Communications
in Computer and Information Science

1624

More information about this series at http://www.springer.com/bookseries/7899

Beno Csapó · James Uhomoibhi (Eds.)

Computer Supported Education

13th International Conference, CSEDU 2021
Virtual Event, April 23–25, 2021
Revised Selected Papers

 Springer

Editors
Beno Csapó
Institute of Education
University of Szeged
Szeged, Hungary

James Uhomoibhi
School of Engineering
University of Ulster
Newtownabbey, UK

ISSN 1865-0929   ISSN 1865-0937 (electronic)
Communications in Computer and Information Science
ISBN 978-3-031-14755-5   ISBN 978-3-031-14756-2 (eBook)
https://doi.org/10.1007/978-3-031-14756-2

This Springer imprint is published by the registered company Springer Nature Switzerland AG
The registered company address is: Gewerbestrasse 11, 6330 Cham, Switzerland

Preface

The present book includes extended and revised versions of a set of selected papers from the 13th International Conference on Computer Supported Education (CSEDU 2021), held from 23 to 25 April, 2021, as a web-based event, due to the COVID-19 pandemic.

CSEDU 2021 received 143 paper submissions from authors in 43 countries, of which 19% were included in this book. The papers were selected by the event chairs and their selection is based on a number of criteria that include the classifications and comments provided by the program committee members, the session chairs' assessment, and also the program chairs' global view of all papers included in the technical program. The authors of selected papers were then invited to submit a revised and extended version of their papers having at least 30% innovative material.

CSEDU 2021, the International Conference on Computer Supported Education, is a yearly meeting place for presenting and discussing new educational tools and environments, best practices and case studies on innovative technology-based learning strategies, and institutional policies on computer supported education including open and distance education. CSEDU 2021 provided an overview of current technologies as well as upcoming trends, and promoted discussion about the pedagogical potential of new educational technologies in the academic and corporate world. CSEDU sought papers and posters describing educational technology research; academic or business case-studies; or advanced prototypes, systems, tools, and techniques.

The papers selected for inclusion in this book contribute to the understanding of current topics in emerging technologies in education for sustainable development and the next generation of teaching and learning environments. They also contribute to treaties on information technologies supporting learning, artificial intelligence in education, ubiquitous learning, social context, and learning environments, as well as learning and teaching methodologies and assessment. These contributions show the wide spectrum of topics presented at CSEDU 2021, where authors had the opportunity to discuss and expand their ideas with colleagues whilst exploring related issues in addressing challenges, devising solutions, and paving clear ways for computer supported education going forward.

We would like to thank all the authors for their contributions and also all the reviewers, who helped in ensuring the quality of this publication.

April 2021

Beno Csapó
James Uhomoibhi

Organization

Conference Chair

James Uhomoibhi   Ulster University, UK

Program Chair

Beno Csapó   University of Szeged, Hungary

Program Committee

Nelma Albuquerque	Concepts and Insights, Brazil
Eleftheria Alexandri	Hellenic Open University, Greece
Abdulwahab Alharbi	University of Glasgow, UK
Fahriye Altinay	Near East University, Cyprus
Zehra Altinay	Near East University, Cyprus
António Andrade	Universidade Católica Portuguesa, Portugal
Francisco Arcega	Universidad de Zaragoza, Spain
Inmaculada Arnedillo-Sanchez	Trinity College Dublin, Ireland
Juan Ignacio Asensio	University of Valladolid, Spain
Breno Azevedo	Instituto Federal de Educação, Ciência e Tecnologia Fluminense, Brazil
Adriano Baratè	Università degli Studi di Milano, Italy
Jorge Barbosa	Unisinos, Brazil
João Barros	Polytechnic Institute of Beja, Portugal
Patrícia Bassani	Universidade Feevale, Brazil
Brett Becker	University College Dublin, Ireland
Jesús Berrocoso	University of Extremadura, Spain
Andreas Bollin	Klagenfurt University, Austria
Ivana Bosnic	University of Zagreb, Croatia
Federico Botella	Miguel Hernandez University of Elche, Spain
François Bouchet	LIP6, Sorbonne Université, France
Patrice Bouvier	SYKO Studio, France
Krysia Broda	Imperial College London, UK
Manuel Caeiro Rodríguez	University of Vigo, Spain
Renza Campagni	Università di Firenze, Italy
Sanja Candrlic	University of Rijeka, Croatia
Nicola Capuano	University of Basilicata, Italy

Ana Carvalho	Universidade de Coimbra, Portugal
Isabel Chagas	Universidade de Lisboa, Portugal
António Coelho	Universidade do Porto, Portugal
Robert Collier	Carleton University, Canada
Fernando Costa	Universidade de Lisboa, Portugal
Gennaro Costagliola	Università di Salerno, Italy
John Cuthell	Virtual Learning, UK
Rogério da Silva	University of Leicester, UK
Sergiu Dascalu	University of Nevada, Reno, USA
Luis de-la-Fuente-Valentín	Universidad Internacional de la Rioja, Spain
Christian Della	University of Glasgow, Singapore
Tania Di Mascio	University of L'Aquila, Italy
Yannis Dimitriadis	University of Valladolid, Spain
Amir Dirin	Metropolia University of Applied Science, Finland
Danail Dochev	Institute of Information and Communication Technologies, Bulgarian Academy of Sciences, Bulgaria
Juan Manuel Dodero	Universidad de Cádiz, Spain
Toby Dragon	Ithaca College, USA
Adam Dubé	McGill University, Canada
Amalia Duch	Politechnic University of Catalonia, Spain
Mihai Dupac	Bournemouth University, UK
Larbi Esmahi	Athabasca University, Canada
João Esteves	University of Minho, Portugal
Vladimir Estivill	Universitat Pompeu Fabra, Spain
Ramon Fabregat Gesa	Universitat de Girona, Spain
Si Fan	University of Tasmania, Australia
Michalis Feidakis	University of West Attica, Greece
Rosa Fernandez-Alcala	University of Jaen, Spain
Débora Nice Ferrari Barbosa	Feevale University, Brazil
Giuseppe Fiorentino	University of Pisa, Italy
Rita Francese	Università degli Studi di Salerno, Italy
Rubén Fuentes-Fernández	Universidad Complutense de Madrid, Spain
Judith Gal-Ezer	The Open University of Israel, Israel
Francisco García Peñalvo	Salamanca University, Spain
Isabela Gasparini	Universidade do Estado de Santa Catarina, Brazil
Henrique Gil	Instituto Politécnico de Castelo Branco, Portugal
Apostolos Gkamas	University Ecclesiastical Academy of Vella of Ioannina, Greece
Anabela Gomes	Instituto Superior de Engenharia de Coimbra, Portugal

Cristina Gomes	Instituto Politécnico de Viseu, Portugal
Maria Gomes	Universidade do Minho, Portugal
Nuno Gonçalves	Polithecnic Institute of Setúbal, Portugal
Ana González Marcos	Universidad de la Rioja, Spain
Anandha Gopalan	Imperial College London, UK
Christiane Gresse von Wangenheim	Federal University of Santa Catarina, Brazil
Christian Guetl	Graz University of Technology, Austria
Raffaella Guida	University of Surrey, UK
Nathalie Guin	Université Claude Bernard Lyon 1, France
David Guralnick	Kaleidoscope Learning, USA
Antonio Hervás Jorge	Universidad Politécnica de Valencia, Spain
Janet Hughes	The Open University, UK
Dirk Ifenthaler	University of Mannheim, Germany
Tomayess Issa	Curtin University, Australia
Ivan Ivanov	SUNY Empire State College, USA
Malinka Ivanova	Technical University of Sofia, Bulgaria
M. J. C. S. Reis	University of Trás-os-Montes e Alto Douro, Portugal
Stéphanie Jean-Daubias	LIRIS, Université Claude Bernard Lyon 1, France
Mike Joy	University of Warwick, UK
M.-Carmen Juan	Universitat Politècnica de València, Spain
Michail Kalogiannakis	University of Crete, Greece
Atis Kapenieks	Riga Technical University, Latvia
Charalampos Karagiannidis	University of Thessaly, Greece
Ilias Karasavvidis	University of Thessaly, Greece
David Kaufman	Simon Fraser University, Canada
Jalal Kawash	University of Calgary, Canada
Samer Khasawneh	Walsh University, USA
Filiz Köse Kalelioglu	Baskent University, Turkey
Adamantios Koumpis	Berner Fachhochschule, Switzerland
Miroslav Kulich	Czech Technical University in Prague, Czech Republic
Lam-for Kwok	City University of Hong Kong, China
Jean-Marc Labat	Sorbonne Université, France
José Lagarto	Universidade Católica Portuguesa, Portugal
Teresa Larkin	American University, USA
Borislav Lazarov	Institute of Mathematics and Informatics, Bulgarian Academy of Sciences, Bulgaria
José Leal	University of Porto, Portugal
Chien-Sing Lee	Sunway University, Malaysia
Marie Lefevre	University Claude Bernard Lyon 1, France

Neena Thota	University of Massachusetts Amherst, USA
Mario Vacca	Italian Ministry of Education, Italy
Michael Vallance	Future University Hakodate, Japan
Leo van Moergestel	HU Utrecht University of Applied Sciences, The Netherlands
Kostas Vassilakis	Hellenic Mediterranean University, Greece
Carlos Vaz de Carvalho	Polytechnic of Porto, Portugal
Giuliano Vivanet	University of Cagliari, Italy
Harald Vranken	Open Universiteit, The Netherlands
Alf Wang	Norwegian University of Science and Technology, Norway
Leandro Wives	Universidade Federal do Rio Grande do Sul, Brazil
Stelios Xinogalos	University of Macedonia, Greece
Diego Zapata-Rivera	Educational Testing Service, USA
Thomas Zarouchas	Computer Technology Institute and Press "Diophantus", Greece
Iveta Zolotova	Technical University of Kosice, Slovakia

Invited Speakers

Gyöngyvér Molnár	University of Szeged, Hungary
Sanna Järvelä	University of Oulu, Finland
Gwo-Jen Hwang	National Taiwan University of Science and Technology, Taiwan, Republic of China

Contents

Learning/Teaching Methodologies and Assessment

Artificial Intelligence in Education

Who is Best Suited for the Job? Task Allocation Process Between Teachers and Smart Machines Based on Comparative Strengths

Michael Burkhard$^{(\boxtimes)}$, Josef Guggemos, and Sabine Seufert

Institute for Educational Management and Technologies,
University of St. Gallen, St. Jakob-Strasse 21, 9000 St. Gallen, Switzerland
{michael.burkhard,josef.guggemos,sabine.seufert}@unisg.ch

Abstract. Due to advances in machine learning (ML) and artificial intelligence (AI), computer systems are becoming increasingly intelligent and capable of taking on new tasks (e.g., automatic translation of texts). In education, such AI-powered smart machines (e.g., chatbots, social robots) have the potential to support teachers in the classroom in order to improve the quality of teaching. However, from a teacher's point of view, it may be unclear which subtasks could be best outsourced to the smart machine.

Considering human augmentation, this paper presents a theoretical basis for the use of smart machines in education. It highlights the relative strengths of teachers and smart machines in the classroom and proposes a staged process for assigning classroom tasks. The derived task allocation process can be characterized by its three main steps of 1) *break-down of task sequence and rethinking the existing task structure*, 2) *invariable task assignment* (normative and technical considerations), and 3) *variable task assignment* (efficiency considerations). Based on the comparative strengths of both parties, the derived process ensures that subtasks are assigned as efficiently as possible (variable task assignment), while always granting priority to subtasks of normative importance (invariable task assignment). In this way, the derived task allocation process can serve as a guideline for the design and the implementation of smart machine projects in education.

Keywords: Education · Human machine collaboration · Task allocation · Task sharing · Augmentation · Comparative advantage · Role of the teacher · Smart machine

1 Introduction

Society, the economy, and the labor market are on the verge of a major transition phase, which comes with different names: The fourth industrial revolution [1], the second machine age [2], the artificial intelligence (AI) revolution [3], the second wave of digitalization [4], and globotics (globalization and robotics) [5]. Technological developments in robotics, machine learning (ML), and artificial intelligence (AI) make it increasingly

© Springer Nature Switzerland AG 2022
B. Csapó and J. Uhomoibhi (Eds.): CSEDU 2021, CCIS 1624, pp. 3–23, 2022.
https://doi.org/10.1007/978-3-031-14756-2_1

important to gain a better understanding of the human-machine relationship, as humans and smart machines may become partners in learning and problem solving [2, 6].

In education, a teacher has many different tasks to perform in the context of a classroom. According to McKenney and Visscher [7], the teaching profession can be described by the three core tasks of teaching: *design* (e.g., create new learning resources), *enactment* (e.g., observe and assess students' needs), and *reflection* (e.g., consider students' progress). From these three core tasks, many other tasks and subtasks can be derived. These may be for example to plan lessons, to create assignments and homework, to conduct and correct exams, to manage the classroom or to activate and motivate students.

Accomplishing all three core tasks well (often simultaneously) can be overwhelming. Increasing student numbers and the requirement for cost savings make it even more challenging to teach successfully [8]. Against this background, it would be desirable if teachers were relieved of some of their workload by smart machines so that they could concentrate even better on their core tasks with the aim of providing high-quality teaching.

A *smart machine* can be defined as a cognitive computer system that can, to a certain extent, make decisions and solve problems without the help of a human being [9]. This is achieved by advanced technology (e.g., AI, ML), which enables the machine to process a large amount of data. Important manifestations of smart machines can be considered chatbots (e.g., Apple's Siri) or social robots, provided that these intelligent machines are able to learn from the environment and develop capabilities based on this knowledge [9]. In a broader sense, however, a smart machine could also be an intelligent tutoring system (ITS) (see e.g., [10]) or a web-based application such as DeepL (www.deepl. com), as long as the underlying technology is based on ML or AI.

In the context of social robots, Belpaeme et al. [11, p. 7] point out, that smart machines could offer a learning experience tailored to the learner, support and challenge students, and free up precious time for human teachers through ways currently unavailable in our educational environments. They conclude that social robots show great promise when teaching restricted topics, with effect sizes on cognitive outcomes almost matching those of human tutoring [11, p. 7]. According to Reich-Stiebert et al. [12, p. 5] such robots could be used as personal tutor for students or assistants to teachers: "provide information on specific topics, query learned lessons, give advice to the learning process, correct errors, or provide feedback on students' progress" [12, p. 5].

Due to advances in sensor and actuator technology, smart machines are increasingly used in everyday life. Starting with the use as an aid in STEM education, the use of smart machines has been increasingly extended to the field of education during the last ten years [11, 13]. Since then, several studies have used smart machines in large classroom-settings (e.g., [14–17]), but also in smaller workshop-like (e.g., [18]) or one-on-one interactions (e.g., [19]).

Currently, a gap exists between available technological capabilities and the use of this technology for educational purposes [20, p. 3]. There could be a disparity between technological readiness and its application in education [21]. Although various AI applications have been developed by the education industry, they may be insufficiently aligned with theoretical frameworks in education [20, p. 3, 13].

To tackle this issue, it might be important to gain a better understanding how specific classroom tasks can be meaningfully allocated between teachers and smart machines. To this end, we will build on and extend our earlier work published in the *Proceedings of the 13th International Conference on Computer Supported Education* (see Burkhard et al. [22]). There, we already analyzed relative strength profiles of teachers and smart machines on a rather general level. However, since the recommendations for the concrete distribution of tasks in the classroom remained largely at an abstract level and could therefore provide only limited guidance in the design and implementation of own projects, we will elaborate on this. Based on the relative strengths of both parties (teacher and smart machine) a *task allocation process* should be derived that can be used as a guideline for the design and the implementation of smart machine projects in education.

In light of the identified research desideratum, the following research question should be addressed:

- *How can specific classroom tasks be meaningfully allocated between teachers and smart machines in terms of normative as well as efficiency considerations?*

The objectives of the paper at hand are therefore twofold:

- Elaboration of the theoretical foundations for the use of smart machines in education, in order to investigate underlying assumptions, relative strengths between both parties, as well as approaches of reasonable task sharing for the design and evaluation of teaching;
- development of a *classroom task allocation process* between teachers and smart machines, to be used as a guideline for the design and the implementation of smart machine projects in education.

From a theoretical point of view, this paper can act as starting point for future empirical research, as it highlights important concepts and variables related to the relative strengths of teachers and smart machines.

From a practical standpoint, the derived *task allocation process* could be helpful in designing use cases as it may act as a guideline in the allocation of tasks between teachers and smart machines. In light of human augmentation, the derived *task allocation process* ensures that normative considerations always take precedence over mere efficiency considerations, which may be a key aspect for the sustainable integration of AI technology [13].

To this end, we lay the foundation in Sect. 2, where we have a closer look at smart machines in the classroom and their role with regard to the human teacher. Based on existing literature in the field of human-machine collaboration and illustrated by a concrete classroom task example, Sect. 3 then explains the *task allocation process*, which can be characterized by the three main steps: 1) *break-down of task sequence and rethinking the existing task structure*, 2) *invariable task assignment* (normative and technical considerations), and 3) *variable task assignment* (efficiency considerations). To be able to conduct the *variable task assignment*, Sect. 4 discusses relative-strength profiles of teachers and smart machines based on the theory of comparative advantages and evaluates them with regard to specific classroom tasks[1]. Section 5 then uses the relative strength profiles for the variable task assignment on the concrete classroom task example. Section 6 concludes with some final remarks.

2 Increasingly Smart Machines Inside a Classroom

Floridi [23] argues that we are entering a new era in which we will become increasingly reliant on our own technological achievements. Information and communication technology (ICT) is increasingly being used to not only record and transmit information, but also to process it autonomously. Floridi [24, pp. 6–7] came up with the term "infosphere", which refers to an information environment that is similar to but distinct from cyberspace and is becoming increasingly interconnected with our daily lives. All informational entities (biological as well as digital agents/smart artefacts) constitute the infosphere. In the context of education, a digital classroom ecosystem (sometimes also referred to as smart classroom [see e.g., [25]] can also be seen as a manifestation of an infosphere. Biological agents (teachers and students) and digital artefacts (such as interactive whiteboards, laptops, and smartphones) interact according to a syllabus and form a classroom ecosystem.

Digital classroom ecosystems have the potential to enhance knowledge transmission from teacher to students in a variety of ways [25]. They can help the teacher with content production, presentation, and distribution [25, pp. 6–12], stimulate interaction between different biological agents [25, pp. 12–14], and provide automated assessment and feedback as well as some background functions (e.g., temperature management inside the classroom) [25, pp. 15–20].

For now, the digital classroom ecosystem can be viewed of as an advanced tool (similar to a car) that aids teachers in achieving their objectives. It assists teachers in getting from point A to point B more quickly, but teachers still must steer and drive. In the future, further advances in AI could make these machines even smarter who in certain situations could also be sitting in the driver's seat.

Such considerations may not be far-fetched. An example could be OpenAI's language generator GPT-3, an autoregressive language model with 175 billion parameters. GPT-3 is able to create computer code for web-page layouts using prompts like "Give me a button that looks like a watermelon" or "I want a blue button that says subscribe" [26]. In addition to that, it can also generate news articles that human reviewers can hardly distinguish from articles written by humans [27]. In this context, a number of critical voices are being raised, indicating that the role of humans in relation to smart machines may be rethought (e.g., [5, 6, 23, 28–30]). As smart machines are increasingly able to perform white collar activities, this also poses a challenge to education and the teaching profession. It raises the question of how to deal and possibly respond to it.

As a possible answer to this question, Davenport and Kirby [30] highlight the mutual complementation and task sharing that they refer to as "augmentation: people and computers supported each other in the fulfilment of tasks" (p. 2). According to Jarrahi [6], *augmentation* can be thought of as a "Human-AI symbiosis," in which humans and AI

---

[1] Note. Section 4 is an adapted version of "Sect. 3: *Comparative Strengths of Teachers and Smart Machines*" published in Burkhard et al. [22].

interact to make both parties smarter over time (p. 583). In this way, the concept of *augmentation* is similar to the concept of *hybrid intelligence* (see [31]), which combines the complementary intelligence of humans and smart machines. Together, the human and the smart machine "create a socio-technological ensemble that is able to overcome the current limitations of (artificial) intelligence" [31, p. 640]. Dellermann et al. [31, p. 640] points out that smart machines and humans learn from each other through experience and gradually improve over time.

Humans Assisting Machines      **Machines Assisting Humans**

Augmentation

Humans need to perform three crucial roles:

1. They must *train* machines to perform certain tasks;
2. *explain* the outcomes of those tasks, especially when the results are counterintuitive or controversial;
3. *sustain* the responsible use of machines (e.g., by preventing robots from harming humans).

Smart machines are helping humans expand their abilities in three ways:

1. They can *amplify* our cognitive strengths (in particular with analytic decision support);
2. *interact* with other humans (e.g., students, teachers) to free us for higher-level tasks;
3. *embody* human skills to extend our physical capabilities.

Fig. 1. Human augmentation. Source: own representation based on Wilson and Daugherty [32]. Note. This figure has already been published in Burkhard et al. [22, p. 77].

Figure 1 illustrates this relationship. On the one hand, humans have to train machines to perform certain tasks. They must communicate and explain the outcomes of those tasks to other stakeholders and ensure that machines are used responsibly. Smart machines, on the other hand, assist humans by boosting their cognitive strengths, relieving them of repetitive tasks, and enhancing their physical abilities.

In a classroom, a teacher has many different tasks to perform. McKenney and Visscher [7] describe the teaching profession by the three core tasks of *design*, *enactment*, and *reflection*. As *designers*, teachers create new learning resources and tasks, or customize them. During lesson *enactment*, teachers observe and assess students' needs, recognize, judge, decide, and act with meaningful measures. During *reflection*, teachers consider their students' progress and, if necessary, adjust their behavior [7].

From these three overarching core tasks, a multitude of concrete teaching activities can be derived. This includes, for example, scheduling lessons, compiling meaningful learning content, creating assignments and homework, presenting content and answering questions, coaching students individually, conducting and correcting exams, managing the classroom, as well as motivating students during the lessons.

Accomplishing all these tasks well (sometimes simultaneously) can be daunting. From the perspective of a teacher, the presence of a smart machine could be beneficial because the smart machine can engage in task sharing and take over some of the teacher's responsibilities. However, determining which tasks or subtasks should be delegated to a smart machine may be difficult for various reasons:

First, it may be generally difficult from a process point of view to divide specific tasks between the two parties in a meaningful way. We will address this issue in Sect. 3 and propose a staged process for task allocation between teachers and smart machines.

Second, it could be that the relative strength profiles between teacher and smart machines in the classroom are not well enough known, making it difficult to assign subtasks to one side or the other. We will address this issue in Sect. 4 by deriving such relative strength profiles.

Third, the nature of the task itself may change by using automated assistance [33], which might require rethinking the existing task process. To illustrate this point, think about a foreign language class where the teacher gives the task of translating a sentence (see Table 1). If the sentence is translated without automatic assistance, this task is carried out it in a different way than if you have access to a smart machine like DeepL (www.deepl.com).

Table 1. Changing nature of tasks through automatic assistance: an example.

Sentence translation (without automatic assistance)	Sentence translation (using automatic assistance)
- Read the sentence	- Read the sentence
- Split the sentence into individual words	- Copy sentence into translation tool
- Translate words you understand	- Read the translation suggestion
- Look up words in the dictionary that you do not understand	- Display alternative words and translation suggestions proposed by the translation tool
- Decide on the grammatical structure of the sentence based on own knowledge	- Decide on the alternative suggestions
- Write down the sentence in the foreign language	- Write down the sentence in the foreign language

As Table 1 shows, using a smart machine does not necessarily mean that less steps have to be performed. It means the user can do other things as the focus of attention changes. This is important to keep in mind when talking about classroom task allocation between human and smart machine in the next section.

3   Classroom Task Allocation Between Human and Machine

How can we analyze specific classroom tasks and divide them meaningfully between humans and smart machines? In the context of human-machine collaboration, Ranz et al. [34, pp. 185–188] suggest a capability-based task allocation approach, which includes the three main steps of 1) *break-down of process sequence*, 2) *invariable task allocation* and 3) *variable task allocation*. On this basis, we will transfer and extend the approach to the classroom setting. For the purpose of illustration, we will use as an example *the correction of exams*. However, the shown approach is also applicable to many other classroom tasks.

According to Ranz et al. [34, p. 185], we should *in a first step* break-down the process sequence into smaller subprocesses or subtasks to get to our investigation object. Therefore, we break down the relatively large task of *correcting exams* into smaller subtasks and number them according to the order of execution. As can be seen in Table 2, the *correction of exams* may include subtasks like correcting closed-ended questions (e.g., multiple choice questions, matching tasks), building groups of similar answers to open-ended questions (to facilitate the correction process), correcting open-ended questions, adding up the points scored, calculating the distribution of points, selecting a grading scale, calculating grades based on that scale, and communicating grades to students together with formative feedback.

Table 2. The correction of exams: break-down of existing process sequence (step 1).

Potential subtasks during the correction of exams
1. Correct closed-ended questions (for each student)
2. Build groups of similar answers to open-ended questions
3. Correct open-ended questions (for each student)
4. Count total points (for each student)
5. Calculate overall point distribution (of the class)
6. Select grading scale (to define best and worst grade)
7. Calculate grades based on grading scale
8. Communicate grades and provide formative feedback (to each student)

Now that we have decomposed and visualized the main task into individual subtasks, we can ask whether the nature of the task would change if we were to receive automatic support from a smart machine. This may also include sequencing the process in a way that allows for a better division of labor. If necessary, we could further adapt the process or decompose it into further subtasks. However, to avoid making the example in Table 2 even more complex, we will refrain from doing so.

How should we now best divide these individual subtasks between the teacher and the smart machine? According to Bradshaw et al. [33], this issue can be viewed in terms of overlapping capabilities (see Fig. 2), with the caveat that in some cases, task allocation is more complicated than simply transferring responsibilities from one party to another, as synergies and conflicts may be involved [33]. In this case, we should rethink and redesign the existing subtasks (step 1, Table 2).

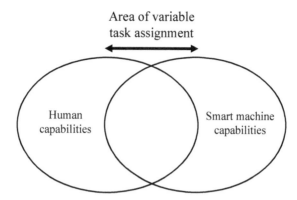

Fig. 2. Adaptive allocation to human and machine. Source: Bradshaw et al. [33].

As we see in Fig. 2, there might be certain *invariable tasks* "that are uniquely suitable for humans and others uniquely suitable for machines" [34, p. 186]. In between, there exist certain *variable tasks* that could potentially be performed by both parties.

In a second step, we should now determine for each of the subtasks identified in step 1 whether they require unique human capabilities and therefore cannot be automated by a smart machine, or vice versa [34, p. 186]. Here, Ranz et al. [34, p. 186] primarily address technical restrictions, but from our point of view, ethical considerations are also highly important. Even tough further advances in AI will make it likely to increase the range of tasks that can theoretically be performed by a smart machine, this does not necessarily mean that we *should* transfer as many tasks as possible to a smart machine. Drawing from the idea of a knock-out list, we can ask "which subtasks should only be performed by humans for ethical reasons?". These subtasks should then always be assigned to humans, regardless of efficiency considerations.

In the context of industrialization, the question of how new technology affects labor and working conditions was already discussed in the 19th and 20th century. Karl Marx's theory of alienation played a substantial role in shaping the discussion [35]. In simple terms, the theory of alienation describes "the estrangement of individuals from aspects of their essence" [36, p. 2]. If a person can no longer identify with their work activity (e.g., due to automatization), this may lead to an alienation between the person and the work performed. Even today, the theory of alienation has some relevance in the context of AI. Wogu et al. [36, pp. 7–8] list several forms of work alienation in relation to recent job automatization tendencies (e.g., low-income service jobs, Gig Economy). While a number of critical voices are being raised, indicating that the role of humans in relation to smart machines may be rethought (e.g., [5, 6, 23, 29, 30]). Acemoglu and Restrepo [28] outline a distinction between a "wrong" and "right" kind of AI that should be promoted. Rather than just replacing labor, AI should serve as a tool for qualified workers to create new application opportunities [28].

For education, this may mean that we should not only "strive for what is technically possible, but always ask ourselves what makes pedagogical sense" [13, p. 21]. As a negative example Zawacki-Richter et al. [13, p. 21] mention so called "Intelligent

Classroom Behavior Management Systems". Those facial recognition systems track student attention with the goal of improving the students' attendance rate and enhancing the classroom discipline [37]. It is questionable whether such AI-systems provide real added value for a good teacher or if it is not just a form of educational surveillance [13, p. 21].

On a general level, an important ethical aspect might be privacy considerations (see e.g., [38]). As smart machines operate (in part) autonomously and collect student data, new strategies may have to be developed to address this issue. Among other things, this may include transparency, student control over data, right of access as well as accountability and assessment [39].

Another important ethical aspect for a sustainable implementation might be technology acceptance (see e.g., [40]). Concerns about replacing teachers with smart machines may be unfounded [41]. Nevertheless, it is important to reassure teachers that the intention is not to replace them with social robots [42]. Use cases should be developed in such a way that teachers do not feel alienated from their work activities, still sit in the driver's seat, and have the final word regarding any decisions.

While some subtasks should only be performed by humans due to ethical reasons, other subtasks can only be performed by smart machines dues to technical reasons. For example, if a planned process is to involve individual coaching with multiple students at the same time, a human teacher is unable to do this because they cannot be in two places at the same time. With a smart machine running on each user's device, this would be theoretically possible.

We can now apply these ethical and technical considerations to the subtasks listed in Table 2. With regard to the smart machine, there are no subtasks that have necessarily to be performed by the smart machine alone (e.g., due to technical reasons). In principle, all subtasks could also be performed by a human teacher. Conversely, however, there may be one subtask that necessarily requires the teacher to perform instead of the smart machine (see Table 3).

We are referring to the selection of the grading scale to be used (subtask 6), which determines how the points achieved in the exam (subtask 5) are translated into grades (subtask 7) (see Table 3). The decision about the grade should be made solely by the teacher, as they will ultimately be held responsible for the grades. The assignment of all other tasks to a certain party (teacher, smart machine) can be decided in step 3: *variable task allocation*.

In a third step, we decide on the assignments of all subtasks that could theoretically be performed by both humans and smart machines (*variable task allocation*). For this purpose, Ranz et al. [34, pp. 186–188] use the concept of *capability indicators*, which were first introduced by Beumelburg [43].

The goal of a *capability indicator* is to objectify *variable task allocation* decisions [34, p. 186]. For each subtask, a capability indicator can be calculated for both the teacher and the smart machine. The capability indicator takes on a value between 0 and 1, where higher values indicate a better relative capability. A particular subtask is then assigned to the party that has the higher capability indicator for a specific subtask. In a

Table 3. The correction of exams: invariable task allocation (step 2).

Subtasks during exam correction	Reasoning	Assigned to
1. Correct closed-ended questions	- Step 3 decision	- Variable task all
2. Build groups of similar answers to open-ended questions	- Step 3 decision	- Variable task all
3. Correct open-ended questions	- Step 3 decision	- Variable task all
4. Count total points	- Step 3 decision	- Variable task all
5. Calculate overall point distribution	- Step 3 decision	- Variable task all
6. Select grading scale	- The teacher should finally decide on grading (normative argument)	- Teacher
7. Calculate grades based on grading scale	- Step 3 decision	- Variable task all
8. Communicate grades and provide formative feedback	- Step 3 decision	- Variable task all

very simplistic form, the capability indicator for a given subtask t can be written as:

$$Capability\ indicator_{t,\ teacher} = \frac{p_t + i_t + q_t}{3} \tag{1}$$

and

$$Capability\ indicator_{t,\ smart\ machine} = 1 - Capability\ indicator_{t,\ teacher}, \tag{2}$$

where p is the process time, i the additional invest, and q the process quality of a given subtask t. The variables p, i and q each can take on the cardinal values of 1 (better), 0.5 (equal), or 0 (worse) when comparing the relative strengths between human and smart machine for a given subtask [34, p. 187]. Depending on the context, the weighting of each variable could be adjusted, or additional variables could be added to calculate the *capability indicators*.

While it can be valuable to quantify relative strengths as objectively as possible using *capability indicators*, it can also be useful to use faster, more intuitive approaches for task allocation [34, p. 188]. To this end, we will derive in the next section relative strength profiles of teacher and smart machine in the context of the classroom in order to guide the *variable task allocation* in an intuitive and efficient way. Figure 3 summarizes the proposed process in this section.

As we see in Fig. 3, *in a first step*, we should break-down the process sequence into smaller subtasks, rethink the existing task process and if necessary, adapt and redesign it. *In a second step*, we can then check whether certain tasks need to be performed only by humans or only by smart machines? In addition to technical considerations regarding smart machines, this step also includes normative questions to ensure that the teacher retains the upper hand and is not degraded to a mere assistant. After completing the invariable task assignment, we can then carry out *in a third step*, the variable task assignment to assign the remaining subtasks to either the teacher or the smart machine.

To this end, we can use either the concept of capability indicators (explained earlier in this section) or the complementary concept of comparative strengths of capabilities, which we discuss in detail in the next section.

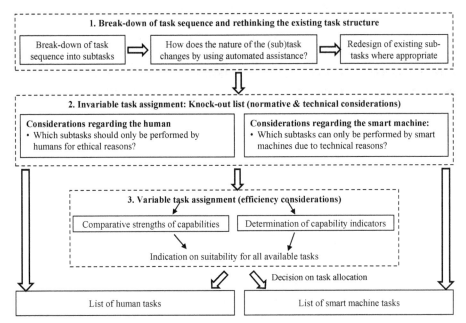

Fig. 3. Staged process for task allocation based on comparative strengths. Source: Bradshaw et al. [33], Ranz et al. [34] and own contributions.

4 Comparative Strengths of Teachers and Smart Machines

To examine the relative strengths of the teacher and the smart machine, we rely on the theory of comparative advantage, a concept that originated in the field of economics [44, 45][2]. Ricardo [46] was the first to show in the late nineteenth century why two countries A and B trade, even if one country is in absolute terms superior to the other in the production of all goods in the economy. He explained why countries specialize in the production of certain goods and trade them. He showed that what matters is not absolute advantage (being better at producing all goods) but relative advantage (having lower opportunity costs).

Applying the concept of comparative advantages to the classroom, the teacher and the smart machine can be seen as countries A and B. In the classroom, various tasks need to be performed (production of goods). For example, two of these tasks could be *to provide feedback* on homework or *to individually support* students. To illustrate the

---

[2] Note. Section 4 is an adapted version of "Sect. 3: *Comparative Strengths of Teachers and Smart Machines*" published in Burkhard et al. [22].

comparative advantage between the two parties, we assume different task times for the teacher and the smart machine.

Table 4. Comparison of task time (absolute) and opportunity costs (relative).

Task time	Teacher (1:1 setting)	Smart machine (1:1 setting)	Smart machine (1:n setting)
Individual coaching	10 min	15 min	15 min
Provide feedback	5 min	15 min	15 min
Opportunity costs			
Individual coaching	10/5 = 2 feedback	15/15 = 1 feedback	0 feedback
Provide feedback	5/10 = 0.5 coaching	15/15 = 1 coaching	0 coaching

As can be seen in Table 4, the teacher is, in absolute terms, better at both individual coaching and providing feedback (lower task times). Note that for the sake of simplicity, differences in task quality are also implicitly reflected in longer task times. The question is now: Even though the teacher is better at both tasks in absolute terms, can it still be beneficial for the teacher to delegate tasks to the smart machine?

According to the theory of comparative advantage, this is possible because it is not the absolute but the relative advantages that count. Table 4 also displays the opportunity costs for our scenario depending on different settings. If we consider the teacher, they will always interact in a 1:1 setting because they cannot split up and do two tasks at the same time. Therefore, if the teacher does individual coaching, they cannot provide two units of feedback during that time (opportunity costs). Conversely, when giving one unit of feedback, the teacher forgoes 0.5 units of coaching.

Similar considerations apply to the smart machine. Let us first think about a setting, where the smart machine like the teacher also cannot act simultaneously (1:1). This might be the case for smart machines with a physical embodiment like social robots, or applications who only run on a specific device that cannot be used by two students simultaneously. In this case, since the smart machine is equally fast in both tasks, the smart machine cannot produce one unit of feedback for each unit of coaching. Therefore, the opportunity costs for both coaching and feedback are 1.

In a setting where a smart machine can interact with multiple students simultaneously (1:n) the opportunity costs even become zero, because the smart machine can perform both tasks at the same time. This might for example be the case for a smart machine application that runs on individual students' smartphones (provided that every student has their own smartphone).

Each party (teacher and smart machine) should now carry out the tasks where they have lower opportunity costs compared to their counterpart. In the comparison with the 1:1 smart machine setting, the teacher will specialize in giving feedback (0.5 < 1) and the smart machine will do the coaching (1 < 2). In the comparison with the 1:n smart machine setting, the smart machine has lower opportunity costs for both tasks, hence it might make sense from an efficiency perspective to even outsource both tasks to the smart machine.

Our example shows that smart machines can be useful even if they are inferior to humans on an absolute level. At a more general scale, smart machines have comparative advantages over the teacher in certain areas. Therefore, it might be advantageous from an efficiency perspective to have them take over certain tasks from the teacher. This means that a certain set of tasks could be completed in a shorter time or within a certain time the number of completed tasks (quality) could be increased.

As the opportunity cost comparison showed, the relative strengths of smart machines could be particularly evident in 1:n settings, where smart machines can interact with multiple students simultaneously. Unlike social robots, which can (like the teacher) only interact in 1:1 settings due to their physical presence, the lower opportunity costs in 1:n settings reflect the potential for scalability of such tools. However, since humans and machines are different in terms of their characteristics, there will always be tasks where humans are better suited from a relative perspective.

As it has been shown, the crucial point is the relative strengths of teachers and smart machines. But how can these relative strengths be characterized? Jarrahi [6, p. 583] constructed relative strength profiles of humans and AI with respect to their *core skill set* along three dimensions: *uncertainty*, *complexity*, and *equivocality* (see Fig. 4).

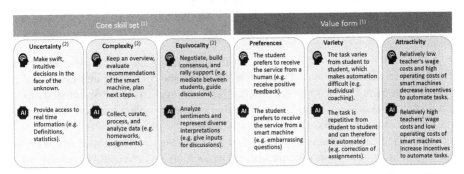

Fig. 4. Relative strength-profiles of teachers and smart machines. Source: (1) Latham and Humberd [47], (2) Jarrahi [6] and own contributions. Note. This figure has already been published in Burkhard et al. [22, p. 78].

Latham and Humberd [47, p. 12] further argue that it is important to look at the *skill set*, but also at how the value of the core skill set is conveyed (*value form*), when assessing the threat of technology to a particular occupation. They ground the *value form* in consumer preferences, task diversity, and wage differentials [47, p. 13], based on which we created the three dimensions of *preferences*, *diversity*, and *attractivity*. Figure 4 summarizes the created relative strength profile of teachers and smart machines along the two categories *core skill set* and *value form*, as well as the six dimensions of *uncertainty*, *complexity*, *equivocality*, *preferences*, *variety* and *attractivity*.

Uncertainty is described by Jarrahi [6, pp. 581–581] as a lack of information about all alternatives or their consequences, which makes interpreting a situation and making a decision more challenging. He states that in situations where there is no precedent, an intuitive decision-making style may be more useful. According to Jarrahi [6, pp. 580–581], humans have a relative advantage over AI in the dimension of *uncertainty* due to their ability to make intuitive decisions (e.g., [48]). Smart machines can still help decrease *uncertainty* by offering access to real-time information, but because machines are largely unable to comprehend human intuition's internal logic and subconscious patterns, humans tend to keep their comparative advantage in situations which require holistic and visionary thinking [6, p. 581].

In the classroom, *uncertainty* can occur through a variety of channels. For example, students may ask unexpected questions or give inputs that require some form of intuitive thinking or creativity to react to it. In addition, the dynamics of the classroom itself are unpredictable and uncertain to some degree because students are individuals with their own needs. Students do not act the same way every day, and sometimes they may not appear at all. How to respond to such situations requires intuition and cannot be solved by a fixed rule alone.

Complexity can be described by its abundance of elements or variables that require the processing of a large amount of information. AI has a comparative advantage in dealing with complexity due to its ability to collect, curate, process, and analyze large amounts of data [6, p. 581].

In the classroom, the *complexity* increases with the number of students because the same assignments, exercises, and exams are given to more people. More students make it harder to get an overview of each student's learning success. Here, smart machines can provide valuable support, especially in large classes, if they assist the teacher in automated feedback for homework and exams.

Equivocality involves the occurrence of multiple simultaneous but divergent interpretations of a decision domain and often arises due to the conflicting interests of stakeholders, customers, and policy makers [6, p. 581]. This suggests that an objective solution to a problem is not always available, but rather that the problem has several different and subjective views. Although smart machines may be able to analyze emotions and represent diverse interpretations of the world, humans have a comparative advantage in dealing with *equivocality* as they are better at negotiating and establishing coalitions [6, p. 582].

In the classroom, *equivocality* may arise due to a variety of circumstances. On the one hand, the curricula of certain school subjects can be more subjective than others. While there are clear guidelines for "true" and "false" in subjects such as accounting or mathematics, it is more difficult to draw a clear line in subjects such as history or literature. On the other hand, students often have different views, and a teacher is needed to guide discussions towards a consensus.

By nature, a human teacher and a smart machine are very different. Therefore, for certain tasks, it will simply depend on the *preferences* of the students as to whom they turn to with their problems.

In the classroom, *preferences* primarily depend on social norms and informal social rules between people. In human conversations, there are informal rules that must be followed (e.g., be friendly), which consumes time and can make communication inefficient. Because those rules do not necessarily apply to smart machines, students can ask any question they want and do not have to worry about asking a "silly" question or behaving in a socially inappropriate way. In addition, smart machines have the ability to repeat answers as many times as needed without becoming tired (e.g., in language learning), which makes them a cooperative learning partner. Teachers, in contrast, may be reluctant to answer the same question multiple times.

For other tasks, the *variety* is the deciding factor. If a task varies from student to student, it is trickier to automate and harder for a smart machine to solve. If a task is repetitive, however, it can be automated more easily because the smart machine can be better trained on it.

In the classroom, *variety* depends on the tasks at hand. In particular, the correction of written work has repetitive elements, since the same subtasks have to be performed for each student. Other tasks, such as conducting individual conversations with students about their research project, differ from student to student and from project to project and cannot simply be taken over by a smart machine.

At last, the attractivity to automate tasks also has an influence on whether a certain task is transferred from a human teacher to a smart machine. The attractiveness depends on the wage costs of the human teacher in relation to the training and operating costs of the smart machine. If a smart machine can be trained for a certain (sub)task with relatively little effort and also incurs few operating costs, it is attractive to outsource partial (sub)tasks to the smart machine. However, if the training costs and the ongoing operating costs are high in relation to the teacher's wage costs, automation is not worthwhile from this perspective.

Training and operating costs vary greatly depending on the type of smart machine and the context in which it is embedded. While existing web-based smart machine applications (e.g., www.deepl.com) can often be used with little or minimal preparation effort and do not need to be trained by teachers themselves, the training effort of own smart machine applications (e.g., in a chatbot or social robot project) is often still very high for the teacher or not yet manageable for the teacher due to the technical challenges. However, as the GPT-3 example shown in Sect. 2 demonstrates (see [26, 27]), it will probably become easier in the future to train smart machines (even without having programming skills). As a result, the *attractivity* and the potential range of tasks to be taken on may increase.

To summarize, both the teacher and the smart machine possess comparative strengths in the classroom. While the teacher's strength lies mainly in situations that involve *uncertainty*, *equivocality*, and high task *variety*, smart machines have a comparative advantage in situations that can be characterized by a high degree of *complexity*, as well as low task *variety*. Depending on the situation and the context, *preferences* (e.g., from students) or the financial *attractivity* to automate may favor one side or the other.

5 Using Relative Strengths for Variable Task Allocation

The relative strength profiles derived in the last section, can now be applied to our example of *correcting exams* (see again Table 1) to conduct the variable task allocation. Table 5 depicts the subtasks and characterizes them with respect to the derived strength dimensions. Based on that process, a specific subtask was either assigned to the teacher or the smart machine.

As can be seen in Table 5, the smart machine received all subtasks that can be described by relatively high *complexity*, low *uncertainty*, low *equivocality*, and low task *variety*. This includes the correction of the closed-ended questions, building groups of similar answers to open-ended questions, the counting of the total points (for each student), the calculation of the overall point distribution (for the class), and the calculation of the grades based on the applied grading scale (for each student).

Technology acceptance is important with regard to smart machines [40]. The teacher, on the contrast, therefore, focuses on the subtasks that have a certain potential of *equivocality* as well as on subtasks where students might prefer the personal contact or opinion of the teacher. This includes the correction of open-ended questions as well as the communication of the grades and the provision of formative feedback. Because human teachers have more empathy than smart machines [31, p. 640], they can perform these tasks in a better way. While the open-ended questions are still corrected by the teacher, smart machines can provide valuable assistance in preparation for the correction of the open-ended questions (e.g., by forming clusters of similar student answers). Particularly, questions that require relatively short answers may be increasingly evaluated automatically in the future (see e.g., [49]). Nevertheless, among other things, the degree to which a smart machine can support the correction of open-ended questions also depends on the subject area of the questions asked.

The subtask of selecting the grading scale was already assigned to the teacher during the invariant task allocation in step 2 for ethical reasons unrelated to efficiency considerations (step 3).

Our example has shown how task processes in the classroom can be meaningfully divided between teachers and smart machines from a theoretical point of view. Nevertheless, from a practical point of view, this does not yet mean that the teacher will actually strive for such a distribution of tasks. Only if the teacher's perceived benefits are greater than the perceived costs of implementation the teacher will outsource (sub)tasks to the smart machine. The perceived benefits can be both quality improvements in terms of teaching as well as time savings due to automatic assistance. The perceived costs include time and money spent on the smart machine's acquisition, training and operation.

Examples of smart machine applications that are already being used today are often web-based applications (e.g., www.deepl.com), since they can be used with only little or no additional effort on the part of the teacher (little or no setup costs).

On the one hand, a barrier towards practical implementation could be that the benefits of smart machines in the classroom are perceived by teachers as too little added value. However, with further progress in the field of AI, it is likely that the perceived benefits will increase.

On the other hand, it could also be that the costs associated with the implementation of smart machine projects are perceived as too high. In many cases, a considerable initial effort would be required for implementation, whose benefits, however, would only be apparent when the number of students is large enough (economies of scale). It is not surprising that teachers who generally teach in smaller groups of less than 30 students and are completely on their own, are not in a position to undertake this initial effort. IT teachers could form an exception here, as using smart machines in the classroom

Table 5. The correction of exams: variable task allocation (step 3) based on relative strength profiles between teacher and smart machine.

Subtasks during exam correction	Relevant relative strength dimensions	Assigned to
1. Correct closed-ended questions	- *Complexity*, no *uncertainty*, no *equivocality*, low task *variety*	- Smart machine
2. Build groups of similar answers to open-ended questions	- *Complexity*, low task *variety*	- Smart machine
3. Correct open-ended questions	- Potential of *equivocality*, potential for student *preferences*	- Teacher
4. Count total points	- *Complexity*, no *uncertainty*, no *equivocality*, low task *variety*	- Smart machine
5. Calculate overall point distribution	- *Complexity*, no *uncertainty*, no *equivocality*, low task *variety*	- Smart machine
6. Select grading scale	- See Table 3: invariable task all. (step 2)	- Teacher
7. Calculate grades based on grading scale	- *Complexity*, no *uncertainty*, no *equivocality*, low task *variety*	- Smart machine
8. Communicate grades and provide formative feedback	- Potential for student *preferences* (empathy), medium task *variety*	- Teacher

might offer them relatively more benefits (e.g., as an object to program) and because IT teachers may also have lower set-up costs due to the related subject matter.

Since in this paper we have focused on the individual perspective of the teacher, these cost-benefit considerations may be very different for each teacher depending on their own prior knowledge as well as the context of application. For example, depending on the country, school system, and discipline, the government may or may not decide to pay for certain costs, leading to different perceived teachers' costs. In a next step, this issue could be explored and addressed in more detail.

6   Conclusion

In this paper, we proposed a staged process for assigning classroom tasks between teachers and smart machines. The process ensures that normative considerations (*invariable task allocation*) are prioritized over efficiency considerations (*variable task allocation*). Drawing from the work of Bradshaw et al. [33] and Ranz et al. [34], we break-down, in a *first step*, the classroom task into smaller subtasks, rethink the existing task structure and redesign it if appropriate. In a *second step*, we carry out the *invariable task assignment*, where subtasks are assigned to one party because there is a normative or technical reason behind it. In a *third step*, we can then carry out the *variable task assignment* to assign the remaining subtasks in an efficient way. In order to do that, we used the theory of comparative advantage and highlighted the relative strengths between teachers and smart machines in the classroom. In this way, we aim to foster and theoretically ground the division of tasks between the teacher and the smart machine in the classroom.

Ultimately, it is also an empirical question of how task sharing in the classroom will look like in the future, as the relative strengths heavily depend on the students' perceptions (reflected in the dimension *preferences*). The *preference* dimension could in a next step be empirically investigated in more detail to gain a better understanding of what tasks could be assigned to smart machines.

Currently, the use of smart machines in educational settings may be limited due to technical and logistical challenges [11, p. 7], but as technology becomes cheaper and better, the use of smart machines in education is likely to increase (reflected in the dimension *attractivity*). From a practical perspective, this may be a big hurdle to sustainable implementation, because teachers need among other things lucrative incentives (e.g., in the form of time savings to be realized) to strive for change (see e.g., [50]). In addition, as many pre-service and in-service teachers are not ready to support and adopt new technologies related to AI, effective teacher education and continuing education programs have to be designed and offered to support the adoption of these new technologies [20, p. 7].

AI transformation does not mean that less teachers are needed [51]. However, the role of the teacher may change. With this paper, we aim at a better understanding of the digital transformation from a teacher perspective in order to provide a guideline and a toolkit when engaging in task sharing with smart machines.

References

1. Braga, C.P., et al.: Services Trade for Sustainable, Balanced, and Inclusive Growth (2019). https://t20japan.org/policy-brief-services-trade-sustainable-inclusive-growth/
2. Brynjolfsson, E., McAfee, A.: The Second Machine Age: Work, Progress, and Prosperity in a Time of Brilliant Technologies. Norton, New York (2014)
3. Makridakis, S.: The forthcoming Artificial Intelligence (AI) revolution: its impact on society and firms. Futures **90**, 46–60 (2017). https://doi.org/10.1016/j.futures.2017.03.006
4. Wahlster, W.: Künstliche Intelligenz als Treiber der zweiten Digitalisierungswelle. IM+io Das Magazin für Innovation Organisation und Management **2**, 10–13 (2017)
5. Baldwin, R., Forslid, R.: Globotics and development: when manufacturing is jobless and services are tradable (Working paper 26731). National Bureau of Economic Research (2020). https://doi.org/10.3386/w26731
6. Jarrahi, M.H.: Artificial intelligence and the future of work: human-AI symbiosis in organizational decision making. Bus. Horiz. **61**(4), 577–586 (2018). https://doi.org/10.1016/j.bushor.2018.03.007
7. McKenney, S., Visscher, A.J.: Technology for teacher learning and performance. Technol. Pedagogy Educ. **28**(2), 129–132 (2019). https://doi.org/10.1080/1475939X.2019.1600859
8. Oeste, S., Lehmann, K., Janson, A., Söllner, M., Leimeister, J.M.: Redesigning university large scale lectures: how to activate the learner. In: Academy of Management Proceedings, vol. 2015, no. 1, p. 14650. Academy of Management, Briarcliff Manor, NY 10510 (2015). https://doi.org/10.5465/ambpp.2015.14650abstract
9. Pereira, A.: What are smart machines? Career in STEM (2019). https://careerinstem.com/what-are-smart-machines/

10. Strobl, C., et al.: Digital support for academic writing: a review of technologies and pedagogies. Comput. Educ. **131**, 33–48 (2019). https://doi.org/10.1016/j.compedu.2018.12.005

11. Belpaeme, T., Kennedy, J., Ramachandran, A., Scassellati, B., Tanaka, F.: Social robots for education: a review. Sci. Robot. **3**(21), 1–9 (2018). https://doi.org/10.1126/scirobotics.aat5954

12. Reich-Stiebert, N., Eyssel, F., Hohnemann, C.: Exploring university students' preferences for educational robot design by means of a user-centered design approach. Int. J. Soc. Robot. **12**, 1–11 (2019). https://doi.org/10.1007/s12369-019-00554-7

13. Zawacki-Richter, O., Marín, V.I., Bond, M., Gouverneur, F.: Systematic review of research on artificial intelligence applications in higher education – where are the educators? Int. J. Educ. Technol. High. Educ. **16**(1), 1–27 (2019). https://doi.org/10.1186/s41239-019-0171-0

14. Abildgaard, J.R., Scharfe, H.: A geminoid as lecturer. In: Ge, S.S., Khatib, O., Cabibihan, J.-J., Simmons, R., Williams, M.-A. (eds.) ICSR 2012. LNCS (LNAI), vol. 7621, pp. 408–417. Springer, Heidelberg (2012). https://doi.org/10.1007/978-3-642-34103-8_41

15. Cooney, M., Leister, W.: Using the engagement profile to design an engaging robotic teaching assistant for students. Robotics **8**(1), 21–47 (2019). https://doi.org/10.3390/robotics8010021

16. Guggemos, J., Seufert, S., Sonderegger, S.: Humanoid robots in higher education: evaluating the acceptance of Pepper in the context of an academic writing course using the UTAUT. Br. J. Educ. Technol. **51**(5), 1864–1883 (2020). https://doi.org/10.1111/bjet.13006

17. Masuta, H., et al.: Presentation robot system with interaction for class. In: 2018 Symposium Series on Computational Intelligence (SSCI), pp. 1801–1806. IEEE (2018). https://doi.org/10.1109/SSCI.2018.8628804

18. Bolea Monte, Y., Grau Saldes, A., Sanfeliu Cortés, A.: From research to teaching: integrating social robotics in engineering degrees. Int. J. Comput. Electr. Autom. Control Inf. Eng. **10**(6), 1020–1023 (2016). https://doi.org/10.5281/zenodo.1124667

19. Gao, Y., Barendregt, W., Obaid, M., Castellano, G.: When robot personalization does not help: insights from a robot-supported learning study. In: 2018 27th International Symposium on Robot and Human Interactive Communication, pp. 705–712. IEEE (2018). https://doi.org/10.1109/ROMAN.2018.8525832

20. Luan, H., et al.: Challenges and future directions of big data and artificial intelligence in education. Front. Psychol. **11**, 1–11 (2020). https://doi.org/10.3389/fpsyg.2020.580820

21. Macfadyen, L.P.: Overcoming barriers to educational analytics: how systems thinking and pragmatism can help. Educ. Technol. **57**, 31–39 (2017). https://www.jstor.org/stable/44430538

22. Burkhard, M., Seufert, S., Guggemos, J.: Relative strengths of teachers and smart machines: towards an augmented task sharing. In: Proceedings of the 13th International Conference on Computer Supported Education (CSEDU), vol. 1, pp. 73–83 (2021). https://doi.org/10.5220/0010370300730083

23. Floridi, L.: Hyperhistory, the emergence of the MASs, and the design of infraethics. In: Information, Freedom and Property: The Philosophy of Law Meets the Philosophy of Technology, vol. 153 (2016)

24. Floridi, L.: The Ethics of Information. Oxford University Press, Oxford (2013). https://doi.org/10.1093/acprof:oso/9780199641321.001.0001

25. Saini, M.K., Goel, N.: How smart are smart classrooms? A review of smart classroom technologies. ACM Comput. Surv. (CSUR) **52**(6), 1–28 (2019). https://doi.org/10.1145/3365757

26. Heaven, W.D.: OpenAI's new language generator GPT-3 is shockingly good—And completely mindless. MIT Technology Review (2020). https://www.technologyreview.com/2020/07/20/1005454/openai-machine-learning-language-generator-gpt-3-nlp/

27. Brown, T.B., et al.: Language models are few-shot learners. arXiv preprint (2020). https://arxiv.org/abs/2005.14165
28. Acemoglu, D., Restrepo, P.: The wrong kind of AI? Artificial intelligence and the future of labour demand. Camb. J. Reg. Econ. Soc. **13**(1), 25–35 (2020). https://doi.org/10.1093/cjres/rsz022
29. Aoun, J.E.: Robot-Proof: Higher Education in the Age of Artificial Intelligence. MIT Press, Cambridge (2017)
30. Davenport, T.H., Kirby, J.: Only Humans Need Apply: Winners and Losers in the Age of Smart Machines. Harper Business, New York (2016)
31. Dellermann, D., Ebel, P., Söllner, M., Leimeister, J.M.: Hybrid intelligence. Bus. Inf. Syst. Eng. **61**(5), 637–643 (2019). https://doi.org/10.1007/s12599-019-00595-2
32. Wilson, H.J., Daugherty, P.R.: Collaborative intelligence: humans and AI are joining forces. Harvard Bus. Rev. **96**(4), 114–123 (2018)
33. Bradshaw, J.M., Feltovich, P.J., Johnson, M.: Human–agent interaction. In: The Handbook of Human-Machine Interaction, pp. 283–300. CRC Press (2011). https://www.taylorfrancis.com/chapters/edit/10.1201/9781315557380-14/human%E2%80%93agent-interaction-jeffrey-bradshaw-paul-feltovich-matthew-johnson
34. Ranz, F., Hummel, V., Sihn, W.: Capability-based task allocation in human-robot collaboration. Proc. Manuf. **9**, 182–189 (2017). https://doi.org/10.1016/j.promfg.2017.04.011
35. Musto, M.: Revisiting Marx's concept of alienation. Socialism Democracy **24**(3), 79–101 (2010). https://doi.org/10.1080/08854300.2010.544075
36. Wogu, I.A.P., et al.: Artificial intelligence, alienation and ontological problems of other minds: a critical investigation into the future of man and machines. In: 2017 International Conference on Computing Networking and Informatics (ICCNI), pp. 1–10 (2017). https://doi.org/10.1109/ICCNI.2017.8123792
37. GETChina Insights. Schools using facial recognition system sparks privacy concerns in China. Medium (2019). https://edtechchina.medium.com/schools-using-facial-recognition-system-sparks-privacy-concerns-in-china-d4f706e5cfd0
38. Sharkey, A.J.C.: Should we welcome robot teachers? Ethics Inf. Technol. **18**(4), 283–297 (2016). https://doi.org/10.1007/s10676-016-9387-z
39. Pardo, A., Siemens, G.: Ethical and privacy principles for learning analytics. Br. J. Educ. Technol. **45**(3), 438–450 (2014). https://doi.org/10.1111/bjet.12152
40. Sohn, K., Kwon, O.: Technology acceptance theories and factors influencing artificial Intelligence-based intelligent products. Telematics Inform. **47**, 101324 (2020). https://doi.org/10.1016/j.tele.2019.101324
41. Frey, C.B., Osborne, M.A.: The future of employment: how susceptible are jobs to computerisation? Technol. Forecast. Soc. Change **114**, 254–280 (2017). https://doi.org/10.1016/j.techfore.2016.08.019
42. Mubin, O., Stevens, C.J., Shahid, S., Mahmud, A.A., Dong, J.-J.: A review of the applicability of robots in education. Technol. Educ. Learn. **1**(1), 1–7 (2013). https://doi.org/10.2316/Journal.209.2013.1.209-0015
43. Beumelburg, K.: Fähigkeitsorientierte Montageablaufplanung in der direkten Mensch-Roboter-Kooperation. Doctoral dissertation. Institut für Industrielle Fertigung und Fabrikbetrieb (IFF), University of Stuttgart (2005)
44. Ruffin, R.: David Ricardo's discovery of comparative advantage. Hist. Polit. Econ. **34**(4), 727–748 (2002)
45. Landsburg, L.F.: Comparative Advantage. The Library of Economics and Liberty (n.d.). https://www.econlib.org/library/Topics/Details/comparativeadvantage.html
46. Ricardo, D.: Principles of Political Economy and Taxation. G. Bell and Sons, London (1891)
47. Latham, S., Humberd, B.: Four ways jobs will respond to automation. MIT Sloan Manag. Rev. **60**(1), 11–14 (2018)

48. Harteis, C., Billett, S.: Intuitive expertise: theories and empirical evidence. Educ. Res. Rev. **9**, 145–157 (2013). https://doi.org/10.1016/j.edurev.2013.02.001

49. Süzen, N., Gorban, A.N., Levesley, J., Mirkes, E.M.: Automatic short answer grading and feedback using text mining methods. Proc. Comput. Sci. **169**, 726–743 (2020). https://doi.org/10.1016/j.procs.2020.02.171

50. Mumtaz, S.: Factors affecting teachers' use of information and communications technology: a review of the literature. J. Inf. Technol. Teach. Educ. **9**(3), 319–342 (2000). https://doi.org/10.1080/14759390000200096

51. Dillenbourg, P.: The evolution of research on digital education. Int. J. Artif. Intell. Educ. **26**(2), 544–560 (2016). https://doi.org/10.1007/s40593-016-0106-z

Prediction of Students' Performance Based on Their Learning Experiences and Assessments: Statistical and Neural Network Approaches

Ethan Lau[✉], Vindya Wijeratne, and Kok Keong Chai

School of Electronic Engineering and Computer Science, Queen Mary University of London,
10 Godward Square, Mile End Road, London E1 4FZ, UK
{e.lau,vindya.wijeratne,michael.chai}@qmul.ac.uk

Abstract. The education approaches in the higher education have been evolved due to the impact of covid-19 pandemic. The predicting of students' final performance has become more crucial as various new learning approaches have been adopted in the teaching. This paper proposes a statistical and neural network model to predict students' final performance based on their learning experiences and assessments as the predictor variables. Students' learning experiences were obtained through educational data analytic platform on a module that delivered the mixed-mode education strategy using Flipped classroom, asynchronous and cognitive learning in combination with the revised Bloom's taxonomy. Statistical evaluations including multiple regressions, ANOVA correlations are performed to evaluate the appropriateness of the input variables used for the later Neural Network output prediction. The Levenberg-Marquardt algorithm is employed as the training rule for the Neural Network model. The performance of neural network model is further verified to prevent the overfitting issue. The Neural Network model has achieved a high prediction accuracy justifying the students' final performance through utilising the aforementioned pedagogical practises along with limitations.

Keywords: Pedagogic practise · Mixed-mode learning · Statistical evaluations · Neural network modelling

1 Introduction

Ever since the COVID-19 pandemic, many students in the world are still affected by normal way of educational learnings. The high level world ministerial meeting back in March 2021 highlighted the importance of prioritising the COVID-19 education recovery to avoid substantial generational catastrophe [1]. Consequently, many universities change their conventional education to mixed-mode learning (MML). MML is getting prevalent ever since the global pandemic, where various pedagogic methods within the MML are rigorously reviewed to apply the best theory and practise of learning among students. However, the evaluation methods to quantify the effectiveness of the MML towards students' learning experiences and their exam performance

© Springer Nature Switzerland AG 2022
B. Csapó and J. Uhomoibhi (Eds.): CSEDU 2021, CCIS 1624, pp. 24–39, 2022.
https://doi.org/10.1007/978-3-031-14756-2_2

are extremely limited at present. Moreover, the conventional questionnaire and feedback surveys from students unfortunately are not sufficiently enough to identify the critical factors attributed to students' experiences in learning.

To evaluate the effectiveness of MML towards students' learning experiences and their final exam, this paper presents the statistical and neural network modelling approach for such evaluations. We present how the modular programming module trialled had successfully implemented the MML, where students' learning experiences and the final examination could be justified and predicted using neural network modelling in addition to the conventional statistical analysis. The MML is entirely based on the Bloom's taxonomy in combination with the flipped classroom, asynchronous and cognitive learning. The delivery of the module comprises of online/offline teaching, learning and assessments. The data metric information about students' learning experiences apart from the assessments are obtained through the educational data analytic software platform. Apart from that, the assessment scores such as the lab and the class test are extracted for statistical evaluations and the later Neural Network (NN) development to predict students' final examinations. The NN model serves as an important tool to bridge students' learning experiences and assessments with their final exams, and finally improve the students' learning outcomes.

The organisation of the paper is as follows: Sect. 2 presents the reviews of mixed-mode learning the NN and the summary, Sect. 3 presents the methodology of conducting the module utilising the MML approach, education dataset information and retrieval, statistical and NN modelling, Sect. 4 presents the results of the statistical evaluations, NN configuration settings and the performance, Sect. 5 concludes the findings.

2 Overview of the Teaching Practises

Bloom's taxonomy is the hierarchical-based pedagogical tool to help educators to classify and specify the education learning objectives [2]. The taxonomy has been applied in software engineering and computer science streams [3–5] and also has improved the assessment methods in science, technology, engineering and mathematics (STEM).

Ever since the COVID-19 pandemic, the pedagogy in the educational field is changing continuously, as hybrid pedagogic strategies are bringing into the balance of stratifying students' learning experiences and interactions, whether there are both offline or live lectures with other assessments such as exercises and labs. Moreover, lectures nowadays have the trends of allowing more spaces for students to elicit the information. The revised Bloom's Taxonomy was proposed to classify the learning content that allows the content to fit into the student-centred learning domain that will further promote critical learning skills among students [6]. The end of the classification provides the best teaching practise model for educational delivery. Apart from that, each of the category within the Bloom's Taxonomy is reviewed extensively to strive for realistic but outstanding students' academic performance, where such encompasses partial or completely redesigning the delivery of lectures, lab and tutorials. Additionally, students are enthused especially for virtual hands-on online education by fully leveraging the realistic exercises that further encourage students thinking at upper levels of Bloom's taxonomy [7].

Several recent papers paid a huge emphasis on revising Bloom's taxonomy to cope with the pandemic. A paper by [8] presented the steps using metacognitive-skills for continuous learning in combination with Bloom's taxonomy to enforce students' learning capability in the normal, online and even in the disrupted settings. This is because the pandemic impacts the usual way of face-to-face teaching, thus hinders the interactivity needed for students to gain the sufficient knowledge. Moreover, the pandemic has dramatically changed the way where the good performance among students should not be measured just in the ability to remember and memorise but more importantly, to focus on students' dissemination of knowledge [8]. The revised Bloom's taxonomy guidance from [9] was followed by the author [10] to mitigate students' mental health issues related to the pandemic.

In summary, Bloom's taxonomy, although being existed for decades, is still strongly inaugurated in the current era. To accommodate the students' learning needs during and beyond COVID-19, for this reason, the revised Bloom's taxonomy is applied this paper that conforms to the MML methods considered – flipped classroom, asynchronous and cognitive learning. Such intends to create a balance of joint teaching elements while ensuring students' learning outcomes are met.

The revised Bloom's taxonomy with MML practices is presented in Fig. 1. With flipped classroom, students were at first being introduced the conceptual information at off-campus (pre-classroom activities, watching pre-recorded videos, offline quizzes and take home exercises). Later, problem-based solving approach is applied where students corporate with the lecturer to solve the exercises during the live lectures (virtual classroom activities). Then, the later assessments include the practical lab exercise, class test and the final exam. Such will later be assessed that concludes the part of '*Evaluating*' and '*Creating*' in Bloom's taxonomy where students' learning experiences and their assessment results will be used later for the NN modelling. The MML model in this case serves as a strategy to circumvent the traditional approach of lecture-centred style of module delivery, but instead strengthen students' conceptual understanding to solve problems and be able to enact the situations [11]. The author [11] affirmed that conceptual teaching approach improved students' comprehension than the traditional procedural-based approach. This is because students like to link their study into the practicality of the real applications. In addition, several existing benefits of using flipped classroom have also already been reported in some of the research works [12–14].

The cognitive learning is the second part of the MML. Such activates students' cognitive-based prior knowledge. Flipped-classroom is tightly related with activating prior cognitive knowledge on knowledge and comprehension understanding within Bloom's taxonomy to enable students to learn actively [16]. Such mechanism fosters the conceptual understanding of STEM and increases students' self-managed responsibility in connecting to the lecture.

The author [17] concluded that the Bloom's taxonomy especially in the pandemic helps to describe the cognitive demand needed by lecturers from '*Testing, Evaluating*', and finally transform into '*Generating, Planning and Producing*'. Lectures are not only being guided by the Bloom's taxonomy but also making the teaching, delivery and knowledge transfer easier. With the activation of students' cognitive prior knowledge of the materials studied before, this allows the creation of ideas and new information. In the next level, the cognitive approach covers the high-order of thinking skill by requiring

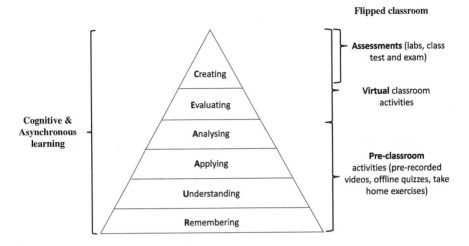

Fig. 1. The revised Bloom's taxonomy. Adapted and modified from the author [15].

students to gather information, apply knowledge, solve or improve a problem. In contrast, the implication of Bloom's Taxonomy in STEM courses was discovered by the author [18] where lower-order cognitive skills increase the overall average exam marks through Bloom's taxonomy and in contrast, decrease in marks for subjects requiring higher-order cognitive skills.

The third part of the pedagogy strategy in MML is the asynchronous learning. Asynchronous in this case enables flexible way of learning as it focuses on "*Anytime, Anywhere*" basis. Such encompasses lecturer-led virtual/remote meetings, module forums, chats, recorded lectures, and tutorials [19]. The author [20] concluded that the asynchronous strategy embedded with cognitive learning can even outperform the traditional face-to-face lecturing and learning approach.

In summary, the revised Bloom's taxonomy in combination with the MML approach intends to motivate and give more spaces for students to learn on/off-line continuously while ensuring students' interactions during virtual meetings, and also, allows student to stay resilient when working alone or with group members during or beyond the COVID-19.

2.1 Neural Network in Educational Context

At present, it is challenging to utilise the conventional methods such as questionnaire feedback to quantify the effectiveness of students' learning experience. Moreover, feedback questionnaire survey data may not help to improve students' performance [21], as some of the questions may not truly reflect the students' needs in improving their learning experiences. Therefore, it is paramount to predict students' academic performance through students' information about their academic learning behaviour from the online-learning platforms, and not depending solely on the reflection on teachings.

Currently, various MML can actually be modelled and further quantified using machine learning techniques for educational improvement purposes. The author [21] applied machine learning in predicting students' successfulness in blended learning by collecting 13 distinct datasets from Moodle based on different aspects of student interactions such as cognitive presence, social presence and teaching presence within the virtual learning. The prediction helped to detect earlier at-risks students so that further negative consequences among students' can be mitigated. The author [18] applied Bayesian linear mixed effects model to analyse and compare COVID-19 effects on students' examination marks. Additionally, the author [22] applied machine learning techniques to classify the low-engagement students, intervening their learning behaviour and improving the accuracy of the prediction that will improve students' learning engagements.

The neural network (NN) model is the fundamental machine learning mechanism and yet has achieving lots of attention for grabbing the relationship between the influencer (the factor variables) and the result of the prediction (the target goal). In terms of the educational domain, it has the capability of modelling students' information into an abstract domain, and to further predict students' performance though the influencer variables (i.e. their learning experiences and interactions). NN is made up by collections of artificial neurons that are trained to mirror and simulate the connection geometry of neurons in biological brains. With the advancement of computational resources and large variety of datasets in computer science, NN is used in numerous applications to improve performance through series prediction, classification and data processing [23, 24]. Impressively, NN modelling has been well received in educational research field especially in prediction students' academic performance [25, 26] and students' courses selection [24].

Therefore, this paper applies NN modelling to predict of students' exam performance using their learning experiences and module engagements as the predictor variables for the statistical and NN modelling. The modelled NN serves as the tool predictor to forecast students' academic performance through the data obtained from students' learning experiences and their engagements to the module, as well as their assessment results to further quantify the prediction on their academic performance.

3   Methodology

This section presents the methodology of acquiring the educational dataset information through the MML approach utilising the revised Bloom's taxonomy, the statistical and NN modelling methods, and performance evaluation of NN model.

3.1   Module Overview

The modular programming module is considered for this study where it aims to emphasises heterogenous functionality of several programs into a loosely-coupled, independent and modular applications. The main learning outcome is to expose students with different modular programmings available and to further design and implement the modular applications.

A total of 180 Year 3 undergraduate students are enrolled to the module. The module is delivered based on the mixture of MML with live (virtual) and pre-recorded lectures, virtual tutorials, offline quizzes, online class test quizzes, and offline lab exercises.

There are four sessions of lab exercises (progressive-based) in 2th, 4th, 6th and 8th of the teaching weeks. Lab task difficulties are increased where for the last two lab exercises students will attempt the lab tasks remotely using the cloud-computing web services. Overall, for all the summative and formative assessments (except the final exam), students accomplish all the tasks through the Moddle learning portal. The class test is held in Week 9th. All assessments conform to the aspects of the UK Quality Code for Higher Education.

3.2   The Mixed-Model Learning

The revised Bloom's taxonomy in this paper includes the MML practises – flipped classroom, cognitive and asynchronous learning. An example of asynchronous learning schedule utilising the *anytime, anywhere* is shown in Table 1. Students have more flexibility to complete the offline tasks (e.g. offline quizzes, homework, readings and watching recorded videos) in their own time or, complete the tasks within the recommended slots. It is not the scope of this paper to improve the Bloom's taxonomy based on the MML strategy but takes students to a higher cognitive level. Microsoft Teams and the Moodle are used as the core lecturing, learning and the course management platform to enable students to access all the required resources. The module team channel is also created using Microsoft Teams to promote better interactions with students to have more spaces for them to ask questions related to their learnings.

Table 1. Timetable schedule.

Session	Week 1	Week 2	Week 3	Week 4
Thursday, 08:00–09:00 am	Intro-Live	Offline	Live	Tutorial
Thursday, 09:00–10:00 am	Live	Live	Offline	Tutorial

Flipped classroom, the second part of MML is implemented by having students to achieve tasks during offline classes and utilise the problem-solving exercises during the live interactions. Even though this is a reverse of the common pedagogical practise, it becomes more prevalent for the effectiveness of students' interactions especially in the pandemic. By combining the asynchronous learning practise, students can accomplish the offline tasks in the manner of the time as they wish or, based on the recommended timetable for the offline slots as shown in Table 1. Students' accomplishment of offline tasks are critically monitored so as not to overwhelm them with their study loads.

Finally, for the cognitive learning, all the questions and tasks covering the tutorials, quizzes, labs and class test are carefully designed to activate students' prior knowledge in programming concepts and previously learned modules (e.g. Java programming module) that are the requisite for the students to study the modular programming module. Those serves as a pre-cognitive domain to activate students' prior knowledge, and to further apply the cognitive knowledge that facilitates the conceptual understanding of the module during the live lecture session. Of course, this practise has to be fitted into the flipped-classroom and asynchronous strategy to realise students' learning experiences and their self-regulated responsibility in engaging the module.

Overall, the module is assessed by 15% coursework (four labs and a class test) and a 85% closed-book final examination.

3.3 Educational Dataset

Summative assessments from the labs, class test and final exam are collected for the later statistical and NN modelling. Data information about students' learning experiences is obtained through the Smart video capture and data analytic platform. It is not intended to perform questionnaire and feedback survey in this case as the main intention is to see how students' interaction with the online learning platform relates to their overall learning experience, where such represent students' behaviour of interaction in virtual classes. The education dataset of students' learning experience is extracted from the Echo360's lecture recording platform [27]. The platform rolls the course data analytics for enrolled students (e.g. engagement, slide views, video views, polling responses, video viewing and attendances apart from facilitating the lecture recordings).

For simplicity purposes, the data metrics *Total engagement* and *Attendance* are extracted from Echo360 platform. The *Total Engagement* metric is the cumulative total of data points such as Video Views, Q&A Entries and Polling Responses. On the other hand, *Attendance* metric contains the attendance where students attempt the online classroom during the class time and also watch the pre-recorded videos. All students' information are kept anonymous and also in line with the General Data Protection Regulation (GDPR)'s data protection policy [28].

3.4 Statistical Testing

Statistical modelling is fundamental not only for the evaluation of relationship between students' learning experiences and their assessment results, but also as a requisite for later NN modelling. Statistical analysis in this paper includes multiple regression analysis that predicts a vector variable based on two or more different vectors of variables. In other words, multiple regression analysis is useful for the NN modelling verification as it determines whether the target (dependent) variable could be predicted through the independent variables. Analysis of Variance (ANOVA) is also used in conjunction with multiple regression analysis to check whether the independent variables influence the dependent variable in a regression model.

Apart from that, Pearson correlation coefficients are applied to measure the linear relationships of the independent variables (four lab assessments, the metrics *Total engagement* and *Attendance* from Echo360, the class test score) towards the final exam score (as the dependent variable). Similar with multiple linear regression, the correlated independent variables are used as the inputs for the NN model.

3.5 Modelling of Neural Networks

The NN is applied to predict the students' academic performance as the target through the input information of students' learning experiences and engagement metrics from Echo360's data analytic platform, and as well the summative assessments labs and the class test. Such scheme is performed based on the supervised machine learning. The NN model is developed using the earlier research works [15,23,29–31]. It is not the scope of this paper to determine the performance of different supervised learning model.

By default, the NN mathematical model is formulated as:

$$B = f(A, W),$$ (1)

where B and A are the output and input vectors. W is the weight vector that learns the behaviour of the data.

The input layer contains the initial data for the NN and is fed to the hidden layer. The hidden layer sits between the input and the output layer and it is where the main computation is done here. Then, the weighted sum of input elements a and w contributes to the output neuron values b_j in for the vector B, where w is updated recursively.

$$b_j = \theta \left(\sum_{i=1}^{N_i} w_{ij} a_i \right).$$ (2)

The i refers tho the current ith connection line to jth neuron. The a_i is the output value from the previous ith neuron. Then, θ is the activation function. N_i is the total number of connection lines.

The hyperbolic tangent in this case is the activation function (θ) that transfers the weighted sum of inputs to the output layer b_j. The node is then fed to the next input layer a_j as:

$$a_j = \theta(b_j).$$ (3)

For training, the supervised learning with back-propagation (BP) approach is applied. BP adjusts w_{ij} from the computed errors to iteratively correct the desired outputs. The error function (E) is the sum of square difference between the target values and the desired outputs for jth neuron:

$$E = \frac{1}{2} \sum_{j}^{N_j} (b_j - t_j)^2,$$ (4)

where t_j is the target value and N_j is the total number of output neurons.

The optimisation algorithm is further applied that performs the training and learning process in a NN. The Levenberg-Marquardt optimisation algorithm is used as the BP-based optimiser. It is a training method using the gradient descent and Gauss-Newton method. It speeds up the convergence to an optimal solution for non-linear problems [32,33]. The algorithm does not introduces the exact Hessian Matrix. Instead, it works on the Jacobian matrix and the gradient vector [33,34]:

$$w_{ij+1} = w_{ij} - [J'J + \zeta I]^{-1} J' \omega_k,$$ (5)

where J is the Jacobian matrix, ω_k is the error [34], w_{ij} and w_{ij+1} are the current and updated weight, ζ is the damping factor.

The Levenberg-Marquardt training algorithm in Eq. 5 is switched between the Gauss-Newton (when ζ is small) and gradient descent algorithm (when ζ is large). The ζ has to be adjusted at every iteration that further guides the optimisation process in the NN training model.

Finally, the output layer will consist the final decision of the prediction based on NN [35]. The output layer is the collection of vector B, that is the predicted students' final examination.

3.6 NN Performance Evaluation Criteria

The Mean Square Error (MSE) is used to evaluate the performance of NN. It is also prevents the arising of over-fitting issues. A properly-trained NN model should result in low MSE value. It is formulated as:

$$MSE = \frac{1}{N_i N_j} \sum_{j=1}^{N_j} \sum_{i=1}^{N_i} (b_{ij} - t_{ij})^2. \tag{6}$$

Even though the low MSE value may denote the good performance of NN, the over-fitting of the NN is still possible. A low MSE value but with the over-fitting scenario implies the good NN model in the training stage, but not in validation and testing phase. This means that the NN model is 'memorising' the historical data but will not respond well to the changes of the observation data.

In order to prevent the issue of over-fittings, a regression is performed to compute the R-value that further validates the goodness-of-fit between the predicted and the desired outputs and the overall fitting performance. In contrast, further trainings are required if low R-value is obtained (poor fitting) by tuning the hidden layers and neurons.

Lastly, the error histogram is also used to validate the NN performance where the histogram illustrates the error distributions with most errors are occurred at near zeroth point. The error value is difference between the targeted t_{ij} and the predicted outputs b_{ij}.

4   Results

This section at first presents the statistical evaluation using multiple regression, ANOVA and Pearson Correlation coefficient. The later includes the NN configuration, its results and the performance verification. All the modelling here are based on the extracted datasets of students' engagement and the attendance metrics from Echo360, and the assessment results include the labs, class test, and the final exam.

Figure 2 shows the overview distributions of the students' final examination grades for the academic year 2019/20 (pre-pandemic) and 2020/21. The breakdown of the results showed that for the 2020/21 academic year, approximately 25% got As, 32% Bs, 16% Cs and unfortunately 12% failed the exam (Fs). Compared with 2019/20 results, they were fewer A's (44.94%) and more F's 8 (4.49%). Apart from that the distribution of B C D and Es' are similar. The overall distribution of the year 2020/21 exam result is better than the year 2019/20. More practical typed exam questions are introduced for the 2020/21 exams and impressively 45 first sit students received grade A in the exam along with the outstanding assessment marks. This shows that the revised Bloom's Taxonomy with MML has successfully improved students' comprehension and creation of knowledge at upper level, thus students scored well for the examinations.

In addition, the average class test score for the year 2020/21 is 78.19 ± 10.10 compared with 81.02 ± 11.18 for the previous academic year, which stays fairly consistent when the teaching model was very different.

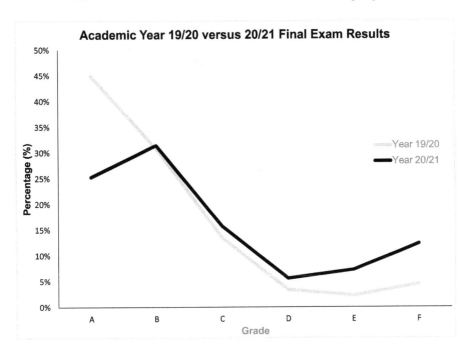

Fig. 2. Academic performance result for the year 19/20 (pre-pandemic) and 20/21.

4.1 Statistical Evaluations

Multiple regression analysis and ANOVA are applied in this paper. The multiple regression analysis validates whether the final exam (the dependent variable) could be predicted for later NN modelling using the independent variables – Lab and class test scores, metrics of students' learning experience such as *Total Engagement* and the *Attendance* from the Echo360 platform. Outcomes of the statistical analysis are presented three main outputs: 1) the regression summary of the multiple regression analysis; 2) ANOVA and 3) statistical significance analysis of the independent variables.

Table 2 reveals the R, R^2, adjusted R^2 and estimations of the standard error. The R value of 0.871 denotes that the independent variables are indeed the good predictors against the dependent variable (the final exam score). Additionally, the adjusted R^2 value of 0.512 for the independent variables contribute to 52.1% variability of the dependent variable.

The ANOVA tests the goodness of fits for the multiple regression analysis. The significant level (α) is set as $\alpha = 0.05$. Table 3 shows the ANOVA result. As ($\alpha < 0.05$), the variables considered are a good fit for the NN model.

The second ANOVA test is performed to check examine the difference in means of all four lab assessment scores. The result is shown in Table 4. As ($\alpha > 0.05$), there is no significant difference of all four lab assessment mean scores among students. Generally, the results show that students' lab performance scores are not influenced by the order of lab assessments and their performances in individual lab assessments are independent with each other labs being carried out.

The final statistical testing analysis for each of the independent variables in relation to the dependent variable is shown Table 5. As $\alpha < 0.05$, all independent variables are statistically significant to the dependent variable (the final exam score).

Table 2. Summary of the regression model.

Multiple R	0.871
R^2	0.521
Adjusted R^2	0.512
Standard error	5.765

Table 3. ANOVA result.

Source	Sum of square	d.f.	Mean square	F-cal	α
Regression	690.47	6	90.24	1.563	0.01
Residual	11600.45	176	55.40		
Total	12512.87	182			

Table 4. ANOVA lab assessment results.

Source	Sum of square	d.f.	Mean square	F-cal	α
Between groups	484.74	3	161.58	0.79	0.50
Within groups	89213.76	436	204.62		
Total	89698.50	439			

Table 5. Statistical significance of the independent variables.

Model	Coefficients	Standard error	t Stat	α
(Intercept)	89.93	6.210	11.212	0.001
Lab 1	−0.333	0.075	−3.24	0.001
Lab 2	−0.218	0.052	−3.808	0.005
Lab 3	−0.442	0.043	12.112	0.001
Lab 4	−0.134	0.059	−2.796	0.004
Total engagement	−0.132	0.232	−3.232	0.002
Attendance	13.1345	1.222	6.933	0.002
Class Test	5.136	1.54	4.456	0.002

Pearson correlation coefficients are computed to evaluate the relationship among the students' engagement metrics (*Total engagement* and *Attendance*), the assessments such as the labs and class test towards the resultant final exam score. The correlation coefficients between lab scores, total engagement, attendance, class test score and the final exam score are shown in Table 6. The degree of closest relationship with ascending order are: Lab 1, Lab 2, Lab 3, Lab 4, Attendance, Total Engagement, and the Class Test. Unsurprisingly, it can be seen that the correlation coefficient value increases as students progress themselves from the labs to the class test. Apart from that, both the Class Test and Engagement metric show the highest relationship with the final exam score. This shows the adopted MML strategy with revised Bloom's Taxonomy successfully guides the students to step forward to the final exam. The finding supports the results from the author [36] where students could perform well even for programming courses in both online and offline classes, as long as students' cognitive levels are prioritised as high as application and analysis.

These correlated variables are used further as input neurons for the NN modelling.

Table 6. Correlation coefficient of four lab scores, total engagement and attendance, the class test and the final exam score.

	Lab 1	Lab 2	Lab 3	Lab 4	Total engagement	Attendance	Class test	Final exam
Lab 1	1							
Lab 2	0.84	1						
Lab 3	0.70	0.70	1					
Lab 4	0.85	0.84	0.80	1				
Total engagement	0.31	0.30	0.28	0.31	1			
Attendance	0.34	0.31	0.28	0.32	0.92	1		
Class test	0.36	0.37	0.40	0.41	0.36	0.42	1	
Final exam	0.35	0.43	0.48	0.49	0.68	0.63	0.70	1

4.2   NN Configuration Settings

The MathWorks MATLAB software is used for the modelling of NN and its evaluations. The input layer consists of seven variables containing all the four lab assessment score, the Engagement and Attendance metrics and the class test score. The NN model contains two hidden layers. Each of the hidden layer comes with 15 neurons where setting was tested and produced the best result throughout several iterations with different settings. The hidden layers are fed into a final single output neuron that brings the decision of the variable (final exam score prediction).

The hyperbolic tangent is used as the activation function. Data samples of 180 students are randomly mixed into the training, validation and test ratio. Therefore in this case the training, validation and the testing ratio is 0.7, 0.15 and 0.15.

For the training and learning phase, Levenberg-Marquardt algorithm (Eq. 5) is applied find the optimal weights. The damping factor ζ is 0.001 and the training epoch is set to 2,000. The NN training runs and stops when the error validation failed to decrease for six iterations consecutively. Typically, the validation process is important that protects over-training of NN.

4.3 NN Simulation Results

The regression plots of NN model is shown in Fig. 3. In this case, the regression plots achieve good fits with an overall R-value of 0.8773. This proves a good data fitting where the predicted outputs b_{ij} converge closely with the target outputs t_{ij}.

The MSE of $\approx$ 42.2 or 6.9% ($<$10%) signals the good performance of NN in completing training and validation runs.

The final error histogram evaluates the error distributions after the NN predictions. The error distributions are shown in Fig. 4. Most errors occurred near zeroth point at horizontal axis are gradually decreasing when shifting away from zeroth point. This further conjectures that NN has successfully performs the predictions with acceptable error distributions.

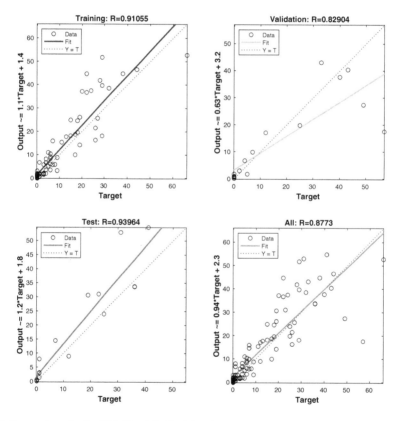

Fig. 3. Regression plot for NN: Top left: Training data; Bottom left: Test data; Top right: validation data; Bottom right; overall regressions.

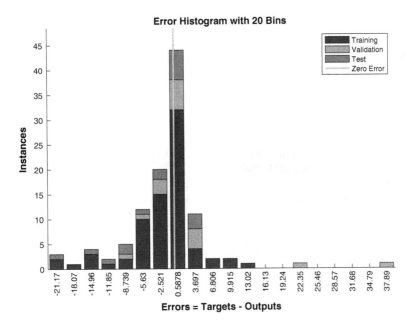

Fig. 4. Error histogram plot.

5 Conclusions

The modular programming module considered in this paper applied the pedagogical practise of revised Bloom's taxonomy in combination with the MML strategy – flipped classroom, asynchronous and cognitive learning throughout the pandemic period. The effectiveness of the module delivery especially the students' final exam performance is evaluated, modelled and predicted using the statistical and NN modelling. Students' learning experiences and assessments were used as the independent variables to predict the students' final exam. Summative assessments from the labs, class test and final exam are collected. The data information about students' learning experiences is obtained from the Echo360's data analytic platform that include the metric *Total engagement* and *Attendance*.

Through the multiple regression analysis and ANOVA, all the independent variables showed statistically significance to the dependent variable. Additionally, the correlation analysis showed the positive relationships of all the independen t variables to the dependent variable, where the Class Test followed by the Engagement metric showed the positive correlation with students' final exam performance. These correlated variables were used for the NN modelling and verification.

Overall, the NN modelling showed a good prediction against students' final exam performance. However, the notable limitation in this study is low sample data sizes as the participation in the study is voluntary. Nevertheless, the education modelling settings using machine learning would further assist the fellow educators to understand students' learning experiences and their learning outcome, apart from improving students' academic performance. The immediate work could include the inclusion of other independent variables. This could be putting lecturers' experiences in the module together with the students as a new perspective. The proposed model is highly modularised in fitting into other courses and thus enabling a smart educational pathway system education for the future.

References

1. UNESCO: Education: From disruption to recovery (2021). https://en.unesco.org/covid19/educationresponse. Accessed 08 Nov 2021
2. Granello, D.: Promoting cognitive complexity in graduate written work: using bloom's taxonomy as a pedagogical tool to improve literature reviews. Couns. Educ. Superv. **40**(4), 292–307 (2001)
3. Fuller, U., et al.: Developing a computer science-specific learning taxonomy. ACM SIGCSE Bull. **39**(4), 152–170 (2007). https://doi.org/10.1145/1345375.1345438
4. Britto, R., Usman, M.: Bloom's taxonomy in software engineering education: a systematic mapping study. In: 2015 IEEE Frontiers in Education Conference (FIE), pp. 1–8 (2015). https://doi.org/10.1109/FIE.2015.7344084
5. Peter, D., Leth, T., Bent, T.: Assessing problem-based learning in a software engineering curriculum using bloom's taxonomy and the IEEE software engineering body of knowledge. ACM Trans. Comput. Educ. **16**(3), 1–41 (2016)
6. Shi, H., et al.: Educational management in critical thinking training based on bloom's taxonomy and solo taxonomy. In: 2020 International Conference on Information Science and Education (ICISE-IE), pp. 518–521. IEEE (2020). https://doi.org/10.1109/ICISE51755.2020.00116
7. Luse, A., Rursch, J.: Using a virtual lab network testbed to facilitate real-world hands-on learning in a networking course. Br. J. Educ. Technol. (BERA) **52**, 1244–1261 (2021)
8. Qadir, J., Al-Furqaha, A.: A student primer on how to thrive in engineering education during and beyond COVID-19. Educ. Sci. **10**(9), 236–258 (2020)
9. Krathwohl, D., Anderson, L.: A Taxonomy for Learning, Teaching, and Assessing: A Revision of Bloom's Taxonomy of Educational Objectives. Longman, London (2001)
10. Sheth, S., et al.: Development of a mobile responsive online learning module on psychosocial and mental health issues related to covid 19. Asian J. Psychiatr. **54**, 102248 (2020)
11. Joffrion, H.: Conceptual and procedural understanding of Algebra concepts in the middle grades. Master's thesis, Office of Graduate Studies of Texas A&M University (2005)
12. Peterson, D.: The flipped classroom improves student achievement and course satisfaction in a statistics course: a quasi-experimental study. Teach. Psychol. **43**, 10–15 (2016)
13. Foldnes, N.: The flipped classroom and cooperative learning: evidence from a randomised experiment. Act. Learn. High. Educ. **17**, 39–49 (2016)
14. Shraddha, B.H., et al.: Enhanced learning experience by comparative investigation of pedagogical approach: flipped classroom. Procedia Comput. Sci. **172**, 22–27 (2020)
15. Lau, E., Chai, K., Goteng, G., Wijeratne, V.: A neural network modelling and prediction of students' progression in learning: a hybrid pedagogic method. In: 13th International Conference on Computer Supported Education (CSEDU), pp. 84–91. Scitepress (2021). https://doi.org/10.5220/0010405600840091
16. Kostons, D., Werf, G.: The effects of activating prior topic and metacognitive knowledge on text comprehension scores. Br. J. Educ. Psychol. **85**(3), 264–275 (2015)
17. Setyowati, Y., Heriyawati, D., Kuswahono, D.: The implementation of 'test of evaluating' and 'test of creating' in the assessment of learning by EFL lectures in pandemic era. J. Lang. Teach. Learn. Linguist. Lit. **8**(2), 578–587 (2020)

18. Tomal, J., Rahmati, S., Boroushaki, S., Jin, L., Ahmed, E.: The impact of COVID-19 on students' marks: a Bayesian hierarchical modelling approach. J. Lang. Teach. Learn. Linguist. Lit. **79**, 57–91 (2021)

19. Wu, D., Bieber, M., Hiltz, S.: Engaging students with constructivist participatory examinations in asynchronous learning networks. J. Inf. Syst. Educ. **19**(3), 321–330 (2008)

20. Michalsky, T., Zion, M.: Developing students' metacognitive awareness in asynchronous learning networks in comparison to face-to-face discussion groups. J. Educ. Comput. Res. **36**(4), 395–424 (2007)

21. Macarini, L., Cechinel, C., Machado, M., Ramos, V., Munoz, R.: Predicting students success in blended learning-evaluating different interactions inside learning management systems. Appl. Sci. **9**, 1–23 (2019)

22. Hussain, M., Zhu, W., Zhang, W., Abidi, S.: Student engagement predictions in an e-learning system andtheir impact on student course assessment scores. Comput. Intell. Neurosci. **6**, 1–21 (2018)

23. Vandamme, J., Meskens, N., Superby, J.: Predicting academic performance by data mining methods. Educ. Econ. **15**(4), 405–419 (2007)

24. Kardan, A.A., Sadeghi, H., Ghidary, S., Sani, M.: Prediction of student course selection in online higher education institutes using neural network. Comput. Educ. **65**, 1–11 (2013)

25. Isljamovic, S., Suknovic, M.: Predicting students' academic performance using artificial neural network: a case study from faculty of organizational sciences. In: ICEMST 2014: International Conference on Education in Mathematics, Science and Technology, pp. 68–72. ISRES Publishing (2014)

26. Okubo, F., Yamashita, T., Shimada, A., Ogata, H.: A neural network approach for students' performance prediction. In: LAK17 - The Seventh International Learning Analytics and Knowledge Conference, pp. 598–599. Association for Computing Machinery (ACM) (2017). https://doi.org/10.1145/3027385

27. Echo360: Echo360- definitions of course analytics metrics (2020). https://learn.echo360.com/hc/en-us/articles/360035037312. Accessed 25 Nov 2020

28. Echo360: The GDPR is coming, and it's a good thing (2018). https://echo360.com/gdpr-coming-good-thing/. Accessed 03 Nov 2021

29. Zhang, Q., Kuldip, C., Devabhaktuni, V.: Artificial neural network for RF and microwave design - from theory to practice. IEEE Trans. Microw. Theory Tech. **51**(4), 1339–1350 (2003)

30. Rashid, T., Ahmad, H.: Using neural network with particle swarm optimization. Comput. Appl. Eng. Educ. **24**, 629–638 (2016)

31. Lau, E.T., Sun, L., Yang, Q.: Modelling, prediction and classification of student academic performance using artificial neural networks. SN Appl. Sci. **1**(9), 1–10 (2019). https://doi.org/10.1007/s42452-019-0884-7

32. Wilson, P., Mantooth, H.: Model-based engineering for complex electronic systems. Newnes, March 2013

33. Yu, H., Wilamowski, B.: Levernberg Marquardt training industrial electronic handbook, intelligent systems, vol. 5, 2 edn. CRC Press, January 2011

34. MathWorks: trainlm - Levernberg-Marquardt backpropagation (2019). http://uk.mathworks.com/help/nnet/ref/trainlm.html. Accessed 17 June 2019

35. Özçelik, S., Hardalaç, N.: The statistical measurements and neural network analysis of the effect of musical education to musical hearing and sensing. Expert Syst. Appl. **38**, 9517–9521 (2011)

36. Othman, M., Zain, N.: Online collaboration for programming: assessing students' cognitive abilities. Turk. Online J. Distance Educ.-TOJDE **16**(4), 84–97 (2015)

The Time Travel Exploratory Games Approach: An Artificial Intelligence Perspective

Oksana Arnold[1] and Klaus P. Jantke[2(✉)] (iD)

[1] Erfurt University of Applied Sciences, Altonaer Str. 25, 99085 Erfurt, Germany
oksana.arnold@fh-erfurt.de
[2] ADICOM Software, Frauentorstr. 11, 99423 Weimar, Germany
klaus.p.jantke@adicom-group.de

Abstract. The authors deploy Artificial Intelligence techniques for the design of digital games aiming at affective and effective education. There is a case of environmental education studied in some detail. The games in focus are time travel exploratory games that enable learners to find data from the past. The narrative is traveling back in time – virtually – and exploring the past – really. Successful learners return with valuable findings. But sometimes, human learners fail. Not everyone is familiar with time travel. How might a learner behave when finding herself back in time in a foreign virtual world? Preparing the exploratory digital game for unforeseeable learners' behavior is an involved planning task. For this purpose, advanced technologies of Artificial Intelligence for planning in dynamic environments such as complex industrial processes are adopted and adapted. This leads to storyboarding of learners' experiences. From the Artificial Intelligence perspective, the time travel exploratory game itself becomes an AI system. It adapts to the learners needs and desires aiming at fun when playing the game and at success and effectiveness.

Keywords: Technology-enhanced learning · Artificial intelligence in education · Game-based learning · Didactic design · Game design · Artificial intelligence design · Design of experience · Dynamic planning · Digital storyboarding · Time travel exploratory games · Exploratory learning · Collaborative learning · Environmental education

1 Introduction

This is a substantially revised and extended version of the authors' contribution [6] to the 2021 Conference on Computer Supported Education (CSEDU).

1.1 The Educational Perspective

The introduction of *time travel exploratory games* and the dissemination of ideas how to deploy those games for *exploratory learning* and for *collaborative learning* may be considered the key contribution of the authors' approach.

© Springer Nature Switzerland AG 2022
B. Csapó and J. Uhomoibhi (Eds.): CSEDU 2021, CCIS 1624, pp. 40–54, 2022.
https://doi.org/10.1007/978-3-031-14756-2_3

1.2   The Artificial Intelligence Perspective

From the point of view of AI, the authors consider two different aspects relevant.

First, system design according to the authors' concepts of storyboarding is an Artificial Intelligence approach of dynamic plan generation. It dates back to 2005, when the principles of storyboarding underlying the present work have been introduced by [19]. This paper, in turn, has adopted and adapted the dynamic plan generation concepts from [4]. The authors' CSEDU contribution [6] has an exclusive AI focus on this first aspect.

Second, what the authors design when storyboarding a time travel exploratory game, seen in the right light, is a certain Artificial Intelligence functionality that aims at pedagogical adaptivity. This second aspect goes beyond the limits of [6].

Putting both aspects in context, one might say that the overall approach is characterized by the deployment of AI plan generation for the design of highly adaptive AI behavior in educational time travel exploratory games.

1.3   The Flying Classroom Background

To make the present contribution self-contained, the background of the so-called *Flying Classroom* project is briefly revisited (Fig. 1).

Fig. 1. The permanently grounded aircraft IL-18 hosting "The Flying Classroom" (figure 1 from section 2 of [6], © 2020, the authors).

A permanently grounded Russian-made aircraft IL-18 at the airport in Erfurt, Germany, is remodeled to become a destination that offers unique experiences of digital game play including virtual journeys and even exploratory time travel.

School classes can come and visit the Flying Classroom to get a guided tour around the airport and to experience educational game play in the aircraft.

In the Flying Classroom, games are played in Web browsers. In this way, they are prepared for a rollout to reach all interested audiences. The Flying Classroom serves as an attractor and as a multiplier.

2   A Case of Environmental Education: Ocean Warming

The environmental education case study from [6] is revisited as well to provide content underlying the elaboration throughout the following sections.

There are available several thousands IR pictures of the ocean world-wide that cover observations over a period of 60 years from 1960 to 2019 (Fig. 2).

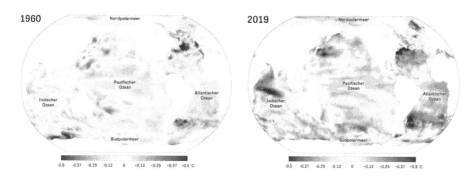

Fig. 2. Exemplified data of ocean warming kindly provided by © Lijing Cheng [11].

Ocean warming invigorates tropical cyclones. On the other hand, the strong winds of tropical cyclones keep the ocean cooler by causing stronger evaporation. This interference is rather intricate [26].

Playing a digital time travel game through which players get information about the process of ocean warming cannot cover all the intriguing phenomena related to climate change in general and to ocean warming in particular. The authors have developed a complete storyboard for such a game putting emphasis on the following effects that are grouped here, perhaps a bit formally, into effects of exploratory learning and effects of collaborative learning.

– Young learners shall become aware of the problem of ocean warming around the globe.
– Inspecting visualizations of real data, they discover an undeniable trend of warming.
– Inspecting visualizations in more detail, they discover the knottiness of the process that shows cooling effects as well.

When returning from the virtual time travel, learners bring their findings with them. They inform each other and work together aiming at a presentation that allows for conveying to others like their parents and other relatives and to their peer group the insights they gained through the experience of game play.

– As a group of learners, they experience the enormous potential of joining their individual findings – the whole is more than the sum of its parts.
– They collaboratively develop a concept of how to disseminate their results.

3   Didactic Game Design from a Bird's Eye Perspective

By way of illustration,

One of the many advantages of digital storyboarding is its modularity and its flexibility. Interdisciplinary teams of educators, psychologists, domain experts, VR technicians, and others negotiate details of anticipated learner experiences. The design can proceed top down, bottom up, or both at one, in the latter case sometimes deeply dovetailed.

Storyboards are hierarchically structured finite families of finite pin graphs. The graphs are usually kept as small as possible to lucidly reflect patterns of game design as well as didactic concepts (see [22] and, especially, [18] vs. [10]).

This approach has been derived from the dynamic plan generation approach of [4] as discussed in some detail in the present paper's predecessor conference contribution [6], section 6. There is no need of a repetition here.

Technicalities will be explained whenever they occur. Real storyboard graphs from the application case study surveyed in Sect. 2 will be used for illustration.

A group of players engage in game play in which

- players undertake independently of each other a virtual journey back in time,
- explore the past virtually,
- find artefacts that carry information about ocean warming,
- bring these artefacts back into the present time,
- compare, discuss, and interpret their individual findings,
- and collaborate toward a presentation of the aggregate of their findings.

From a bird's eye perspective, the didactic approach assumes that there is a first phase of exploratory learning that precedes a second phase of collaborative learning. The explorations take place individually and are experienced by the human players/learners as an adventurous time travel. When returning from their journeys, the team of players/learners engages in collaborative work.

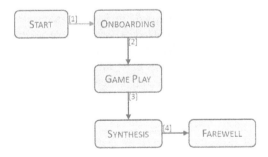

Fig. 3. Top level storyboard graph specifying the macroscopic structure of game play (figure 4 from section 5 of [6]).

Notice that Onboarding means the intuitive exploration of the game mechanics.

In a storyboard graph like the one in Fig. 3, the nodes are called *episodes*. Other nodes called *scenes* will be introduced later. Episodes are placeholders for more detailed descriptions of anticipated experience. In slightly more technical terms, the episode nodes determine places for graph substitution. Replacement of episodes by other graphs of the storyboard takes place at execution time, an issue that is essential to the dynamics of interaction and to the adaptivity of the system's behavior. The discussion of these technicalities is postponed.

Edges from one node to the other specify the flow of interaction. Every edge is annotated by a logical condition of execution. In this way, it is determined in which conditions game play may proceed from one episode to the other. The logical conditions seen as formulas contain variables. These variables refer to environmental data – game play may depend on the day time and even on the outside weather conditions – and on data from the user/learner/player model including the interaction history.

For simplicity, one may assume that the conditions [1], ..., [4] in Fig. 3 have constantly the value *true*.

The design of interaction, game play, and learning proceeds via a step by step introduction of storyboard graphs admissible for the expansion of the episodes. Storyboarding means *the organization of experience* (see [19]. p. 25)

In Fig. 3, **Game Play** denotes the episode of exploration and **Synthesis** is the episode of collaboration. For every episode node in the storyboard graph of Fig. 3, there may be available several other storyboard graphs for expansion. By way of illustration, one may implement the phase of collaborative learning differently in dependence on the players' findings in the virtual past. These data are not yet known at planning time, but occur as late as at execution time.

Fig. 4. A particular storyboard graph for the expansion of the **Start** episode in Fig. 3.

For the first time, there is in use a collapsed box flatter than the boxes on display so far. Such a box indicates a *scene*. According to [19], scenes have a semantics in the domain. They are not subject to substitution by graphs from the storyboard.

Nevertheless, notice that there may exist alternative semantics implementing a scene. For the **Introduction** episode, e.g., implementations may be an audio file, a video file, a cutscene, a text on the screen, or even – offline – a teacher's speech.

Fig. 5. A storyboard graph for expanding the **Registration** episode shown in Fig. 4.

Alternative semantics of a scene implementation have conditions of execution.

The **Book Keeping** scene in Fig. 5 denotes an action of the digital system usually unobservable by the human player/learner. The system needs to process the registration data.

Take alone the issue of book keeping throughout game play, this turns out to be a relevant part of didactic and game design. Negotiations among the designers of what to record, when to record, why to record, and how to use records later on are decisive to the future behavior of the system, especially to its adaptivity.

We continue with another expansion of an episode in the top level storyboard graph on display in Fig. 3. For the first time, there occur branching conditions. This phenomenon will be discussed in more detail in the subsequent section.

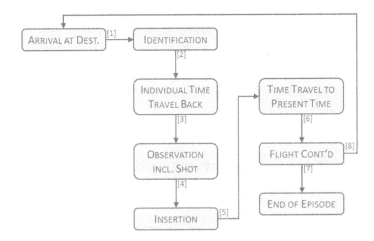

Fig. 6. A particular storyboard graph for the expansion of the **Game Play** episode (figure 5 from section 5 of [6]).

Every storyboard graph such as the one on display in Fig. 6 has its substitution condition. Only if the condition is valid, the graph may be selected to expand an episode. Among others, these conditions control the game's intelligent behavior.

Members of the design team may anticipate different variants of game play, of pedagogical patterns, and of interaction. They represent their ideas by means of storyboard graphs. During the design process, they negotiate in which conditions the one or the other idea should be reasonably applied.

Interested readers are directed to prior applications of the present approach such as [7, 8, 12, 20–23] and [27].

One may even experiment with varying ideas of game play, of pedagogy, and of interaction by, so to speak, *plug and play*. Insert or remove graphs into resp. from a storyboard or modify conditions when a system is in use. One may observe the emergence of varying game play and, hence, varying learner experiences.

This flexibility becomes more comprehensible when stepping down from the bird's eye perspective to, so to speak, a frog's eye view. This corresponds to *layered languages of ludology*, a term coined in [15] and studied in detail in [24].

4   The AI Approach to Didactic and Game Design

Intelligence – human as well as artificial – shows in the details of behavior, in originality, in empathy, in adaptivity, in variety, and the like.

The Artificial Intelligence behavior of an interactive digital system, especially a
digital game, is determined on finer levels of granularity. This section is aimed at a
demonstration of the design technology by means of the authors' application case study
of environmental education.

Artificial Intelligence is a feature of a digital system. The implementation of AI has
a digital representation – it is formally represented. The storyboarding approach explicates, so to speak, where the intelligence of the system is sitting. By way of illustration,
larger storyboard cutouts demonstrate the adaptivity of a practical implementation for
training executive staff of professional disaster management [8].

The intelligence of a time travel exploratory game results from an interference

[1] of the structure of storyboard graphs,
[2] of the execution conditions of edges in graphs,
[3] of, in particular, the branching conditions,
[4] of the assignment for substitution of storyboard graphs to nodes,
[5] of the substitution conditions of graphs.

These five parameters controlling AI are negotiated by the interdisciplinary design
team when creating an educational game. In this way, the emerging AI reflects principles of pedagogy, of psychology, of game design, and takes the potential of the involved
technical disciplines such as VR into account.

The resulting storyboard graphs including all their logical annotations are instances
of the principles invoked. They are instances of more general patterns.

The pattern concept became explicit in science and technology almost half a century
ago by Christopher Alexander's related work in architecture [1, 2].

At almost the same time, Dana Angluin developed a very precise and lucid approach
to patterns and their instances [3]. Trendsetting is Angluin's insight that there might be
patterns in science and technology, but what we can perceive are only their instances.
This leads directly to the pattern inference problem. Seeing a few or, perhaps, a large
amount of instances, what is the pattern behind? This problem is usually an intriguing
one.

In some areas, the pattern concept is rather vague and the borderline between patterns and instances gets blurred [10].

Surprisingly, despite the appealing title of the book [9], this work does not contribute
much to a dovetailed didactic and game design, as the majority of the concepts discussed
are far from being patterns.

That patterns are a key concept of game-based learning is illustrated by [17] where
the occurrences of patterns–more correctly, of instances of patterns–may be interpreted
as indicators of mastery. Thus, patterns and instances are key to assessment. Consequently, systematic storyboarding is setting the stage for assessment, unfortunately an
issue beyond the limits of the present contribution.

4.1   The Pattern-Instance Relation

Because there is a lot of confusion about patterns, instances and their relationship, it is worth to invest a few lines into a clarification.

Some authors claim that they work on finding patterns. Even Angluin [3] used this terminology in the title of her seminal paper: *Finding Patterns Common to a Set of Strings*. A closer look reveals that patterns are something quite general. They are principles behind their instances and, thus, remain under the hood, so to speak. What we find in structures, in behavior, and the like are only instances.

As a consequence, the practical observation of several patterns brings with it the question for the underlying pattern. This exactly is a key problem of assessment [17]. Loosely speaking, you see what a learner – who, in our case, is also a player – is doing and you may ask yourself for the deeper reasons behind.

Having an instance and a candidate pattern possibly underling the instance, it is not easy to decide whether or not this is the case. Angluin [3] demonstrates that this decision problem may be of extraordinary computational complexity – it may be NP-complete [13]. We drop further details of discussing this issue, because it relies on a quite intricate basis in computation theory [25].

Let us return to the didactic and game design by means of storyboarding, along the way keeping an eye on instances and patterns.

4.2   Instances of Artificial Intelligence Design

Because we deal with time travel exploratory games, in every expansion of the Game Play episode such as the one on display in Fig. 6, one finds an episode of Individual Time Travel Back. The design of storyboard graphs for the expansion of this node is a crucial step of didactic and game design.

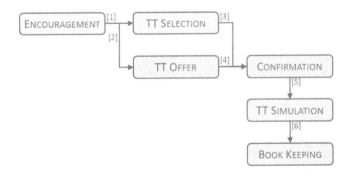

Fig. 7. An expansion of the Individual Time Travel Back episode shown in Fig. 6 before (figure 7 from section 7 of [6]).

The storyboard graph of Fig. 7 is just one of several alternative design proposals. It exclusively consists of scenes. That means that no further expansion is intended.

Everything about time travel back in time is specified in this graph (including all the logical conditions).

There are the scenes Encouragement, TT Selection, TT Offer, Confirmation, TT Simulation, and Book Keeping.

Encouragement means to prompt a player to go for a journey back in time. This may be implemented in different ways by an audio file, by a video, or anything like this. Similarly, Confirmation means to accept the chosen destination in time and to start the journey. TT Simulation means the way of how to present the journey. There are hundreds of ideas how to show this on a screen, perhaps, accompanied by sound. In a project on *time travel prevention games* for the purpose of accident prevention in the industries [5], the authors have a time tunnel as illustrated in Fig. 8. Last but not least, BOOK KEEPING is needed as usual.

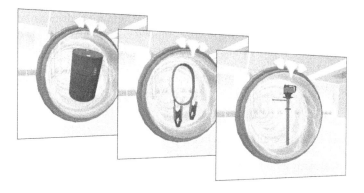

Fig. 8. A time tunnel in which iconic objects represent events of the past; clicking the left or the right button moves the objects back in time or forward in time, respectively, and a click to the button in the middle selects an event and, thus, a destination in time.

In the present case study of environmental education, a time tunnel may be less appropriate, because there are no iconic objects available. But for other time travel exploratory games, this is definitely an option worth to be considered. Therefore, the authors decided to include it here.

The illustrated concept of a time tunnel integrates the meaning of the scenes TT Selection, TT Offer, and TT Simulation. This explicates the interference of different contributions to the interdisciplinary design process. Technical features may either undercut or support other ideas from pedagogy, from game design, or from other disciplines and vice versa.

The two alternative scenes TT Selection and TT Offer are more interesting. In the TT Selection scene, the player has a choice from a set of destinations, perhaps a list, turning a wheel of time, or anything like this.

There is some Artificial Intelligence in the control of switching between the scenes TT Selection and TT Offer. The logical conditions may reflect the designers aim at guidance, an attempt of treating the player with empathy, and may be other aspects brought in by the team of designers from their varying topical perspectives. The behavior unfolds at play time.

For simplicity, the authors present two alternatives to the storyboard graph that is on display in Fig. 7.

This graph, apparently, is a simplification. It reflects the intention of never taking away the player's choice of alternatives. The scene TT Offer that offers only a single destination, possibly in response to earlier mistakes, is removed. Making TT Selection an episode allows for more player-system interactions when selecting the destination of time travel. This may include more player guidance.

The storyboard graphs of the Figs. 7, 9, and 10 have a general idea in common. They may be seen as instances of a certain pattern. In the process of game design, one may select one or two of them, but one might also insert all three in the storyboard and equip every graph with a substitution condition, accordingly. There is another approach briefly investigated in Sect. 4.3 that follows immediately.

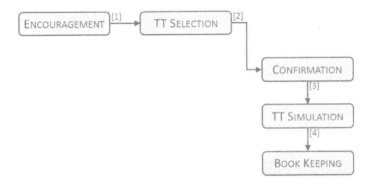

Fig. 9. An simplified expansion of the Individual Time Travel Back compared to Fig. 7.

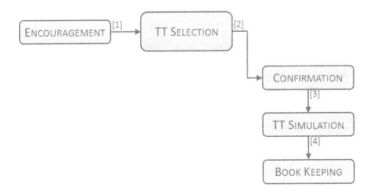

Fig. 10. A generalized expansion of the Individual Time Travel Back compared to Fig. 7.

4.3 Patterns of Artificial Intelligence Design

There is a large variety of pattern concepts appropriate to the design and to the analysis of digital games and to the understanding of the impact of game playing [16]. For the present contribution, the authors confine themselves to a particularly lucid, but practically useful approach.

Patterns and their instances are investigated within the framework of some storyboard. This is a decisive assumption, but it is appropriate for considerations within a particular design process. In case a storyboard is under development, at any point of time there exists the current collection of storyboard graphs to which the two related concepts pattern and instance refer.

A pattern is a storyboard graph with a finite number of distinguished episodes as indicated in Fig. 11. Distinguished nodes including the execution conditions of outgoing edges serve as variables exactly in the same way as variables in Angluin's approach [3]. In contrast to [3], the present approach is more flexible, because the admissible substitutions for a variable node as well as the validity of logical conditions may change over time when the storyboard is growing.

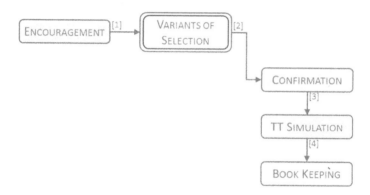

Fig. 11. A most simple storyboard pattern with an episode serving as a variable node.

If the storyboard under construction is sufficiently rich, the pattern on display in Fig. 11 has the storyboard graphs on display in the Figs. 7, 9, and 10 among its instances.

Patterns of this type play a decisive role in storyboarding. For problems within the design process that are relevant to anticipated experiences of play and to the effectiveness of playing, designers are frequently coming up with generic ideas on a somehow abstract level. It is helpful to try a generic formalization as a storyboard pattern like this in Fig. 11. This is directing the focus of design to details and to the interference of the five parameters listed at the beginning of Sect. 4. What are suitable substitution alternatives and in which conditions do they apply? How to continue at play time afterwards in which conditions? Apparently, this is a top-down approach.

The other way around, members of design teams, according to their different topical backgrounds, have varying ideas of how to affect the human players aiming at effectiveness and sustainability. Searching for common patterns behind their ideas brought in bottom-up is a way toward a holistic design.

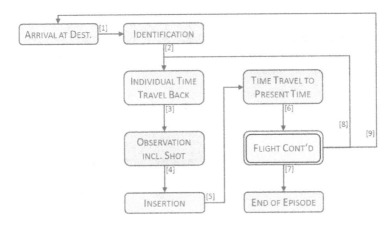

Fig. 12. A one-variable storyboard pattern with variable conditions of execution.

The variable, so to speak, of the pattern on display in Fig. 12 is the episode node Flight Cont'd together with the three conditions [7], [8] and [9].

Looking back at Fig. 6, readers may easily recognize that the storyboard graph of this earlier figure is an instance of the pattern in Fig. 12. The episode variable is substituted by a graph that consists of one scene named Flight Cont'd and the condition [8] is substituted by the logical constant 'false'. Accordingly, [9] is renamed to [8].

Seen from the viewpoint of Artificial Intelligence design, the instantiation of the pattern's conditions [7] and [9] is of particular interest. It is assumed that the meaning of the scene Flight Cont'd within the narrative of the game is that after the virtual return from an exploration in the past the player flies virtually back home (if condition [7] is true at play time) or goes for another adventure.

During the design process, one needs to decide whether or not it is possible to assess the player's satisfaction with the mini-game established by the two subsequent scenes Arrival at Dest. and Identification. In this little game, players get information – say pictures taken from some height showing something distinctive such as the Amazon delta – and are asked to identify the area where they did arrive virtually. If such an assessment is possible, results are recorded.

If a record says the a player did enjoy the mini-game and if there are no critical limits of time to play, designers can arrange condition [9] to outstrip [8]. In case it is known to the system that for the next destination there do not exist any sufficiently good visual representations, the condition [8] should guarantee to skip the mini-game to avoid frustration, hence, designing empathy of the AI.

However practically useful, readers might correctly complain that literally every storyboard graph becomes a pattern by just indicating variable episodes. On the one hand, this is true. It fits the experience that a storyboard is a usually big collection of graphically represented ideas, concepts, principles – let's just say, patterns – of pedagogy, psychology, and game play. But on the other hand, the authors admit that it is a methodological flaw just to rename a concept.

Consequently, the authors prefer to consider *patterns in the strict sense* that obey the additional requirement that all nodes that are not indicated to represent variables are simply scenes, not episodes. Notice that the graph in Fig. 11 meets this condition, whereas the one in Fig. 12 does not.

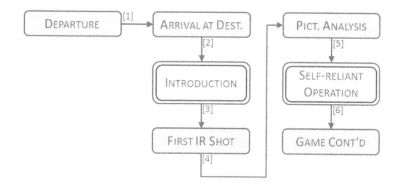

Fig. 13. A storyboard pattern in the strict sense to expand the **Onboarding** episode.

Patterns in the strict sense direct the team of designer's attention to episodes that are embedded in a somehow predefined flow of game play, of interaction, and of exploratory learning, in particular. This resembles Angluin's approach of embedding variables among strings of ground text [3]. Nevertheless, in graphs the string concept is generalized.

By way of illustration, let us briefly investigate expansions of the **Onboarding** episode (revisit Fig. 3). There is such a pattern in the strict sense in Fig. 13. To create instances of this pattern, the designers need to balance the interplay – in the full sense of the word – of the **Introduction** of the game mechanics and the player's opportunity to try it out in a **Self-Reliant Operation**. Results of the **Introduction** episode determine what the human player may possibly master in the **Self-reliant Operation** episode. What is required in the latter must be prepared in the introduction before.

Features of Artificial Intelligence may be designed by means of patterns in the strict sense. Adaptive behavior in a later episode may unfold in response to the player/learner achievements in one or even more preceding episodes of play. Guidance can be provided to learners in response to problems in earlier episodes. To those players who are strong and demonstrate their mastery of game play, later episodes can offer challenges that make interactions attractive to them. Storyboarding Artificial Intelligence via pattern concepts is the key technology.

Acknowledgement. The two authors gratefully acknowledge the patience of all the members of the editorial team during the period of time in which this contribution was in progress and, apparently, Hofstadter's Law ([14], p. 152) was ruling.

References

1. Alexander, C.: A The Timeless Way of Building. Oxford University Press, New York (1979)
2. Alexander, C., Ishikawa, S., Silverstein, M.: A Pattern Language: Towns, Buildings, Construction. Oxford University Press, New York (1977)
3. Angluin, D.: Finding patterns common to a set of strings. J. Comput. Syst. Sci. **21**, 46–62 (1980)
4. Arnold, O.: Die Therapiesteuerungskomponente einer wissensbasierten Systemarchitektur für Aufgaben der Prozeßführung, DISKI, vol. 130. Infix, St. Augustin (1996)
5. Arnold, O., Franke, R., Jantke, K.P., Wache, H.-H.: Dynamic plan generation and digital storyboarding for the professional training of accident prevention with time travel games. In: Guralnick, D., Auer, M.E., Poce, A. (eds.) TLIC 2021. LNNS, vol. 349, pp. 3–18. Springer, Cham (2022). https://doi.org/10.1007/978-3-030-90677-1_1
6. Arnold, O., Jantke, K.P.: AI planning for unique learning experiences: the time travel exploratory games approach. In: 13th International Conference on Computer Supported Education (CSEDU), vol. 1. pp. 124–132. SCITEPRESS (2021)
7. Arnold, S., Fujima, J., Jantke, K.P.: Storyboarding serious games for large-scale training applications. In: Foley, O., Restivo, M.T., Uhomoibhi, J., Helfert, M. (eds.) Proceedings of the 5th International Conference on Computer Supported Education, CSEDU 2013, Aachen, Germany, 6–8 May 2013, pp. 651–655 (2013)
8. Arnold, S., Fujima, J., Jantke, K.P., Karsten, A., Simeit, H.: Game-based training for executive staff of professional disaster management: storyboarding adaptivity of game play. In: Tan, D. (ed.) Proceedings of the International Conference on Advanced Information and Communication Technology for Education (ICAICTE 2013), Hainan, China, 20–22 September 2013, pp. 68–73. Atlantis Press (2013)
9. Björk, S., Holopainen, J.: Patterns in Game Design. Charles River Media, Hingham (2005)
10. Board, P.P.A. (ed.): Advice for Educators. Joseph Bergin Software Tools (2012)
11. Cheng, L., Abraham, J., Zhu, J., Trenberth, K.E., Fasullo, J.: Record-setting ocean warmth continued in 2019. Adv. Atmos. Sci. **37**, 137–142 (2020)
12. Fujima, J., Jantke, K.P., Arnold, S.: Digital game playing as storyboard interpretation. In: Proceedings of the 5th International Games Innovation Conference (IGIC), Vancouver, BC, Canada, 23–25 September 2013, pp. 64–71. IEEE Consumer Electronics Society (2013)
13. Garey, M.R., Johnson, D.S.: Computers and Intractability: A Guide to the Theory of NP-Completeness. Freeman, San Francisco (1979)
14. Hofstadter, D.R.: Gödel, Escher, Bach: An Eternal Golden Braid. Basic Books (1979)
15. Jantke, K.P.: Layered Languages of Ludology: The Core Approach. Diskussionsbeiträge 25, TUI IfMK, November 2006
16. Jantke, K.P.: Patterns in Digital Game Playing Experience Revisited: Beiträge zum tieferen Verständnis des Begriffs Pattern. Diskussionsbeiträge 22, TU Ilmenau, IfMK (2008)
17. Jantke, K.P.: Patterns of game playing behavior as indicators of mastery. In: Ifenthaler, D., Eseryel, D., Ge, X. (eds.) Assessment in Game-Based Learning: Foundations, Innovations, and Perspectives, pp. 85–103. Springer, New York (2012). https://doi.org/10.1007/978-1-4614-3546-4_6
18. Jantke, K.P.: Pedagogical patterns and didactic memes for memetic design by educational storyboarding. In: Arnold, O., Spickermann, W., Spyratos, N., Tanaka, Y. (eds.) WWS 2013. CCIS, vol. 372, pp. 143–154. Springer, Heidelberg (2013). https://doi.org/10.1007/978-3-642-38836-1_12
19. Jantke, K.P., Knauf, R.: Didactic design through storyboarding: standard concepts for standard tools. In: Baltes, B.R., et al. (eds.) First International Workshop on Dissemination of

E-Learning Technologies and Applications, DELTA 2005, in: Proceedings of the 4th International Symposium on Information and Communication Technologies, Cape Town, South Africa, 3–6 January 2005, pp. 20–25. Computer Science Press, Trinity College Dublin, Ireland (2005)

20. Jantke, K.P., Spundflasch, S.: Storyboarding pervasive learning games. In: Tan, D. (ed.) Proceedings of the International Conference on Advanced Information and Communication Technology for Education (ICAICTE 2013), 20–22 September 2013, Hainan, China, pp. 42–53. Atlantis Press (2013)

21. Knauf, R., Jantke, K.P.: Storyboarding - an AI technology to represent, process, evaluate, and refine didactic knowledge. In: Kreutzberger, K.P.J.G. (ed.) Knowledge Media Technologies. First International Core-to-Core Workshop, pp. 170–179. No. 21 in TUI IfMK Diskussionsbeiträge, TU Ilmenau (2006)

22. Knauf, R., Sakurai, Y., Tsuruta, S., Jantke, K.P.: Modeling didactic knowledge by storyboarding. J. Educ. Comput. Res. **42**(4), 355–383 (2010)

23. Krebs, J., Jantke, K.P.: Methods and technologies for wrapping educational theory into serious games. In: Zvacek, S., Restivo, M.T., Uhomoibhi, J., Helfert, M. (eds.) Proceedings of the 6th International Conference on Computer Supported Education, CSEDU 2014, Barcelona, Spain, 1–3 May 2014, pp. 497–502. SCITEPRESS (2014)

24. Lenerz, C.: Layered languages of ludology. In: Beyer, A., Kreuzberger, G. (eds.) Digitale Spiele - Herausforderung und Chance, pp. 39–52. Boizenburg, VWH (2009)

25. Rogers, H., Jr.: Theory of Recursive Functions and Effective Computability. McGraw-Hill, New York (1967)

26. Trenberth, K.E., Cheng, L., Jacobs, P., Zhang, Y., Fasullo, J.T.: Hurricane Harvey links to ocean heat content and climate change adaptation. In: Environmental Science, pp. 730–744 (2018)

27. Winter, J., Jantke, K.P.: Formal concepts and methods fostering creative thinking in digital game design. In: 3rd Global Conference on Consumer Electronics (GCCE 2014), Makuhari Messe, Tokyo, Japan, 7–10 October 2014, pp. 483–487. IEEE Consumer Electronics Society (2014)

Guidelines for the Application of Data Mining to the Problem of School Dropout

Veronica Oliveira de Carvalho[1]([⊠]), Bruno Elias Penteado[2],
Leandro Rondado de Sousa[1], and Frank José Affonso[1]

[1] Instituto de Geociências e Ciências Exatas, Universidade Estadual Paulista (Unesp),
Rio Claro, Brazil
{veronica.carvalho,leandro.rondado,f.affonso}@unesp.br
[2] Instituto de Ciências Matemáticas e de Computação, Universidade de São Paulo (USP),
São Carlos, Brazil
bruno.penteado@alumni.usp.br

Abstract. Dropout is a complex phenomenon based on interrelated factors such as personal, institutional, structural, sociocultural, among other ones. It represents a waste of resources for students, their families, schools and society, and continues to be a challenge for educational institutions. In the last decade, the growing amount of data from educational institutions and the emergence of data science have led to data mining methodologies to explore this problem empirically. In this work, we map the literature on how data mining has been addressed face-to-face dropout. We synthesize different aspects, all of them related to steps of a generic data mining process. Our findings reveal a low level of formalism in theories, methodologies and pre-processing steps, with most papers making comparisons of different algorithms and features on the data available in the institution's information system. Finally, we present some guidelines that can be used to improve the research on this topic.

Keywords: School dropout · Data mining · Systematic Literature Mapping · Revisited

1 Introduction

Dropout is a problem widely studied worldwide [17], as it leads to economic and social problems [26]. Due to this fact, many works were carried out to understand the subject. These works can be divided into two approaches [8]: (a) survey-based, in which theoretical models are developed (such as Tinto's work [32]); (b) data-driven, in which institutional data are used to understand the problem through analytical methods.

Regarding the second approach (data-driven), data mining is one of these analytical solutions. *Data mining* is an iterative process that makes it possible to extract knowledge from a set of historical data. Process steps can generally be divided into pre-processing, pattern extraction, and post-processing (details in Sect. 2). The knowledge extracted can predict whether a student will drop out, understand why students drop out, etc. To understand how dropout and data mining are related, some reviews/surveys were carried

© Springer Nature Switzerland AG 2022
B. Csapó and J. Uhomoibhi (Eds.): CSEDU 2021, CCIS 1624, pp. 55–72, 2022.
https://doi.org/10.1007/978-3-031-14756-2_4

out [2, 7, 17, 22]. However, few of them follow a structured research protocol and cover all aspects of the process (usually, only one step is handled (pattern extraction)). In order to contribute to this gap, in [28], we present a Systematic Literature Mapping (SLM) [13] covering all steps of the data mining process regarding the dropout problem in face-to-face education[1].

This current work is an extension of our previous work [28]. In [28], we covered six research questions. In this paper, we intend to revisit these questions, correlating each one with the data mining steps, and present guidelines to highlight the aspects that must be considered when data mining is applied to the face-to-face school dropout problem. Besides, two other general questions were addressed regarding educational theories and data mining methodologies applied to research. We hope to complement the findings discussed in [28]. In this way, everyone (researchers, students, etc.) who aims to apply data mining to address the dropout problem can be aware of the aspects that must be thought.

This work is structured as follows: Sect. 2 describes the data mining process. Section 3 discusses how data mining has been applied to address the dropout problem in face-to-face education. For this, the six research questions presented in [28] are revisited, along with two new general questions. Section 4 presents the proposed guideline. Finally, Sect. 5 concludes the paper.

2   The Data Mining Process

First, it is essential to discuss some concepts that are currenlty used. In this new era, lots of data are constantly generated - this can be seen by some infographics available on the internet[2]. In this context, many concepts emerge. Data science, the most general, can be considered a multidisciplinary area that includes statistics, visualization, machine learning, etc. Figure 1 presents an interesting relationship between the concepts, trying to explain them like a puzzle. It is available on KDnuggets[3]. To be able to develop a data science project, a process must exist. Data mining is the process. Several methodologies execute the process, such as KDD, CRISP-DM, or SEMMA [23]. Finally, machine learning allows the process, through its algorithms, to obtain the patterns in one of the steps.

With the above discussion in mind, the SLM we presented in [28] was made considering all the general steps of a data mining process. As many methodologies exist, we consider in this work the steps proposed by [25] and presented in Fig. 3. We believe that these steps, described below, are the ones that commonly appear in most methodologies. Therefore, from now on, data mining will be understood as a process containing these steps.

As mentioned before, data mining is an iterative process that allows the extraction of knowledge from a set of historical data. However, what is knowledge? Figure 2 gives an example: the data is raw and meaningless; information gives meaning to data. Based

---

[1] Only face-to-face dropout was addressed in [28], since all institutions have data on the trajectory of their students in their academic systems.

[2] Some of them: https://www.statista.com/chart/25443/estimated-amount-of-data-created-on-the-internet-in-one-minute/, https://lorilewismedia.com/.

[3] https://www.kdnuggets.com/.

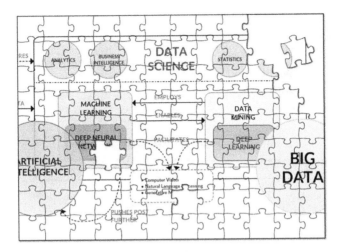

Fig. 1. Puzzle explained [https://www.kdnuggets.com/2016/03/data-science-puzzle-explained.html] and Puzzle [https://www.kdnuggets.com/2017/01/data-science-puzzle-revisited.html].

on data and information, it is possible to extract patterns (knowledge). Many works are discussing this, with the work of [1] being widely known. It is this kind of pattern that can be extracted through data mining.

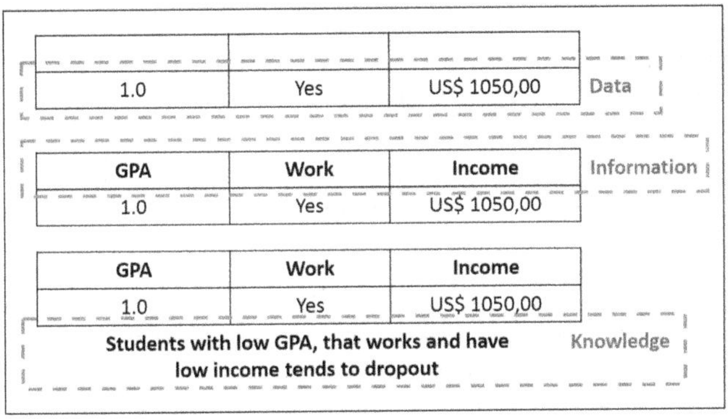

1.0	Yes	US$ 1050,00	Data

GPA	Work	Income	Information
1.0	Yes	US$ 1050,00	

GPA	Work	Income	
1.0	Yes	US$ 1050,00	

Students with low GPA, that works and have low income tends to dropout   Knowledge

Fig. 2. Data, information and knowledge.

As seen in Fig. 3, the process contains three main steps (pre-processing, pattern extraction, post-processing), described here, along with two complementary steps.

Problem Identification. The process begins by defining the aim and goals to be achieved, identifying and selecting the data to be used. Also, it is crucial to understand the domain.

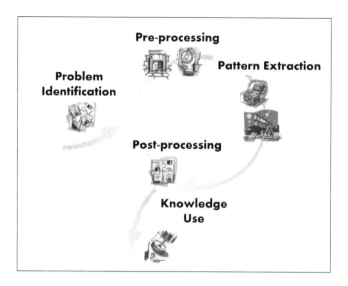

Fig. 3. The steps of a data mining process [25].

Pre-processing. In this step, the data is prepared to be used in the next step, and many aspects can be considered [4, 30][4]:

– data cleaning: missing, redundant and inconsistent values; noise/outlier detection;
– data transformation: discretization; normalization; standardization; encoding;
– data reduction: instance reduction (sampling); feature reduction (feature selection; feature transformation);
– data balance [10]: oversampling, undersampling and hybrid techniques.

Pattern Extraction. In this step, the patterns are obtained through machine learning algorithms. For this, algorithms are selected, set (with/without hyperparameter tuning), and executed.

Post-processing. The patterns that were obtained are validated. This validation can be in terms of validation measures (accuracy, precision, recall, etc.), interpretability (through the model itself (interpretable models[5]) or XAI techniques [6, 20]).

Knowledge Use. After post-processing, it is verified whether the aim and the goals were achieved. If not, the process is re-executed (it is iterative); otherwise, the knowledge and/or patterns can be incorporated, for example, into an intelligent system.

3   Dropout Through the Data Mining Perspective

In [28], we covered six research questions through an SLM. The aim here is to extend these questions and correlate them with the data mining steps previously discussed. The

---

[4] There is no consensus on the division presented here.
[5] White-box algorithms produce interpretable models, i.e., models that can be understood by humans, such as decision trees [20].

protocol used to conduct the SLM is reported elsewhere [28]. The SLM was carried out to retrieve and analyze primary studies that use data mining in the face-to-face dropout context considering 10 years (01/01/2010 to 31/12/2020), and 118 papers were selected. Figure 4 presents the correlation between the research questions and the data mining steps. The six research questions addressed in [28] are:

- RQ1. What are the levels of education explored?
- RQ2. Considering the samples (datasets) used, how big are they, and how are they generated?
- RQ3. What aspects (features, attributes) have been used to model the dropout problem?
- RQ4. What kind of pre-processing has been applied to the samples?
- RQ5. What algorithm families have been used?
- RQ6. What measures have been used to validate the extracted patterns?

As mentioned previously, we also extend these questions with the following to provide a richer context on the research maturity level:

- RQ7. What theoretical frameworks have been used to deal with the dropout problem? This question relates to the "Problem Identification" step (see Fig. 4).
- RQ8. What data mining methodologies have been used to analyze the dropout problem?
  This question is general and does not appear associated with a specific step (as seen in Fig. 4).

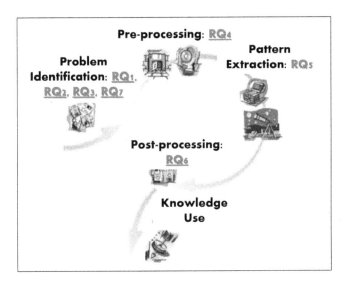

Fig. 4. The research questions and the steps of a data mining process. Adapted from [25].

The results for the above questions will be presented and discussed in the following sections, considering their place in the data mining process. Since RQ8 is a general one, i.e., it is not related to a specific step, we will to present its results as a methodological aspect.

3.1 Methodological Aspect

This section presents the results associated with question RQ8, as it is not related to a specific step. This new research question was elaborated to identify the methodologies that have been used to analyze the dropout problem.

A data mining methodology specifies steps, inputs, outputs and provides guidelines and instructions on how the steps should be performed to achieve the aim/goals of the investigation. Several methodologies were developed to model the necessary steps that make up the data mining pipeline, and their use has grown substantially since 2007 [23]. Many of these methodologies provide similar general phases based on the conceptual process model proposed by [9] for knowledge discovery in databases, differing in aspects such as specific recommended steps, particularities of an area of application, integration with methodologies from other disciplines (e.g., quality management, industrial engineering, software engineering), or in relation to the linearity of the process [23].

Existing research suggests that when a data mining methodology is selected, it is generally applied 'as-is', with specific patterns of adaptation [23]. The absence of a formal methodology implies an ad-hoc strategy for the data mining process. Although this is not an issue per se, the formal and explicit use of existing methodologies can help in the analysis, reproducibility and understanding of the limitations of an investigation, as it has been previously tested in different applications or even domains.

Table 1 summarizes the frequencies of the methodologies that were applied by the literature. We note that most papers do not adopt a formal data mining methodology, using *ad-hoc* steps to execute the mining process, even resembling existing ones. As pointed out by [23], their adoption is increasing, as is the customization according to contextual factors. However, in this study, we only investigate the methodologies that are explicit in the papers. Two data mining methodologies, both domain agnostic, stood out. The KDD process is a general methodology, commonly recognized as the basis of structured data mining methodologies and adopted in academic and industrial domains [16]. Since the beginning of the KDD process, several methodologies were proposed to address the gaps and limitations of this process, and some of them converged to the CRISP-DM model [23, p. 6]. The SEMMA methodology was also derived from the KDD process in 2005 based on the tools and practices of the SAS Enterprise Miner solution.

Table 1. Data mining methodologies applied in the dropout research.

Methodology	Freq.	%
CRISP-DM	16	13.56%
KDD	14	11.86%
SEMMA	2	1.69%
Not specified	86	72.88%

3.2 Problem Identification

Research questions RQ1, RQ2, RQ3, and RQ7 are related to this step, as seen in Fig. 4. Before answering these questions, we must draw attention to some aspects of this data mining step, considering the 118 selected papers.

Concerning *domain understanding*, a gap found in the papers was a formal dropout conceptualization without presenting its definition of what was understood by dropout. As is known, there is no consensus on the definition of dropout. In [28], it was considered as the students who interrupt the course for any reason (course transfer, registration locking, etc.) and do not finish their studies with their cohorts. This definition is an important aspect, as the interpretation of the results will be related to it.

About *aim and the goals*, it was observed that they are not explicitly presented. In [28], the 118 selected papers were classified into three categories (created by the authors)[6]: "Comparative Analysis", "Case Study", and "Solution Proposal". The Comparative Analysis (56.78%) includes papers that performed comparative analyzes between techniques and/or algorithms based on one or more datasets. The Case Study (50%) includes papers that presented an analysis on a specific dataset using one or more techniques and/or algorithms. The Solution Proposal (7.63%) includes papers that proposed a new solution to the dropout problem through data mining. Thus, most papers perform exploratory analysis considering different algorithms, but the aim and the goals, like the ones described below, are not explicitly defined (hypothetical examples elaborated by the authors):

- Aim: predict whether a student has a tendency to dropout during the current semester; Goals: the model must be interpretable (must use white-box algorithms), as it is necessary to understand the tendency; the model must achieve an accuracy above 80%;
- Aim: predict whether a student has a tendency to dropout after a certain semester or after some semesters; Goals: the model doesn't need to be interpretable (can use white-box or black-box algorithms); the model must achieve an accuracy above 85%;
- Aim: understand the profile of students who usually dropout of a given course considering theirs trajectories; Goals: understand the profile by evaluating the features using statistics (correlation with heat maps and scatter plots) and feature selection (filter, wrapper and embedded approaches [30]); validate the important features through model generation;
- Aim: understand the profile of students who usually dropout of a given course considering theirs trajectories; Goals: understand the profile through model generation (the model selects the features and, therefore, must be interpretable);
- and so on...

The papers generally focus on evaluating a sample of data using different algorithms without explicitly explaining the aim and goals. This fact has some consequences: it is difficult to identify the effective contribution of the paper regarding dropout; a concrete

---

[6] Sometimes a publication is counted more than once (for example, it is classified into more than one group). Thus, the sum of some percentages can be greater than 100%.

advantage of understanding dropout ends up being limited; the interpretation of the results is affected; recommendations about the findings, depending on the aim/goals, do not appear (a list of strategies that the institution could follow).

RQ1. What are the Education Levels Explored? This question was elaborated to identify which educational levels (university, high school, etc.) the papers focused on: 86.44% were related to universities and 16.10% with high school. The higher level is a concern in several countries, including theoretical models of study in this context [21].

Unlike the discussions presented in [28] regarding each research question, in this work considerations are made to complement the previous findings and relate them to the data mining steps. The same reasoning is valid for the other questions presented in the rest of the paper. In this sense, how does this question relate to and/or contribute to the data mining process? It can be used to refine the aim and the goals as below (some examples):

– Aim: predict whether an **undergraduate** student tends to drop out during the current semester; Goals: the model must be interpretable (must use white-box algorithms), as it is necessary to understand the tendency; the model must achieve an accuracy above 80%;
– Aim: understand the profile of **high school** students who usually drop out of a given course considering their trajectories; Goals: understand the profile through model generation (the model selects the features and, therefore, must be interpretable).

RQ2. Considering the Samples (Datasets) Used, How Big are They, and How are They Generated? Questions RQ2 and RQ3 are related to identifying and selecting data that will be used in the process. RQ2 was elaborated to identify the sample sizes and the cut made in the data to obtain the sample (by year, by course, etc.).

It was noticed that 47.46% of the works used small samples compared to the educational context; in this case, the sizes were less than or equal to 5,000 instances (]1, 5,000]). Considering this range, 60.70% use sizes smaller than 1,000 (28.81% in relation to the total of papers (118)). Besides, regarding the strategies used to obtain the samples, it could be noticed that the samples include data from periods longer than two years, and data from more than two courses: 56.78% of the studies used samples covering two or more courses considering more than two years.

How does this question relate to and/or contribute to the data mining process? To identify aspects that should be considered and that may or may not be observed, as the ones below:

– 83.05% of the studies do not mention the moment when dropout is analyzed. This information is essential since it can be used to refine the aim and the goals: Aim: predict whether a student tends to drop out **during the current semester**; Goals: the model must be interpretable (must use white-box algorithms), as it is necessary to understand the tendency; the model must achieve an accuracy above 80%;
– samples sizes impact generalization and overfitting [30]. [18] present an interesting discussion regarding the Law of Large Numbers, Empirical Risk, Expected Risk and Generalization, Bias-Variance Dilemma and its correlation with underfitting and overfitting. The size of the samples used was small, and aspects related to generalization were not considered, which can hamper the analysis of the results;

- the data used to carry out the experiments are not available, which hinders the reproducibility of the research (more on reproducibility, see [31]). In general, the data is a sample of a database from a specific institution;
- a positive aspect of the samples is that there is a concern about diversification (several courses) considering a more extended period. This information can also be used to refine the aim and the goals: <u>Aim</u>: predict whether a student tends to drop out during the current semester; <u>Goals</u>: the model must be interpretable (must use white-box algorithms), as it is necessary to understand the tendency; the model must achieve an accuracy above 80%; sample must present **diversification (several courses) considering a more extended period**.

RQ3. What Aspects (Features, Attributes) Have Been Used to Model the Dropout Problem? This question was elaborated to identify the features that were used to induce the models. As many distinct features were found, they were combined into groups (details in [28]). Table 2 summarizes the results. The first column refers to the name of the group, the second to the most representative features of the group, the third to the number of works that used features of that group, and the fourth to the percentage in relation to the total of works.

How does this question relate to and/or contribute to the data mining process? To identify aspects that should be considered and that may or may not be being observed. As seen, different features are used, but the reasons for the choices are not clarified, indicating a gap to be explored. However, as many works are categorized as "Comparative Analysis" and "Case Study", many use a sample of the data stored in the institution's database. Due to this fact, features from academic, demographic, and finance groups end up being the most used. This selection is a point that differs from the survey-based solutions, in which many groups of features, i.e., many aspects, are considered, not just those stored in the database.

Efforts must be made to find ways to collect other characteristics of the students to try to obtain models that best express the world. A compelling work is to correlate the features that are available inside the institution's database with those found in theoretical models. Perhaps, a hybrid approach can bring good results.

Table 2. Feature groups. Adapted from [28].

Feature group	Most representative features	Freq.	%
Academic	Course, yield, year ticket and GPA, course area, admission note; conclusion year, credits per semester and admission form and native language note	85	72.03%
Demographic	Gender, age, has work, marital status, schooling of the father, mother's schooling, address, mother has work and father has work and ethnicity	77	65.25%
Finance	Familiar income, financing type	28	23.73%
Social	Relationship with friends	12	10.17%
Psychological	Interest in studies, personality	10	8.47%
Not specified	N/A	26	22.03%

RQ7. What Theoretical Frameworks Have Been Used to Deal with the Dropout Problem? This section presents the results associated with this new research question, elaborated to identify the theoretical frameworks that have been used to deal with the dropout problem. A theoretical framework specifies which key variables influence a phenomenon of interest and highlights the need to examine how these key variables may differ and under what circumstances. Also, theories can help design a research question, guide the selection of relevant data, interpret the data, and propose explanations of causes or influences. They provide a comprehensive conceptual understanding of the phenomenon, providing researchers with multiple "lenses" to look at complex problems and social issues, providing a framework for conducting their analyses [24].

With the increasing availability of digital data and the development of data mining methodologies, there is a tendency to rely solely on the convenience of data to select variables to describe the phenomenon and provide explanations, generalization or even limitations [5]. This 'data deluge' brings many opportunities to leverage valid ecological data into the research process. However, this may cause troubles that could be avoided by integrating theoretical models into the pipeline [15, 19].

Only one study presented an explicit theoretical framework. In [3], the authors explored the factors as stated in Tinto's Theory of Dropout and selected the variables and analysis technique (discrete-time hazard) based on its assumptions to predict the graduation time of students from a large American research university. The results corroborated the importance of the main factors - social integration and college participation - as stated by the theory. Five other studies (4%) analyzed a slice of the problem, testing hypotheses relative to only one or a few factors related to dropout. For instance, paper [11] studied the importance of extracurricular activities in predicting school dropout rates for students at a Chilean university; paper [29] investigated the impact of individuals' technological habits (e.g., social media and internet addiction) in predicting school dropout at a university in Ecuador. These studies can serve as extensions to theories about school dropout, providing complementary perspectives on the problem. Thirteen papers (11%) based their theoretical underpinnings on previous generic studies, not specific theoretical sources. Most papers ($n = 100$, 85%) used data available, mainly from individual proprietary academic information systems, to build the prediction with different algorithms, aiming to achieve the greatest accuracy with this data.

In this sense, how does this question relate to and/or contribute to the data mining process? It can be used to refine the aim and the goals: <u>Aim</u>: correlate the features (profile) identified by higher education dropout models (survey-based) with the ones obtained through data mining process (data-driven); <u>Goals</u>: select an appropriate theoretical framework, considering the context in which it will be analyzed, to guide feature selection, techniques for analysis, and interpretation of results; hypothesize about likely extensions that could be made to explain results not explained by the selected theory (understanding its limitations).

Thus, data mining's empirical research should integrate and extend existing theories or help build new ones, providing alternative explanations for understanding the dropout phenomena.

3.3 Pre-processing

Question RQ4 is related to this step: what kind of pre-processing has been applied to the samples? This question was elaborated to identify the techniques that have been applied to prepare the samples for model induction.

Table 3, presented in [28], is here reintroduced (some counts were corrected - the ones on the lines marked with "*"). Based on this table, Table 4 was created and presents the techniques grouped according to the aspects presented in Sect. 2 - the first column refers to the name of the technique, the second to the number of works that used the respective technique, and the third to the percentage in relation to the total of works.

How does this question relate to and/or contribute to the data mining process? First, to identify aspects that should be considered and that may or may not be being observed. While some techniques are general, in the sense that they can be applied to many problems, such as data cleaning, data reduction and data transformation, others are more domain dependent, such as data balance. Dropout is a typical problem where data balance must be taken into account. But, surprisingly, only 11.86% of the works commented on this. Besides, in these cases, only general data balance techniques are used (such as random oversampling, random undersampling), but many of them exist (see [10]). This is a gap that must be addressed. Another thing is that almost 50% of the works do not specify whether the data was pre-processed.

Table 3. Studies by pre-processing techniques [28]: revised version.

Technique	Study ID	Freq.	%
Missing values*	2, 5, 10, 16, 18, 21, 22, 23, 24, 29, 30, 32, 37, 38, 41, 45, 54, 58, 67, 68, 99, 100, 103, 108, 113	25	21.19% [↑ 0.85%]
Attribute selection	1, 2, 18, 21, 28, 30, 35, 41, 49, 53, 59, 63, 77, 80, 84, 88, 89, 93, 99, 105, 110, 112	19	16.10%
Descriptive statistic	2, 5, 7, 10, 11, 18, 19, 23, 24, 29, 37, 41, 42, 45, 46, 48, 50, 56	18	15.25%
Data balance*	30, 31, 38, 41, 60, 61, 62, 63, 64, 79, 81, 85, 99, 111	14	11.86% [↓ 7.63%]
Data normalization*	1, 9, 18, 23, 28, 33, 47, 49, 50, 107	10	8.47% [↑ 5.93%]
Attribute reduction	2, 4, 30, 33, 58, 69, 103	7	5.93%
Discretization	10, 24, 29, 30, 32, 40, 109	7	5.93%
Outlier	1, 58, 93	3	2.54%
Attribute creation	1	1	0.85%
Not specified	3, 6, 12, 13, 14, 15, 17, 20, 25, 26, 27, 34, 36, 39, 43, 44, 51, 52, 55, 57, 65, 66, 70, 71, 73, 74, 75, 76, 78, 82, 83, 86, 87, 90, 91, 92, 94, 95, 96, 97, 98, 101, 102, 104, 106, 114, 115, 116, 117, 118	50	42.37%

Note: Nine (9) works on "data normalization" were counted on "data balance" and two (2) works on "missing values" were counted on "data normalization". The revised spreadsheet is available at https://bit.ly/dropout-slm-2021.

3.4 Pattern Extraction

Question RQ5 is related to this step: what algorithm families have been used? This question was elaborated to identify the algorithm families that have been explored for model induction and whether they are all predictive or whether there are solutions using descriptive tasks.

Table 4. Pre-processing techniques, presented in Table 3, grouped according to the aspects presented in Sect. 2.

Technique	Freq.	%
Data cleaning	28	23.73%
Data reduction	26	22.03%
Data transformation	17	14.41%
Data balance	14	11.86%
Not specified	50	42.37%
Others		
Descriptive statistic	18	15.25%
Attribute creation	1	0.85%

As many distinct algorithms were found, they were grouped into families (details in [28]). Table 5 summarizes the results. The first column refers to the name of the family, the second to the most representative algorithm of the family, the third to the number of works that used algorithms of that family, and the fourth to the percentage regarding the total of works.

How does this question relate to and/or contribute to the data mining process? To identify aspects that should be considered and that may or may not be being observed, like the ones below:

– the task that stood out was classification. Clustering, association, and sequential pattern are rarely used. Thus, the works are focused on predictive models, which may or may not be interpretable. However, depending on the aim/goals, other techniques could be interesting. This gap can be explored, depending on the defined aim/goals;
– the algorithm family that stood out was decision tree with almost 70% (69.49%) (Table 5). This family may have stood out because that its algorithms are interpretable (white-box) [6,20] since it is possible not only to generate a predictive model but also to understand the model. Another aspect is that, in general, nothing is mentioned about hyperparameter optimization. This is an important aspect as validation measures vary when the hyperparameter is changed;
– since many of the works are categorized as "Comparative Analysis" and "Case Study", many of them use different algorithms to seek diversity. However, as the aim and goals are not explicitly defined, the choice is not justified. Is the aim to predict? Is the aim to predict and understand the patterns obtained? Is the aim to understand the domain? Algorithms must relate to aim and goals.

Table 5. Algorithm families. Adapted from [28].

Algorithm family	Most representative algorithm	Freq.	%
Decision tree	J48/C4.5	82	69.49%
Ensemble	Random forest	61	51.69%
Regression	Logistic regression	38	32.20%
Bayesian	Naive bayes	36	30.51%
Neural network/Deep neural network	MLP	33	27.97%
Support vector machine	SVM	30	25.42%
Instance-based	KNN	21	17.80%
Rule-based	OneR	13	11.01%
Clustering	K-means	7	5.93%
Association	Apriori	5	4.24%
Discriminant analysis	Linear Discriminant Analysis (LDA)	3	2.54%
Nature-inspired	Bacterial Foraging Optimization (BFO)	1	0.85%
Sequential pattern	PrefixSpan	1	0.85%

3.5 Post-processing

Question RQ6 is related to this step: what measures have been used to validate the extracted patterns? This question was elaborated to identify the measures that have been used in the post-processing step to validate the extracted patterns (knowledge).

How does this question relate to and/or contribute to the data mining process? First, to identify aspects that should be considered and that may or may not be being observed, as the ones below:

- as classification was the prevalent task, measures such as accuracy, precision, recall, and f-measure were frequently used, with accuracy appearing in almost 82% of the studies (81.36%). However, one must observe the choice of appropriate measures. For imbalanced data, for example, other measures can be used, as G-mean (geometric mean of true positive and true negative rates) (only 4.24% of the works used it) and balanced accuracy (none of the works used), as stated by [10]. Thus, for the dropout problem, complementary measures must be used;
- as mentioned in Sect. 2, validation can be done in terms of validation measures or interpretability. Regarding interpretability, it was observed that few papers discuss and/or present the best predictive features in an attempt to understand the problem, even when white-box models are used. Again, as the aim and goals are not explicitly defined, even the preference for interpretable models (obtained, for example, through decision tree family algorithms) is not taken into account to complement the discussion of the results nor suggest actions to avoid dropout.

4 Dropout Methodological Guideline: A Data Mining Perspective

As seen so far, many aspects must be considered when data mining is used to address the dropout problem and, sometimes, gaps emerge. Therefore, in this section, we present a guideline that can be used by those looking to apply data mining to address the dropout problem.

The guideline can be seen in Table 6. The first column refers to the general data mining step and the second to the aspects that must be considered regarding the respective step.

Table 6. Data mining methodological guideline to dropout in education.

Data mining step	Aspects to be considered
Methodological aspect	- Explicitly define which formal methodology will be chosen to guide the research process or justify the reasons for using *ad-hoc* approaches: customization or limitations of existing methodologies?
Problem identification	General
	- Explicitly define how dropout is understood
	- Explicitly define how dropout will be detected:
	• before the student leaves: current data as well as historical data can be used. The detection can consider a specific semester (first semester, for example), a specific subject (one with a high number of failures), etc.
	• after the student leaves: only historical data is used. The detection can consider all semesters of a specific course or specific semesters, all subjects of a specific course or specific subjects, etc.
	- Explicitly define the level of education that will be explored
	- When possible, choose the theoretical lenses that will guide the research, stating its assumptions and limitations
	Aim/Goals
	- Explicitly define the aim, that, in general, can be:
	• Obtain a predictive model
	• Obtain a predictive model and understand the dropout problem
	• Perform a domain exploration
	• Perform a domain exploration and understand the dropout problem
	- Explicitly define the goals
	- Explicitly define the effective contribution of the work regarding dropout
	Data
	- When possible, use as much data we can to avoid overfitting
	- When possible, use a sample that contains several subjects and/or courses considering a longer period of time to take diversification into account
	- When possible, use features from different groups (categories) to try to obtain models that best express the world. In this case, features not available in the institutional database should be collected by other means (this complements what is considered in the survey-based approaches). Besides, explicitly justify the choices
	- When possible, use FAIR principles in your data and, if feasible, make them open. This ensures reproducibility
Pre-processing	- When possible, explicitly mention the treatments that were done to the data, especially if you use FAIR principles and make them open
	- When applied, use general techniques such as data cleaning, data reduction and data transformation
	- Balance your data: in general, dropout samples will be imbalanced

(continued)

Table 6. (*continued*)

Data mining step	Aspects to be considered

Pattern extraction	- Considering the previously defined aim/goals, choose the appropriate tasks/algorithms:
	  • Obtain a predictive model
	    * Use classification task
	    * Any family of classification algorithms can be used: it is important to evaluate the validation measures (see post-processing) along with algorithms' bias/variance regarding generalization [18]
	  • Obtain a predictive model and understand the dropout problem
	    * Use classification task
	    * Use white-box classification algorithms or black-box algorithms with XAI approaches [6, 20]: it is important to evaluate the validation measures (see post-processing) along with algorithms' bias/variance regarding generalization [18]. However, we think that white-box classification algorithms bring more clarity, as they are inherently interpretable
	  • Perform a domain exploration
	    * Different tasks can be used, such as classification, clustering and association
	    * Statistical analyses can complement the results, as correlation with heat maps and scatter plots
	  • Perform a domain exploration and understand the dropout problem
	    * Different tasks can be used, such as classification, clustering and association
	    * Statistical analyses can complement the results, as correlation with heat maps and scatter plots
	    * In applying classification, use white-box algorithms or black-box algorithms with XAI approaches [6, 20]: it is important to evaluate the validation measures (see post-processing) along with algorithms' bias/variance regarding generalization [18]. However, we think that white-box classification algorithms bring more clarity, as they are inherently interpretable
	  • When necessary, perform hyperparameter optimization and explicitly describe how this was done

Post-processing	- Use, among others, the following validation measures, as dropout problem is inherently imbalanced: balanced accuracy, precision, recall, f-measure, G-mean and AUC-ROC (see [10]) - Compute generalization (difference of Empirical Risk and Expected Risk) if a small sample be used. It provides a more reliable estimate [18]. Besides, analyze the bias/variance of the selected algorithms to make a better decision [18] - When interpretability is required, explicitly describe how the analysis is done. When interpretability is not required, it is interesting, when possible, to carry out additional analyzes regarding interpretability. Interpretability allows the elaboration of strategies that could be followed by the institution to support students regarding dropout
Knowledge-use	- Explicitly describe what will be done with the discovered patterns (new knowledge): will they be incorporated into a system? Will they be used to develop strategies that could be followed by the institution to support students regarding dropout? Will they help to better understand cases not explained by previous theories? Who will use the new knowledge: tutors, teachers, course administrators, policymakers?

5   Concluding Remarks

This work presented a synthesis of how dropout is being investigated by the data mining community and also a guideline to emphasize the aspects that must be considered when data mining is applied to the face-to-face school dropout problem. In this way, anyone (researches, students, etc.) who intends to apply data mining to address the dropout problem can be aware of the aspects that must be considered.

The guideline was elaborated considering the findings of our previous work [28] and here extended with two additional research questions. Besides, we also revisited these questions, correlating each one with the general data mining steps.

We found that most of the literature works adopt *ad-hoc* processes and lack theoretical background to conceptualize and guide the process. They are highly concentrated at the university level with relatively small datasets (a few thousand records) covering multiple courses and academic years. A wide range of features is explored, mainly centered on well-studied general factors, but with new and interesting hypotheses. This type of academic data is often very noisy and pre-processing steps are required, in particular on dealing with imbalanced datasets. However, we found that almost half of the papers do not cover this step. Most papers apply a computer science approach to the problem, conducting experiments evaluating the effectiveness of different algorithms in the accuracy of the classification task.

These characteristics may be attributed to the recent study of the dropout phenomenon by the data mining community. Although dropout is a topic studied for many decades in educational literature, the emergence of digital technologies has brought new potential, recording several aspects related to this problem and making them available for analysis. As the field of research matures, it is likely that new levels of formalism will occur in the scientific method. An additional explanation may be due to a paradigm shift, as in Kuhn's perspective [14] as a consequence of using big data as a fourth scientific paradigm [12]. Previous research on education has developed robust theories, but data-intensive applications and instruments are increasingly being deployed and used for analytical purposes. Thus, new methods and constructs are being tested to help better understand this social phenomenon and extend the educational literature on the subject.

Considering the discussion presented and the suggested guideline, we also list some works that could be done or some suggestions that could be though by the research community:

- the works should more carefully describe the details of the process when data mining is applied. This improves readability, enables reproducibility and makes it easier to understand the contribution of the work. To help with this, the suggested guideline can be used;
- as mentioned before, when possible, use FAIR principles in your data and, if feasible, make them open. This ensures reproducibility and open the possibility to create public repositories on the topic[7]. Besides, this can contribute to the improvement of research in this area as many samples could be available to enable (some examples): (a) broader comparisons between algorithms, as well as the adaptation and/or proposal of specific solutions to the problem; (b) studies on data balance and its impacts;
- an interesting effort would be studies that attempt to correlate the survey-based and data-driven approaches. Perhaps a hybrid solution can improve the understating of the problem.

---

[7] Like the ones cited and used in [27] for software defect.

References

1. Ackoff, R.L.: From data to wisdom. J. Appl. Syst. Anal. **16**, 3–9 (1989)
2. Agrusti, F., Bonavolonta, G., Mezzini, M.: University dropout prediction through educational data mining techniques: a systematic review. J. e-Learn. Knowl. Soc. **15**, 161–182 (2019)
3. Aiken, J.M., Bin, R.D., Hjorth-Jensen, M., Caballero, M.D.: Predicting time to graduation at a large enrollment american university. PLoS One **15**(11) (2020). https://doi.org/10.1371/journal.pone.0242334
4. Alexandropoulos, S.A.N., Kotsiantis, S.B., Vrahatis, M.N.: Data preprocessing in predictive data mining. Knowl. Eng. Rev. **34**(e1), 1–33 (2019)
5. Anderson, C.: The end of theory: the data deluge makes the scientific method obsolete (2008). https://www.wired.com/2008/06/pb-theory/
6. Burkart, N., Huber, M.F.: A survey on the explainability of supervised machine learning. CoRR abs/2011.07876 (2020)
7. Cardona, T., Cudney, E.A., Hoerl, R., Snyder, J.: Data mining and machine learning retention models in higher education. J. Coll. Student Retention Res. Theor. Pract. 25p. (2020)
8. Delen, D.: Predicting student attrition with data mining methods. J. Coll. Student Retention Res Theor. Pract. **13**(1), 17–35 (2011)
9. Fayyad, U., Piatetsky-Shapiro, G., Smyth, P.: The KDD process for extracting useful knowledge from volumes of data. Commun. ACM **39**(11), 27–34 (1996). https://doi.org/10.1145/240455.240464
10. Fernández, A., García, S., Galar, M., Prati, R.C., Krawczyk, B., Herrera, F.: Learning from Imbalanced Data Sets. Springer, Cham (2018). https://doi.org/10.1007/978-3-319-98074-4
11. Hasbun, T., Araya, A., Villalon, J.: Extracurricular activities as dropout prediction factors in higher education using decision trees. In: 2016 IEEE 16th International Conference on Advanced Learning Technologies, pp. 242–244 (2016). https://doi.org/10.1109/ICALT.2016.66
12. Hey, T., Tansley, S., Tolle, K.: The fourth paradigm: data-intensive scientific discovery. Microsoft Research (2009). https://www.microsoft.com/en-us/research/publication/fourth-paradigm-data-intensive-scientific-discovery/
13. Kitchenham, B., Charters, S.: Guidelines for performing systematic literature reviews in software engineering. Technical report EBSE 2007-001, Keele University and Durham University Joint Report (2007)
14. Kuhn, T.: The Structure of Scientific Revolutions. University of Chicago Press, Chicago (1962)
15. Lazer, D., Kennedy, R., King, G., Vespignani, A.: Big data. The parable of google Flu: traps in big data analysis. Science **343**, 1203–1205 (2014). https://doi.org/10.1126/science.1248506
16. Marbán, O., Segovia, J., Menasalvas, E., Fernández-Baizán, C.: Toward data mining engineering: a software engineering approach. Inf. Syst. **34**(1), 87–107 (2009). https://doi.org/10.1016/j.is.2008.04.003, https://www.sciencedirect.com/science/article/pii/S0306437908000355
17. Mduma, N., Kalegele, K., Machuve, D.: A survey of machine learning approaches and techniques for student dropout prediction. Data Sci. J. **18**(14), 10p. (2019)
18. Mello, R.F., Ponti, M.A.: Machine Learning: A Practical Approach on the Statistical Learning Theory. Springer, Cham (2018). https://doi.org/10.1007/978-3-319-94989-5
19. Metaxas, P.T., Mustafaraj, E., Gayo-Avello, D.: How (not) to predict elections. In: 2011 IEEE Third International Conference on Privacy, Security, Risk and Trust and 2011 IEEE Third International Conference on Social Computing, pp. 165–171 (2011). https://doi.org/10.1109/PASSAT/SocialCom.2011.98

20. Molnar, C.: Interpretable Machine Learning (2019). https://christophm.github.io/interpretable-ml-book/
21. Nicoletti, M.C.: Revisiting the Tinto's theoretical dropout model. High. Edu. Stud. **9**, 52–64 (2019)
22. Pedroza, K.Y.D., Chasoy, B.Y.C., Gómez, A.: Review of techniques, tools, algorithms and attributes for data mining used in student desertion. In: Sixth International Meeting of Technological Innovation. Journal of Physics: Conference Series (2019)
23. Plotnikova, V., Dumas, M., Milani, F.: Adaptations of data mining methodologies: a systematic literature review. PeerJ Comput. Sci. **6**, e267 (2020). https://doi.org/10.7717/peerj-cs.267
24. Reeves, S., Albert, M., Kuper, A., Hodges, B.D.: Why use theories in qualitative research? BMJ **337** (2008). https://doi.org/10.1136/bmj.a949, https://www.bmj.com/content/337/bmj.a949
25. Rezende, S.O.: Sistemas Inteligentes: fundamentos e aplicações. Editora Manole Ltda (2003)
26. Rumberger, R.W.: The economics of high school dropouts. In: Bradley, S., Green, C. (eds.) The Economics of Education: A Comprehensive Overview, 2nd edn., chap. 12, pp. 149–158. Academic Press (2020)
27. Shao, Y., Liu, B., Wang, S., Li, G.: A novel software defect prediction based on atomic class-association rule mining. Expert Syst. Appl. **114**, 237–254 (2018)
28. Sousa, L.R., Carvalho, V.O., Penteado, B.E., Affonso, F.J.: A systematic mapping on the use of data mining for the face-to-face school dropout problem. In: Proceedings of the 13th International Conference on Computer Supported Education CSEDU, pp. 36–47 (2021)
29. Taipe, M.S.A., Sánchez, D.M.: Prediction of university dropout through technological factors: a case study in Ecuador. Rev. Espacios **39**(52) (2018)
30. Tan, P.N., Steinbach, M., Karpatne, A., Kumar, V.: Introduction to Data Mining, 2nd edn. Pearson, London (2018)
31. Tatman, R., VanderPlas, J., Dane, S.: A practical taxonomy of reproducibility for machine learning research. In: Reproducibility in Machine Learning Workshop at ICML 2018, p. 5p. (2018)
32. Tinto, V.: Leaving College: Rethinking the Causes and Cures of Student Attrition. University of Chicago Press, Chicago (1993)

Existing Machine Learning Techniques for Knowledge Tracing: A Review Using the PRISMA Guidelines

Sergio Iván Ramírez Luelmo$^{(\boxtimes)}$ (ID), Nour El Mawas (ID), and Jean Heutte (ID)

CIREL – Centre Interuniversitaire de Recherche en Éducation de Lille, Université de Lille,
Villeneuve d'Ascq, France

{sergio.ramirez-luelmo,nour.el-mawas,jean.heutte}@univ-lille.fr

Abstract. Nowadays, Machine Learning (ML) techniques play an increasingly important role in educational settings such as behavioral academic pattern recognition, educational resources suggestion, competences and skills prediction, or clustering students with similar learning characteristics, among others. Knowledge Tracing (KT) allows modelling the learner's mastery of skills and to predict student's performance by tracking within the Learner Model (LM) the students' knowledge. Based on the PRISMA method, we survey and describe commonly used ML techniques employed for KT shown in 51 articles on the topic, among 628 publications from 5 renowned academic sources. We identify and review relevant aspects of ML for KT in LM that contribute to a more accurate panorama of the topic and hence, help to choose an appropriate ML technique for KT in LM. This work is dedicated to MOOC designers/providers, pedagogical engineers and researchers who need an overview of existing ML techniques for KT in LM.

Keywords: Machine Learning · Knowledge Tracing · Learner Model · Technology enhanced learning · Literature review · PRISMA

1 Introduction

Evidence from several studies has long linked having a Learner Model (LM) can make a system more effective in helping students learn, and adaptive to learner's differences [1].

LMs represent the system's beliefs about the learner's specific characteristics, relevant to the educational practice [2], encoded using a specific set of dimensions [3]. Ultimately, a perfect, ideal LM would include all features of the user's behavior and knowledge that effect their learning and performance [4]. Modelling the learner has the ultimate goal of allowing the adaptation and personalization of environments and learning activities [5] while considering the unique and heterogeneous needs of learners. We acknowledge the difference between Learner Profile (LP) and LM in that the former can be considered an instantiation of the latter in a given moment of time [6].

Knowledge Tracing (KT) models students' knowledge as they correctly or incorrectly answer exercises [7], or more generally, based on observed outcomes on their previous

© Springer Nature Switzerland AG 2022
B. Csapó and J. Uhomoibhi (Eds.): CSEDU 2021, CCIS 1624, pp. 73–94, 2022.
https://doi.org/10.1007/978-3-031-14756-2_5

practices [8]. KT is one out of three approaches for student performance prediction [9]. In an Adaptive Educational System (AES), predicting students' performance warrants for KT. This allows for learning programs recommendation and/or level-appropriate, educational resources personalization, and immediate feedback. KT facilitates personalized guidance for students, focusing on strengthening their skills on unknown or less familiar concepts, hence assisting teachers in the teaching process [10].

Machine Learning (ML) is a branch (or subset) of Artificial Intelligence (AI) focused on building applications that learn from data and improve their accuracy over time without being programmed to do so [11]. To achieve this, ML algorithms build a model based on sample data (a.k.a. input data) known as 'training data'. Once trained, this model can then be reused with other data to make predictions or decisions.

ML techniques are currently applied to KT in vast and different forms. The goal of this literature review is to survey all available works in the field of "Machine Learning for Knowledge Tracing used in a Learner Model setup" in the last five years to identify the most employed ML techniques and their relevant aspects. This is, in general terms, what common ML techniques and aspects, designed to trace a learner's mastery of knowledge, also account for the creation, storage, and update of a LM. Moreover, we aim to identify relevant ML aspects to consider insuring KT in a LM. The motivation behind this work is to present a comprehensive panorama on the topic of ML for KT in LM to our target public. To our knowledge, currently there is no research work that addresses the literature review of ML techniques for KT accounting for the LM.

Thus, we decided to focus our literature review on the terms "machine learning", "knowledge tracing" and "learner model", a.k.a. "student model" (SM). Using the PRISMA method [12], we performed this research in the IEEE, Science Direct, Scopus, Springer, and Web of Science databases comprising the 2015–2020 period. The thought behind these choices is to obtain the most recent and high-quality corpus on the topic.

This work differs from other literature reviews [13–15] on two accounts. First, we focus exclusively on ML techniques for KT accounting for the LM. That is, we do not cover pure Data Mining (DM) techniques, nor AI intended for purposes other than KT, such as Natural Language Processing (NLP), gamification, computer vision, learning styles prediction, nor any processes that make pure use of LP data (instead of LM data), nor other User Model data, such as sociodemographic, biometrical, behavioral, or geographical data[1]. Second, we do not compare the mathematical inner workings of ML techniques: we feel that our target public might be unable to exploit appropriately such complex form results. Instead, we shift the focus to a pragmatic report on ML for KT in a LM application and purpose(s).

The remainder of this article is structured as follows. Section 1 of this paper oversees the theoretical framework concerning this paper, namely the definition of ML and its categorization. Section 3.3 details the methodology steps taken. Section 5 presents the findings of this research, Sect. 5 discusses the results and, finally Sect. 5 concludes this paper and presents its perspectives.

---

[1] Please note that we did include such works, if they also employed ML for KT (the core of this paper).

2   Theoretical Background

In this section we present the theoretical background put in motion behind this research, namely the definition of ML and how it is categorized.

2.1   Machine Learning

ML is a branch (or subset) of AI focused on building applications that learn from data and improve their accuracy over time without being programmed to do so [11]. Additional research [16, 17] to this definition allows us to present Fig. 1 to illustrate and discern the situation of ML against other common terms used in the field.

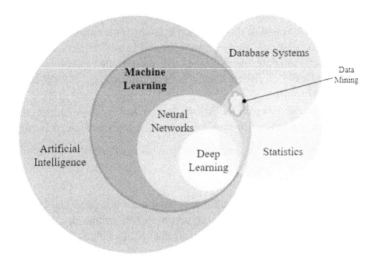

Fig. 1. Situational context of ML [18].

2.1.1   ML Methods/Styles/Scenarios

Although some authors [13, 19] admit several more ML methods (or styles or paradigms or scenarios), we retain the following categorization: Supervised ML, Semi Supervised ML, Unsupervised ML, Reinforcement Learning, and Deep Learning [11]. The first three differentiate each other on the labelling of the input training data while creating the model. The two latter constitute special cases altogether [11, 19, 20].

First, in **Supervised Learning (SL)** labels are provided (metadata containing information that the model can use to determine how to classify it). However, properly labelled data is expensive[2] to prepare, and there is a risk of creating a model so tied to its training data that it cannot handle variations in new input data accurately ("overfitting") [20].

Second, **Unsupervised Learning (UL)** must use algorithms to extract meaningful features to label, sort and classify its training data (which is unlabeled) without human

---

[2] Mostly in terms of computational resource allocation.

intervention. As such, it is usually used to identify patterns and relationships (that a human can miss) than to automate decisions and predictions. Because of this, UL requires huge amounts of training data to create a useful model [20].

Third, **Semi Supervised Learning (SSL)** is at the middle point of the two previous methods: it uses a smaller labelled dataset to extract features and guide the classification of a larger, unlabeled dataset. It is usually used when not enough labelled data is made available (or it is too expensive) to train a preferred, Supervised Model [21].

Fourth, **Reinforcement Learning (RL)** is a behavioral machine learning model akin to SL, but the algorithm is not trained using sample data but by using trial and error. A sequence of successful outcomes will be reinforced to develop the best recommendation or policy for a given problem. RL models can also be deep learning models [11].

Lastly, **Deep Learning (DL)** is a subset of ML (all DL is ML, but not all ML is DL). DL algorithms define an artificial neural network[3] that is designed to learn the way the human brain learns. DL models require a large amount of data to pass through multiple layers of calculations, applying weights and biases in each successive layer to continually adjust and improve the outcomes. DL models are typically unsupervised or semi-supervised [11]. For clarity reasons, the figure illustrating this ML categorization is available in the Appendix.

In this subsection we covered the ML definition and a categorization of ML techniques. In the following subsection we deepen into the relevant aspects in ML for KT in LM.

2.1.2   ML for KT in LM

An overwhelming number of ML techniques have been designed and introduced over the years [13]. They usually rely on more common ML techniques, within optimized pipelines. As such, we identify the **ML techniques** (or algorithms) upon which any new research is based.

Additionally to performing KT in LM, researchers have acknowledged that ML techniques can reliably determine the initial parameters when instantiating a LM [23, 24]. This led us to consider this purpose when reviewing ML techniques. Different ML techniques are applied at different stages of the ML pipeline, and not all stages are responsible for KT (other applications can be NLP, computer vision, automatic grading, demographic student clustering, mood detection, etc.) We differentiate purposes related to KT and/or learner modelling, specifically if the ML technique is used for (1) either grade, skills, or knowledge prediction (and hence later, clustering, personalizing, or suggesting resources), (2) either for LM creation (or instantiation), or (3) both.

Studies highlight the importance of justifying the **rationale** when choosing a ML technique [25–27]. We note such rationale, when made explicit, and contrast it to other authors' rationale for commonalities, on the same technique. This allows us to weigh and present known, favorable, and unfavorable features specific to ML techniques applied to KT accounting for the LM.

Research studies stress the ultimate importance of the input data (**dataset**) and the effects of the chosen programming language **software** employed for ML [25, 28]. Indeed,

---

[3] A quite complete and updated chart of many neural networks was made available by [22].

ML techniques require input data for creating a model. The feature engineering of this input data (dataset) might be determinant for a ML project to succeed or fail [25]. We list and verify the availability of all public datasets presented in the reviewed articles. Furthermore, the choice of the programming language for ML plays a role in collaboration, licensing, and decision-making processes: it helps to determine the most appropriate choices for ML implementation (purchasing licenses, upgrading hardware, hiring a specialist, or considering self-training). Hence, we highlight the family of ML programming languages used by researchers on their proposals.

Thus, based on this state-of-the-art, we identify relevant aspects to consider in ML for KT in LM: the **ML technique** employed, its **purpose**, the contextual, known **rationale** for choosing it, the **programming language software** used for ML, and the **dataset**(s) employed for KT. We consider that these aspects are relevant for our target public when choosing a ML technique for KT in LM.

3 Review Methodology

This review of literature follows the PRISMA [12] methodology, comprising: Rationale, Objectives & Research questions, Eligibility criteria, Information sources & Search strategy, Screening process & Study selection, and Data collection & Features.

3.1 Rationale, Objectives and Research Questions

The goal of this literature review is to present a comprehensive panorama on the topic of ML for KT in LM. This is, in general terms, what ML techniques designed to trace a learner's mastery of skill also account for the creation, storage, and update of the LM.

This article aims thus to answer the following two research questions (RQ):

- RQ1: What are the most employed ML techniques for KT in LM?
- RQ2: How do the most employed ML techniques fulfil the considered relevant aspects to insure KT in LM?

3.2 Eligibility Criteria, Information Sources and Search Strategy

In this section we describe the inclusion and exclusion criteria used to constitute the corpus of publications for our analysis. We also detail and justify our choice of in-scope publications, the search terms, and the identified databases.

In this research, we focus on recent ML techniques (and/or algorithms) that explicitly "learn" (with minimal or no human intervention) from its data input to perform KT, while accounting for the LM. Thus, we do not cover all predictive statistical methods (as they are not all ML), nor pure DM techniques, nor AI intended for purposes other than KT (such as NLP, gamification, computer vision, learning styles prediction, etc.), nor any processes that make pure use of LP data (instead of LM data), nor other User Model data, such as sociodemographic, biometrical, behavioral, or geographical data.

On one hand, our Inclusion criterion are: Works that present a ML technique for KT while accounting for the LM, in the terms presented in the previous paragraph. On the other hand, our chosen Exclusion criterion consist of: Works written not in English, under embargo, not published or in the works. We choose to keep subsequent works on the same subject from the same research team because they represent a consolidation of the techniques employed.

We performed this research at the end of October 2020 in the following scientific databases: IEEE, Science Direct, Scopus, Springer, and Web of Science, comprising 2015–2020. The thought behind these two choices is to have the most recent and quality-proven scientific works on the subject. Our general search terms were:

```
(("learner model" OR "student model" OR "knowledge trac-
ing") AND "machine learning")
```

They were declined for the specificities of each scientific database (search engines parse and return verbal, noun, plural, and continuous forms of search terms). We used their 'Advanced search' function, or we queried them directly, if they allowed it. Some direct queries did not allow for year filtering, so we applied it manually on the results page. For accessibility reasons, we explicitly selected "Subscribed content" results for the scientific databases supporting it.

3.3   Screening Process and Study Selection

The paper selection process happened as follows: First, we gathered all the results in two known Citation Manager programs to benefit from the automatic metadata extraction, the report creation, and duplicate merging. We also used a spreadsheet to record, based on Sect. 2.1.2, the following information: doi, title, year, purpose, ml_method, method_rationale, software, data_source, and observations. Second, we screened the abstracts of all 708 results: three categories appeared: obvious Out-of-scope results, clear Eligible results, and Pending (verification needed) results. Third, using the institutional authentication, we downloaded all the papers in the Eligible and Pending categories. Fourth, we read the full papers in the Eligible and Pending categories and re-classified them as Eligible or Out-of-scope, as needed (Fig. 2).

3.4   Data Collection and Features

In this section we review the relevant features of interest described in Subsect. 2.1.2 found in the reviewed literature.

During the full text read, we extracted the following information from the selected papers: (1) ML technique employed; (2) purpose of the ML technique; (3) rationale for employing that specific ML technique; (4) software employed for ML; and (5) dataset employed for KT, if any.

We note here that rarely a single, known technique ML is employed, but it is rather implemented in a pipeline, connected with another secondary ML (probabilistical, or DM) techniques. In such cases, we focused on the technique(s) employed for KT and on the reasons given for choosing it over other techniques acknowledged by the authors.

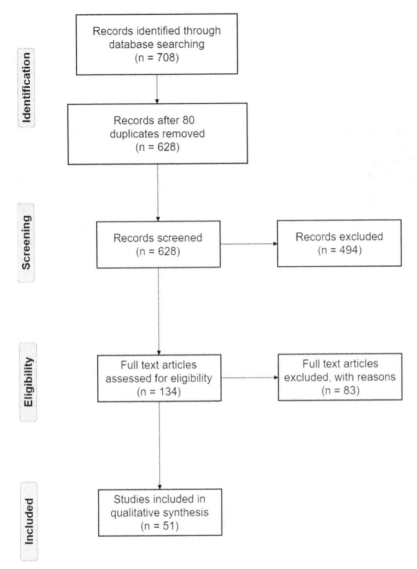

Fig. 2. PRISMA Flow diagram of the publication screening process [18].

We surveyed the software used to perform the calculation of ML and we grouped them by programming language, which is a rather meaningful description, compared to combinations of libraries and platforms. We think this result shows a clear tendency on the necessary requirements to implement and perform ML for KT in LM.

We surveyed all datasets presented in the 51 reviewed papers and checked for their existence. We understand that our target public may not have data made available to perform ML for KT accounting for the LM and we feel that this resource may be invaluable when evaluating their results.

In this section we presented our literature review methodology, the considered features, and the train of thought behind them. The following section details our literature review results.

4 Results

We aggregated the data collected (described in the previous section) to make it easier to digest.

First, we present the seven most employed[4] ML techniques for KT in LM found in the reviewed publications. These comprise based-upon techniques for the paper proposal, techniques used as starting bases within a pipeline, and techniques employed when comparing ML techniques.

Bayesian Knowledge Tracing (BKT) [8] is the most classical method used to trace students' knowledge states. It is a special case of a Hidden Markov Model (HMM) [29]. In BKT, skill is modeled as a binary variable (known/unknown) and learning is modeled by a discrete transition from an unknown to a known state. The basic structure of the model, as well as its update and prediction equations are depicted in Fig. 3, the probability of being in the known state is updated using a Bayes rule based on an observed answer. The basic BKT model uses the following data [9, 29, 30]:

- Global learner data: P_i is the probability that the skill is initially learned (also known as *p-init*, or $P(L_0)$), P_g is the probability of a correct answer when the skill is unlearned (a guess, a.k.a. *p-guess* or $P(G)$), P_s is the probability of an incorrect answer when the skill is learned (a slip, a.k.a. *p-slip* or $P(s)$), and P_l is the probability of learning a skill in one step (a.k.a *p-transit*, $P(T)$), assumed constant over time.
- Local learner data: probability θ that a learner is in the known state.
- Global domain data: a definition of Knowledge Components (KCs) (sets of items). There are no relations among KCs, i.e., the parameters for individual KCs are independent.

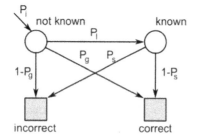

update equation:

if $c = 1$: $\theta' := \dfrac{\theta(1-P_s)}{\theta(1-P_s)+(1-\theta)P_g}$

if $c = 0$: $\theta' := \dfrac{\theta P_s}{\theta P_s+(1-\theta)(1-P_g)}$

$\theta := \theta' + (1 - \theta')P_l$

prediction equation:

$P_{correct} = P_g \cdot \theta + (1 - P_s) \cdot (1 - \theta)$

Fig. 3. Basic structure and equations for the BKT model [29] (*c* is the correctness of an observed answer).

---

[4] With more than five applications in the last five years.

Parameter fitting for the global learner parameters (the tuple P_i, P_l, P_s, P_g) is typically done using the standard expectation-maximization algorithm, alternatively using a stochastic gradient descent or discretized brute-force search. The specification of KC is typically done manually, potentially using an analysis of learning curves [9, 29].

Deep Knowledge Tracing (DKT) was proposed by [31] to trace students' knowledge using Recurrent Neural Networks (RNNs), achieving great improvement on the prediction accuracy of students' performance. It uses a Long Short-term Memory (see next item) to represent the latent knowledge space of students dynamically. DKT uses large numbers of artificial neurons for representing latent knowledge state along with a temporal dynamic structure and allows a model to learn the latent knowledge state from data [32]. It is defined by the following equations [31, 32]:

$$h_t = \tanh(W_{hx}x_t + W_{hh}h_{t-1} + b_h) \tag{1}$$

$$y_t = \sigma\left(W_{yh}h_t + b_y\right) \tag{2}$$

In DKT, both tanh and the sigmoid function (σ) are applied element wise and parameterized by an input weight matrix W_{hx}, recurrent weight matrix W_{hh}, initial state h_0, and readout weight matrix W_{yh}. Biases for latent and readout units are represented by b_h and b_y [31, 32].

Long Short-Term (LSTM) is a variant of RNN, effective in capturing underlying temporal structures in time series data and long-term dependencies more effectively than conventional RNN [33]. LSTM builds up memory by feeding the previous hidden state as an additional input into the subsequent step. While typical RNN consist of a chain of repeating modules of NN, in LSTM, instead of having a single NN layer, there are three major interacting components: forget, input, and output (i_t, f_t, and o_t, respectively) [33]. In LSTM, latent units retain their values until explicitly cleared by the action of the 'Forget gate'. Thus, they retain more naturally information for many time steps, which is believed to make them easier to train. Additionally, hidden units are updated using multiplicative interactions, and they can thus perform more complicated transformations for the same number of latent units. This makes the model particularly suitable for modeling dynamic information in student modeling, where there are strong statistical dependencies between student learning events over long-time intervals. The equations for an LSTM are significantly more complicated than for an RNN [31]:

$$i_t = \sigma\left(W_i\left[x_t, h_{t-1}\right] + b_i\right) \tag{3}$$

$$f_t = \sigma\left(W_f\left[x_t, h_{t-1}\right] + b_f\right) \tag{4}$$

$$o_t = \sigma\left(W_o\left[x_t, h_{t-1}\right] + b_o\right) \tag{5}$$

$$c\_in_t = \tanh\left(W_{c\_in}\left[x_t, h_{t-1}\right] + b_{c\_in}\right) \tag{6}$$

$$c_t = f_t \odot c_{t-1} + i_t \odot c\_in_t \tag{7}$$

$$h_t = o_t \odot \tanh(c_t) \tag{8}$$

$$k_t = \sigma(W_{kh}h_t + b_k) \tag{9}$$

Although LSTM has a certain capability of learning relatively long-range dependency, it still has trouble remembering long-term information [34].

Bayesian Networks (BNs) are graphical models designed to explicitly represent conditional independence among random variables of interest and exploit this information to reduce the complexity of probabilistic inference [35]. They are a formalism for reasoning under uncertainty that has been widely adopted in AI [36]. Formally, a Bayesian network is a directed acyclic graph where nodes represent random variables and links represent direct dependencies among these variables.

If we associate to each node X_i in the network a Conditional Probability Table (CPT) that specifies the probability distribution of the associated random variable given its immediate parent nodes parents (X_i), then the BN provides a compact representation of the Joint Probability Distribution (JPD) over all the variables in the network:

$$P(X_1, \ldots, X_n) = \prod_{l=1}^{n} P(X_i | Parents(X_i)) \tag{10}$$

A few examples of simple BN and their associated equations are shown in Fig. 4.

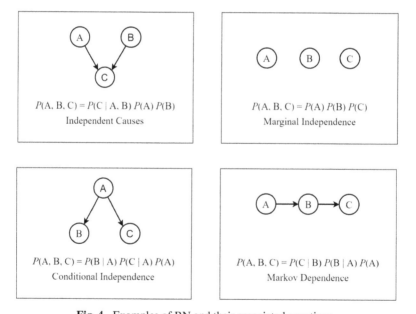

Fig. 4. Examples of BN and their associated equations.

Static BNs track the belief over the state of variables that don't change over time as new evidence is collected, i.e., the posterior probability distribution of the variables given the evidence. Dynamic BNs on the other hand, track the posterior probability of variables whose value change overtime given sequences of relevant observations [36].

Support Vector Machines (SVM) are one of the most robust prediction methods, based on statistical learning frameworks [37]. The primary aim of this technique is to map nonlinear separable samples onto another higher dimensional space by using different types of kernel functions. The underlying idea is that when the data is mapped to a higher dimension, the classes become linearly separable [38]. SVM try to reduce the probability of misclassification by maximizing the distance between two class boundaries (positive vs. negative) in data [39]. Assume that a dataset used for training is represented by a set $j = \{(x_i, y_i)\}_{i=1}^l$, where $(x_i, y_i) \in R^{n+1}$, l is the number of samples, n is the number of features and a class label $y_i = \{-1, 1\}$. The separating hyperplane, defined by the parameters w and b, can be obtained by solving the following convex optimization problem [40, 41]:

$$\min \frac{1}{2} \|w\|^2 \tag{11}$$

$$s.t. \ y_i\left(w^T \varphi(x_i) + b\right) \geq 1 \qquad i = 1, 2, \ldots\ldots, l \tag{12}$$

For actualizing SVM for more than two classes two possible strategies can be used: One-Against-All (OAA) and One-Against-One (OAO). In OAA, to unravel an issue of n classes, n binary problems are solved rather than fathoming a single issue. Each classifier is basically used to classify one single class; that is why values on that specific class will grant positive response and [data] points on other classes will give negative values on that classifier. In the case of OAO, for n course issues, $\frac{n(n-1)}{2}$ SVM classifiers are built and each of them is prepared to partition one class from another. When classifying an unknown point, each SVM votes for a class and the class with most extreme votes is considered as the ultimate result [41].

They key advantage of SVM is that they always find the global minimum because there are no local optima in maximizing the margin. Another benefit is that the accuracy does not depend on the dimensionality of data [38]. This is a clear advantage when the class boundary is non-linear, as other classification techniques will produce too complex models for non-linear boundaries [38]. They distinctively afford balanced predictive performance, even in studies where sample sizes may be limited.

Dynamic Key Value Memory Network (DKVMN) is a Memory Augment Neural Network-based model (MANN), which uses the relationship between the underlying knowledge points to directly output the student's mastery of each knowledge point [42]. DKVMN uses *key-value* pairs rather than a single matrix for the memory structure. Instead of attending, reading, and writing to the same memory matrix in MANN, the DKVMN model attends input to the *key* component, which is immutable, and reads and writes to the corresponding *value* component [42]:

At each timestamp, DKVMN takes a discrete exercise tag q_t, outputs the probability of response $p(r_t|q_t)$, and then updates the memory with an exercise-and-response tuple (q_t, r_t). Here, q_t also comes from a set with Q distinct exercise tags and r_t is a binary value. [42] affirms that there are N latent concepts $\{c^1, c^2, \ldots, c^N\}$ underlying the exercises, which are stored in the *key* matrix M^k (size $N \times d_k$) whereas the students' mastery levels of each concept $\{s_t^1, s_t^2, \ldots, s_t^N\}$ (concept states) are stored in the *value* matrix M_t^v (size $N \times d_v$), which changes over time [42]. Thus, DKVMN traces the knowledge of a student by reading and writing to the *value* matrix using the correlation *weight* computed from the input exercise and the *key* matrix. Equations for the read and write process can be found in detail in [42].

Performance Factor Analysis (PFA) [43] is one specific model from a larger class of models based on a logistic function [29]. In PFA, the data about learner performance are used to compute a skill estimate. Then, this estimate is transformed using a logistic function into the estimate of the probability of a correct answer. The update and prediction equations are depicted in Fig. 5. The PFA model uses the following data [29]:

- Global learner data: parameters γ_k, δ_k specifying the change of skill associated with correct and wrong answers for a given KC_k.
- Local learner data: a skill estimate θ_k for each KC_k.
- Global domain data: a KC difficulty parameter β_k, a Q-matrix Q specifying item-KC mapping; $Q_{ik} \in \{0, 1\}$ denotes whether an item i belongs to KC_k.

$$\theta_k := \theta_k + Q_{ik}\big(\gamma c + \delta(1 - c)\big) \text{ for each } k. \tag{13}$$

$$P_{correct} = \frac{1}{(1+e^{-m})}, \text{ where } m = \sum_k Q_{ik}(\beta_k + \theta_k). \tag{14}$$

Fig. 5. Update (θ_k) and prediction ($P_{correct}$) equations for the PFA model, according to [29] (c is the correctness of an answer, i is the index of an item).

Parameter fitting for parameters β, γ, δ is usually done using standard logistic regression. The Q-matrix is also usually manually specified, but can be also fitted using automated techniques like matrix factorization [29].

This list answers then RQ1. "What are the most employed ML techniques for KT in LM?". Figure 6 shows a yearly heatmap of the most used techniques: the number indicates the total number of applications[5] in all 51 combined-and-reviewed papers, per year. DKT was applied eight times in 2019 (emerging from two consecutive zero years) while BKT was mostly applied in 2016 and 2017, five and six times respectively, decreasing since. LSTM peaked in 2017, with 7 applications, and has decreased since. BN remains with a steady application since 2017. For clarity reasons, the 29 ML techniques found in the 51 papers issued from this study are available in the Appendix.

Second, we noted the rationale (if any) given by authors when choosing a ML technique. We do not account for the rationale of the paper's unique ML proposal if its improvements are related to parameter fine-tuning, or if the justification is à posteriori. Instead, we account rationale for the general application of the original, unmodified technique. Also, very few publications detail the shortcomings of their choice. We grouped these rationales (Fig. 7) in the following categories:

---

[5] Programming and teaching the ML model with input data.

ML technique	2015	2016	2017	2018	2019	2020	Cumulative
BKT	1	5	6	0	3	3	18
DKT	0	2	0	0	8	3	13
LSTM	0	0	7	1	1	3	12
BN	1	0	2	3	3	2	11
SVM	2	1	1	0	2	1	7
DKVMN	0	0	0	0	4	3	7
PFA	0	1	2	0	0	3	6

Fig. 6. Yearly heatmap of the most employed ML techniques [18].

R1-Uses Less Data and/or Metadata. These techniques handle sparse data situations better compared to others, according to the authors, e.g. DKT [44].

R2-Extended Tracing. These techniques provide additional attributes and/or dimensional tracing with ease when compared to other techniques, according to authors, e.g. LSTM [45].

R3-Popularity. These techniques were chosen because of their popularity, e.g. BN [24].

R4-Persistent Data Storage. These techniques explicitly save their intermediate states to long-term memory, e.g. DKVMN [46].

R5-Input Data Limitations. These techniques have been acknowledged to lack when the number of peers is "too high", e.g. BN [47].

R6-Modelling Shortcomings. Techniques in this category face difficulties when modelling either forgetting, guessing, multiple-skill questions, time-related issues, or have other modelling shortcomings, e.g. BKT [48].

	ML technique	R1	R2	R3	R4	R5	R6
RL	BKT	0	0	1	0	0	4
	DKT	5	0	1	0	0	0
DL	LSTM	0	4	0	0	0	1
	DKVMN	0	0	0	1	0	1
SL	BN	0	1	3	0	2	0
	SVM	0	0	0	0	0	0
UL	PFA	0	1	0	0	0	1

Fig. 7. Heatmap of most employed ML techniques, categorized by Method (SL, UL, SSL, RL, DL) and number of publications sharing a Rationale (R1–R6) [18].

A heatmap illustrating the number of publications mentioning each of these rationales, for each of the most common ML techniques, is shown in Fig. 4. This heatmap includes the ML categorization presented in Sect. 2.1.1 (SL, UL, SSL, RL, DL).

BKT faced mostly R6 rationales (four occurrences) and it alone conformed all of the RL techniques found in this study. DKT and BKT were mostly commented on R1 and R2, with five and four occurrences respectively. This leads to the DL categorization (DKT + LSTM + DKVMN) to be extensively justified in the literature, while UL (PFA) is sparsely commented, and SVM not at all, despite its non-negligeable number of applications (seven). BN had the highest R3 count of all (three occurrences) and because of the absence of SVM comments, it carries all the justifications related to SL.

Third, we looked over the intended purpose of the ML implementation, besides the intended KT. In one hand, out of the 51 publications reviewed, seven (~15%) employ ML for initializing the LM (e.g., for another course, academic year, or for determining the ML parameters in a pipeline) by accounting previous system interactions, grades, pretests, or other data. In the other hand, most publications (44–85%) perform some form of prediction. Finally, only one proposal (~2%) incorporates both a prediction and/or recommendation mechanism as well. A pie chart of ML techniques' purpose distribution is presented in Fig. 8.

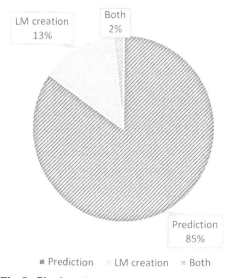

Fig. 8. Pie chart distribution of ML purpose [18].

Fourth, we surveyed the software used to perform the ML calculations. Note that many publications (~50%) do not mention their software of choice. Python (comprising Keras, TensorFlow, PyTorch and scikit-learn) is the largest group, with 13 papers. *Ad-hoc* solutions follow with five papers, and finally C, Java (-based), Matlab and R solutions, with 2 publications each. Outliers were SPSS and Stan, with one (1) paper each. A pie chart illustrating the distribution of programming languages is shown in Fig. 9.

Fifth, we highlighted (and checked for existence) the public datasets employed, shown in Table 1. All the datasets we found in the literature were online and accessible when reviewed. We made the version distinction (yearly or by topic) of datasets from the same source (such as DeepKnowledgeTracing and ASSISTments, respectively) because

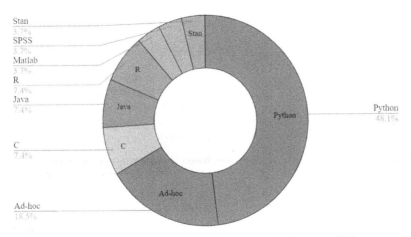

Fig. 9. Pie chart distribution of ML programming language [18].

they differ on either the number of features, or the dimensioning, or dataset creation method.

We make the distinction from our previous paper [18] in that the datasets "MOOC [Big Data and Education on the EdX platform]" and "Hour of Code" are not available anymore[6] online and thus, do not appear anymore in Table 1. Moreover, the "DataShop" dataset (https://pslcdatashop.web.cmu.edu/) was removed as well because it points to a repository of learning interaction data with no specific dataset.

Moreover, for each available singular dataset we catalogued the number of files it encompasses, the file format in which it is saved, its total size (in multiples of bytes), and lastly, the list of available features (the mixed uppercase and lowercase features' labels are 'as found' within the datasets):

- ASSISTments2009. 1 CSV file (61.4 MB) with the following features:
    order_id, assignment_id, user_id, assistment_id, problem_id, original, correct, attempt_count, ms_first_response, tutor_mode, answer_type, sequence_id, student_class_id, position, type, base_sequence_id, skill_id, skill_name, teacher_id, school_id, hint_count, hint_total, overlap_time, template_id, answer_id, answer_text, first_action, bottom_hint, opportunity, opportunity_original
- ASSISTments2013. 1 CSV file (2.8 GB) with the following features:
    problem_log_id, skill, problem_id, user_id, assignment_id, assistment_id, start_time, end_time, problem_type, original, correct, bottom_hint, hint_count, actions, attempt_count, ms_first_response, tutor_mode, sequence_id, student_class_id, position, type, base_sequence_id, skill_id, teacher_id, school_id, overlap_time, template_id, answer_id, answer_text, first_action, problem-logid, Average_confidence(FRUSTRATED), Average_confidence(CONFUSED), Average_confidence(CONCENTRATING), Average_confidence(BORED)
- ASSISTments2015. 1 CSV file (17.4 MB) with the following features:
    user_id, log_id, sequence, correct

---

[6] As of end of October 2021.

Table 1. Public datasets found.

Name	URL
ASSISTments2009	https://sites.google.com/site/assistmentsdata/home/assistment-2009-2010-data/skill-builder-data-2009-2010
ASSISTments2013	https://sites.google.com/site/assistmentsdata/home/2012-13-school-data-with-affect
ASSISTments2015	https://sites.google.com/site/assistmentsdata/home/2015-assistments-skill-builder-data
KDD Cup	https://pslcdatashop.web.cmu.edu/KDDCup/downloads.jsp
DataShop: OLI Engineering Statics - 1.14 (Statics2011)	https://pslcdatashop.web.cmu.edu/DatasetInfo?datasetId=507
The Stanford MOOCPosts Data Set	https://datastage.stanford.edu/StanfordMoocPosts/
DeepKnowledgeTracing dataset	https://github.com/chrispiech/DeepKnowledgeTracing
DeepKnowledgeTracing dataset - Synthetic-5	https://github.com/chrispiech/DeepKnowledgeTracing/tree/master/data/synthetic

- KDD Cup. 6 TXT files (8.4 GB) with the following features:
  Row, Anon Student Id, Problem Hierarchy, Problem Name, Problem View, Step Name, Step Start Time, First Transaction Time, Correct Transaction Time, Step End Time, Step Duration (sec), Correct Step Duration (sec), Error Step Duration (sec), Correct First Attempt, Incorrects, Hints, Corrects, KC(Default), Opportunity(Default)
- DataShop: OLI Engineering Statics – (Fall 2011). 1 TXT file (171 MB) with the following features:
  Row, Sample Name, Transaction Id, Anon Student Id, Session Id, Time, Time Zone, Duration (sec), Student Response Type, Student Response Subtype, Tutor Response Type, Tutor Response Subtype, Level (Sequence), Level (Unit), Level (Module), Level (Section1), Problem Name, Problem View, Problem Start Time, Step Name, Attempt At Step, Is Last Attempt, Outcome, Selection, Action, Input, Input, Feedback Text, Feedback Classification, Help Level, Total Num Hints, KC (Single-KC), KC Category (Single-KC), KC (Unique-step), KC Category (Unique-step), KC (F2011), KC Category (F2011), KC (F2011), KC Category (F2011), KC (F2011), KC Category (F2011), School, Class, CF (oli:activityGuid), CF (oli:highStakes), CF (oli:purpose), CF (oli:resourceType)
- The Stanford MOOCPosts Data Set. 11 CSV files (3.28 MB) without headers within the files.
- DeepKnowledgeTracing dataset. 2 CSV files (2.58 MB) without headers within the files.
- DeepKnowledgeTracing dataset - Synthetic-5. Over 40 CSV files (15.64 MB) without headers within the files.

Thus, the elements presented here-in, namely the ML techniques, their chosen rationale, their KT in LM purpose, the most usual programming language software employed, and the subsequent required datasets, found in the 51 reviewed publications constitute the answer to "RQ2: How do the most employed ML techniques fulfil the considered relevant aspects (identified in Sect. 2.1.2) to insure KT in LM?".

5   Discussion

In this section we present our observations on the ML techniques addressed in the precedent section, issued from the 51 reviewed publications. This discussion covers the five elements mentioned in Subsect. 2.1.2.

ML Technique: We begin by noting that, in the reviewed papers, rarely a clear, well-defined, single ML technique is employed: very often additions or variants are employed (which make the point of the paper). Research teams seem to focus their attention on fine-tuning parameters (to improve prediction) rather than on expanding the application of ML for KT to other educational data sources or contexts. Authors recognize that additional features (or dimensions) would encumber the learning phase for limited gains, compared to parameter fine-tuning. As such, many papers propose pipelines ('chains') of ML techniques to optimize the process without increasing the calculation load. Performance improvements aside, this brings up two inconveniences: the difficulty of identifying the ML technique suitable for KT, and the difficulty to evaluate and compare any two papers employing different pipelines, as the intermediary inputs and outputs of the chain elements are quite different between papers.

ML Purpose: We distinguish two families of stated purposes in the reviewed ML techniques for KT: prediction and LM creation. Prediction is often portrayed as a probability, which can be interpreted as a mastery (or degree) of a skill (0–100), a grade (0–10), or a likelihood (0–1) of getting the answer right (in binary answers). In LM creation, ML predicts parameters for initializing the LM. We noticed that clustering, personalization, and/or resource suggestion (or other ML techniques, such as NLP) were performed once the predicting phase had taken place.

ML Choice Rationales: We condense the rationales exposed by the authors when choosing a ML technique. We omit rationales based on novelty, status-quo, or vague generalities, e.g., "nobody had done it before", "the existing system already uses this mechanism", "because it helps predict students' performance", respectively. The choice of BKT's was mostly driven by popularity, although it had issues on learners' individuality, multi-dimensional skill support and modelling forgetting. BN also seemed to be a common, popular choice. Its main advantage was its ability to model uncertainty, although it seems to reach its limits if the number of students is kept "relatively low". On the contrary, DKT may benefit from large datasets and has proven being able to model multi-dimensional skills, although lacking in consistent predicted knowledge state across time-steps. DKVMN (based on LSTM) can model long-term memory and mastery of knowledge at the same time, as well as finding correlations between exercises and concepts, although it does not account for forgetting mechanisms. LSTM appears to additionally handle tasks other than KT satisfactory. It also models forgetting mechanisms over long-term dependencies within temporal sequences. It is then well suited for time series data with unknown time lag between long-range events. PFA does not consider answers' order (which is pedagogically relevant), nor models guessing, nor multiple-skills questions. Finally, RNNs are well suited for sequential data with temporal relationships, although long-range dependencies are difficult to learn by the model, hence the resurgence of LSTM.

Software for ML: Python (all frameworks and libraries merged) is the most common programming language employed for ML, more than doubling the number of papers employing *ad-hoc* languages. In this subject, we think that employing platform-specific programming languages for ML assures a lack of code portability (implying licensing issues, steep learning curve, little replicability, code isolation, and other situations) and thus, little to no adoption of research proposals adopting this approach. However, specialized ML software, designed by worldwide experts on the field, with a large user base maintaining it, backed up by large and specialized ML corporations, tends to be performance-optimized for diverse hardware and software and quasi bug-free. An *ad-hoc* solution developed in-house by a comparatively small team of developers cannot compete with such an opponent. We were taken aback by two facts: the sparse use of specialized mathematical software (Matlab, R, SPSS) in ML, and to learn that about 50% of all reviewed publication do not specify what software was employed for their ML calculations, leaving little room for independent replication, results verification, and additional development.

Datasets: We noticed that frameworks proposal papers aim to prove the performance of their approach using publicly available datasets. An overview of the found public datasets can be found in Table 1, in the previous section. The datasets found in the publications (chosen by papers' authors) are static (*i.e.*, they are not part of a "live" system), mostly contain grades or other similar evaluation measurements (but no behavioral or external sensor data) and provide the non negligible advantages of being explained in detail (their data structure) and having their data already labelled, often by experts. This contrasts with the "organic" data employed in publications where ML is addressed for an existing, live system, even if it is for testing purposes. Both variants could benefit from each other's approaches, but this would require diverse, detailed, copious high-quality data that many institutions simply cannot afford to generate nor stock, let alone analyze.

One of most recurring datasets is the ASSISTment [49] (employed in 11 publications), of which there are different versions. A noteworthy fact is that this dataset has been acknowledged to have two main kind of data errors: (1) duplicate rows (which are removed if acknowledged by the authors) and (2) "misrepresented" skill sequences. Drawbacks of the latter issue (2) have been discussed: while this does not affect the final prediction, it nevertheless might conduce the learner to being presented with less questions on one of the merged skills (the less mastered) because the global (merged) mastery of skill is achieved mainly through the mastery of the most known skill [50, 51]. This raises the importance of the data cleaning process [25], which processing time is not negligible and should be accounted at early data mining stages.

6   Conclusion and Perspectives

The aim of this research work is to present existing ML techniques for KT in LM employed during the last five years. It also helps to prepare our target public to the complex task of choosing a ML for KT technique by outlying the current trends in the research field. To reach this objective, we used the PRISMA guidelines for systematic reviews methodology, which led us to conclude that the following five ML for KT techniques were the most employed in the State-of-the-Art during the last five years: BKT (18 applications), DKT (13 applications), LSTM (12 applications), BN (11 applications), and SVM (7 applications). However, the reasons behind choosing any given ML technique were not really detailed in reviewed publications. We also noticed that combinations of ML techniques arranged in pipelines are a common practice, and that the most recent research (2019–2020) favored pipeline and/or parameter optimization over new techniques implementation. The use of public datasets is recurrent: they contain usually grades or other similar evaluating metrics, but no other pedagogical relevant data. On this subject, we insist that extensive data cleaning and other pre-treatments are highly recommended before using these public datasets. Finally, our results show that the ML programming language of choice is Python (all libraries & frameworks combined).

This review of literature is inscribed in the context of the "Optimal experience modelling" research project, conducted by the University of Lille. This research project [52] aims to model and trace the Flow psychological state, alongside KT, via behavioral data, using the generic Bayesian Student Model (gBSM), within an Open Learner Model.

The current challenge is to incorporate the ML relevant aspects highlighted in this study, and the behavioral and psychological aspects (log traces and Flow state determination) specifically linked to the project. Namely, a ML technique supporting the gBSM, capable to initialize the LM and perform KT, supported by the most common programming language for ML, based on a sound rationale. The originality of such research lies in the use of live, behavioral, Flow-labelled data issued from the French-spoken international MOOC "Project Management"[7].

Acknowledgements. This project was supported by the French government through the Programme Investissement d'Avenir (I-SITE ULNE/ANR-16-IDEX-0004 ULNE) managed by the Agence Nationale de la Recherche.

Appendix

The Appendix is composed of: (a) the ML categorization figure, (b) the summary table of ML for KT in LM (for clarity reasons, the extensive column 'rationale' has been removed), and (c) the full table of the 29 ML techniques.

It can be found at the following address:

https://nextcloud.univ-lille.fr/index.php/s/DpJwFtRHg399pXm.

---

[7] https://moocgdp.gestiondeprojet.pm/.

References

1. Corbett, A., Anderson, J., O'Brien, A.: Student modeling in the ACT programming tutor, Chap. 2. In: Nichols, P., Chipman, S., Brennan, R. (eds.) Cognitively Diagnostic Assessment. Lawrence Erlbaum Associates, Hillsdale (1995)
2. Giannandrea, L., Sansoni, M.: A literature review on intelligent tutoring systems and on student profiling. Learn. Teach. Med. Technol. **287**, 287–294 (2013)
3. Nakić, J., Granić, A., Glavinić, V.: Anatomy of student models in adaptive learning systems: a systematic literature review of individual differences from 2001 to 2013. J. Educ. Comput. Res. **51**(4), 459–489 (2015)
4. Wenger, E.: Artificial Intelligence and Tutoring Systems: Computational and Cognitive Approaches to the Communication of Knowledge. Morgan Kaufmann, Burlington (2014)
5. El Mawas, N., Gilliot, J.-M., Garlatti, S., Euler, R., Pascual, S.: As one size doesn't fit all, personalized massive open online courses are required. In: McLaren, B.M., Reilly, R., Zvacek, S., Uhomoibhi, J. (eds.) CSEDU 2018. CCIS, vol. 1022, pp. 470–488. Springer, Cham (2019). https://doi.org/10.1007/978-3-030-21151-6_22
6. Martins, A.C., Faria, L., De Carvalho, C.V., Carrapatoso, E.: User modeling in adaptive hypermedia educational systems. J. Educ. Technol. Soc. **11**(1), 194–207 (2008)
7. Swamy, V., et al.: Deep knowledge tracing for free-form student code progression. In: Penstein Rosé, C., et al. (eds.) AIED 2018. LNCS (LNAI), vol. 10948, pp. 348–352. Springer, Cham (2018). https://doi.org/10.1007/978-3-319-93846-2_65
8. Corbett, A., Anderson, J.: Knowledge tracing: modeling the acquisition of procedural knowledge. User Model. User-Adap. Inter. **4**(4), 253–278 (1994)
9. Yudelson, M.V., Koedinger, K.R., Gordon, G.J.: Individualized Bayesian knowledge tracing models. In: Lane, H.C., Yacef, K., Mostow, J., Pavlik, P. (eds.) AIED 2013. LNCS (LNAI), vol. 7926, pp. 171–180. Springer, Heidelberg (2013). https://doi.org/10.1007/978-3-642-39112-5_18
10. Zhang, J., Li, B., Song, W., Lin, N., Yang, X., Peng, Z.: Learning ability community for personalized knowledge tracing. In: Wang, X., Zhang, R., Lee, Y.-K., Sun, L., Moon, Y.-S. (eds.) APWeb-WAIM 2020. LNCS, vol. 12318, pp. 176–192. Springer, Cham (2020). https://doi.org/10.1007/978-3-030-60290-1_14
11. IBM: What is Machine Learning?. IBM Cloud Learn Hub, 18 déc. 2020. https://www.ibm.com/cloud/learn/machine-learning. (consulté le 23 déc. 2020)
12. Moher, D., Liberati, A., Tetzlaff, J., Altman, D.G.: The PRISMA Group: Preferred reporting items for systematic reviews and meta-analyses: the PRISMA statement. PLOS Med. **6**(7), e1000097 (2009). https://doi.org/10.1371/journal.pmed.1000097
13. Das, K., Behera, R.N.: A survey on machine learning: concept, algorithms and applications. Int. J. Innov. Res. Comput. Commun. Eng. **5**(2), 1301–1309 (2017). https://doi.org/10.15680/IJIRCCE.2017.0502001
14. Olsson, F.: A literature survey of active machine learning in the context of natural language processing. SICS Technical report, p. 59 (2009)
15. Shin, D., Shim, J.: A systematic review on data mining for mathematics and science education. Int. J. Sci. Math. Educ. **19**, 639–659 (2020). https://doi.org/10.1007/s10763-020-10085-7
16. Chakrabarti, S., et al.: Data mining curriculum: a proposal (version 1.0). In: ACM SIGKDD, 30 avr. 2006 (2006). Consulté le: 24 déc. 2020. [En ligne]. Disponible sur: https://www.kdd.org/curriculum/index.html
17. Schmidhuber, J.: Deep learning in neural networks: an overview. Neural Netw. **61**, 85–117 (2015). https://doi.org/10.1016/j.neunet.2014.09.003

18. Ramírez Luelmo, S.I., El Mawas, N., Heutte, J.: Machine learning techniques for knowledge tracing: a systematic literature review. In: Proceedings of the 13th International Conference on Computer Supported Education, vol. 1, pp. 60–70 (2021). https://doi.org/10.5220/001051 5500600070

19. Mohri, M., Rostamizadeh, A., Talwalkar, A.: Foundations of Machine Learning, 2nd edn. MIT Press, Cambridge (2018)

20. Brownlee, J.: A tour of machine learning algorithms. Machine Learning Mastery, 11 août 2019. https://machinelearningmastery.com/a-tour-of-machine-learning-algorithms/ (consulté le 28 déc. 2020)

21. van Engelen, J.E., Hoos, H.H.: A survey on semi-supervised learning. Mach. Learn. **109**(2), 373–440 (2020). https://doi.org/10.1007/s10994-019-05855-6

22. van Veen, F., Leijnen, S.: The Neural Network Zoo. The Asimov Institute (2019). https://www.asimovinstitute.org/neural-network-zoo/ (consulté le 23 déc. 2020)

23. Eagle, M., et al.: Estimating individual differences for student modeling in intelligent tutors from reading and pretest data. In: Micarelli, A., Stamper, J., Panourgia, K. (eds.) ITS 2016. LNCS, vol. 9684, pp. 133–143. Springer, Cham (2016). https://doi.org/10.1007/978-3-319-39583-8_13

24. Millán, E., Jiménez, G., Belmonte, M.-V., Pérez-de-la-Cruz, J.-L.: Learning Bayesian networks for student modeling. In: Conati, C., Heffernan, N., Mitrovic, A., Verdejo, M.F. (eds.) AIED 2015. LNCS (LNAI), vol. 9112, pp. 718–721. Springer, Cham (2015). https://doi.org/10.1007/978-3-319-19773-9_100

25. Chicco, D.: Ten quick tips for machine learning in computational biology. BioData Min. **10**(1), 35 (2017). https://doi.org/10.1186/s13040-017-0155-3

26. Wen, J., Li, S., Lin, Z., Hu, Y., Huang, C.: Systematic literature review of machine learning based software development effort estimation models. Inf. Softw. Technol. **54**(1), 41–59 (2012). https://doi.org/10.1016/j.infsof.2011.09.002

27. Winkler-Schwartz, A., et al.: Artificial intelligence in medical education: best practices using machine learning to assess surgical expertise in virtual reality simulation. J. Surg. Educ. **76**(6), 1681–1690 (2019). https://doi.org/10.1016/j.jsurg.2019.05.015

28. Domingos, P.: A few useful things to know about machine learning. Commun. ACM **55**(10), 78–87 (2012). https://doi.org/10.1145/2347736.2347755

29. Pelánek, R.: Bayesian knowledge tracing, logistic models, and beyond: an overview of learner modeling techniques. User Model. User-Adap. Inter. **27**(3–5), 313–350 (2017). https://doi.org/10.1007/s11257-017-9193-2

30. van de Sande, B.: Properties of the Bayesian knowledge tracing model. J. Educ. Data Min. **5**(2), 1–10 (2013)

31. Piech, C., et al.: Deep knowledge tracing. In: Advances in Neural Information Processing Systems, vol. 28, pp. 505–513 (2015). [En ligne]. Disponible sur: https://proceedings.neu rips.cc/paper/2015/file/bac9162b47c56fc8a4d2a519803d51b3-Paper.pdf

32. Minn, S., Yu, Y., Desmarais, M.C., Zhu, F., Vie, J.-J.: Deep knowledge tracing and dynamic student classification for knowledge tracing. In: 2018 IEEE International Conference on Data Mining (ICDM), pp. 1182–1187 (2018). https://doi.org/10.1109/ICDM.2018.00156

33. Mao, Y., Lin, C., Chi, M.: Deep learning vs. Bayesian knowledge tracing: student models for interventions. J. Educ. Data Min. **10**(2), 28–54 (2018)

34. Daniluk, M., Rocktäschel, T., Welbl, J., Riedel, S.: Frustratingly short attention spans in neural language modeling. arXiv preprint arXiv:1702.04521 (2017)

35. Pearl, J.: Probabilistic Reasoning in Intelligent Systems: Networks of Plausible Inference, 2nd edn. Morgan Kaufmann, Los Angeles (1988)

36. Conati, C.: Bayesian student modeling. In: Nkambou, R., Bourdeau, J., Mizoguchi, R. (eds.) Advances in Intelligent Tutoring Systems, pp. 281–299. Springer, Heidelberg (2010). https://doi.org/10.1007/978-3-642-14363-2_14

37. Vapnik, V.: Statistical Learning Theory. Wiley, Hoboken (1998)
38. Hämäläinen, W., Vinni, M.: Classifiers for educational data mining, Chap. 5. In: Handbook of Educational Data Mining, pp. 57–74. CRC Press, Boca Raton, USA (2010)
39. Lee, Y.-J.: Predicting students' problem solving performance using support vector machine. J. Data Sci. **14**(2), 231–244 (2021). https://doi.org/10.6339/JDS.201604_14(2).0003
40. Salzberg, S.L.: Book Review: C4.5: Programs for Machine Learning by J. Ross Quinlan. Morgan Kaufmann Publishers, Inc. (1993). Kluwer Academic Publishers (1994). [En ligne]. Disponible sur: http://server3.eca.ir/isi/forum/Programs%20for%20Machine%20Learning.pdf
41. Janan, F., Ghosh, S.K.: Prediction of Student's Performance Using Support Vector Machine Classifier (2021)
42. Zhang, J., Shi, X., King, I., Yeung, D.-Y.: Dynamic key-value memory networks for knowledge tracing. In: Proceedings of the 26th International Conference on World Wide Web, pp. 765–774 (2017)
43. Pavlik, P.I., Cen, H., Koedinger, K.R.: Performance factors analysis – a new alternative to knowledge tracing. Online Submission, p. 8 (2009)
44. Zhang, J., King, I.: Topological order discovery via deep knowledge tracing. In: Hirose, A., Ozawa, S., Doya, K., Ikeda, K., Lee, M., Liu, D. (eds.) ICONIP 2016. LNCS, vol. 9950, pp. 112–119. Springer, Cham (2016). https://doi.org/10.1007/978-3-319-46681-1_14
45. Sha, L., Hong, P.: Neural knowledge tracing. In: Frasson, C., Kostopoulos, G. (eds.) BFAL 2017. LNCS, vol. 10512, pp. 108–117. Springer, Cham (2017). https://doi.org/10.1007/978-3-319-67615-9_10
46. Trifa, A., Hedhili, A., Chaari, W.L.: Knowledge tracing with an intelligent agent, in an e-learning platform. Educ. Inf. Technol. **24**(1), 711–741 (2019). https://doi.org/10.1007/s10639-018-9792-5
47. Sciarrone, F., Temperini, M.: K-OpenAnswer: a simulation environment to analyze the dynamics of massive open online courses in smart cities. Soft. Comput. **24**(15), 11121–11134 (2020). https://doi.org/10.1007/s00500-020-04696-z
48. Crowston, K., et al.: Knowledge tracing to model learning in online citizen science projects. IEEE Trans. Learn. Technol. **13**(1), 123–134 (2020). https://doi.org/10.1109/TLT.2019.2936480
49. Razzaq, L., et al.: Blending assessment and instructional assisting. In: Proceedings of the 2005 Conference on Artificial Intelligence in Education: Supporting Learning through Intelligent and Socially Informed Technology, NLD, pp. 555–562 (2005)
50. Pelánek, R.: Metrics for evaluation of student models (2015). https://doi.org/10.5281/zenodo.3554666
51. Schatten, C., Janning, R., Schmidt-Thieme, L.: Vygotsky based sequencing without domain information: a matrix factorization approach. In: Zvacek, S., Restivo, M.T., Uhomoibhi, J., Helfert, M. (eds.) CSEDU 2014. CCIS, vol. 510, pp. 35–51. Springer, Cham (2015). https://doi.org/10.1007/978-3-319-25768-6_3
52. Ramírez Luelmo, S.I., El Mawas, N., Heutte, J.: Towards open learner models including the flow state. In: Adjunct Publication of the 28th ACM Conference on User Modeling, Adaptation and Personalization, Genoa Italy, pp. 305–310 (2020). https://doi.org/10.1145/3386392.3399295

Information Technologies Supporting Learning

Learning Analytics Metamodel: Assessing the Benefits of the Publishing Chain's Approach

Camila Morais Canellas[1,2(✉)] ⓘ, François Bouchet[1] ⓘ, Thibaut Arribe[2], and Vanda Luengo[1] ⓘ

[1] Sorbonne Université, CNRS, LIP6, 75005 Paris, France
{camila.canellas,francois.bouchet,vanda.luengo}@lip6.fr
[2] Société Kelis, Thourotte, France
thibaut.arribe@kelis.fr

Abstract. In this work, we propose a learning analytics implementation based on a model-driven engineering approach. It aims at assessing the benefits that could arise from such an implementation, when pedagogical resources are produced via publishing chains, that already use the same approach to produce documents. Previously, we have discussed these potential benefits from a more theoretical point of view. In the present work, we present a concrete implementation of a metamodel to integrate a learning analytics system closely linked to the knowledge of the semantics and structure of any document produced, natively. Finally, we present an initial evaluation of this metamodel by modelers and discuss the limits of this metamodel and the future changes required.

Keywords: Learning analytics (LA) · Model-driven engineering (MDE) · Metamodel · Publishing chains

1 Introduction

In this work, we describe a metamodel for a learning analytics (LA) solution into a platform framework that uses a model-driven engineering (MDE) approach to design and publish pedagogical resources via publishing chains, as well as a first evaluation of the metamodel proposed.

Learners' interactions with courses and resources offered by learning platforms on the Web have generated a vast amount of learning-related data. For a decade now, the interdisciplinary field of Learning Analytics (LA) has focused its attention on the collection, processing, and analysis of such data. A common definition of learning analytics is: "Learning analytics is the measurement, collection, analysis, and reporting of data about learners and their contexts, for the purposes of understanding and optimizing learning and the environments in which it occurs" [16]. According to [17], when applying learning analytics, one can focus on what e-learning interaction traces need to be captured, how to process them and how to present them to stakeholders in a useful way. Although many times the analyses carried out in these systems are based on traces already collected [14], there is currently no consensus on the interactions actually

© Springer Nature Switzerland AG 2022
B. Csapó and J. Uhomoibhi (Eds.): CSEDU 2021, CCIS 1624, pp. 97–114, 2022.
https://doi.org/10.1007/978-3-031-14756-2_6

relevant for effective learning [1] and, subsequently, on the traces to be recorded and analyzed.

Our work takes place in a context where detailed knowledge of the structure and semantics of the documents learners interact with is known a priori. We posit that this knowledge represents an asset which value has to be assessed in the context of an LA implementation. Our interest relies on how to natively include this detailed knowledge of the structured document in the trace analysis cycle. The idea is to illustrate how this knowledge a priori can be used to enhance educational resources produced by a set of MDE authoring tools. The ultimate aim is to create solutions that support stakeholders in making data-driven decisions in order to improve learning, while also taking advantage of the existing context of publishing chains to do so.

The rest of this work is structured as follows: in Sect. 2 we situate our research with related literature regarding models for learning analytics, in Sect. 3 we define the model-driven engineering approach we use and in Sect. 4 we describe how the same approach is already used to create pedagogical materials—resulting in the semantics and structure knowledge of such documents to be known beforehand, as illustrated in Sect. 5. In Sect. 6 we present the details of the proposed metamodel dedicated for the application of LA processes. In Sect. 7 we describe the first evaluation of the metamodel proposed. A discussion can be found in Sect. 8. This work summarizes parts of [4] as the present work is an extension of this previous work. More particularly, it adds a more detailed description of the proposed metamodel, as well as an initial evaluation based on its use by two expert modelers. It relies on the same context and illustration of the processes already described, which are necessary for understanding of the stakes at hand. On the other hand, the present work is less focused on the potential benefits of our approach and on the learning analytics processes, previously discussed in [4].

2 Related Work

The underlying assumption [14] of a number of LA solutions is that the interaction data is either already at hand, or recordings of all student interactions with a given system will start to be collected, usually then it is decided to use one of the available standards, such as IMS Caliper or Experience API (xAPI), among others. Moreover, one may argue that a result of the "lack of staff and technology available for learning analytics projects" [12, p. 366] would favor an outsourcing of the issue of traces collected, and a standard would be used, without consulting a local expert for the establishment of these.

Besides the fact of having competing specifications, it is not clear at this time whether one or the other works better [14]. One of the aspects related to these specifications that has received criticism is the lack of relevant information if one follows these specifications by implementing only the mandatory aspects (*required*). The same authors [14] point out that this lack of relevant information leads to situations where we consider the system to conform to one or another specification via certification tests, but in practice it produces data streams with largely redundant events of one type, or that do not describe any user behavior that could be useful for an intended analysis.

An effort on enriching traces via models found in the literature is Trace-Based Reasoning (TBR) [8]. The so-called obsels are formally described in the trace model and each is characterized by a name, a timestamp, and a set of properties, usually attribute-value pairs. The authors briefly mention that a metamodel should define these properties in order to ensure interoperability between different traces.

Researchers behind TBR mention that the name "obsel" is chosen—instead of event—in order to emphasize that an obsel is recorded with a purpose (defined a priori). This idea to think about the traces beforehand and enrich them with context information is a common point with our approach here. However, a difference lays in the fact it seems interesting to us to explore how to guide this modeling of traces not only a priori and via modeling, but also starting from a specific question, translated into one or more indicators. In other words, it is because we are trying to answer a question that we record interactions, but also, traces are potentially enriched with some information about the documents read.

Among the works going towards models more closely related to learning analytics, in [5] the authors present a dedicated metalanguage defining both data and needs: *Usage Tracking Language* (UTL). Its components allow defining information to enrich the collected traces and to structure the information through the definition of intermediate data and indicators. This model has the advantage of adapting to the elements it describes via the addition/modification of information already available, and its focus is to enable stakeholders to take educational scenarios into account in the process of collecting and analyzing traces.

The metalanguage is made of three parts. First, its UTL/T (traces) part makes it possible to represent the transformation of the traces generated by describing them as a set of data including information on their source. One may also define the relationship of the latter with the expected primary data in order to create an indicator, as well as define the information extraction method from the traces. Second, its patron part (UTL/P) is based on the DGU (Defining, Getting, Using) model and proposes the usage trace before the learning session [6] and not once the traces have been collected. This aims to allow a future comparison of descriptive scenarios with predictive scenarios of the educational situation, and therefore make it possible to describe the structure of an observable. Finally, the UTL/S (pedagogical scenario) part makes it possible to semantically link the pedagogical scenario to a given indicator.

This work shows us the benefit of enriching the analyses (and traces to do so) with context information that can be modeled, in this case, from the educational scenario, in our case, with the detailed information from the various documents consulted, and natively in a previous phase by the modeler (what could relieve the author from defining the traces).

To conclude, here we see an effort made to model the traces and indicators, but also the educational scenarios, so that one can enrich the other during the analyses. Our approach is therefore close to those cited, but we intend to explore an approach guided by educational documents in order to complete and enrich traces. As we have mentioned, documentary production in our context already relies on modeling, which allows us to explore this fine knowledge of the document in the context of learning analytics.

We sustain that, by allowing to specifically define, at the same time, both the desired LA indicators and the educational resources' production, we could enable both the document and its usage analysis to be closely linked. In order to do so, we use an MDE approach both to produce documents—already happening in our context—and analyze their usage—also via an MDE approach—, that has been used in other domains such as automotive, banking, printing, web applications, among others [11], yet not as much in LA systems.

3   Methodology: Models

When developing complex systems, model-driven engineering is an approach that allows one to focus on a more abstract level than classical programming [7]. It allows a description of both the problem at hand and its solution. The focus is to create and leverage domain models—in our case, the LA one—as a conceptual model taking into account all the topics related to this specific task/domain. Model-driven engineering is considered as a form of generative engineering [7] as some or all of a computer application is generated from models. The procedure includes simplifying aspects of reality (or a solution to a problem) by creating a model or a set of models. Thus, a model is a simplified representation of a concept or object, which can be more or less abstract/precise. Likewise, a metamodel is a model which defines the expression language of a model, i.e., the modeling language [13].

Digital publishing chains are an example of systems based on MDE, as they use this approach to create document models that will then be used in document production. In our context, all pedagogical documents are produced using digital publishing chains; we provide the details of this publishing process in the next sections.

4   Models for Publishing Educational Documents

Some documents, such as pedagogical resources, often have a well-known consistency in structure, i.e., the parts that compose the document, and in semantics, i.e., what type of content (a definition, an example) corresponds to each part. Digital publishing chains can assist the production and publication of such structured documents, especially in their design processes [2]. A publishing chain can be defined as "a technological and methodological process consisting in producing a document template, assisting in content creation tasks and automating formatting" [9, p. 2, our translation].

One of the benefits stemming from the use of digital publishing chains is the documents' homogeneity, even when editing and publishing large amounts of documents. The models' definition that documents are based on is strongly linked to the profession or associated context. The document needs are assessed and then formalized in the appropriate model, therefore when an author decides to use one document model over another, he or she reflects a given intention.

Moreover, such systems have the advantage of separating content and form [3], allowing for authors to focus on the content itself. The form is later applied, automatically, when the document is generated. Thus, the same content could have different forms of presentation according to the publication format, which can include PDF, Web, among others [3]. Other features of such systems are making teaching practices explicit, sharing of practices, optimizing production management, and reducing costs in document production [10].

Typically, a document will be based on a document model, the document model is itself based on a metamodel, the metamodel can be based on a metametamodel and so on, depending on the level of abstraction needed. In our context, we consider "document primitive" which is a computer code abstracting documentary objects that later in the chain will allow the generation of specific code instantiating multiple document models [2]. This process is illustrated in the next section. Therefore, we propose to follow a similar approach for our LA modeling. For our metamodel, we propose to abstract the various elements related to implementing a learning analytics solution ("analytical primitives") via a metamodel, in order to design learning indicators. Unlike the approaches described in the state of the art, we suggest that the definition of indicator models should be done at the same time as the definition of documentary models (and not a posteriori), which makes it possible to take into account detailed information of the structured documents in question. This approach has the advantage of allowing the definition of the necessary traces (and their contents), thus reversing the usual approach [14] of starting from the traces to arrive at indicators.

5   The Publishing Processes

As mentioned before, in our particular context, **developers** have defined "building bricks" called document primitives. They are the building blocks that will be used as the basis for the creation of a document's model.

In the next phase, those bricks are made available to a modeling tool that allows the creation of any document model desired using and combining those initial bricks. In other words, the **modeler** defines a document **model** using the available document primitives.

When the document model is ready, it will be made available to a writing tool dedicated to structured documents. An **author** (a teacher or a pedagogical team, for example) uses this document model in order to create his or her **document** according to the model chosen (a course module, a case study game, etc.). The writing tool (s)he uses includes the transformation algorithms that automatically publish the document at the desired format. We are obviously interested in the **document** in its Web format and in the ways in which it will later be used by its final **users** (learners).

The challenge of choosing such an approach is to be able to abstract the technical aspects as much as possible—specially at the metamodel level—in order to reduce the work and facilitate the creation processes as the phases at the end are more numerous and stakeholders' roles are less technical.

5.1   Illustrating Our Context

As briefly described before, a number of phases take place and several stakeholders may intervene. In this section, in order to better illustrate these processes taking place when designing documents for educational purposes via model-driven engineering, we describe next a—highly simplified—practical example. Note that we first reported this example in our previous work (cf. [4]). In this example, we rely on the vocabulary of the existing Opale document model[1], used for creating linear academic courses (on-site training, distance or blended learning). It is worth noting that other document models can be used, for pedagogical purposes or not, such as to carry out case studies with a gamified twist, create exercisers with different modes of execution (self-learning, evaluation), build question banks for evaluations, among others.

First Phase: The Developer. The first phase consists in providing the so-called document primitives and is done by the developers of the tool(s). Those could be, for our example, "text", "multimedia", "quiz" and "organization". Each of these bricks (see Fig. 1) will be used on the next phase to build the document model.

Fig. 1. First phase: metamodel with the document primitives for our illustrative document modeling, from [4].

Second Phase: The Modeler. A modeler then uses the available primitives on the design tool to define document models. The challenge of this phase is to understand the needs of a given community of users, and translate them into a document model. The elements created, their organization and the language used are therefore appropriate for this group of users and the domain in question. Certainly, the same applies to the available transformation functions that translate the models into actual documents in various formats.

In our example, the modeler uses the organization primitive to define a "learning activity" as being made of:

– "Introduction" (text primitive)
– "Concept" (text or multimedia primitives)
– "Content" made up of parts (text and multimedia primitives) titled "Information" and "Example"
– "Conclusion" (multimedia primitive)
– "Practice" made up of quizzes (quiz primitive).

As seen in Fig. 2, the modeler also defines that a learning module can have one or more "learning activities". A "learning activity" must have exactly one "introduction", one "conclusion" and a "practice" part at the end, and include between "introduction" and "conclusion" one or more "concept" and/or one or more "content" parts. A "practical" part must have one or more quizzes.

Regarding the generation of the document in a Web format, the modeler may choose to have a page created for each "learning activity". He or she could also define a menu

---

[1] Available at: https://doc.scenari.software/Opale/en/.

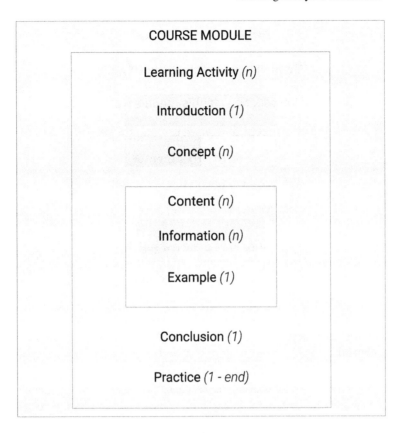

Fig. 2. Second phase: metamodel used to define the model for our illustrative document modeling, adapted from [4].

reflecting the structure of the document, allowing users to browse the module, either by the pages names or by jumps to each internal part of the pages, the blocks mentioned above. The document model is at last made available for authors to use it on the editing tool.

It is important to observe that it is during this phase that the modeler defines both the structure (each block and its optional/mandatory components) and the semantics (the nature of what each block is supposed to contain) for each document based on this particular document model. Note that this information will also be used by the publishing algorithm. In other words, the different possible "parts" of the document are pre-established at the time of modeling and the content type of each part is defined by the chosen blocks.

Third Phase: The Author. An author, such as an instructor, uses this document model to create his or her course. Its course consists of four "learning activities", each with an "introduction", two "concepts", four "contents", a "conclusion", followed by a "practice" activity to check the understanding of the theoretical content (see Fig. 3). Once the course has been created, the instructor can publish it, for example in a Web format.

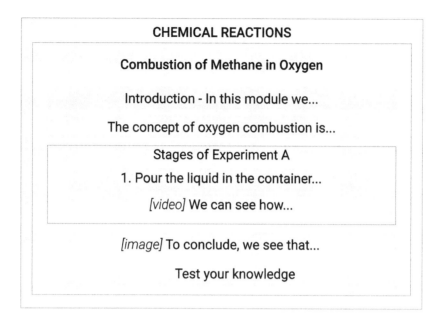

Fig. 3. Third phase: document model used in the instantiating process of our illustrative document modeling, from [4].

Choosing a document model over another reflects the intentions of the author. Moreover, when creating his or her document, each choice regarding each block used also represents the intention regarding its *content*: he or she decides it is time to introduce a "practical activity" to check the understanding of the new "concept" just introduced— and not simply: (s)he adds a quiz after a subsection.

Thus, 1/the author must choose the model corresponding to the needs of his or her profession (or type of course, in our case) and 2/this will also have consequences in trace analysis phase, in particular regarding semantics analysis.

Fourth Phase: The Learner. Finally, learners access the published content, open pages in whatever order they want, scroll through them, take quizzes, etc. See Fig. 4 for an example of a published document.

6   Models for Learning Analytics

We firstly decided to have indicators as the core of the LA metamodel. So this concept was isolated and defined as "a learning indicator used in a learning analytics approach is a calculated measure [computability property] linked to a behavior or an activity instrumented by digital technology [traceability property] of one or more learners, given to a user [visibility property] and which can be used in the calculation of other indicators".

More specifically, according to this definition, the following elements are not learning indicators:

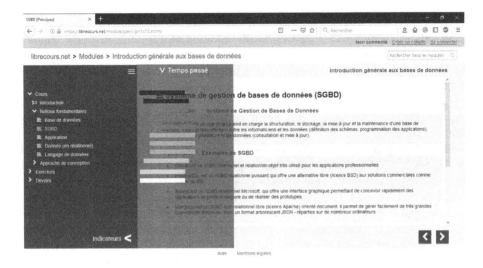

Fig. 4. Fourth phase: document transformed into a Web format, ready to be used by learners. We added as an example the visualization of an indicator following the menu (structure) of the course, from [4].

- Page X was seen or not by learner Y [not respecting the property of computability, as it is directly traced this way].
- The gender of a learner [not respecting the properties of traceability and computability].
- The motivation if it comes only from declarative data, such as a questionnaire [not respecting the property of traceability].
- The number of pages viewed by a learner in a system using this data to calculate a dropout risk indicator, if only the latter is reported to the instructor (because the number of page views is then only a variable internal to the system, not accessible by a user) [not respecting the visibility property].

These three key properties will impact the proposed metamodel below, as it represents the first element of it from which the others derive. Additionally, we used this definition to conduct a systematic review of indicators used by the learning analytics community [to be published soon]. The goal of this work was to ensure that our metamodel could indeed be used to transpose the most commonly used indicators from the LA community, and to avoid defining a metamodel that is too specific to some particular indicators. Indeed, the metamodel proposal must make it possible, once instantiated, to provide the essential specifications to the various transformation and generation processes. The designed metamodel is made up of three main dimensions:

- Interactions with the document: these are the potential sources of traces of interactions that can be recorded.
- Indicators: the core of the metamodel, with the primitives used to define an indicator.
- Visualization modes: allowing to decide how the visualization of the results obtained will be offered to the different stakeholders.

The core of the metamodel is clearly the "indicators" dimension, as it is the selection of an indicator which will trigger or not the addition of a trace to trace log (among all the possible ones associated to each interaction) and the visualizations depend on the nature of the selected indicators (e.g. a bar chart can be preferable to a pie chart) and they can be seen as more general parts of a visualization system (e.g. deciding to group all indicators on a dashboard page or embedding them into the learning content as shown in Fig. 4). Thus, we will describe in detail here only the "indicators" dimension (see Fig. 5).

The main primitive allowing to model the indicator (indicator on Fig. 5) has some traditional attributes regarding its identification (code, name and description), as well as the attribute used to indicate who has access to this primitive (the type of user). This last aspect is already present and modeled in the system in question, and therefore a link between these primitives will be enough concerning the proposed primitive. The other three attributes refer to the inputs (inputs), analyses (analysis) and outputs (output) required to define the indicator that will be modeled. The calculation relating to the indicator must also be indicated in this primitive.

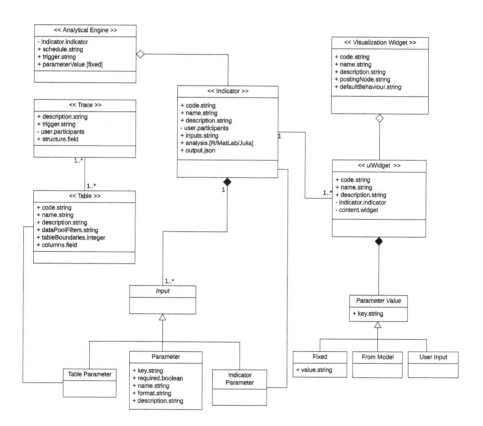

Fig. 5. Metamodel: creating indicators.

Several types of inputs can be provided:

- A parameter: this is to define a necessary entry that is flexible enough for the modeler, for example a start and end date that will be used to perform the analysis from the log traces, or a number which will be used as a threshold to trigger a message to a user, etc.
- A table: an entry can also be a table with data more or less ready to be used by the analysis in question.
- An indicator: some indicators can have as input another indicator calculated beforehand. For example, an indicator aiming to display the involvement of a learner could have as input the time spent on content or the number of connections, etc.

The traces that will be produced from the interaction with the documents are defined in the corresponding primitive (`trace`). It is above all a matter of determining the type of interaction (opening a page, clicking on the play button of a video, etc.) triggering the tracing, but also the documentary components to which this rule must be applied, such as "trace all openings of pages that include a "conclusion" tag". This is done through the documentary model, via an identification key. Note that some information for each trace will be added by default and therefore does not need to be modeled each time. These are mainly elements either not necessarily related to the analyses but technologically necessary or a convention always needed, for example the address of the page, project identification on the server, or the timestamp.

Another primitive (`table`) aims to allow modelers to define tables from raw traces' logs, which would be ready to use for the calculation of certain indicators. This primitive must make it possible to determine the filtering of the traces taken into account to fill the table in question. A Boolean system allows this selection, for example by choosing (`typeBlock = concept AND (focus = started OR focus = ended)`). It is also possible to define calculations to be performed during filling, for each column, if necessary. Then, the modeler must be able to define the columns of this table and the information available in the traces which will fill each of these them, as well as the corresponding type of data. It is also possible to set the retention time of such data, for technical purposes (freeing up memory space) but also for compliance to data protection regulations, such as the European GDRP[2]).

The `Analytical Engine` primitive is used to model the triggers for indicator calculations. It is with this primitive that the modeler determines the scheduling of the actual calculations for each indicator. Parameters, such as start and end dates, can also be defined here. This should allow, for example, to gain in performance for indicators where an incremental calculation avoids recalculating it from all the data (e.g., an indicator showing the number of new concepts seen daily by a learner could be computed every day at midnight and only the traces for the last 24 h would need to be analyzed and added to the existing indicator).

7 Testing and Evaluation

Once the primitives of the metamodel have been implemented in the tool dedicated to modeling (an example is given in Fig. 6), we sought to verify the correct understanding

---

[2] https://gdprinfo.eu/.

and handling on the part of its future users, the modelers. Note that the different trans-formations were not all yet implemented at the time, so the study focuses on getting started with the metamodel. However, understandability by the first-level users (model-ers) is key to ensure the proposed LA metamodel will be actually used.

Fig. 6. Metamodel: initial implementation in the dedicated tool, from [4].

7.1   Statement of the Problem

The main issue is to understand the usage by the modelers of the proposed LA meta-model by focusing on how they go about creating indicators using the proposed primi-tives. It entails understanding several aspects—utility, advantages/disadvantages, under-standing, sticking points, etc.—of the use of the metamodel by experienced modelers. We were particularly interested in observing the following points:

- How they get started: what actions are taken, with what goal and in what order (planned vs. actual actions).
- Blocking or improvement points: anything that could lead to a hesitation or possible improvements when modeling an indicator.
- Their perceptions on several aspects: the advantages/disadvantages of our approach.

Research Questions. More precisely, the questions we seek to answer with this study are:

1. How is the handling of the metamodel by modelers?
2. Are the primitives used as expected?
3. What are the blocking points during this first usage?
4. Are there any improvements to be expected, which ones?
5. Is the vocabulary used clear and understandable to a modeler?
6. What is the perceived usability of the metamodel by modelers?
7. How are the proposed primitives used to encourage maintainability, reuse, and cus-tomization of indicators?

Delimitations and Limitations. Modelers are experts in document modeling and so quite familiar with metamodels in general, however they are not necessarily experts in LA. The first phase of the study therefore had to be conducted with care in order to help them understand these new concepts and vocabulary.

Qualitative Research Approach Used. This study is based on a triangulation comprising three phases:

1. Presentation of the model with a concrete example of an indicator.
2. Modeling (by experienced modelers) of an indicator in autonomy and response to a questionnaire (System Usability Scale and phrase completion).
3. Explanatory interview on the modeling of an indicator in autonomy (done in phase 2). At the end, modelers are asked to change the indicator that was modeled, creating a new one.

Researcher's Role. The role of the researcher in this study is that of a guide.

Sampling Method. A targeted sampling method [15] is used, that is to say the sampling by criterion—which consists in selecting cases who satisfy a predefined criterion—in our case, experienced modelers.

7.2   Results: Interviews with Modelers

Each interview with the two participating modelers lasted an average of one hour. They did the exercise of telling their actions, primitive by primitive, in order to create the model of the proposed indicator. During the interview, the discussion was based on this modeling, with screen sharing.

The first modeler started by setting the primitive allowing to define the traces to be recorded (`Trace`) and the corresponding table (`Table`). He stated he understood how to define what will be added into the published document and then will feed the trace logs. He had questions about some information, such as the time stamp, which was not to be modeled, since it was part of the information automatically inserted in the traces. The researcher confirmed that some information were automatically collected for all traces, information transmitted during the model presentation session but not stored. A list with this information should be provided in a guide to modelers to help them better understand this point and which information does not need to be modeled. Other than that, the modeler claimed to be confident in what he was doing. A suggestion made by the modeler consisted in adding the possibility of indicating on which element(s) of the interface the traces should be created. For example, if we want to target only the clicks on one button of a page and not all, by indicating the identifier of this button coming from the documentary model (at the moment there is no unique identifier created automatically for each button, therefore changes would have to be considered in the documentary model first).

Then the first modeler affirmed to have set the primitive corresponding to the (`Table`) that is going to be created from the traces: "That, too, is a part that I have fully understood". Regarding the filters making it possible to choose the information from the traces in order to fill the rows and columns that correspond to the need, the logic seemed to be well understood and easy to use. Here, often, the information inserted automatically in the traces must be used, which reinforces the previous remark. The modeler advised that these options could be offered by the system in the form of lists, so that the modeler would only choose from the ones that already exist—the benefit being that the information of what exists by default would also be presented that way.

The next primitive by the modeler was the one that allowing to set information about the indicator itself (`Indicator`). The modeler linked the table necessary as input to the computation carried out subsequently. The output is conditioned by this calculation, and it should be noted that the corresponding typology should result in a defined list, the values of which have not yet been decided and that therefore this field remained open for the moment. The objective is to have, in this field, the type of output expected, which will be used as information for the visualization chosen below (a list, a matrix, etc.). Because of that open field, the modeler asked for details, but understood the purpose of the field without problem.

The widget which allows for preparing the visualization of these results (*Visualization Widget*) was used later. For the indicator that served as an example, the input parameters (start date and end date) came from the instructor, but this could also be set by the modeler at the time of modeling, etc. The modeler did not understand that this is where the origin of this information is defined. With the explanation in this interview, the modeler understood how to define these inputs in several ways using this primitive.

Then, the (`uiWidget`) primitive was used by the modeler, for modeling in terms of the chosen visualization means (dashboard, alongside the menu, as a prompt message, etc.) for each type of user. The modeler suggested changing the naming of the primitive in order to have the term "visualization" appearing in it, making it easier to find/understand among all the available primitives. As a reminder, during this study, the details concerning visualizations type of graph, etc. were not yet implemented.

Finally, the modeler used the (`Analytical Engine`) primitive in order to define the periodicity of the calculations.

Overall, the modeler mentioned that the use of the primitives of the metamodel required some back and forth actions for certain parts, where he realized a need and had to return to a previous primitive before continuing. These actions are natural and even expected, they prove a step-by-step understanding on the one hand of the metamodel, and on the other hand of the indicator itself.

Nonetheless, according to the first modeler, the part of the metamodel corresponding to the visualization seemed a little more complicated, in particular the definition of the inputs as the parameters, but has nevertheless been understood. This difficulty could be caused by the fact that the metamodel corresponding to this dimension could not be implemented in its entirety at the time of the tests.

Regarding the naming, the other primitives—apart from (`uiWidget`) already mentioned—were easy to navigate. Finally, the modeler said:

> [...] I found it to be concise as a metamodel anyway, there weren't that many [primitive] items and that's not bad.

About the change requested by the client (last task in phase 3), the modeler was able to make the necessary changes: "I would start by changing the traces". Subsequently, he affirmed:

[...] If I have to change from one event to another, yes, the fact that it's taken care of by the metamodel is pretty quick.

The second modeler reported that he started with the (Indicator) primitive. Using it, he realized which inputs he was going to need. Thus, he then took in hand the primitive relative to the traces (Trace), so that these entries were traced. He suggested changing the internal term for this primitive from (fromModel) to (callModel) in order to keep the nomenclature already existing in the tool regarding these situations.

Next, he determined the table (Table) built from the traces. The modeler noted that this primitive is not explicitly related to the traces. This type of link is common among the primitives of the tool used, making it possible to determine the network of the different components and thus to not forget them (the information is also present in the item networks, which makes it possible to check these links at the end of the modeling).

When defining the elements necessary to insert data into tables, the modeler questioned whether a primitive can have one or more values (focus in and focus out) or that these two values are each set in a component. This implies that the link between primitives just mentioned is made in two ways, as needed. We noted that a reflection on this topic should be carried out to see whether the complexity brought by this flexibility is justified—which could be the case for certain indicators—or whether only one value can be assigned to this primitive. This thought led the modeler to make two possible changes:

1. having two items from this primitive, each with a unique value: then link each one into the component to create the table (and create these links);
2. having an item from this primitive with several values: link the item in question, then choose the value(s) concerned.

The modeler stated that the primitive dedicated to the indicator itself is clear, and that the (Analysis) part will depend on the technologies chosen.

Regarding the primitive used for modeling the visualization of the indicator, the second modeler questioned the chosen names, but found the primitive clear:

The item itself is super clear, it's just the naming system compared to the existing ones...

Finally, the modeler stated that the primitive (Analytical Engine) seemed less clear, because it is "a little more difficult to visualize without the context that might arise").

Regarding the last task of phase 3, the change supposedly requested by the client, the modeler was able to perform it without any issue.

To conclude, the second modeler stated that he would need a practical application and a real context for even more precise opinions, because it is with use and time that we will be able to find other improvements.

7.3   Discussion: Summary of the Feedbacks

In order to synthesize the results of this analysis, we propose to come back to the research questions asked.

How is the Handling of the Metamodel by Modelers? The modelers seem to handle the metamodel without much difficulty. Some areas for improvement were suggested, some parts were less clear, but in general the primitives were understood and used correctly for a very first modeling of an indicator.

Are the Primitives Used as Expected? Yes, they are used as intended. The order of use of the components is logical and sometimes with back and forth actions, which is expected, especially for first-time use.

What are the Blocking Points During this First Usage? A visualization-related primitive was less clear to one of the modelers; for the other modeler, it was a primitive linked to the calendar of calculations. In the first case, this difficulty could be related to the fact that the visualizations were implemented with less detail at the time of the tests. In the second case, the modeler asserted that the difficulty was linked to the abstract project/need as opposed to a practical situation, in which it would probably be easier to perceive the use of the primitive.

Are There Any Improvements to be Expected, Which Ones? Yes, the two main improvements are related to the naming system (see below) and the fact that some primitives are not explicitly linked.

Is the Vocabulary Used Clear and Understandable to a Modeler? Most of the vocabulary seems to be understood without problem, especially the one relating to the field of LA. However, some suggested changes are: 1/add the term (*Visualization*) to (`uiWidget`), or do not consider this primitive as a widget, because it is too different from the existing ones; 2/In (`Trace`), change the term of (`fromModel`) to (`callModel`) which already exists, for consistency.

What is the Perceived Usability of the Metamodel by Modelers? Perceived usability (via the questionnaire on phase 2) does not have very high scores, but during interviews it was possible to understand that this was due to the fact that it is a metamodel, which is inherently perceived as complex even by experts used to manipulate them. The fact that the modelers were able to model a first indicator autonomously, and the interviews, allow us to put this initially perceived complexity into perspective.

How are the Proposed Primitives Used to Encourage Maintainability, Reuse, and Customization of Indicators? This aspect was analyzed through the request to change an aspect of the indicator at the end of phase 3. The modelers were able to understand

and perform the requested task, creating a new indicator by changing a few aspects of the metamodel. In addition, the maintainability, reuse, and customization of the use of an indicator via the metamodel seem to be well received considering the latest responses to the questionnaire (and certainly also the modeling experience).

8   Conclusion

In this work, we proposed and evaluated a learning analytics metamodel that uses a model-driven approach within digital publishing chains based on the same MDE approach. The metamodel aims at being sufficiently abstract to allow the implementation of the vast majority of learning analytics indicators, and includes the possibility to enrich them natively with the prior knowledge of documents' semantics and structure.

Results show that modelers seem to take in hand the metamodel without significant difficulties: they used the primitives in a logical order, sometimes with back and forth actions, which is expected for a metamodel and especially for a first use.

Most of the vocabulary seems to be understood and some suggestions for changes have been made, which are easy to fix. During the interviews, it was possible to understand that usability was rated low given the inherent complexity of use. It should be noted that the modelers were able to model a first indicator after being shown only one example and that especially some aspects such as maintainability, reuse, and customization of an indicator via the metamodel seemed to be well perceived. This could be verified via the request to change an aspect of an indicator (phase 3), where the modelers could change only part of a primitive to adapt it according to new needs or to create a new one from the primitives modeled previously.

Although this work is instanciated in the context of a particular model-driven tool, we believe the work presented here is generic enough to be replicated in a similarly designed environment. Further work would involve a deeper analysis of the modelers' work on a real task once all the elements will be fully implemented into the system. Although a larger sample size would be appreciated, the very specialized nature of the modelers' work does not make it realistic to imagine having enough participants for a quantitative analysis.

References

1. Agudo-Peregrina, Á.F., Iglesias-Pradas, S., Conde-González, M.Á., Hernández-García, Á.: Can we predict success from log data in VLEs? Classification of interactions for learning analytics and their relation with performance in VLE-supported F2F and online learning. Comput. Hum. Behav. **31**(1), 542–550 (2014). https://doi.org/10.1016/j.chb.2013.05.031
2. Arribe, T., Crozat, S., Bachimont, B., Spinelli, S.: Chaînes éditoriales numériques: allier efficacité et variabilité grâce à des primitives documentaires. In: Actes du colloque CIDE, Tunis, Tunisie, pp. 1–12 (2012)
3. Bachimont, B., Crozat, S.: Instrumentation numérique des documents: pour une séparation fonds/forme. Revue I3 - Inf. Interact. Intell. **4**(1), 95 (2004)
4. Canellas, C., Bouchet, F., Arribe, T., Luengo, V.: Towards learning analytics metamodels in a context of publishing chains. In: Proceedings of the 13th International Conference on Computer Supported Education, pp. 45–54. SCITEPRESS - Science and Technology Publications, San Francisco (2021). https://doi.org/10.5220/0010402900450054

5. Choquet, C., Iksal, S.: Usage tracking language: a meta-language for modelling tracks in tel Systems. In: International Conference on Software and Data Technologies (ICSOFT), pp. 133–138 (2006). https://doi.org/10.5220/0001312701330138

6. Choquet, C., Iksal, S.: Modélisation et construction de traces d'utilisation d'une activité d'apprentissage?: une approche langage pour la réingénierie d'un EIAH. Sci. Technol. l'Inform. Commun. pour l'Éduc. Form. **14**(1), 419–456 (2007). https://doi.org/10.3406/stice. 2007.968

7. Combemale, B.: Ingénierie Dirigée par les Modèles (IDM) - État de l'art (2008)

8. Cordier, A., Lefevre, M., Champin, P.A., Georgeon, O., Mille, A.: Trace-based reasoning - modeling interaction traces for reasoning on experiences. In: FLAIRS 2013 - Proceedings of the 26th International Florida Artificial Intelligence Research Society Conference, pp. 363–368 (2013)

9. Crozat, S.: Scenari, la chaîne éditoriale libre. Eyrolles (2007)

10. Guillaume, D., Crozat, S., Rivet, L., Majada, M., Hennequin, X.: Chaînes éditoriales numériques (2015)

11. Hutchinson, J., Rouncefield, M., Whittle, J.: Model-driven engineering practices in industry. In: Proceedings - International Conference on Software Engineering, pp. 633–640. ACM Press, USA (2011). https://doi.org/10.1145/1985793.1985882

12. Ifenthaler, D.: Are higher education institutions prepared for learning analytics? TechTrends **61**(4), 366–371 (2016). https://doi.org/10.1007/S11528-016-0154-0

13. Jézéquel, J.M., Combemale, B., Vojtisek, D.: Ingénierie dirigée par les modèles - Des concepts à la pratique. Ellipses, Paris (2012)

14. Kitto, K., Whitmer, J., Silvers, A.E., Webb, M.: Creating data for learning analytics ecosystems. SOLAR Position Paper (September), pp. 1–43 (2020)

15. Patton, M.Q.: Qualitative Evaluation and Research Methods, 2nd edn. SAGE Publications, Thousand Oaks (1990)

16. Siemens, G.: Learning and Academic Analytics (2011). https://www.learninganalytics.net/uncategorized/learning-and-academic-analytics/

17. Wise, A.F., Vytasek, J.: Learning analytics implementation design. In: Lang, C., Siemens, G., Wise, A., Gašević, D. (eds.) Handbook of Learning Analytics, 1st edn., chap. 13, pp. 151–160. SoLAR (2017). https://doi.org/10.18608/hla17.013

Developing an Interest in Mathematics with Occupational Exemplars

Päivi Porras[1](✉) ⓘ and Johanna Naukkarinen[2] ⓘ

[1] LAB University of Applied Sciences, Lappeenranta, Finland
paivi.porras@lab.fi
[2] LUT University, Lappeenranta, Finland

Abstract. This paper introduces four different ways for developing upper secondary school students' interest in and motivation toward mathematics by connecting mathematics topics and contextualizing mathematical problems to working life scenarios in different occupations. They include materials for teachers, a self-study online course for students, a virtual reality environment with embedded mathematics problems, and the use of near-peer role models (university students) as math ambassadors that visit upper secondary schools. All four activities were created in project TyöMAA with the general objectives of informing upper secondary school students where and how mathematics is used in their future careers, enhancing students' mathematical skills to enable them to progress in their postgraduate studies, and infusing the students with enthusiasm and self-confidence in mathematics. Although the COVID-19 pandemic has inevitably slowed the implementation and dissemination of the created tools, we have observed some encouraging indicators of the need for and effectiveness of the solutions.

Keywords: Mathematics · Work-life · MOOC · GeoGebra · Near-peer role models · Contextual framing

1 Introduction

A decreasing interest in mathematics is a problem in math-intensive fields, such as engineering, but also in fields like physiotherapy, nursing, tourism, and hospitality, where mathematics proficiency links to many professional tasks. As noted in [1], many first-year college and university students are unprepared and struggling in disciplines in which mathematics is applied, such as business and economics. Basic calculation skills are important in nursing and medical care [2], and processes should be automatic so that they can be applied easily in conjunction with other required professional skills. Although the mathematical skills of young Finnish people are relatively high in international comparison, their attitudes toward mathematics are alarmingly negative—57% of eighth graders do not like mathematics, and only a quarter of pupils appreciate mathematics to a large extent [3]. In the past couple of years, the number of students taking matriculation exams in the long syllabus of mathematics (explained below, in contrast with the short syllabus) has increased [4], probably due to changes in university admission criteria.

© Springer Nature Switzerland AG 2022
B. Csapó and J. Uhomoibhi (Eds.): CSEDU 2021, CCIS 1624, pp. 115–129, 2022.
https://doi.org/10.1007/978-3-031-14756-2_7

However, in addition to external motivators, the inner motivation of students studying mathematics should also be aroused for better learning outcomes in both long and short syllabus studies.

Gender segregation in the Finnish labor market is very strong. Much of this is due to a strong gender segregation of educational choices at the tertiary level, which, in turn, is affected by the gender segregation of subject choices at the secondary level. The gendered subject choices in upper secondary school seem to have created a situation in which it is relatively easier for men than women to be admitted to university studies [5]. This can be explained by the large number of study placements in STEM (science, technology, engineering, and mathematics) fields in comparison to less math-intensive fields, such as in the humanities or behavioral science, which are popular among women [5].

The contextual framing of mathematics problems has been suggested as one means to motivate and engage students, even though more empirical evidence is still needed [6]. Appropriate contextual framing depends on the topics and level of mathematics taught, and on the age and interests of the learners. A great majority of Finnish upper secondary school students have been indecisive of their occupational interests and have had related worries during their first year of studies [7]. Thus, it appears that future working life is a meaningful context for many upper secondary students and has the potential to act as a fruitful starting point for contextualizing mathematics.

Traditionally, Finnish upper secondary teachers in mathematics have studied mathematics at the university in a special teacher program. After graduating, they have started their career either in a lower secondary school or in an upper secondary school. Mathematics studied at the university level is usually more theoretical than applied. This kind of education provides teachers with excellent theoretical knowledge (which is also needed), but unfortunately, it does not expand their knowledge of how and where mathematics is used in one's working life.

The occupations with which mathematics teachers are usually most familiar are related to STEM fields. Therefore, they may unintentionally neglect the application of mathematics in other fields. However, expecting teachers to be able to provide a wide array of mathematics examples related to different occupations is not realistic, and teachers have to be supported in this by other professionals [8]. Creating examples of mathematical problems related to all fields covered at our campus, such as technology, nursing and health care, business, tourism and hospitality, and the arts, is hoped to help teachers to motivate students, as they can illustrate a multitude of contexts where and how mathematics is needed.

This paper introduces four different ways of developing upper secondary school students' interest in and motivation toward mathematics by connecting mathematics topics and contextualizing mathematical problems to working life scenarios in different occupations. They include developing materials for teachers, constructing an open online course for students, creating a virtual reality environment, and starting peer role model visits to schools. The practices have been created and tested in a European Social Fund-supported project, with gender equality as a central aspect of the project. An outline of the project is presented in [9]. Next, the issues of gender equality and contextual framing in mathematics will be discussed further. The four practices will then be introduced and discussed in more detail. Finally, some conclusions and implications will be presented.

1.1 Mathematics and Gender Equality

Currently, girls form a majority of Finnish upper secondary school students and generally outperform boys in matriculation exams [5]. However, boys choose more mathematics courses and typically get better grades in them [10]. In upper secondary education in Finland, all students must choose between a long and short syllabus in mathematics. Students choosing the long syllabus must take a matriculation exam in mathematics, but for the students taking the short syllabus, the exam is optional. Choosing the short syllabus in mathematics and not taking a matriculation exam in mathematics is more common among girls than boys [10]. Studies have shown that the mathematics competence level of students completing only the minimum number of courses remains at the level they had achieved during compulsory basic education [11]. As a result, the average competence in mathematics for girls trails boys at the end of upper secondary school by approximately one year [11].

In the Netherlands, girls often choose science and mathematics at the secondary level to keep their options open for tertiary-level studies [12]. However, in Finland, girls commonly seem to make choices that rule out the possibility of studying in math-intensive fields, such as engineering, at university [5]. The subject choices of girls—psychology, health education, and religion—often relate to plans to study educational sciences, humanities, or health sciences [5]. Although mathematics is also needed in these disciplines, this is not recognized by upper secondary school students. This can create difficulties in tertiary-level studies, for example, when a nursing student has to pass an exam involving medical calculations.

1.2 Contextual Framing of Mathematics

In the contextual framing of mathematics, a mathematical problem is connected to a real or imaginary problem in some other area of expertise and used to solve it. Although the usefulness of mathematics in solving many kinds of problems is widely acknowledged, many students still experience mathematics as not being relevant [13]. This is under-standable if usefulness is regarded as a property of mathematics, whereas relevance is defined as a connection between mathematics, its usefulness, and the learner [13]. Hence, students may perceive mathematics as generally useful but not as personally relevant. In order to transform the perception of usefulness to the experience of relevance, mathematics needs to somehow connect to students' personal lives in the past, present, or future. The experience of using engineering problems in upper secondary school mathematics teaching has shown that these kinds of exemplars can increase student perception of practicality and usefulness of mathematics even if the examples are not taken from students' everyday lives. In this case, the relevance of mathematics was enhanced through a meaningful connection of mathematics to scientific problems and working life [8]. Earlier studies have also shown that the perception of relevance of mathematics to real life and a future career increases student motivation toward the subject [14]. Although the utility value of mathematics can be expected to increase the extrinsic motivation of the students, it has also been noted to not decrease the intrinsic motivation to learn [15].

Despite the potential positive relationships among contextual framing, experience of relevance, and motivation to learn mathematics, possible dilemmas in the use of contextual framing have also been pointed out [16]. The first dilemma concerns the perceived utility of mathematics. If in making a connection between the abstract world of mathematics and other contexts, the utility of mathematics as a language for explaining patterns in the real world is reinforced by manipulating and "sanitizing" the real-world experiences to enable them to be modeled by a pre-ordained set of mathematical techniques, then the result can appear to be artificial and thus not credible or relevant. Another dilemma relates to the cognitive effect of framing the questions in real-life contexts. On one hand, this can be argued to help to solve the mathematical task by providing mental scaffolding; on the other hand, it may complicate the thinking process by making assumptions of certain context knowledge outside of mathematics [16].

We address both dilemmas by offering the teachers and students working-life problems that are as genuine as possible without extensive "sanitation," but with prerequisite context knowledge. We believe that the high school students' motivation toward mathematics can be enhanced by presenting them mathematical tasks contextually framed in working-life problems from different occupations as career considerations, and the role of mathematics in them invites the students to personally relate to the problems in their fields of interest. In addition to evoking an interest in mathematics, this is also hoped to provide students with support in the development of their professional identities.

2   Teaching Materials in GeoGebra

"Word problems" are part of teaching mathematics, and they are also included in mathematics books. Vos [17] explains the problem of "word problems" widely used in school mathematics: in many cases, the solution is found easier without mathematics; there is only "one" correct method for solving; and the solution is only one number. These kinds of "sanitized" problems prevent students from experiencing the usefulness of mathematics, and, therefore, "many students perceive mathematics as alien, that they don't understand it, and that they drop it as soon as possible" [17, p. 2]. Although traditional word problems in mathematics may not have the value hoped, connecting mathematics to working life provides problems a context with which students can personally relate and hence can increase their perception of its relevance. The more practical work-life examples from different fields that we can show, the more we, hopefully, can motivate students to study mathematics at all levels.

The teaching materials created in the project are organized based on the mathematics curriculum in the upper secondary schools. Because these are real- and work-life examples, they often can be used under different topics. The teacher's expertise is used to select examples best for the group and the case. Examples represent all fields of our campus: engineering, social and health care, business, tourism, and hospitality. The material is available in GeoGebra, as it is well-known by teachers, and it also enables interactivity.

Students who have selected the long syllabus are interested in mathematics at least at some level. They also study mathematics much further than students in the short syllabus. Exemplars given for first courses handle mostly "non-mathematical" fields, such as nursing, tourism, business, and hospitality. These fields represent additions to the graduate studies of students in humanism (which is not studied at our campus) in the short syllabus. Raising their interest in mathematics at an early stage is very important, and, therefore, we have exemplars of these fields at the beginning. Of course, advanced mathematics is not studied in the short syllabus, so the challenging exemplars are mainly for students in the long syllabus.

Corresponding courses in both syllabuses have common exemplars. As these exemplars are highly connected to occupations, it is important to show the reality: sometimes you need more challenging mathematics, sometimes not. One exemplar concerning vectors is shown in [18]. Although vectors are not studied in the short syllabus, they have a big role, for example, in physiotherapy. If a student has thoughts of his or her future career, it is good to already know in upper secondary school the mathematics that will help or are required in the field (when they still have time to acquire the knowledge).

3   MOOC—Mathematics of Working Life

This section presents a brief description of the MOOC (massive open online course) offered to students. More detailed information of the framework and about STACK (system for teaching and assessment using a computer algebra kernel) questions can be found in [18]. The technical issues of exercises are handled in more detail in [19].

3.1   The Framework of Learning Outline

Motivation is the key element in learning [20, 21]. As mentioned in [20], a person modifies his or her motivation from external settings and, little by little, intrinsic motivation may arise. However, a student's extrinsic motivation for a reward (such as a course mark) is probably not enough for the longer-term experience with a MOOC. It is our responsibility to stir intrinsic motivation by promoting personal achievement through exercises, the use of interesting exercises, and good supporting feedback.

Self-efficacy describes the confidence of a student when completing a task successfully, and it is correlated with academic achievement [22]. E-learning self-efficacy [23] implies that students are more motivated and will benefit from e-learning if they feel confident and are capable of learning. There is evidence that digital learning materials have a motivational impact on engagement [24].

This MOOC of the mathematics of working life is directed at Finnish upper secondary school students (later, only students). This creates different kinds of frames, which must be considered in planning and in execution. For example, students are normally in upper secondary high school for three years. Although the topics to be handled are the same in all schools, the timing during the school year may change. Students can also select either the short or long syllabus in mathematics, so the topics studied are somewhat different. In addition, the MOOC should not be confined to a specific time and place, and it should appeal to both female and male students.

The MOOC was planned in cooperation with local upper secondary schools. The teachers in upper secondary schools know the timing of the courses, which topics are considered difficult, and which topics do not receive as much time as desired. The teachers also pointed out topics that students do not necessarily connect to their future careers. For example, vectors and forces are rarely connected with bodily movements, although they have a big role in physiotherapy and in athletics, as shown in [18, p. 210].

Teachers recommended separating the MOOC into two parts: Part A for first-year mathematics and Part B for the last two years. Both parts are also separated by the syllabus, so students can easily find exercises suitable for them. As upper secondary school education normally lasts three years, students can complete this MOOC at the same time. This enables students to study in the MOOC in parallel with courses at their schools.

Although the exercises are separated by short and long syllabus, the students can select any to attempt. Some students in the short syllabus may have good calculation skills and would want to challenge themselves with exercises from the long syllabus. Also, those struggling in the long syllabus may want to try exercises from the short syllabus to gain additional practice and improve their self-esteem. The scores are weighted by their difficulty level to motivate students in selecting more challenging exercises. Applied problems are weighted twice as much as mechanical exercises, and exercises from the long syllabus are worth more points than corresponding exercises in the short syllabus.

Students do not need to do all the exercises from the selected package at once: they can always continue from the point that they solved previously. They can also do exercises in an order of their own selection, as packages also may contain question topics that have not yet been handled by their teacher. Furthermore, a student may notice the need to review earlier topics before continuing in exercises. Most of the questions have several (but limited) trials with hints. Quizzes can be attempted as many times as a student wants. Exercises are randomly selected from folders, and all mathematical information (variable names and values) is randomized, so the quizzes look different every time.

3.2 Technical Framework

Technical resources have a major effect on what and how the material can be produced. If the MOOC is not working properly from a student's point of view, it may also negatively influence the student's motivation and self-regulation levels. The Moodle learning management platform [25] is used at our universities and was an obvious choice for this MOOC: it enables automatic grading, has good analytic tools, and is closely connected to our administrative systems.

STACK [26] is a computer-aided assessment package for mathematic questions on the Moodle platform. It enables the randomizing of variables, as well as provides feedback based on a student's answer. Various question types can be formed, for example, algebraic, numerical, multiple choice (radio button or checkbox), and equivalence reasoning. When combining different kinds of question types, understanding can be reviewed in addition to calculation skills. STACK supports JSXGraph [27], which makes graphing possible. In [18], it is also mentioned that GeoGebra [28] is supported by STACK for graphing. It was later noted that JSXGraph is better to use inside questions, as no connection to outside services is required.

Some of the exercises require more knowledge than what is offered at courses in schools. The needed extra material is provided in an interactive GeoGebra book available in the MOOC. There are also videos (H5P and mp4) in Moodle that are available that clarify topics and provide correct answering techniques for questions. H5P [29] is an interactive video format that enables the clarification of notes and examination of understanding.

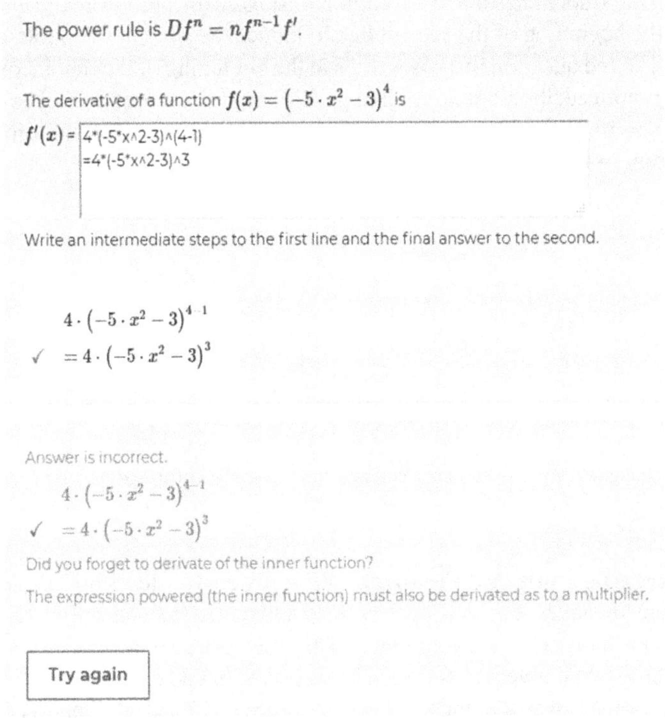

The power rule is $Df^n = nf^{n-1}f'$.

The derivative of a function $f(x) = (-5 \cdot x^2 - 3)^4$ is

$f'(x) =$ | 4*(-5*x^2-3)^(4-1)
=4*(-5*x^2-3)^3

Write an intermediate steps to the first line and the final answer to the second.

$$4 \cdot \left(-5 \cdot x^2 - 3\right)^{4-1}$$
$\checkmark \quad = 4 \cdot \left(-5 \cdot x^2 - 3\right)^3$

Answer is incorrect.

$$4 \cdot \left(-5 \cdot x^2 - 3\right)^{4-1}$$
$\checkmark \quad = 4 \cdot \left(-5 \cdot x^2 - 3\right)^3$

Did you forget to derivate of the inner function?

The expression powered (the inner function) must also be derivated as to a multiplier.

Try again

Fig. 1. Feedback given on derivatives when the inner derivative is missing. (Color figure online)

3.3 Exercises

One part of increasing motivation is to provide proper feedback and a chance to correct mistakes. The Moodle activity called "Quiz" has different behavioral modes. The mode "Interactive with Multiple Tries" enables extra trials if there are hints used in an exercise. When forming a STACK question, an author can add in "Options" as many hints as needed. Every hint adds one trial to the exercises without changing the randomized values. Hints are general and cannot be customized based on a student's answer. A small deduction to a student's score can be made whenever the answer is incorrect.

The feedback can be given in a potential response tree (PRT) or, in some cases, it is automatically given by STACK. Good and motivating feedback requires understanding the most common mistakes, as they must be added one by one to the PRT. As shown in Fig. 1. Feedback given on derivatives when the inner derivative is missing, when a derivative with intermediate steps is written, the program automatically checks the consistency of the consecutive steps. In this case, lines one and two are different forms of the same expression, so a green check mark is given. At this point, STACK does not check the correctness of the answer. When an answer is locked for checking, its correctness is checked with the PRT. In this case, it was noticed that the inner derivative was missing, and the student was informed about it.

In Fig. 2, the student has made a common mistake with signs. The red question mark is shown at the beginning of the second line to induce the student to double-check his or her answer. The red question mark is written at the beginning of the first inconsistent row as soon as it is noticed: the student can change this answer before checking its correctness. This reduces writing mistakes and guides students toward a correct solution. Learning is the key point of this course, not assessment.

Solve $10 - 7 \cdot s = -s - 3$ with all steps.

$$10 - 7 \cdot s = -s - 3$$
$$?\quad -7 \cdot s - s = -3 + 10$$

Fig. 2. A consistency problem is noticed and the student is informed. (Color figure online)

Issues Concerning Hints and Feedback. With extensive teaching experience, the most common mistakes are well known by mathematics teachers. Because programming many questions in a short amount of time is quite stressful, the most frequent calculation hints were given as common hints (ones that added extra trials to the same exercise). A problem was that these hints sometimes confused students, as they had not made that specific mistake. Thus, these common hints were reformed as "Try again" and "You still have one trial."

The previously mentioned changes meant that we had to reform the PRT. We had to program the most common mistakes and check them in the PRT. It was not a difficult task but very time-consuming, and it is still partly under construction. It is quite an easy thing to do for mechanical exercises, where the procedure is fixed. In applied exercises in which only the final numerical answer is given, the programming is not always an easy task. Figure 3 shows an applied exercise that requires extra knowledge, and it is part of the challenging exercises of the long syllabus. The PRT checks and also provides hints for units if the given number is incorrect. In this case, the PRT noticed that the answer is in kilograms, not Newtons. One branch of the PRT provides hints if it assumes that the upper and lower bounds are not given from the water level toward the bottom. As the calculations are not written, the mistake can also be something else, but at least, the student is notified to stay on the correct path. Hopefully, his/her self-esteem is not lowered because of an incorrect answer.

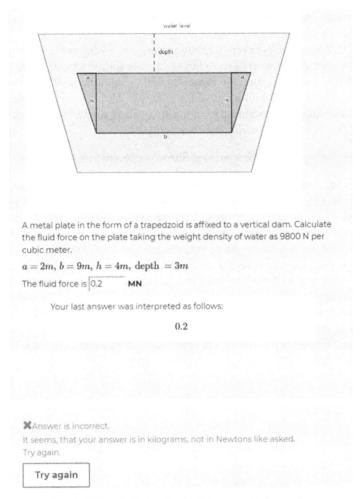

A metal plate in the form of a trapedzoid is affixed to a vertical dam. Calculate the fluid force on the plate taking the weight density of water as 9800 N per cubic meter.

$a = 2m, b = 9m, h = 4m, \text{depth} = 3m$

The fluid force is [0.2   MN

Your last answer was interpreted as follows:

0.2

✖Answer is incorrect.
It seems, that your answer is in kilograms, not in Newtons like asked.
Try again.

[**Try again**]

Fig. 3. The calculation feedback in an applied exercise.

Issues on the Technical Part of Exercises. Several technical problems were found during the first round. Some of them were small and easy to solve, but some prevented even opening the package. STACK programming is very easy if you have even basic knowledge in programming and writing MathJax/LaTeX [30]. The normal procedure— (1) plan an exercise, (2) program it, (3) check with a preview two or three times, and (4) publish the exercise—is not usually enough with STACK. A mentioned in [19], judging the correctness of expressions is sometimes problematic. Therefore, several kinds of correct answers may be required to enable the judging to work correctly.

Despite previews and pretrials, students sometimes struggled in opening the packages. There were at least two major reasons for this: (1) because of randomization, discontinuous cases occurred, and STACK could not solve them; and (2) when there were many students opening packages at the same time, STACK overloaded and would become stuck.

The STACK community is very friendly and helpful. Professor C. Sangwin provided personal guidance that helped in recognizing reasons for the technical issues and also the possibilities of this program. For example, all the exercises should have a "Question Note." This note should be informative for a teacher, and randomized values make it unique for each case. For example, for a linear equation where two linear expressions (yht1, yht2) are equal, the question note was written as

$$\[\{@yht1@\} = \{@yht2@\}, \; \{@vast[5]@\}\].$$

In this case, "yht1" and "yht2" refer to the expressions that are equal, and "vast[5]" refers to the final solution in the teacher's answer. STACK creates different variants of this exercise and uses those variants every time the question is used. It decreases the resources needed when the packages are opened. Figure 4 shows some of the variants created with the previous note. As shown, the teacher can easily check what kinds of exercises are included and what the answer is. If some of the variants look too easy or difficult compared to the others, they can be excluded from the list, and they are no longer available. Of course, all this should be done before the exercise is published for students. Just doing this helped for both problems. First, the students get only exercises that are tested and working. Second, STACK does not form questions for all students from scratch but takes one from the list, which was an essential improvement.

$$3 - 4 \cdot s = \frac{s}{2} - 4, \; s = \frac{14}{9}$$

$$5 \cdot x + \frac{5}{2} = \frac{x}{8} + \frac{39}{4}, \; x = \frac{58}{39}$$

$$6 - \frac{t}{2} = -\frac{5 \cdot t}{4} - \frac{5}{4}, \; t = -\frac{29}{3}$$

$$s + \frac{1}{4} = \frac{7 \cdot s}{10} + \frac{31}{4}, \; s = 25$$

$$s + \frac{9}{4} = -s - \frac{35}{4}, \; s = -\frac{11}{2}$$

$$s + 6 = \frac{23}{4} - \frac{s}{2}, \; s = -\frac{1}{6}$$

Fig. 4. Variants used in one linear equation.

4   Mathematics in Virtual Reality

Already in the twentieth century, virtual reality (VR) was seen as having enormous educational potential [31], as it allows an individual to experience situations that are very close to reality [32]. The idea of this project was to motivate students with examples from different professional fields. It would be best to experience them in authentic surroundings. However, there are typically not opportunities to take all students to a variety of workplaces. A virtual reality experience was produced to help with this, as students can solve applied problems related to working life in an almost authentic surrounding. All the activities in the virtual reality will be connected to the exercises solved in the MOOC, but they will be extended versions of them.

Fig. 5. Hospitality exercise in VR [18].

As we have VR/AR knowledge at our campus, we decided to take advantage of it. We were able to liaise with Pulsan Asema, which is a very popular resort in Lappeenranta, Finland. An inventory model of the resort was digitalized some years ago by Saimaa UAS, Finland, and our project can use that digital information in the VR model. The VR model is so accurate that even the location of the plumbing can be checked. This allows us to add actual working life exercises from civil engineering through hospitality in the same VR model. In Fig. 5 is shown an exercise concerning hospitality [18]. An order made by a customer is given on the screen to the right, a recipe is to the left of the window, and a cake tin needs to be found to see measurements. The volume of this "old" cake tin is not known, so the student must calculate it to know how much pastry is needed. Students must also price the order to cover all costs, like material, work, VAT.

The plan is that students, who actively take part in this course, will be asked to visit campus as a final activity. At this point, it seems that COVID-19 restrictions are cancelled at the time and this final activity could be arranged during the spring 2022. Due to Covid-19, the number of active students is quite low, and VR exercise may be used as an enticement for new students. In that case, the level of exercises must be rethought.

5 University Students as Math Ambassadors

Murphey [33] defines near-peer role models as peers who are close to one's social, professional, and/or age level and who one may respect and admire. College-level near-peer mathematical mentors giving outreach presentations to high school students were noted to have a positive impact on their sense of enjoyment in mathematics, sense of value of mathematics in their lives, and motivations to pursue higher mathematics [34]. Curiously, female near-peer role models in mathematics were observed to strengthen the sense of belonging and self-efficacy in mathematics for women with high mathematics identification and males with low mathematics identification [35].

In the project, the principle of peer role models was employed in the form of math ambassadors. Math ambassadors are students from the university or university of applied sciences who visit the upper secondary schools and present a roughly 30-min talk that introduces the mathematical aspects of several occupations and study areas (health care, business, engineering, tourism and hospitality, arts, social work, education, and psychology), as well as talking about their personal experiences of studying mathematics at the tertiary level. Some ambassadors also act as assistants in the online course, in which they have consultation hours to answer upper secondary school students' questions via chat.

The first five visits by the math ambassadors were conducted as physical visits to the schools in autumn 2020. Based on immediate feedback from the teachers, the upper secondary school students found the ambassador visits interesting and encouraging. In autumn 2021, the visits were conducted virtually due to the COVID-19 situation to an audience of fourteen classes of 25–30 students in two upper secondary schools. After these visits, feedback was collected from the teachers and the students.

The feedback from students (N = 39) indicated that the idea of near-peer role models worked, as approximately 64% of the respondents somewhat or fully agreed with the statement "it is more interesting to hear about [the use of mathematics in different fields and occupations] from a university student than from my own teacher." About 30% of the respondents had no opinion on this statement, so only 5% disagreed with the statement. In addition, almost 80% of the students stated that they received new information from the math ambassadors, and 46% agreed that the presentation motivated them to study mathematics, as opposed to 20% who disagreed with the statement regarding motivation (a third of the respondents had no opinion on this). The topics that interested respondents the most were the need for mathematics in different fields and the need for mathematics in everyday life. Although the latter topic was not the core content of the presentation, it appears that it was either an issue that is also not covered in a mathematics lesson, or that the near peers can address this issue better than the teachers.

The teachers agreed that secondary school students received new information from the ambassadors and that this kind of information would not otherwise be visible in mathematics teaching. The teachers were also of the opinion that the message from the math ambassadors was received better than from the teacher and that it helped in increasing the student motivation. However, the number of teachers providing feedback was rather small (N = 4), so these results should be considered with caution.

6   Discussion and Implications

The whole project, including the piloting of the four presented forms of action, was heavily affected by the COVID-19 pandemic, which forced all Finnish schools to use distance learning in March 2020. It is understandable that in this new situation, the teachers had no time to familiarize themselves with the GeoGebra materials or advertise the MOOC to the students. Even though some students enrolled in the MOOC, after all day in distance learning, they had no interest in continuing with online learning. Nevertheless, we believe that although the pandemic has momentarily decreased teachers' and students' interest in online tools, in the long run, the online learning experiences gained during the pandemic will lower the threshold for engaging in online activities and increase the attractiveness of both the GeoGebra materials and the online course.

This project has indicated that both GeoGebra and STACK provide good tools for contextualizing mathematics problems in working life scenarios without too much simplification or "sanitizing," which could obscure their efficiency in supporting the learning of mathematics [16]. We strongly believe that illustrating the need for mathematics in different occupations through different means developed in the project will help to affect the upper secondary school students' actions when making study choices both at the secondary level and in their further education. However, this belief still needs to be verified, and we are currently looking for possibilities to look deeper into this issue during and after the project.

One of the remarks we received so far based on the piloting of this MOOC is that not all upper secondary students are ready for self-learning at the level required in it. Currently, we have also employed some of the math ambassadors as MOOC assistants, who hold regular consultation hours and help the students with STACK answering techniques and motivate them to keep up with the course. Whether this increased interaction in the MOOC proves to be an efficient tool for better student engagement remains to be seen.

Currently, there are more girls than boys enrolled in the MOOC. This suggests that this kind of course could especially help in supporting and motivating girls to study mathematics in upper secondary school; they may increasingly choose mathematics-intensive fields in tertiary education and thus reduce gender segregation in education and occupations. However, the numbers of students in the MOOC are small due to the reasons explained earlier, and all the conclusions are preliminary and must be considered very carefully as well as requiring further investigation.

When the project finishes, the GeoGebra material will be published for the free use of all interested teachers, and the online course will become part of the regular open university offering at LAB University of Applied Sciences. The visits by math ambassadors will also be available for local upper secondary schools. As all the activities will continue, they will also provide a venue for more research on the effectiveness of the actions than is possible during the project lifetime. Increasing interest in the activities from the upper secondary schools is also expected due to the requirement for the upper secondary schools to strengthen their co-operation with higher education institutions included in the legislation. The requirement has initiated many co-operation schemes, such as at Lappeenranta Junior University [36], which provides an excellent channel for further marketing of the activities, dissemination of the results, and actions for research.

Acknowledgements. We thank European Social Fund for funding this project (S21637).

References

1. LeSage, A., Friedlan, J., Tepylo, D., Kay, R.: Supporting at-risk university business mathematics students: shifting the focus to pedagogy. Int. Electron. J. Math. Educ. **16**(2), em0635 (2021)
2. Zwart, D., Norooz, O., Van Luitc, J., Goeia, S., Nieuwenhuis, A.: Effects of digital learning materials on nursing students' mathematics learning, self-efficacy, and task value in vocational education. Nurse Educ. Pract. **44**, 102755 (2020)
3. Vettenranta, J., et al.: Tulevaisuuden avaintaidot puntarissa: Kahdeksannen luokan oppilaiden matematiikan ja luonnontieteiden osaaminen: kansainvälinen TIMSS 2019 -tutkimus Suomessa [Key skills for the future: Mathematics and science skills of eighth-graders: International TIMSS 2019 study in Finland]. Koulutuksen tutkimuskeskus, Jyväskylä (2020)
4. YTL. www.ylioppilastutkintolautakunta. Accessed 24 Sept 2021
5. Pursiainen, J., Muukkonen, H., Rusanen, J., Harmoinen, S.: Lukion ainevalinnat ja tasa-arvo [Upper secondary school subject choices and equality], Oulun yliopisto (2018). http://urn.fi/urn:nbn:fi-fe201803135965
6. Beswick, K.: Putting context in context: an examination of the evidence for the benefits of 'contextualised' tasks. Int. J. Sci. Math Educ. **9**, 367–390 (2011)
7. Kärkkäinen, J., Luojus, L.: Ammatillinen identiteetti ja koulumotivaation toisen asteen opiskelijoilla [Professional identity and study motivation at upper secondary school]. Jyväskylän yliopisto (2019)
8. Nieminen, I.: Practical Mathematics in High School. Tampere University (2015)
9. Naukkarinen, J., Porras, P.: Motivation for upper secondary school mathematics through working life connection. In: Proceedings of the SEFI 2021 Annual Conference, Berlin, 13–16 September 2021 (2021)
10. Vipunen. Vipunen – Education statistics Finland. https://vipunen.fi/en-gb/. Accessed 24 Sept 2021
11. Metsämuuronen, J., Tuohilampi, L.: Matemaattisen osaamisen piirteitä lukiokoulutuksen lopussa 2015 [Mathematical competence at the end of upper secondary education]. Finnish Education Evaluation Centre (FINEEC), Publications 3:2017 (2017)
12. Yazilitas, D., Saharso, S., de Vries, G.C., Svensson, J.S.: The postmodern perfectionist, the pragmatic hedonist and the materialist maximalist: understanding high school students' profile choices towards or away from mathematics, science and technology (MST) fields in the Netherlands. Gender Educ. **29**(7), 831–849 (2016)
13. Hernandez-Martinez, P., Vos, P.: "Why do I have to learn this?" A case study on students' experiences of the relevance of mathematical modelling activities. ZDM Mathematics Education **50**, 245–257 (2018). https://doi.org/10.1007/s11858-017-0904-2
14. Summala, T.: Ensimmäisen ja toisen vuoden lukio-opiskelijoiden motivaatio matematiikassa [The first and the second year upper secondary school students' motivation in mathematics]. Itä-Suomen yliopisto (2020)
15. Porras, P.: Utilising student profiles in mathematics course arrangements. Yliopistopaino, Lappeenranta (2015)
16. Little, C., Jones, K.: The effect of using real world contexts in post-16 mathematics questions. In: Jourbert, M., Andrews, P. (eds.) Proceedings of the British Congress for Mathematics Education (2010)
17. Vos, P.: "How real people really need mathematics in the real world"—Authenticity in mathematics education. Educ. Sci. **8**, 195 (2018)
18. Porras, P., Naukkarinen, J.: Motivating upper secondary students to learn mathematics with working life exercises. In: Proceedings of the 13th International Conference on Computer Supported Education (CSEDU 2021), vol. 2, pp. 208–215. SCITEPRESS, Portugal (2021)

19. Porras, P.: Some Stack issues faced during programming exercises for TyöMAA project. In: International Meeting of the STACK Community 2021 (2021)

20. Legault, L.: Intrinsic and extrinsic motivation. In: Zeigler-Hill, V., Shackelford, T.K. (eds.) Encyclopedia of Personality and Individual Differences, pp. 1–4. Springer, Cham (2016). https://doi.org/10.1007/978-3-319-28099-8_1139-1

21. Porras, P.: Enthusiasm towards mathematical studies in engineering. In: Tso, T.Y. (ed.) Proceedings of the 36th Conference of the International Group for the Psychology of Mathematics Education, Taipei, Taiwan, vol. 3, pp. 313–320 (2012)

22. Bandura, A.: Social Foundations of Thought and Action: A Social Cognitive Theory. Prentice-Hall, Englewood Cliffs (1986)

23. Wu, J.-H., Tennyson, R.D., Hsia, T.-L.: A study of student satisfaction in a blended e-learning system environment. Comput. Educ. **55**, 155–164 (2010)

24. Passey, D., Goodison, R., Machell, J., McHugh, G.: The motivational effect of ICT on pupils. Department of Educational Research, Lancaster University (2004)

25. Moodle homepage. www.moodle.org. Accessed 31 Aug 2021

26. STACK. https://moodle.telt.unsw.edu.au/question/type/stack/doc/doc.php/About/. Accessed 27 Aug 2021

27. JSXGraph. http://jsxgraph.uni-bayreuth.de/wiki/. Accessed 17 Aug 2021

28. GeoGebra. https://www.geogebra.org/. Accessed 26 Aug 2021

29. H5P homepage. https://h5p.org/. Accessed 27 Aug 2021

30. MathJax. https://www.mathjax.org/#features. Accessed 30 Aug 2021

31. Hoffman, H., Vu, D.: Virtual reality: teaching toll of the twenty-first century? Acad. Med.: J. Assoc. Am. Med. Coll. **71**(12), 1076–1081 (1997)

32. Chavez, B., Bayona, S.: Virtual reality in the learning process. In: Rocha, Á., Adeli, H., Reis, L.P., Costanzo, S. (eds.) WorldCIST'18 2018. AISC, vol. 746, pp. 1345–1356. Springer, Cham (2018). https://doi.org/10.1007/978-3-319-77712-2_129

33. Murphey, T.: Motivating with near peer role models. In: Proceedings of JALT 1997: Trends & Transitions (1997)

34. Wilson, A., Grigorian, S.: The near-peer mathematical mentoring cycle: studying the impact of outreach on high school students' attitudes toward mathematics. Int. J. Math. Educ. Sci. Technol. **50**(1), 46–64 (2019). https://doi.org/10.1080/0020739X.2018.1467508

35. Nickerson, S., Bjorkman, K., et al.: Identification matters: effects of female peer role models differ by gender between high and low mathematically identified students. In: Proceedings of 20th Annual Conference on Research in Undergraduate Mathematics Education, pp. 1367–1372 (2018)

36. Lappeenranta Junior University. https://www.uniori.fi/. Accessed 6 Oct 2021

Moroccan Higher Education at Confinement and Post Confinement Period: Review on the Experience

H. Akhasbi[✉], N. Belghini, B. Riyami, O. Cherrak, and H. Bouassam

Institut Supérieur du Génie Appliqué (IGA),
Place de la Gare Voyageur, 20300 Casablanca, Morocco
hassna.akhasbi@iga.ac.ma

Abstract. The Moroccan university like universities all over the world will long remember the year 2020 marked by the spread of the Coronavirus. The imposed closures of schools and universities, and the switch to distance learning in uncertainty have disrupted the lives and habits of students overnight.

None of the universities were prepared for disruptions of this magnitude. However, learning had to continue regardless of the COVID-19 pandemic.

The results recently published in the article [1] revealed that, despite the efforts made, the Moroccan university was not really prepared to continue these online distance learning, and that the decision to go online was made hastily and without any preparation. In addition, the lack of access to the necessary infrastructure, the lack of technical support, and the digital divide that exists among students, have provoked stress, anxiety, and uncertainty among students. Unfortunately, this can generate negative effects on the quality of students' motivation [2].

To this end, it was crucial to answer the following question: What was the impact of distance learning during the confinement period on the motivational state of young learners? How did the students experience this period of unusual teaching?

Based on the Moroccan university context, the literature, and a study via a questionnaire, this article discusses the higher education response to COVID-19, presents lessons learned, and provides recommendations that could help universities in the future. It reports on a previous study that analyzed the situation of the Moroccan university regarding COVID-19, collected over the two periods: Confinement and post-confinement.

Keywords: COVID-19 · Higher education · Distance learning · Hybrid learning · Design thinking · Human development · Human interaction · Confinement · Post confinement · Motivation · Self-determination · Mental health

1 General Context

"In crisis, it is acceptable to have more questions than answers. In crisis, there's no room for 'not-invented-here'. In crisis, we should all be learners". H.R.H. Princess Laurentien of the Netherlands".

© Springer Nature Switzerland AG 2022
B. Csapó and J. Uhomoibhi (Eds.): CSEDU 2021, CCIS 1624, pp. 130–164, 2022.
https://doi.org/10.1007/978-3-031-14756-2_8

The COVID-19 pandemic (also called CoronaVirus Disease) [3] outbreak in Wuhan, China, spread nationwide between December 2019 and early 2020. Weeks after weeks, the virus spread rapidly across the World [4]. The World Health Organization declared it a Public Health Emergency of International Concern on 30th January 2020 [4,5]. Since that, the pandemic has continued to affect people's physical as well as mental health and lives.

The COVID-19 pandemic has had several repercussions on the economic and social sectors. This has prompted governments around the world to implement a series of urgent decisions to fight its spread. They mainly imposed a total confinement, which impacted social, economic and educational sectors. For the historian [6], pandemics have played a very important role in societal, political, economic and sociological developments throughout history. Covid-19 coupled with considerable advances in information technology have led to a revolution that is reinventing most of the foundations already established [7]. Indeed, in educational institutions, the'contract' that is defined by the unit of time, place and subordination is being reinvented. Currently, new technologies make it possible to work synchronously and asynchronously in different contexts. The notion of place is no longer the same. It has been mutated into new spaces. Hence, human and professional relationships may be shaped by new management and leadership styles.

In these contexts, a wide variety of issues were triggered:

– What impact of COVID-19 pandemic on education sector?
– How student's life has been influenced?
– How mental health has been significantly affected by COVID-19
– How virtual learning has impacted student motivation and engagement during the pandemic period.
– What lessons to learn after this difficult period of confinement and post-confinement.

This article discusses issues cited above, related to the impact of COVID 19 pandemic on the education sector. Our contribution is divided into four parts. First, we present the impact of the COVID19 pandemic on Moroccan higher education sector. Second, we explore some new challenges facing higher education in Morocco. In the third part, we present the results of surveys collecting students' feedbacks on Confinement and post confinement time. The last section is reserved to discuss results and propose some recommendations.

2   Impact of COVID-19 on Higher Education Sector

In the light of the restrictions imposed by the pandemic and state of health emergency, UNESCO estimates that "191 countries have closed all their schools and more than 1.5 billion students from pre-school to tertiary level have been affected" because of COVID-19 [8].

According to [9], this new dynamism will surely lead to unprecedented changes in the education system. [7] asserts in a recent article published in the newspaper "Le Monde" that "when faced with a disaster, one is reassured by considering it as a parenthesis rather than a warning". For Claire Marin, the experience of the pandemic and its implications (e.g., fear of the epidemic and confinement), have a destructive potential at all levels (psychic, moral, social, economic and political). They deconstruct the foundations of society and the rules by which they function.

Academic community around the world was obliged to explore new teaching and learning methods. This has proved difficult for both students and teachers, who must, not only cope with the emotional [10], psychological [11] and economic [12] difficulties posed by the pandemic, but also do their best to adapt to these new changes and ensure "pedagogical continuity".

The Moroccan Ministry of National Education, Vocational Training, Higher Education and Scientific Research has taken a set of measures. Educational continuity for the year 2019/2020 was ensured, adopting a remote mode by broadcasting online courses in synchronous/asynchronous modes.

As a consequence, several challenges have appeared:

– psychological symptoms among students, professors and staff: depression, panic disorder, anxiety, etc.;
– impact of self-Determination motivation to deal with crisis and diseases;
– how to provide theoretical and practical courses;
– how to find internships and ensure professional integration for students in the COVID era.

The following paragraphs cover some aspects of these challenges:

2.1 Mental Health and Motivation

The impact of the COVID-19 pandemic on mental health is complex, diverse and wide-ranging, affecting all sections of societies and populations. Studies carried out in USA have revealed that the COVID-19 pandemic has been associated with mental health, mainly challenges related to the morbidity and mortality caused by the disease and to mitigation activities, including the impact of physical distancing and stay-at-home orders. Symptoms of anxiety disorder and depressive disorder increased considerably during April-June of 2020, compared with the same period in 2019 [13, 14]. The prevalence of symptoms of anxiety disorder was approximately three times than reported in the second quarter of 2019 (25.5% versus 8.1%), and prevalence of depressive disorder was approximately four times than those reported in the second quarter of 2019 (24.3% versus 6.5%) [14].

In addition, fears about the risks of infectious pandemics from COVID-19 and the rapid transition to distance have highlighted the imperative need to involve all actors in the higher education sector in understanding the impact of these events on the teaching and learning process of learners.

Therefore, we assume that this may affect self-determined motivation according to the causal chain presented in the following hypothesis [15].

From this hypothesis we think that the pandemic situation and the confinement, may give students a feeling of isolation, anxiety and uncertainty. In addition, the switch to another mode of learning fully online, with technical and logistical conditions (e.g.

Internet connection problems, unsuitable platforms, lack and/or absence of communication, etc.) judged to be poorly prepared [16], may strongly influence the quality of the teaching during this period. Other conditions leading to further decreased motivation among learners are related to the various difficulties that were observed during the pandemic period (Fig. 1):

Fig. 1. The impact of covid-19 pandemic on self-determined motivation [15].

- The massive use of exchange and sharing tools (Zoom, Skype, Google Meet, G-suite, or Teams, etc.), whose characteristics have a number of limitations, such as: the number of users; security problems; or problems related to connection time (meeting time is limited to 40 min for Zoom) [17];
- A number of students are not familiar with learning in such a distance learning situation;
- Young students have problems with self-discipline, so in a distance learning context they will be less motivated to attend classes on a regular basis in the absence of supervision and coaching;
- Others do not always have the means to acquire a smartphone, tablet, computer or Internet connection, which can influence learning continuity and consequently lead to anxiety and worry about not being able to pass exams;
- Cancellation of a number of practical tasks that require the presence and handling of materials and the work on models in the workshops and laboratories of their establishments;
- Some teachers have not been sufficiently engaged or do not regularly integrate ICT into their teaching practices;
- Most students were bombarded by the workload of assignments and digital resources, which can increase stress levels;
- In addition, students may perceive a lower quality of interaction with their professors (especially in large groups);
- and so on.

2.2 Theoretical and Practical Courses

During this pandemic, universities and colleges have opted for online lectures in the hope of limiting the spread of the COVID-19. Students from various private and public higher education institutions were obliged to follow their courses at distance. They used several digital media: PPT presentations, Word or PDF documents, recorded video and interactive courses via internet platforms (university website, Moodle, Zoom, Meet, Google Classroom, Teams, etc.). In the light of the organizational and personal constraints resulting from the pandemic that affected teachers, the courses were not constructed in an optimal way, or even in a home-made way. In developing countries (including Morocco), the lack of network infrastructure in rural areas has prevented several thousands of students from pursuing programs online.

Tutorials are a form of teaching that allows you to apply the knowledge learned in the classroom or to introduce new concepts. Students work individually on application or discovery exercises in the presence of the teacher, who intervenes to help and correct the exercises. The tutorials are done in a small group, so that the teacher can more easily help the students and adapt his interventions to their difficulties.

Practical work is a type of teaching based on practical learning with, in particular, the realization of experiments to verify and complete the knowledge given in the theoretical courses. Therefore, Practical Work (TP) is considered an essential part of science teaching and learning [18–20].

In the context of distance learning, and if dispensing of lectures and tutorials were possible thanks to the sharing of digital media, it was more delicate to simulate the materials virtually to carry out practical work. In many cases, several hours of practical work had to be cancelled.

2.3   Evaluation and Exams

The objective of the tests is to evaluate the knowledge acquired during a training period. The context of the pandemic raises the question of the evaluation method based on numerical scoring. This issue is of concern during this period of crisis in the light of the difficulty that teachers may encounter in evaluating their students at distance. It necessitates the obligation to rethink alternative forms to guarantee equal opportunities for all students and to apply measures to control and reduce fraud, plagiarism and cheating [10].

2.4   Internships

It is important for every student to obtain a first professional experience. This contributes to the acquisition of skills related to the discovery of the company's context. Thus, internships are presented as a first experience and a means of acquiring skills, complementary to those obtained in training. Also, they give young graduates access to a professional network and facilitate integration into the job market [21].

In these uncertain times, and especially for millions of students who have to graduate during the Covid period, they will be confronted with an economically almost paralyzed world in certain fields (especially tourism, transport, etc.).

In Morocco, the results of a study under the theme "the impact of COVID-19 on the employability of young people in Morocco" showed that 67% of recruitments and 64% of internships were postponed or suspended during the crisis. Only 2% of companies plan not to suspend the recruitment of young graduates in 2020 [22].

3   Situation of Higher Education in Morocco

University represents a primary place of socialization for individuals [23]. Its role is to complete the educational institutions (school and high school), and to permit students to acquire useful knowledge. Beyond the intellectual formation, the university contributes actively in the development of critical thinking and in the preparation of students for professional life [24]. It provides a passion for learning and builds the skills necessary to make these new generations future active citizens in their economy.

Moreover, the role played by the university is changing. Dedicated to transmit knowledge to students, this purely academic role is now combined with a social role aimed at integrating new graduates into the professional world. We can thus think of the university as "the place of intellectual and civic training par excellence"[1]. This is why we often talk about the role of the university in the production of information and training that are central to the subsequent professional exercise.

The professional world has often evolved in a rapid and often unpredictable manner. At the same time, education has always known and followed important evolutions and revolutions that have marked it throughout human history. Currently, the significant advances that have taken place in Information and Communication Technologies (ICT) have created a favorable and enriching environment for developing new and innovative education and research approaches in higher education [25,26], since they facilitate more efficient ways of managing the work of teachers [27]. Increasingly, questions are being asked about the content learned and the way it is acquired and transmitted: is it adapted to new socio-economic realities?

Moreover, could the COVID-19 pandemic be a "godsend" for higher education in Morocco? Certainly, the Moroccan educational landscape, and higher education in particular, is undergoing a serious transformation brought about by the integration of ICT as innovative tools to improve performance and quality, and to harmonize with international standards. More or less lagging behind in this area, has higher education in Morocco been able to catch up several years of delay, moving from 1.0 to 4.0 education? And have initiatives had anything to do with it?

In what follows, we consider higher education under the following headings: Education in the age of digital transformation;

- Digital transformation;
- Digital native.

3.1   Digital Transformation

Over the last few years many changes have transformed society: the arrival of the television, the new ways of transport and the new Information and Communication Technologies known as ICT. These changes, coupled with the changes in family customs and social values, have had a special impact on the young students who have grown up in the heart of this technological and societal revolution. The new generations, in contrast to the old generations, have new expectations and new requirements that seem to be especially present in educational environments such as universities.

---

[1] http://www.unistra.fr/index.php?id=20878.

In addition, the world that is taking shape in the 4.0 perspective of this digital transformation, is reinventing the way businesses operate in virtually all areas and especially with the arrival of the "fourth industrial revolution" called industry 4.0 which promises to impact, in a remarkable way, the way goods and services are produced and sold.

The field of education has also been affected by this transformation through the integration of ICT into the educational process. However, for educational institutions, as for companies in other sectors, the digital transformation is taking place at the strategic and organizational level as well as at the human and technological levels.

Thus, in this context of global digital transformation, the landscape of teaching and education (especially higher education) in Morocco is then directly concerned by the requirements of international competitiveness which forces the university to change its paradigm. To this end, it is crucial to train profiles with a strong potential capable of meeting the challenges of a developing country such as Morocco, or even on a continental scale, that of Africa.

In order to place education at the heart of the digital transformation and help young graduates to become lucid citizens in a changing world, the teacher must be supported in the integration of new technologies so as to develop new curricula that respond effectively to the new requirements of today's world.

3.2 Digital Native

Technological progress has influenced the way we acquire knowledge and learn. On the other hand, the Internet provides fast access to information technology in different fields and thus, improves efficiency and saves time. The importance of online technology is especially emphasized in new methods for learning and education. This is particularly important among Generation Z ("Gen Z"), which derives knowledge from the Internet and is focused on a quick search of information.

Generation Z (Gen.Z) is recognized as the second-wave millennials as they were born between 1995 and 2010 (11–26 years old). They, also known as Generation C for Communication, Collaboration, Connection and Creativity, represents people born in 1997 and beyond (wikipedia). They are considered to be the first generation that is referred to as a digital native [28, 29].

Despite being Digital Natives, specialists talk about some common characteristics of this generation:

– Poor self-branding with instant lifestyle, and unable to survive in a workplace that is not ideal for them [30];
– Gen.Z have high work ambitions and are enterprising (they invest themselves and want to be enterprising) and yet are so easily stressed out that they are difficult to be retained for a long-term commitment. This finding is supported by Steinerowska-Streb and Wziątek-Staśko [31] who stated that Gen.Z do not pay attention to stability at work and easily change workplaces. Thus, Gen.Z employees tend to be less committed [30];
– They are known to be effective employees in the digital age that can undertake multitasking [32];

- They are technology-dependent and the most connected generation. About half of Generation Z reports being connected 10 h a day and using an average of five screens (desktop, laptop, smartphone, TV and tablet).
- It promotes an experiential and interactive learning style, involving collaborative work around projects;
- In their choice of studies, Generation Z adults no longer select their university based on academic performance, but rather seek the one that can offer them a unique learning experience.

Furthermore, the majority of Moroccan Universities students fall within the digital native generation, and thus are assumed to be digital literate. However, many studies found that students over the world are more literate and comfortable with using mobile devices than computers [33, 35]. Furthermore, [36] pointed out that this "knowledge of how to use mobile devices for specific personal or social functions is not always a good indicator of knowledge of educational functions".

It is in these regards that training and technical support are always recommended even for digital natives, with the purpose of not only making a positive impact on the remote online teaching and learning outcomes, but also of addressing the digital divide which was found among the current generation of university students [35, 37].

4   Confinement and Post Confinement Time: Towards a Student Experience

In order to better understand the situation of distance learning in the Moroccan university context during the period of confinement, and to draw up a general assessment of the difficulties encountered by students and their impact on the motivational state and commitment of the learner, two results will be exploited in this article:

- A first survey (already published as [1]), conducted during the confinement period (between April and June 2020), was in the form of a feedback questionnaire intended for a homogeneous sample group of 24 students of the Master 2 "Business Administration" (AE), respecting the ethics of anonymity and confidentiality of their responses;
- A second survey was administered after the confinement and the resumption of classes (between February and June 2021). The latter targeted all IGA School of Engineering students (2nd year through 4th year) more than 132 students. 83 students accepted to answer the questionnaire. The data was collected and held anonymously. It consists of two parts of questions: 1) Students' worries and stress about the health situation and 2) Motivation to Learn.

The empirical results of the two questionnaires, will be exploited to the point of collecting additional arguments contributing to the design of lessons learned feedback on distance education during COVID-19 and to verify the causal chain already proposed [15].

4.1 On Confinement Analysis Survey

In a preceding paper [1], a survey was conducted among a sample group of students of the IGA during the switching period to remote learning. The targeted students are enrolled in the professional Master 2 "Business Administration" for the 2019–2020 academic year. The main objective of this study is to collect a number of information in order to know:

– Impact of Covid19 on the student life and the course of the distance studies;
– To what extent the current distance learning devices have allowed to maintain a link with the educational institution;
– Impact of COVID19 on students' professional life.

In addition to the demographic information of the participants, 9 questions were asked. The synthesis of the results collected is presented in the figure below (Fig. 2).

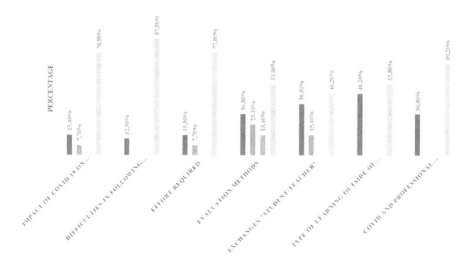

Fig. 2. Results of the first survey.

Analysis of the survey results reveals:

– Over 70% of the students had been negatively affected by the COVID19;
– The Majority of the learners, about 87.5%, had difficulty in attending online courses. This can be ascribed to the feeling of isolation, pressure and the difficulty of working without the direct support of the professor or because of the problems of the internet connection which is sometimes unstable or noisy;
– 38.5% of students say that the transition to distance education requires more effort. They considered as new and stressed context;
– More than half of the participants (53.8%) considered online MCQs (multiple-choice questions) the most suitable form for assessments. We hypothesize that this method of assessment appears to be the simplest, and most comfortable method from the students' perspective;

– Despite a variety of platforms have been put in place to ensure continuity of exchange and communication between teachers and their students 46.2% have expressed a difficulty in communication. This is a significant percentage, can be explained by the difficulty of managing different communication channels at the same time, especially during this special period;
– Between the 100% distance, hybrid and 100% face-to-face teaching mode, the participants chose the hybrid (53.8%) mode. Perhaps with the aim of seeking balance between face-to-face and distance learning;
– The COVID-19 pandemic has had a considerable impact on the Moroccan economy. This explains the high rate of students about 70% confirmed that the crisis of covid-19 will influence their professional plan.

This study highlighted a number of interesting findings related to the negative effects of the COVID-19 pandemic and the shift towards distance education on students' lives.

The table below summarizes students' overall appreciation of the switchover to distance learning during the confinement period.

Some students tried to take advantage of the confinement to acquire other skills online like language or computer programming classes outside of the school's program to occupy themselves.

But for the majority of the participants, the current conditions have strongly impacted the follow-up of courses, the efforts to be made to adapt, to communicate with the teaching and administrative staff, and have strongly favoured the feeling of uncertainty about their professional future and their integration into the job market.

Table 1. Summary of students' overall appreciation of the lockdown period.

Categories	N°	Appreciation
Impact of Covid 19 on students' lives	2	🙁
Difficulties following distance learning courses	3	🙁
Effort & Time required	4	🙁
Evaluation Methods	5	😐
Interaction and Communication	6	🙁
Type of Learning Outside of COVID-19	7	🙂
Learning outside the school program	8	🙂
Learning space in open space	9	🙂
Professional insertion	10	🙁

During confinement, it can be concluded that the system of teaching and learning was disrupted on several levels. Regarding IGA, the switch to planning the video capsules and the virtual classrooms was an unprecedented race. The teaching staff had to upload the course content online, whereas students had to download the online course. Despite the urgency, it was a challenge that proved to be successful, but the feedback revealed several limitations and failures. These reflect the lack of access to the required infrastructure and tools, the lack of technical support and training, and expertise on the part of the professors, the organizers of this crisis situation and even on the part of the students.

To well understand the deep impact of this crisis and its consequences on the learners' lives, a survey of the motivational effects is extremely important, considering the context in which the learners were placed during this unprecedented period.

4.2   Post Confinement Results

Background. Motivation is a postulated structure representing physiological and psychological mechanisms. It is also the motor of internal and external forces (situational, contextual and global), directed or not by a goal, which influence individual on a cognitive, emotional or behavioral level [38].

Academic motivation occurs when these internal and/or external forces are present. It's also requiring a personal investment on the part of the student [39]. The learner must be willing to be involved in his or her studies, to apply himself or herself to his or her schoolwork, and to participate actively in class. If his or her academic motivation is strong and increased, then the student can expect greater academic success, a more effective achievement of learning.

Numerous studies have shown that lack of motivation is one of the main causes of absenteeism, poor grades, and dropping out of school, and a low level of motivation or the absence of any form of motivation is one of the main symptoms of dropping out of university [40]. In this respect, many people note that a demotivated student will eventually fail or will not learn much.

In the usual circumstances (outside of crises), many professors and tutors find it difficult to promote the development and support of their students' academic motivation. So, what impact did distance learning have on the motivational state of young learners during the period of confinement, and how did the students experience in this unusual period of learning?

It is important to remind that learning within the university is specific. Indeed, the learner is called, within the framework of his university studies to be more autonomous, fast and responsible. In this sense, the modification of learning methods requires a considerable effort of adaptation from the new students.

Consequently, and following the crisis of the COVID-19, the universities are faced with the challenge of associating ICT to the university pedagogy in order to consolidate the relationship between the student and his studies. To this end, university is called upon more than ever to learn from these new dynamics and to open up pragmatically to the evolution of training modalities and the accompaniment of learners and the distribution of the roles of teachers, administration and learners.

Theoretical Framework of Investigation. From the theoretical perspective of [41], it seems that the academic motivation of a student is influenced by his or her feelings of self-determination, competence, and affiliation, and that whatever affects these three determinants can also impact motivation.

Feelings of Self-determination. This is an individual's perception of the origin of his or her action. If the person perceives that his or her behavior was by choice, his or her sense of self-determination will be more important. The context in which the task was performed is therefore perceived as supporting autonomy. To this extent, a stronger sense of self-determination will have a positive impact on the development of the student's academic motivation, while its opposite will have a negative impact.

Feeling of Competence. According to [41,42], this element can be defined as a complex, fairly stable, enduring affective state related to an individual's perception of his or her ability, competence in a given activity. Events that help an individual feel competent increase his or her self-determined motivation. In contrast, events that undermine an individual's feelings of competence decrease his or her self-determined motivation. Several contextual factors can affect students' perceptions of competence and, consequently, their academic motivation: "the curriculum, the structure of the classroom and the teacher are all sources of influence that can affect motivation" [43].

Context of Study and Analysis of the Results. After the confinement, the Moroccan Ministry of National Education, Vocational Training, Higher Education and Scientific Research has decided to allow schools and institutes of higher education with limited access to adopt either:

- The face-to-face training approach: for small groups not exceeding twenty students per class, with of course, the implementation of a very strict sanitary protocol (aeration of classes, wearing of masks within the school mandatory, sterilization and cleaning, presence of hydroalcoholic gel, etc.);
- The hybrid approach: which consists in opting for face-to-face courses (with the sanitary measures in force) according to the technicality of the subject and others at a distance for the courses considered less technical;
- Distance learning: For students who live in other regions, because even after the decontamination, the state of emergency is still applied and travel between regions and even between some cities is still prohibited, depending on the state of spread and the number of cases felt.

For open access institutions (faculties and schools with boarding facilities), the teaching approach adopted is distance learning, especially in areas heavily affected by the pandemic.

The Higher Institute of Applied Engineering (IGA) is one of the limited access higher education schools. As a result, it was proposed to students living in Casablanca and its regions, at the beginning of the 2020–2021 academic year, to choose between the two approaches: face-to-face or distance learning.

For students outside of Casablanca or outside of Morocco, it was automatically the distance learning continuity, given that the region of Casablanca was in the red zone (with a very high number of cases of COVID-19).

Students who chose the face-to-face option were responsible for respecting the barrier measures against the pandemic, including social distancing and wearing masks in classrooms. However, students who have chosen the distance mode are required to follow the course via the GoogleMeet platform, during the same period that the course is being facilitated by the face-to-face professor. They are also required to hand in their work and attend the eventual corrections during the lives organized during the week.

The context of the confinement due to the COVID-19 health crisis seemed special to investigate the impact of a virtual course on students' academic motivation. In order to answer to this crucial question, a questionnaire surveying the self-determined motivation of learners was drawn up.

An anonymous online questionnaire (Google Forms) has been proposed at the end of the first semester (between February and June 2021), to 132 students of the IGA School of Engineering (2nd year EISE: Engineering of Information Systems and Electronics, 3rd year EIS/ESE: Engineering of Information Systems and Electronics and 4th year all options combined: EINS "Engineering of Information Networks and Security", ESIP "Engineering of Software and Image Processing", EASQC "Engineering of Automated Systems and Quality Control" and ETNES "Engineering of Telecoms, Networks and Embedded Systems". The questionnaire is composed of a total of 14 questions using different scales according to the questions asked. It includes two parts:

1. Students' anxiety and stress about the health situation, is composed of questions using either a 4-level frequency scale or a 1 to 4 rating scale;
2. Motivation to Learn is composed of simple questions with response's choices; Filtered questions; and an open-ended question giving the student a free choice of expression.

For the development of the questionnaire, and in order to facilitate the analysis of the results, closed questions with check boxes were used or simple check-off answers to facilitate students in expressing their appreciations. The present section is dedicated to the synthetic description of the results obtained from the analysis of the questionnaire.

Regarding the profile of the students participating in the study, the sample consists of three levels (2nd, 3rd and 4th year of the engineering school). They are distributed according to the percentages as shown in Fig. 3.

Second year undergraduate engineering students (EISE): 25% of the participants represent the population of the study. These students have gone through an unprecedented phase, passing through a double transition: from secondary education to higher education and then a switch, forced, ill-prepared and imposed by the advent of the COVID-19 pandemic, from face-to-face to distance learning.

– The students of the license (all options), represent 55% of the participants;
– And 22% of the participants are students of master 1 in engineering (all options).

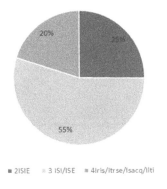

Fig. 3. Profile of participants.

Analysis of Students' Anxieties about the Health Situation. Regarding the first part of the questionnaire, five questions were proposed to the students in order to further clarify the impact of distance education on the psychology and mental health of the students.

For the first question, a rating scale of 1 to 4 ranging from "I am not at all worried" to "I am extremely worried" was used. More than 54% of all participants chose level 3 or 4 (I am very concerned) regarding their personal health concerns, while the other respondents chose level 1 or 2 (I am not at all worried) (Fig. 4).

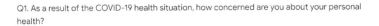

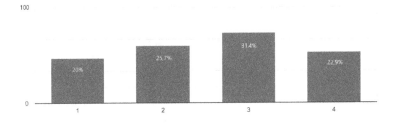

Fig. 4. Students worries about their personal health.

Nevertheless, 80% chose level 3 or 4 (I am very worried) regarding their worries about the health of their loved ones (Fig. 5).

Q2. As a result of the COVID-19 health situation, how concerned are you about the health of your family members?

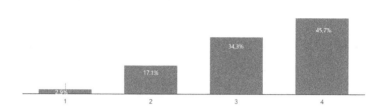

Fig. 5. Students' concerns about their families' health.

The following figure shows that more than 67% of the students chose level 3 or 4 (I am very worried) regarding their worry about their university education in terms of difficulties of organization, level of understanding, etc. (Fig. 6).

Q3. As a result of the health situation related to COVID-19, how concerned are you about your academic learning (educational continuity at a dist...sonal organization, difficulties in understanding)?

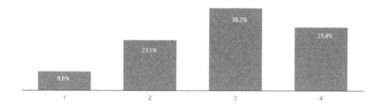

Fig. 6. Students' concerns about their learning.

A rating scale of 1 to 4 ranging from "no, not at all" to "yes, definitely" was used to assess whether students had questioned their choice of study and profession. 67.8% of all students chose level 1 (no, not at all) or 2; 32.2% chose level 3 or 4 (yes, definitely) (Fig. 7).

Q4. As a result of the COVID-19 health situation, have you had to rethink your choice of study and profession?

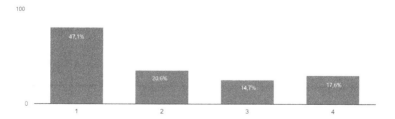

Fig. 7. Choice of study.

A Likert scale with 4 levels (excellent-medium-mediocre-unsatisfactory) was used to obtain the students' feelings about the work environment from home during the confinement period. During this semester of distance learning, 17.6% of all the students felt very dissatisfied with the work environment, 38.2% felt the environment was mediocre, 38.2% felt it was medium and just 6% were satisfied (Fig. 8).

Q5. Following the COVID-19 health situation, do you feel satisfied with the homeschool environment?

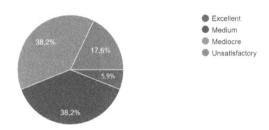

Fig. 8. Satisfaction with the home-school environment.

Analysis of Motivation to Learn. This health crisis has imposed the rapid implementation of a pedagogical continuity at a distance concerning the teachings but also the follow-up of the supervision in mini-projects, tutored projects and also for the internships in companies, for the students of the license and the Master 2. Although distance learning (DE) in initial education has existed for years, and according to the work of [44], a web-mediated course stimulates the feeling of competence, increases the feeling of self-efficacy and competence [45], the training structures as well as the students of the IGA were not specifically prepared for it. Specifically prepared for it.

This second part of the questionnaire allowed students to explore their motives for engagement and their level of motivation during the period of confinement.

There was a significant variation in student engagement in their learning during this period. A large gap was observed in the analysis of responses between those who felt that their engagement had increased and those who felt the opposite. The engagement of all students in their learning was distributed between an increase of 5.9%, a decrease of 70.6%, and no change for 23.5% of students (Fig. 9).

Q6. Your engagement in your learning compared to a face-to-face semester:

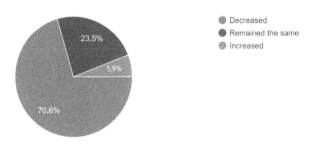

Fig. 9. Learning engagement.

Throughout the distance learning semester, motivation has continually decreased for 55.9% of students and has continually increased for only 14.7% of students. Motivation varied greatly for 8.8% of all students, while it remained the same for 20.6% of them (Fig. 10).

Q7. Throughout your distance learning continuum, your motivation :

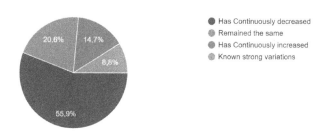

Fig. 10. Learners' motivation variations.

The autonomy was perfectly suitable or appreciated by just 8.8% of the students. 26.5% of the students felt that they were able to organize themselves perfectly or knew how to organize themselves, even if there were some limitations, in the different learning activities, and 29.4% felt that they were able to acquire knowledge perfectly or knew how to acquire knowledge, even if there were some limitations. On the opposite, 23.5% had no opinion because they felt disturbed during this period and 11.8% did not appreciate the autonomy left. The complete results are presented in Fig. 11.

The personal organization in the different learning activities was completely changed for 17.1% of all students, kept with some limitations for 57.2%, and perfectly kept for 14.3% of students. 11.4% had no opinion because of the conditions imposed by the pandemic. The detailed results are presented in the figure below (Fig. 12).

As shown in the Fig. 13, during the distance learning semester, 14.7% of all students felt they had learned almost nothing, 20.6% said they had learned but it has too many limitations, 47.7% had learned, although with some limitations, 11.8% felt they had learned perfectly, and 5.8% had no opinion because this period was very disruptive.

Q8. How would you evaluate the autonomy you have been given in advancing the various courses?

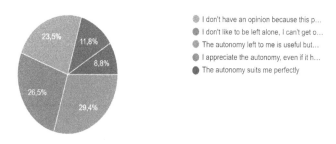

Fig. 11. Evaluation of autonomy.

Q9. How would you rate your personal organization in your various learning activities?

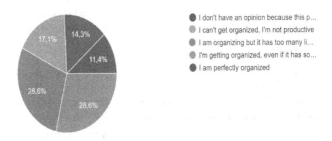

Fig. 12. Personal organization.

Q10. How would you rate your feelings about learning in this period?

Fig. 13. Feeling about learning.

Compared to before the confinement, organizational habits were completely disrupted for 15.2% of all students, slightly disturbed for 63.6%, and maintained for 21.2% of students (Fig. 14).

Q11. Your organizational habits, compared to before the confinement, have been :
33 réponses

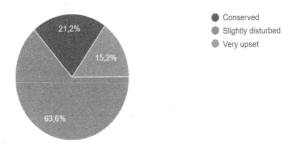

Fig. 14. Organizational habits.

In order to understand the motivational determinants for student engagement during the different pedagogical activities (courses, TD, TP, etc.), the question: How do you explain your level of involvement (participation, engagement, etc.) in your learning during this period? was asked to the students with a series of proposed answers to be checked:

1. I was involved because it allows me to develop my academic and professional skills;
2. I don't feel like I was involved because I didn't like this way of learning;
3. I was involved because I find pleasure in learning new knowledge and overcoming difficulties and in sharing my knowledge;
4. Not being involved would make me feel guilty towards those around me;
5. I was involved because the methods used by the teachers helped keep me interested in learning;
6. I don't have an opinion because this period was very disturbed;
7. I was involved because the learning activities were attractive and diverse;
8. I was involved because the learning environment (platforms and tools) put in place, are attractive;
9. I was involved because I like to get feedback from my teachers and bonus points.

From the Fig. 15, we can see that the motivational motifs of the students vary between: the academic and professional motive (31.4%), the motif of the pleasure of learning new knowledge (20%), the reason related to the methods implemented were attractive (14.3%), or again because of the feeling of guilt towards one's entourage (parents, friends, etc.) (14.3%). The learning environment 5.7%, the activities implemented 5.7%, and the feedback from teachers 5.7% were also chosen by the participants as reasons for their commitment. 22.9% did not feel they were engaged because they did not like this method of distance learning and 11.4% had no opinion because of the disruption caused by the pandemic.

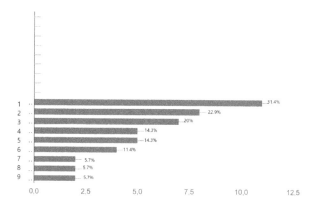

Q12. How would you explain your level of involvement (participation, commitment, etc.) in your learning during this period?

Fig. 15. Motivational determinants for student engagement.

The analysis of the results of the question: Do you attend all the pedagogical activities proposed by your teachers (courses, tutorials and supervision of mini-projects and projects)? showed that if a minority of the participants in this study have never or occasionally attended courses, most of them have always attended all these pedagogical activities (Fig. 16).

Q13. Do you attend all the educational activities provided by your teachers (courses, tutorials and supervision of mini-projects and projects)?

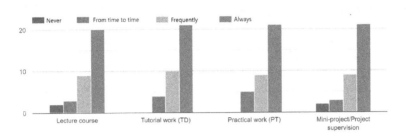

Fig. 16. Attendance rate for pedagogical activities.

The reason presented by 42.9% of students justifying their attendance to the pedagogical activities (Courses, TD, TP and others...) was "To acquire the necessary skills for my academic and professional career", 11.4% "For the pleasure of learning and sharing knowledge with teachers and other students", 8.6% "Because working in a digital learning environment has drawn my curiosity and interest in discovering new learning techniques", and 5.7% had attended pedagogical activities "Because the methods used by the teachers helped keep me interested in learning". The other responses are divided as follows:

- 17.1% attend activities for the risk of being penalized with a malus for absences;
- 2.9% as a result of guilt towards those around them;
- 11.4% don't have an opinion because this period was very disturbed (Fig. 17).

Q14. How do you explain being regularly present at all educational activities provided throughout this confinement period?

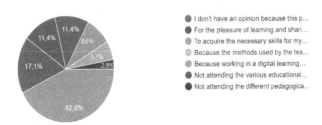

Fig. 17. Motives for attendance at various pedagogical activities.

5   Discussion

Students' Anxieties. The analysis of the empirically collected results of the questionnaire revealed that this health crisis caused a sudden change in the lifestyle of both students and teachers. As this change was made in an emergency, it is not without risk insofar as this unusual situation has caused anxiety and permanent stress in the students' environment (Question 1 "Q1" to Question 5 "Q5" of the questionnaire). The students were in a state of uneasiness, stress and distress, and they were anxious because of the fear of the pandemic and also because of the fear of losing the academic year. This can be explained by the following reasons:

– Development of the feeling of isolation: Alone in front of the screen for several hours, students, cut off from the university environment where they used to meet their classmates and teachers, no longer had this physical contact with which they were familiar, therefore, are taken by difficult and unmanageable psychosociological difficulties;
– Heavy sanitary measures to be applied: This pandemic requires adherence to several measures, some of which are necessary for personal protection against infection (e.g., hand hygiene, avoidance of direct contact with an infected person) and others for the protection of loved ones and society as a whole (e.g. staying at home, social distancing). Things that no one used to apply or assume;
– Health communication "A lot of information and not much credibility": Several studies conducted during the COVID-19 pandemic show that the high prevalence of mental health problems, especially anxiety and depression in the general population, is positively associated with frequent exposure to social media [46]. Others consider that this period was characterized by an abundance of information, considered inconsistent, ambiguous, contradictory and sometimes incredulous [46–49]; contributing to the development of a feeling of uncertainty and invisibility about the future,
– Other conditions, observed in the first study [1], related to difficulties of connection, equipment, ergonomics, as well as an environment that was not conducive to teaching and collective learning and to group dynamics, also participated in affecting the mental health of students.

Motivation to Learn. In view of the specific nature of the health crisis related to the COVID-19 pandemic and its repercussions on the students' life, this study focuses on a major dimension of investment, perseverance and performance in learning situations: motivation [50].

Thus, this section discusses the impact of the period of confinement on the determinants of student motivation: self-determination (working at their own rhythm, structuring and self-managing learning, being active in learning), as well as that of engagement in studies, and attempts to verify the causal chain previously presented (Sect. 2.1).

Commitment and Motivation. In the second part of the questionnaire, the analysis of the results shows that these difficulties (mental fatigue, anxiety, stress and uncertainty) began to be felt by most students through a decrease in their commitment (Q6) and motivation (Q7) towards their studies. This lack of motivation could be explained by the difficulty of projecting themselves into the future for some students. For other students, this lack of motivation may be due to the fact that they do not feel that they are in a situation/environment that allows them to reach the level of learning that meets their expectations (Q10), they think that this new "homeschool" mode of teaching does not promote effective learning in the same way as classroom teaching (Q5). The fear of the digital world can also be one of the main reasons why students are not motivated to study at a distance. This fear can be explained by the lack of basic digital skills.

On the other hand, and to better explain this lack of motivation of students, we must return to the generational phenomenon. Indeed, current students were born around 2000. They belong to the Z generation which is characterized by its connectivity, inter-activity and reactivity. The demand for immediacy leads the generation Z to not stay in the waiting. They are on their own and look for the necessary information. Soliciting the network of friends allows them to achieve their goals. This is why this generation prefers to be in groups rather than alone.

Autonomy. In contrast to a classroom course that imposes the same constraints and deadlines on all students, leaving relatively little opportunity for autonomy, a virtual course has fostered autonomy for a good portion of students (Q8), perhaps because the learning approach proposed in a distance course requires self-management and allows students to work at their own rhythm.

Nevertheless, for some students, the autonomy they were granted during this period seemed difficult to manage, especially during the first few weeks of the switch to distance learning. This can be explained by the lack of experience in self-learning, the lack of monitoring and follow-up, the work schedule, the diversity of channels of communication to manage, stress and anxiety, the decrease in motivation and many other factors.

Personal variables can also influence the degree of students' autonomy. The difficulties in adapting to the learning modalities required by the academic environment are particularly acute for new students who are unable to find a good rhythm at the beginning of their student life. Indeed, the change in subjects and their content between secondary and higher education, and the sometimes, radical change in the study context, creates a certain uneasiness in new students that can lead to a disruption of the work rhythm. To this effect, it suggests that those who appreciated the autonomy granted were the older students, i.e., the students of the Bachelor and Master, who are familiar with the university pedagogy and had already developed the attitude of being more autonomous in class and were more comfortable away from the teacher's support.

Work Organization. It is important to remember that learning at university is specific. Indeed, the learner is called, within the framework of his university studies to be more autonomous, fast and responsible. In this sense, the modification of work methods requires a considerable effort of adaptation on the part of new students. During distance learning, these efforts will be coupled with the efforts to adapt to the new conditions imposed by the COVID-19 (confinement and switch to virtual classes). This explains the responses obtained in the questions (Q9 and Q11). Another reason that could have slightly or strongly disrupted the organization of work for some students is that the teachers were stressed by the volume of the modules that they absolutely had to finish before the end of the year. It should also be noted that the period of confinement and isolation experienced was destabilizing, in terms of organizational habits and work rhythm, for some young students.

Attendance in Pedagogical Activities. During the period of confinement and distance learning, student absenteeism can then be explained by student demotivation, lassitude, and disarray felt during the period of confinement and isolation, and too high workload (while being isolated at home). Other reasons can be added such as technical difficulties related to internet connection problems, distrust of the digital world, health problems caused by screen work, as well as the lack of commitment felt by students towards their studies. Therefore, it is important that teacher-researchers adapt their teaching methods more to students' learning styles. A majority of the respondents considered the professional dimension (development of professional skills) as the driving force behind their motivation to attend the various pedagogical activities. We believe that those who consider this dimension were the older ones, i.e., the Bachelor's and Master's students (diploma years) who had already mapped out their professional goals and orientations than those in the first year.

Reasons for Commitment to Pedagogical Activities. The theory of self-determination, developed in [42], makes it possible to establish the motivational profile of individuals according to their level of self-determination. A high level of self-determination translates into a free engagement of the subject in an activity that he or she favors because of the interest, satisfaction, and even pleasure that it provides. She distinguishes several types of behavior, according to the decreasing level of self-determination:

– intrinsic motivation: characterizes individuals who practice an activity for pleasure,
– extrinsic motivation by identified regulation: concerns subjects who engage in an activity because it will allow them to achieve professional or personal goals, even if the interest in the task itself is weak,
– extrinsic motivation by introjected regulation: distinguishes behaviors dictated by the expectations of the subjects' social environment,
– extrinsic motivation by external regulation: corresponds to behaviors influenced by the search for rewards or the avoidance of sanctions of sanctions,
– amotivation: refers to individuals who are unable to amotivation refers to individuals who are unable to explain the origin of their behavior.

The evaluation of the collected results illustrates the way in which both the conditions imposed by COVID-19 and the changeover to the virtual classroom have affected commitment to their studies. The contribution should contain no more than four levels of headings. Table 1 gives a summary of all heading levels.

The analysis of the results presented in Table 2, highlights the prevalence of the identity motive (30%) (development of academic and professional skills), over the self-determined motive (20%) (the pleasure of acquiring new knowledge). The instrumental motive related to the work methods used by the professors and the guiltless motive are also highlighted with identical rates of 14.3% each. While the rate of students who are in the Amotivation level is very important 34.3%. 30% of the learners are of course interested in the professional dimension of the skills targeted by their training. Thus, the students who considered this motif were the older ones, i.e., the students of the Bachelor's and Master's 1 levels, who already constitute 80% of the surveyed population (see Fig. 3) and who are, perhaps, more mature in terms of their vocational orientation and choice of professional project.

Table 2. Motif of commitment.

Motifs	Rate
Intrinsic motivation	
I was involved because I find pleasure in learning new knowledge and overcoming difficulties	20%
Motif of identity	
Professional Motif: I was involved because it allows me to develop my academic and professional skills	31.4%
Extrinsic motivation by identified regulation	
I was involved because the methods used by teachers kept me interested	14.3%
I was involved because the learning activities were attractive and diverse	5%
I was involved because I like to get feedback from my teachers and bonus points	5%
I was involved because the learning environment (platforms and tools) are attractive	5%
Extrinsic motivation by introjected regulation	
Not being involved would make me feel guilty towards those around me	14.3%
Extrinsic motivation by external regulation	–
Amotivation	
I don't have an opinion because this period was very disturbed	11.4%
I don't feel like I was involved because I didn't like this way of learning	22.9%

The feeling of acting in a self-determined way is one of the most fundamental psychological needs in the causal explanation of human behavior. A change or jostling of the factors (both internal and external) of the learning conditions (perception, activity, interpersonal relationship, place, person, situation, autonomy, obligation, etc.) are likely to have a negative impact on students' self-determined motivation. The fact that students expressed little sense of self-determination thus testifies to the fact that the self-determined motivational profile of learners was negatively impacted.

6   Recommendations

The switching to distance education has permitted to rescue the Moroccan educational system during the confinement insofar as it was a solution not to interrupt the studies. However, this experience has shown that the structures put in place to ensure this transition must offer the same services and advice as in the face-to-face learning environment while considering the mental health and well-being as well as the motivational determinants of students.

This section will contribute to the formulation of a list of recommendations to address and apprehend the expectations of students in such an educational environment and learning situation.

The Mural.co tool was adopted and used for this purpose. It is quick and easy to use visual online collaboration tool that allows teams to think and collaborate together to find innovative solutions to the most complex problems. Users benefit from Mural to create diagrams, which are popular in design thinking and agile methodologies.

For this study, the Lightning Decision Jam (LDJ) model was chosen. It is an animation technique that can be used in all situations that require a group of people (between 2 and 8 people per workshop) to make decisions: to solve problems, discuss challenges or implement quick actions. Indeed, LDJ exploits collective intelligence and allows defining concrete and viable actions, for relevant problems, in record time. The Lightning Decision Jam (LDJ) method was used as a working tool, considering the IGA as a case study [1] (see Fig. 18).

Fig. 18. Lightning Decision Jam (LDJ) [1].

It is imagined by AJ & Smart, a Berlin Sprint Design agency, it takes place in several stages and allows:

– Identify and prioritize the problems to be treated;
– Bring out a maximum number of solutions and define the most relevant ones (with more impact and minimum effort);
– Draw up an action plan to be implemented in the short term.

Below are the steps of the workshop process.

The workshop begins with what's working: In this step, the different participants use sticky notes to write down all possible ideas related to the theme. Here, the participants list the existing elements that can help to reduce the impact of this abrupt changeover on the teaching and learning process of the students. The Fig. 19, shows the first ideas generated by the group of teachers participating in the workshop.

Then, the participants individually capture all the problems (Regardless of the degree of their importance or their priorities). During this phase, a series of difficulties encountered by teachers during this rapid transition to distancing were identified.

The Fig. 20 shows a summary of a set of issues affected by teachers during this period of continuity of distance education courses with their votes.

For 3 min participants vote on the most relevant issues to be prioritized for treatment. The result ranked from least to highest priority is as follows:

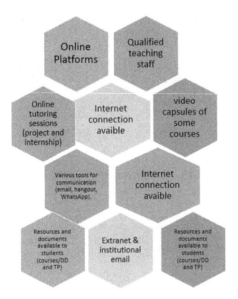

Fig. 19. Ideas working.

Fig. 20. Problems and annoyances with distance learning.

This study has also highlighted a number of interesting findings related to the negative effects of the COVID-19 pandemic and the shift towards distance education on students' lives. For the majority of the participants in the survey, the current conditions have strongly impacted the follow-up of courses and organization habits, the efforts to be made to adapt, their motivation and commitment to learn, and have strongly favored the feeling of anxiety, stress and uncertainty about their study, professional future and their integration into the job market. The next phase consists of rewording the problems in the form of challenges. Four main challenges are taken up (Figs. 21 and 22):

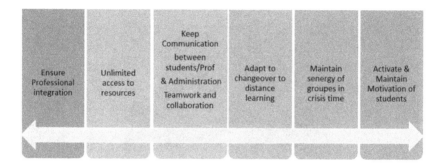

Fig. 21. Prioritizing the tasks to be done.

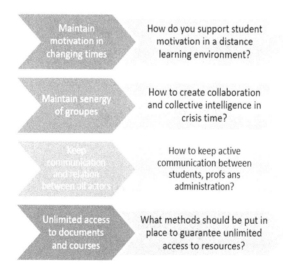

Fig. 22. Rewording the problems in the form of challenges.

Faced with this context, Moroccan higher education institutions find themselves obliged to act to remedy and/or mitigate the negative effects of such a situation on the operation of teaching and learning by adopting a participatory and mobilizing strategy of all actors: Teachers, Students and Administrators, to respond to a massive societal demand and its challenge: how to continue to share knowledge and help learners to be motivate and remain in the dynamics of knowledge building?

In regard to these elements, some proposals for the creation of a student experience adapted to the context of COVID-19 coupled with the digital technologies will be presented.

The step entitled "Generation and generation of ideas/solutions", provides participants (professors) in the workshop to propose a number of ideas and solutions. The results of this step are shown in the Fig. 23.

Fig. 23. Generation of ideas and propositions.

Here, the ideas and solutions proposed by the group of teachers participating in the workshop will be prioritized as illustrated in the Fig. 23 (Fig. 24).

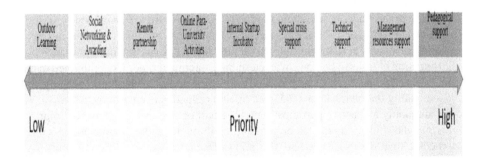

Fig. 24. Prioritize solutions.

This crisis will constitute an opportunity for reflection about solutions to be implemented in the short, medium and long term. To this effect, a number of solutions to be implemented were listed, considering the degree of impact and efforts required for their applicability, taking into account the student as a central actor in the process of learning.

The last step concerns deciding what to execute based on the effort/impact. Solutions are classified on four categories: what to do now, what to make as a task, what to make as a project and what to delay later.

In order to act more efficiently, it is more judicious to start by implementing actions that will anticipate new similar periods of crisis, and to re-think the traditional mode of learning.

Based on the results highlighted in the previous work [1], the contributions of the literature review and the ideas generated by the workshop group of the LDJ workshop, a list of proposals and recommendations was drawn up as follows:

- Special crisis support: This situation may create pressure as well as an anxiety in a large number of individuals, which may affect quality of students' motivation [51]. Hence, educative interventions in crisis situations must be accompanied by coaching or support sessions in the management of crises. The objective is to allow both students and teachers to develop competences that enable them to confront such a crisis situation and to reduce feelings of isolation and stress. Furthermore, school and teacher support can help to motivate the students, despite their anxiety [52];
- Mentoring and accompanying professors and students to speak the same ICT language: Following the COVID-19 crisis, universities are faced with the challenge of combining ICT with university pedagogy in order to consolidate the relationship between the teacher and his or her students [53]. To this end, it is necessary to ensure that professors, who are part of the "generation of digital immigrants" (people born before the 90's) [54], are possessing adequate digital competence in order to satisfy the needs of their students qualified as "digital natives", and to remediate the digital divide that has been observed between generations. In other words, the more teachers and students speak the same language in terms of ICT, the better the academic exchange between the two parties will be;
- Management resources support: One of the difficulties observed in this new experience is that the students are bombarded with digital resources. Consequently, a strategy for managing and sharing resources in an efficient way has to be put in place, sharing resources at the right time so that the learner is not lost in an avalanche of information or tasks to be done in a determined amount of time. As highlighted in the literature, learners need to be provided with open-access, consistent, engaging and good quality online resources [55–57]. Such a pedagogical environment that allows for autonomy and active participation on the part of students. Such an environment permits the student to plan his or her learning through a more flexible and less busy schedule than the usual one and also allows free access to the different courses in asynchronous mode at any time;
- Pedagogical support: Indeed, this period, questioning the practices of pedagogical teams and students, should lead to further reflections on new pedagogical devices that will eventually modify the "classic" organization of training and teaching methods [58]. If the technological and digital evolution has allowed the creation of virtual spaces of exchanges capable of enhancing the students' learning, it is also judicious to implement a rigorous integration of those technologies into teaching processes in

order to allow students to optimize the learning task in an autonomous and organized manner. In fact, specialists affirm that the inverted classroom provides a learning context, focused on more learner-centered pedagogical devices with active and collaborative learning approaches. The aim is to seek possible balances between teaching and learning, between face-to-face and distance work, between information and knowledge, between knowledge and skills [59]. Identifying the motivational profile of students during this crisis period highlighted the strong vocational motive related to the acquisition of academic and professional competences of their motives for engaging in studies. Therefore, it is the role of the professors to include this dimension in the pedagogical reflection associated with the construction of learning sequences. In this context, the use of active teaching and learning methods, based on project and trial-and-error pedagogy are appropriate.

- Provide Technical Support to Students and Professors: Scientific research [60] states that the use of digital tools can have unpleasant side effects, leading to negative emotional states such as frustration, confusion, anger or anxiety, while other researchers have defined a state of mind of fear or apprehension among teachers in the face of the constant use of digital resources in their teaching practice [61, 62] that has been called "technological anxiety". As for digital natives, they are certainly supposed to be digitally literate, but studies have found that these students are more comfortable using mobile devices than computers. Furthermore, Stockwell and Hubbard [36] highlighted that this "knowledge of how to use mobile devices for specific personal or social functions is not always a good indicator of knowledge of educational functions". In this regard, training and technical support is always recommended, both for digital natives, as well as for professors in order to have a positive impact on the quality of online distance teaching and learning;

- A range of communication tools: If in the normal case, studying at a distance, in a virtual way, can generate in the person the feeling of being distant and not very affiliated with his peers of the class group, what can we say in the case of confinement where each of us has to be forced to live for over three months under the constraint of social distancing and isolation at home! Indeed, the course should be designed to promote the maximum exchange of information and ideas, the confrontation of opinions and points of view and the organization of a mutual aid network. To do this, a wide range of communication means (email, instant messaging, chat, forum, discussion group, etc.) must be set up and must be accessible at all times, promoting teamwork and frequent feedback between the professor and his students. Thus, the immediate feedback provided to learners is aimed at supporting motivation through mechanisms similar to those identified by Malone and Lepper [63].

7 Conclusion

The last two years have been marked by the spread of the COVID19 pandemic which has had impacts on the majority of vital sectors (in Morocco as in other countries of the world). Especially, the higher education sector has been affected by the pandemic in terms of mental health, maintaining strong relationships between professors and students, providing courses and practical work 100% remotely, helping students to find a job, and so on.

Certainly, the switching to distance learning has saved the school and university year 2019–2020 of the Moroccan educational system during the confinement insofar as it was a solution to prevent the COVID-19 to spread. Significant efforts have been made by higher education organizations and actors to replace face-to-face classes and ensure pedagogical continuity.

However, this experience has revealed that the structures and platforms put in place to ensure this changeover suffer from a number of technical and pedagogical problems and do not, unfortunately, offer the same conditions and services as in the classroom, while considering the mental health and well-being as well as the motivational determinants of the students.

Indeed, in this new context of crisis, several challenges have emerged: How to deal with the negative impact of COVID-19 pandemic on the education sector? How to attenuate the influence of COVID on students' lives? What are the best ways to increase student motivation and engagement during the pandemic period?

This article discusses issues related to the impact of COVID 19 pandemic on the higher education sector in particularly on the motivation and commitment of students in their studies during the period of confinement. The results of the study developed on the evaluation of the experience of distance learning by students in the Moroccan higher education sector at the time of Covid-19 show that the educational system was facing a great challenge and that it is still and should be able to do so, but the effort required is collective.

This return on experience highlights that distance learning is much more exhausting than face-to-face learning, both physically and mentally. The statistics confirm that most of the students suffered from stress, anxiety and uncertainty about their academic and professional future, and as a consequence a decrease in motivation and commitment to their studies during the period of confinement, which confirms the hypothesis assumed at the beginning.

From these elements, a series of recommendations to improve the quality of the said device have been proposed, because distance learning will constitute, henceforth, a reality to be admitted and an essential lever for the change and the success of Moroccan higher education.

Finally, it is necessary to remain optimistic because at times of crises, lessons have to be learned and many opportunities to be exploited, as the philosopher Edgar Morin [64] said: "In the unknown, everything progresses by trials and errors as well as by deviant innovations first misunderstood and rejected ....". So, we join the voices of most researchers that confirmed that universities need to reinvent their learning environments to remain relevant. And as perspectives and future work, we are working on the conception of a framework that can help the development of competences in times of crisis.

References

1. Akhasbi, H., Belghini, N., Riyami, B., Benitto, M., Gouttaya, N.: Moroccan higher education at the time of Covid-19: issues and challenges: a case study among master students business administration at IGA casablanca. In: Proceedings of the 13th International Conference on Computer Supported Education. 2: CSEDU, pp. 73–85 (2021). https://doi.org/10.5220/0010438700730085. ISBN 978-989-758-502-9; ISSN 2184-5026

2. Liu, W.C., Wang, D.K.J., Kee, Y.H., Koh, C., Lim, B.S.C., Chua, L.: College students motivation and learning strategies profiles and academic achievement: a self-determination theory approach. Educ. Psychol. **34**, 338–353 (2014)
3. World Health Organization: statement on the second meeting of the international health regulations (2005) emergency committee regarding the outbreak of novel coronavirus (2019-nCoV) (2020)
4. Cheng, C., Jun, H., Liang, B.: Psychological health diathesis assessment system: a nationwide survey of resilient trait scale for Chinese adults. Stud. Psychol. Behav. **12**, 735–742 (2014)
5. Sareen, J., et al.: Risk factors for post-injury mental health problems. Depress. Anxiety **30**(4), 321–327 (2013). https://doi.org/10.1002/da.22077
6. Hildesheimer, F.: Entretien. Récupéré du site du quotidien ≪Le Monde≫
7. Marin, C.: Entretien Récupéré du site du quotidien ≪e Monde≫ (2020). https://www.lemonde.fr/
8. UNESCO: COVID-19: une crise mondiale pour l'enseignement et l'apprentissage. https://unesdoc.unesco.org/ark:/48223/f0000373233_freUniversiteLaval,Quelssontlesavantagesdela formationadistance?, https://www.enseigner.ulaval.ca/ressources-pedagogiques/introduction-la-formation-distance. Accessed 17 May 2020
9. Devinney, T., Dowley, G.: Is this the crisis higher education needs to have. Times High. Educ. (2020)
10. Béland, S., Bureau, J. S., Peters, M.: Plagier en temps de pandémie. e-JIREF (1), 35–40 (2020)
11. Wang, C., et al.: Immediate psychological responses and associated factors during the initial stage of the 2019 coronavirus disease (COVID-19) epidemic among the general population in China. Int. J. Environ. Res. Public Health **17**(5), 1729 (2020)
12. Aissaoui, Y., Aissaoui, A.: Brève réflexion sur l'atténuation des impacts économiques et sociaux de la pandémie du (Covid-19). Rev. d'Etudes Manag. Financ. d'Organ. [S.l.] **4**(11) (2020). ISSN 2489-205X
13. CDC, national center for health statistics. : indicators of anxiety or depression based on reported frequency of symptoms during the last 7 days. household pulse survey. US Department of Health and Human Services, CDC, National Center for Health Statistics, Atlanta (2020)
14. CDC, National center for health statistics: early release of selected mental health estimates based on data from the January-June 2019 national health interview survey. US Department of Health and Human Services, CDC, National Center for Health Statistics, Atlanta (2020)
15. Akhasbi, H., Belghini, N., Riyami, B.: Self-determined learner motivation evaluation in distance learning context using web 2.0 tools, during the COVID-19 pandemic period. In: INTED 2021 Proceedings, pp. 2143–2151 (2021)
16. Villiot-Leclercq, E.: L'ingénierie pédagogique au temps de la Covid-19. Distance Mediat. Knowl. (30) (2020)
17. Rechidi, N., et al.: L'intégration pédagogique des TIC à l'épreuve de la crise COVID-19: quels enseignements à tirer?? Revue Int. Chercheur **1**(2), 274–297 (2020)
18. Pekmez, E.S., Johnson, P., Gott, R.: Teachers' understanding of the nature and purpose of practical work. Res. Sci. Technol. Educ. **23**(1), 3–23 (2005)
19. Abrahams, I., Millar, R.: Does practical work really work? A study of the effectiveness of practical work as a teaching and learning method in school science. Int. J. Sci. Educ. **30**(14), 1945–1969 (2008)
20. Abrahams, I., Reiss, M.J., Sharpe, R.M.: The assessment of practical work in school science. Stud. Sci. Educ. **49**(2), 209–251 (2013)
21. Vincens, J.: Définir l'expérience professionnelle. Travail Emploi. (85), 21–34, 83 (2001). ISSN 0224-4365

22. Medias24.com: Crise du Covid : Les 2/3 des recrutements et stages ont été reportés ou sus-pendus (2020). https://www.medias24.com
23. Berthaud, J.: L'intégration sociale étudiante: relations et effets au sein des parcours de réussite en Licence. Education. Université Bourgogne Franche-Comté (2017)
24. Khamassi, M., Decremps, F.: Apprentissage de la demarche scientifique et de l'esprit cri-tique: un enseignement de sorbonne universite pour les etudiants d'aujourd'hui, citoyens de demain (2019)
25. Flores-Lueg, C., Roig Vila, R.: Diseño: y validación de una escala de autoevaluación de competencias digitales para estudiantes de pedagogía. Pixel Bit Rev. Medios Educ. (12), 209–224 (2016)
26. Ocaña-Fernández, Y., Valenzuela-Fernández, L., Morillo-Flores, J.: La competencia digital en el docente universitario. Propósitos Represent **8**(1), e455 (2020)
27. Mimirinis, M.: Qualitative differences in academics' conceptions of e-assessment. Assess. Eval. High. Educ. **44**, 233–248 (2018)
28. Troksa, L.M.: The study of generations: a timeless notion within a contemporary context. University of Colorado Boulder (2016)
29. Twenge, J.M., Campbell, S.M., Hoffman, B.J., Lance, C.E.: Generational differences in work values: leisure and extrinsic values increasing, social and intrinsic values decreasing. J. Manag. **36**(5), 1117–1142 (2010)
30. Bencsik, A., Machova, R.: Knowledge sharing problems from the viewpoint of intergenera-tion management. In: ICMLG 2016–4th International Conference on Management, Leader-ship and Governance: ICMLG2016, no. 42 (2016)
31. Steinerowska-Streb, I., Wziątek-Staśko, A.: Effective motivation of multi-generation teams-presentation of own research results. In: The Proceeding of the Management International Conference, Pula, Croatia (2016)
32. Elmore, T.: How generation Z differs from generation Y, July 2015
33. Grigoryan, T.: Investigating digital native female learners' attitudes towards paperless lan-guage learning. Res. Learn. Technol. **26**, 1–27 (2018). https://doi.org/10.25304/rlt.v26.1937
34. Stockwell, G., Liu, Y.C.: Engaging in mobile phone-based activities for learning vocabulary: an investigation in Japan and Taiwan. Calico J. **32**(2), 299–322 (2015). https://doi.org/10.1558/cj.v32i2.25000
35. Thinyane, H.: Are digital natives a world-wide phenomenon? An investigation into South African first year students' use and experience with technology. Comput. Educ. **55**(1), 406–414 (2010). https://doi.org/10.1016/j.compedu.2010.02.005
36. Stockwell, G., Hubbard, P.: Some emerging principles for mobile-assisted language learning. The International Research Foundation for English Language Education (2013). http://www.tirfonline.org/english-in-the-workforce/mobile-assisted-language-learning
37. Brown, C., Czerniewicz, L.: Debunking the 'digital native': beyond digital apartheid, towards digital democracy. J. Comput. Assist. Learn. **26**(5), 357–369 (2010). https://doi.org/10.1111/j.1365-2729.2010.00369.x
38. Karsenti, T.: Comment le recours aux TIC en pédagogie universitaire peut favoriser la moti-vation des étudiants: le cas d'un cours médiatisé sur le Web. Cahiers Recherche Éduc. **4**(3), 455–484 (1997)
39. Maehr, M.L.: Meaning and motivation: toward a theory of personal investment. In Ames, C., Ames, R. (eds.) Research on Motivation in Education, vol. 1, pp. 115–144. University of Maryland, Baltimore (1984)
40. Vallerand, R.J., Sénécal, C.B.: Une analyse motivationnelle de l'abandon des études. Appren-tissage Soc. **15**(1), 49–62 (1993)
41. Deci, E.L., Ryan, R.M.: A motivational approach to self: integration in personality. In: Dientsbier, R.A. (ed.) Perspectives on Motivation. Nebraska Symposium on Motivation, pp. 237–288. University of Nebraska Press, Lincoln (1991)

42. Deci, E.L., Ryan, R.M.: Intrinsic Motivation and Self-determination in Human Behavior. Plenum, New York (1985)
43. Vallerand, R.J.: La motivation intrinsèque et extrinsèque en contexte naturel: implications pour les secteurs de l'éducation, du travail, des relations interpersonnelles et des loisirs. In: Vallerand, R.J., Thill, E.E. (eds.) Introduction à la Psychologie de la Motivation, pp. 533–581. Éditions Études vivantes, Montréal (1993)
44. Shin, M.: Promoting students' self regulation ability: guidelines for instructional design. Educ. Technol. **38**(1), 38–44 (1998)
45. Christoph, R., Schoenfeld, G.A., Tansky, J.W.: Overcoming barriers to training utilizing technology: the influence of self-efficacy factors on multimedia-based training receptiveness. Hum. Resour. Dev. Q. **9**(1), 25–38 (1998)
46. Gao, J., et al.: Mental health problems and social media exposure during COVID-19 outbreak. Plos One **15**(4), e0231924 (2020)
47. Zarocostas, J. : How to fight an infodemic. Lancet **395**(10225), 676 (2020). World Health Organisation. Infodemic management - Infodemiology. Ad-hoc technical consultation on managing the COVID-19 infodemic. https://www.who.int/teams/risk-communication/infodemic-management
48. World Health Organisation. Non-pharmaceutical public health measures for mitigating the risk and impact of 881 epidemic and pandemic influenza. 882 (2019). https://www.who.int/influenza/publications/public_health_measures/publication/en/
49. Talya, P., Nyrup, R., Calvo, R.A., Paudyal, P., Ford, E.: Public health and risk communication during COVID-19-enhancing psychological needs to promote sustainable behaviour change (2020)
50. Pelaccia, T., Delplancq, H., Triby, E., Leman, C., Bartier, J.-C., Dupeyron, J.-P.: La motivation en formation: une dimension réhabilitée dans un environnement d'apprentissage en mutation. Pédag. Méd. **9**, 103–121 (2008)
51. Liu, W.C., Wang, C.K.J., Kee, Y.H., Koh, C., Lim, B.S.C., Chua, L.: College students' motivation and learning strategies profiles and academic achievement: a self-determination theory approach. Educ. Psychol. **34**(3), 338–353 (2014)
52. Elmelid, A., Stickley, A., Lindblad, F., Schwab-Stone, M., Henrich, C.C., Ruchkin, V.: Depressive symptoms, anxiety and academic motivation in youth: do schools and families make a difference? J. Adolesc. **45**, 174–182 (2015)
53. Paivandi, S.: L'appréciation de l'environnement d'études et la manière d'étudier des étudiants. Mesure Éval. Éduc. **35**(3), 145–173 (2012). https://doi.org/10.7202/1024673a
54. Marjane, D.: Les technologies de l'information et de la communication et l'enseignement des langues: les défis et les opportunités pour le Tamazight. Langues Linguist. **42**(43), 35–45 (2019)
55. Bishop, J.L., Verleger, M.A.: The flipped classroom: a survey of the research. In: ASEE National Conference Proceedings, Atlanta, GA, vol. 30, no. 9, pp. 1–18 (2013)
56. Enfield, J.: Looking at the impact of the flipped classroom model of instruction on undergraduate multimedia students, CSUN. TechTrends **57**(6), 14–27 (2013)
57. Sales, N.: Flipping the classroom: revolutionising legal research training. Legal Inform. Manag. **13**(04), 231–235 (2013). https://doi.org/10.1017/S1472669613000534
58. Zorn, C., Feffer, M.L., Bauer, É., Dillenseger, J.P.: Évaluation d'un dispositif de continuité pédagogique à distance mis en place auprès d'étudiants MERM pendant le confinement sanitaire lié au COVID-19. J. Med. Imaging Radiat. Sci. **51**(4), 645–653 (2020)
59. Lebrun, M.: Les classes inversées: intégration d'idées pédagogiques anciennes pour une réelle innovation. Blog de Marcel (2017). http://lebrunremy.be/WordPress/. Accessed 13 July 2017
60. Saade, R., Kira, D.: The emotional state of technology acceptance. Issues Inf. Sci. Inf. Technol. **3**, 1–11 (2006)

61. Celik, V., Yesilyurt, E.: Attitudes to technology, perceived computer self-efficacy and computer anxiety as predictors of computer supported education. Comput. Educ. **60**, 148–158 (2013)
62. Rogers, R.K., Wallace, J.D.: Predictors of technology integration in education: a study of anxiety and innovativeness in teacher preparation. J. Lit. Technol. **12**, 28–60 (2011)
63. Malone, T., Lepper, M.: Making learning fun: a taxonomy of intrinsic motivations for learning. In: Snow, R.E., Farr, M.J. (eds.) Aptitude, Learning, and Instruction, 3, Conative and Affective Process Analyses, pp. 223–253 (1987)
64. Morin, E.: Le Monde. https://www.lemonde.fr. Accessed 19 Apr 2020

Efficient and Accurate Closed-Domain and Open-Domain Long-Form Question Answering

Rhys Sean Butler, Vishnu Dutt Duggirala$^{(\boxtimes)}$, and Farnoush Banaei-Kashani

University of Colorado, Denver, CO 80204, USA
{rhys.butler,vishnudutt.duggirala,
farnoush.banaei-kashani}@ucdenver.edu
https://cse.ucdenver.edu/~bdlab/

Abstract. We present an efficient and accurate long-form question answering platform, dubbed *iLFQA* (i.e., short for *intelligent Long-Form Question Answering*). iLFQA was created as a continuation of *iTA* (i.e., short for *intelligent Teaching Assistant*). iTA was originally designed as a narrow domain question answering platform that performed generative question answering with a single textbook as a reference. The core purpose of iTA was expanded into iLFQA as we attempted to expand the narrow domain of iTA into an open-domain question answering system. iLFQA functions as a platform that accepts unscripted questions and efficiently produces semantically meaningful, explanatory, and accurate long-form responses. iLFQA uses classification tools as well as Transformer-based text generation modules, and is unique in the question answering space because it is an example of a *deployable and efficient long-form question answering* system. Question Answering systems exist in many forms, but long-form question answering remains relatively unexplored. The source code for both iLFQA and iTA are freely available for the benefit of researchers and practitioners in this field.

Keywords: Long-form question answering · Generalized language models · Text retrieval · Text generation · Natural language processing

1 Introduction

Question answering is a highly studied problem in the domain of natural language processing. However, long-form question answering is only recently coming into focus as a topic of research [12]. Most question answering techniques focus on returning factoid-type answers no longer than a sentence. iTA *iTA* (i.e., short for *intelligent Teaching Assistant*) and iLFQA *iLFQA* (i.e., short for *intelligent Long-Form Question Answering*) are different from other question answering systems because they provide multi-sentence responses to the given questions. Originally iTA was designed as a digital teaching assistant capable of retrieving answers from a single textbook and generating detailed answers. iTA [8] was further expanded into iLFQA to increase efficiency and

© Springer Nature Switzerland AG 2022
B. Csapó and J. Uhomoibhi (Eds.): CSEDU 2021, CCIS 1624, pp. 165–188, 2022.
https://doi.org/10.1007/978-3-031-14756-2_9

accuracy to provide a wide knowledge domain from which to produce answers. Currently, there are only a few other examples of long-form question answering systems available [9, 12].

The application of question answering platforms is wide-ranging. To implement question answering systems, several different methods of machine learning, text retrieval, text generation, and topic classification are used. To generate accurate and meaningful responses, a question answering system must be able to understand each given question and retrieve the answer from unstructured text. There are many techniques to retrieve appropriate documents including calculating Term frequency-inverse document frequency as well as the use of machine learning techniques and architectures that retrieve answers from the source material. To develop the machine learning models which perform text generation, it is necessary to have large data sets with long-form answers that can be used to train models on similar problems.

Question Answering systems could be used to help customers find answers to difficult questions quickly and effectively. The question answering systems could serve as digital teaching assistants, only allowed to answer certain questions, but available at all times. Most question answering systems that currently exist are constrained to a specific domain. The question answering systems can also serve as chatbots that perform support tasks, but an effective one could quickly ascertain when a question might require immediate escalation.

We have to overcome two barriers in order to construct such capable systems that provide thorough responses. The first obstacle is that most Machine Reading Comprehension (MRC) solutions are designed to assess understanding from a passage of little more than 500 words. The second aim was to create a model that offers detailed, easy-to-understand results. The majority of question answering datasets produce extractive and brief replies.

Some of the question answering systems that currently exist do not publish the results of their run-time or efficiency. Therefore, it is not clear whether any of the existing work is deployable. iLFQA has been expanded into a system that performs open-domain and open-source long-from question answering with accuracy, comparably accurate to other existing systems. iTA and iLFQA were developed and evaluated on a set of volunteer-created questions. The volunteers formulated questions based on materials in the text and chose sections that offered the text's replies. All accuracy scores are based on a comparison between the generated answer and the volunteer selected context. To develop iLFQA and iTA, we have taken advantage of different machine learning architectures, including Long-Short Term Memory, Transformers [26], and T5 [22]. We hope that iLFQA serves as an open-source platform that researchers and practitioners in the field of long-form question answering can build on. A brief video demonstration of the iLFQA system is made available for public view[1].

2   Related Work

Long-form question answering can be defined as the task of producing answers by retrieving content from appropriate documents and accordingly generating a multi-

---

[1] https://youtu.be/4rzWrEZWq2E.

sentence response. Generative question answering refers to a process where a response is dynamically generated. This is different than retrieving a span of text for a given question. Knowledge grounded chatbots can be defined as any chatbot that incorporates knowledge from a domain or set of domains when interacting with a user. Generative question answering refers to a process where a response is dynamically generated. This is different than retrieving a span of text for a given question. Knowledge grounded chatbots can be defined as any chatbot that incorporates knowledge from a domain or set of domains when interacting with a user. Few question answering systems use recurrent neural networks. Recurrent neural networks (RNNs) are effective for generating semantically accurate and meaningful responses, some RNNs are auto-regressive. That is to say that each output token in a sequence of text makes up the input for the next token. So, given a long sequence of text to process as input, or to generate as output, each word must be processed one at a time. This makes interpretation and generation of long passages difficult. While there are many existing question answering systems, few long-form question answering systems have been introduced. A relatively new recurrent neural network architecture, namely, Transformer [26], has been introduced that makes producing long-form responses easier.

There has been previous work that focuses on generative question answering [2,4,25], knowledge grounded chatbots [11], and ambiguous question answering [16]. Generative question answering refers to a process where a response is dynamically generated. This is different than retrieving a span of text for a given question. Knowledge grounded chatbots can be defined as any chatbot that incorporates knowledge from a domain or set of domains when interacting with a user. Ambiguous question answering can be defined as answering questions that may contain multiple plausible answers.

One reason for the rarity of long-form question answering systems is a lack of resources necessary to train appropriate models [12]. There are several open-domain sources of data that can be used to produce open-domain question answering platforms; however, ELI5 [9] remains the only appropriate data source to train text-generation models to produce long-form responses. The lack of appropriate data sets remains a critical bottleneck for more advanced research into the field of long-form question answering. The most comprehensive comparable work in long-form question answering to date is a paper produced by a group of Google researches titled, "Hurdles to Progress in Long-form Question Answering" [12]. Their paper provides impressive results on the Kilt leader-board but they fail to provide code, run time performance details, or a deployable version of their system. In this paper, we introduce a deployable and open-source platform for open-domain, long-form question answering trained based on a ELI5-like Wikipedia dataset.

2.1 Question Answering Systems

One project that showcases the capabilities of new technologies for use in the question answering domain is called AmbigQA [16]. In their paper, "AmbigQA: Answering Ambiguous Open-domain Questions", the authors explain that one of the difficulties in generative question answering is that answering dynamically generated questions requires knowledge not necessarily gained from previous user inputs. For instance, if I ask how many movies are in a particular franchise, the system must understand the context of the question, then seek the correct result from a series of documents. The authors of AmbigQA create a database that can be used to train models to disambiguate similar questions within the same passage.

To perform the necessary tasks on the AmbigQA dataset, a model first selects spans of text from its document base (Wikipedia passages). It then ranks those passages and concatenates the most promising ones together up until it has reached a total of 1024 tokens. At that point a sequence to sequence model is used to generated the answer from the concatenated spans. The model then generates a series of questions that are paraphrased from the original input. It tries to match the best spans to the whole series of generated questions and the original to find the most likely answer.

Another system, StockQA [25] attempts to answer questions in the domain of stock trading. StockQA accepts open ended questions about stocks and attempts to return answers based on those questions that could have been generated by professional stock analysts [25]. What makes StockQA so interesting is that stock value and advice generally vary by the moment. A stock that would be recommended for purchase today may be recommended for sale tomorrow. StockQA is of note because it must analyze the current market before returning most answers. Operating in a dynamic environment where the data is constantly changing is something that few question answering systems can claim to have tackled in the same way. Some questions posed to StockQA may require calculations to be performed based on information contained within the database.

As noted earlier, StockQA may generate two different answers to the same question depending on when the question is asked. The knowledge corpus that StockQA uses to generate answers is organized into feature, value pairs [25]. So for any given stock there are a number of feature value pairs that are used to generate answers to a given question. Most of the questions that are asked are questions that will ask for a prediction on a stock or an action recommendation [25] such as to buy, sell, hold etc. The knowledge base for training StockQA consisted of question answer pairs generated from online Chinses stock forums where users could ask questions for free or a very low price. The authors limited the questions selected from the forum by keeping only questions that involve a single stock. Any questions that involved multiple stocks were removed from the training/testing data. All question answer pairs selected involve the same stock and any question answer pairs that had answers that referenced stocks not in the original question were removed [25]. StockQA generates answers by using a memory-augmented encoder-decoder architecture. The StockQA encoder transforms each question into a vector representation. The memory stores the dataset concerning all the stock information as an array of key-value pairs. The decoder uses the information stored in memory, along with the vector representation of the question in order to generate the text that is returned as an answer.

Eli5 [9] is a dataset composed of threads from a subreddit named "Explain Like I'm Five". The ELI5 dataset is the product of authors that included members of Google and Facebook. Explain Like I'm Five is a forum where users can ask questions. Answers to user questions must be understandable by a five year old. This makes ELI5 unique because most of the answers require multi sentence explanations. The questions and answers that were selected from ELI5 all have more upvotes than down votes. For threads that are selected with multiple answers, the highest voted answer is chosen.

This yields 272,000 question and answer pairs. Most answers consist of multi-sentence responses for their given questions.

For each question ELI5 contains a support document, which is text extracted from the web, and likely to contain information that can help answer the question. The support documents are included in the dataset for each question. The text chosen for the support documents are pages from Common Crawl that have the highest TF-IDF similarity to the question. Questions in the validation and test sets of ELI5 are chosen such that they are not similar to questions in the training set.

The creators of ELI5 note that their dataset is one of the first of it's kind in terms of Long-Form question answer pairs. The authors use ELI5 to train multiple Natural Language Processing machine learning models that include, text generation, sequence to sequence, extractive, and multi task models. ELI5 had human evaluations based on three assessments. The human evaluators assess the question answer pairs for fluency (readability). The evaluators make a determination on if the ELI5 answer is correct, they also indicate if they prefer the ELI5 answer or a human generated answer for the same question.

Overall the human evaluators heavily preferred the human generated questions. The creators of ELI5 freely acknowledge that there is room for improvement on this particular dataset. ELI5 remains a highly functional and effective dataset for training machine learning models on long form extraction, text generation, and sequence to sequence tasks.

Researchers at google have recently created a system that is designed specifically to perform the task of Long-Form Question Answering. The authors of Hurdles to Progress in Long-form Question Answering [12], design a Long-Form Question Answering system that is trained with the ELI5 dataset, which also takes advantage of state of the art document retrieval techniques. Currently the system made by the Google Researchers ranks 3rd on the KILT leaderboard [19]. KILT is a benchmark measure used by entities that want to validate knowledge intensive tasks.

The authors of the Google paper uses a dense retriever by the name of contrastive REALM [12]. REALM works by comparing questions to documents in a knowledge corpus made up of information from Wikipedia. Questions are compared to the documents in a 128 dimension embedding space. The documents most similar in content to the question. A vectorized representation of the question is pushed further away from possible answers, and pushed closer to the vector representation of the ground truth answer.

The text generation model used by the Google researches is actually a transformer based model named Routing Transformer. The Routing Transformer used was trained on the PG-19 dataset [21]. PG-19 contains a series of text documents that are much longer than the average reddit response and Wikipedia article. It was created to serve as a benchmark for text generation that can help train and evaluate a model's ability to generate Long-Form responses.

The system was evaluated on the KILT leaderboard mentioned earlier and on a subset of the ELI5 dataset. The ELI5 dataset does not have human annotations that measure or designate "gold answers". So this makes evaluation of generated answers more difficult than it might be on the KILT leaderboard.

One of the most interesting findings in the Google paper is the discovery that their model doesn't utilize the retrieved documents in the process of text generation as originally thought. The authors state that the generated text is similar regardless of the documents that are retrieved for any question. The authors performed experiments where retrieved documents were replaced with random documents for the purpose of text generation.

Despite the fact that the generated text may not be grounded in retrievals, the Google system performs quite well compared to other similar systems. The authors believe this is due to the fact that the Text Generation model was actually trained on next-word prediction on long sequences of text. Normally BART [14] is trained by filling in masked tokens. The google authors believe that by training their model for next-word prediction on long sequences, the model was made to sound more fluent.

2.2 Question Answering Applications

Few of the applications that long-form question-answering focuses on any open-domain questions that may require explanatory answers, particularly those that begin with How. Hugging Face has developed a conversational application that focuses on long-form answers for open-domain questions. They have used wiki40b [10], which are disjointed into 100 words passages. They use elastic search as their information retrieval system, which provides a convenient way to index the documents; it uses the nearest neighbor search using BM25 similarity. They made the demo available for both the retrieval and generative models. Microsoft Canada Research team has developed a long document MRC, they have developed R-Net [28] which uses gated attention-based recurrent networks to gather the question-passage representation and uses a self-attention mechanism for refinement. As of the demonstration, it points out where the answer could be in the document. Most of the educational chatbots focus on the Yes or No type of responses, Adaptive Conversations for Adaptive Learning: Sustainable Development of Educational Chatbots [7], students can use it to understand the course modules and to keep track of their overall progress. As an education chatbot, it can act as a teaching assistant, providing answers to any technical questions posed by the student. iTA: A digital Teaching Assistant [8] that comprehends unstructured text and then responds to questions.

3   System Overview

In this section, we provide an overview of the two long-form question answering systems we have introduced, i.e., iTA and iLFQA. Thereafter, we elaborate on each systems in more detail in subsequent sections.

3.1   iTA

iTA is a digital teaching assistant which can provide extensive responses to questions by effectively locating the most relevant answers in a "long" text source (textbook). iTA

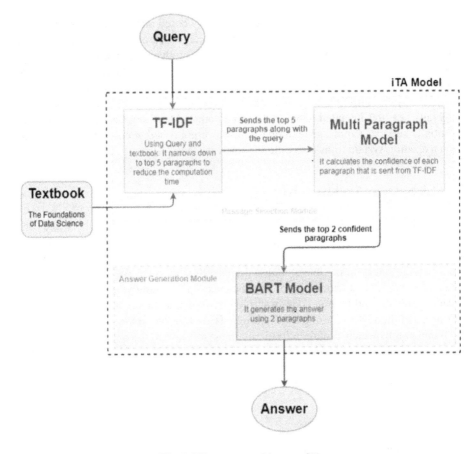

Fig. 1. iTA system architecture [8].

uses a two-stage technique to respond to questions. First, the topmost relevant paragraphs are identified in the selected text source using TF-IDF and passed to the LSTM model which scores each passage according to the question. Second, using a generative model, extracted the relevant content from the top-ranked paragraph to construct the answer. Figure 1 shows iTA overview.

3.2   iLFQA

iLFQA is an open domain question answering platform. This is different from iTA because iLFQA uses an open domain dataset instead of a single textbook as a reference for questions. In order to achieve this functionality, incorporates topic classification. iTA used only a single textbook for reference where iLFQA incorporates multiple textbook and a portion of the Wikipedia dataset. For this reason iLFQA uses classification to constrain the amount of text that must be analyzed for passage selection. In addition to the change in the size of the data set, iLFQA uses a Transformer based question answering models where iTA used a Long Short Term Memory based question answering implementation. The change from LSTM to a Transformer based architecture was made in order to increase the run-time performance. iLFQA produces responses at more than 50% faster per responses when compared with iTA. The changes to, passage selection, and the inclusion of topic classification significantly increased the efficiency and accuracy of the system. iLFQA is also designed to be used for many different applications. Originally iTA was intended to be used as an automated teaching assistant. iLFQA can potentially serve as a more general question answering tool that could be incorporated into an existing system or as a module for a new system.

4 iTA System Architecture

iTA supports multi-paragraph generative-based responses, avoids noisy labels, and uses an unstructured text as its source [8]. We use a Data Science textbook, stored as a .txt file as the source from which answers are generated [8]. iTA is a two-tier model which has a passage selection module and answer generation module. We have scrubbed the data in the text "The fundamentals of Data Science" [1], by removing the mathematical equations, tables, and python code from the text. When a student asks a question, the question and the textbook are passed through TF-IDF in the first module to select the top 5 paragraphs which contain information most relevant to the question. This step is employed to save the computationally expensive step of calculating each paragraph's confidence score in the entire text document. We use the top 2 candidate paragraphs in BART model to generate the response.

4.1 Passage Selection Module

With a user-generated query, the paragraph selection module picks the paragraphs with the shortest TF-IDF cosine distance. For computing TF-IDF, iTA employs a modified approach. iTA now generates answers using just one-long format document. Instead of taking the word-frequency from a vast number of documents, iTA calculates the term-frequency using the frequency of each keyword in the Data Science textbook. This method prioritizes the least common terms in the query and extracts the relevant paragraphs. The implementation of iTA employs an attention mechanism. This mechanism, presented in detail in [5], computes a confidence score for each paragraph containing a relevant text and then chooses the best passages based on the confidence score. The weight in TF-IDF is a statistical metric used to determine the importance of a word in a corpus of texts. The weights are calculated by multiplying TF by IDF. In our technique [5], TF-IDF is defined as the number of times a keyword appeared in a paragraph divided by the total number of paragraphs. GloVe [18] was utilized to vectorize each student question and context. A shared bi-directional GRU [3] mapping connects the context and the question. Attention processes construct a representation of the relationship between the question and the context. Self-attention processes improve machine understanding of complicated environments. The last layer of iTA's first module employs a softmax operation to compute the confidence ratings for the relevant sections. The paragraphs with the highest scores are routed to the iTA response creation module.

4.2   Generative Module

iTA's language module [30] was pre-trained using BART on the ELI5 dataset [9]. BART [14] is a sequence-to-sequence model with a bidirectional encoder similar to BERT [6] BART uses an auto-regressive decoder as with GPT [20]. iTA modules are made up of Transformers [27]. BART employs multi-tasking. This enables reliable sequence-to-sequence modeling. Using multi-headed attention processes improves accuracy in processing context and creating answers. BERT encoder replaces tokens with masks at random, and missing tokens are anticipated separately, therefore it cannot be used for generation. BART decoder employs a transformer decoder block; its decoder has an additional layer of disguised self-attention, which prevents a position from peeking at tokens to its right. The transformer decoder differs from standard language models that employ auto-regression in that it outputs one token at a time. The answer is generated via a beam search. Larger beam widths result in improved iTA performance. The presence of several candidate sequences increases the chance of a better match to a target sequence. Beam-width has an inverse relationship with performance.

5   iLFQA System Architecture

Figure 2 depicts the system architecture of iLFQA. iLFQA leverages and integrates three distinct methods to introduce a generative long-form question answering platform. iLFQA represents the first instance of a long-form question answering system that is deployable, open source, and which freely shares its run-time performance metrics. Currently no other long-form question answering systems share run-time performance details, nor are they deployed in such a way that they can be used by the general public. When a question is submitted to iLFQA, it is first sent to a classifier module. The classifier associates the question with one of the topics in a long list of topics, such as Data Science, Statistics, History, Politics, International Relations, Social Science, and Wikipedia articles. Each topic is associated with a corresponding set of text sources. Once a question is assigned a topic label, only the text sources associated with that label is retrieved. iLFQA then calculates the *TF-IDF* (i.e., short for *Term Frequency–Inverse Document Frequency*) of each paragraph in the given text sources to determine the passages most likely to contain relevant answers. The most likely paragraphs are then scored, and the likely answers from each paragraph are extracted. At this point, the possible answers and the highest scoring paragraph are sent to a text generation module. The text generation module takes this as input and synthesizes a long-form answer for the given question.

5.1   Classification Module

When a question is given to iLFQA, the question is first passed to a classification module. The main purpose of the classification module is to limit the amount of data that needs to be processed for any given question in order to ensure efficient (and deployable) real-time question answering. Our system currently uses a portion of the

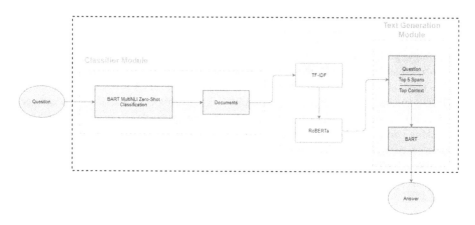

Fig. 2. iLFQA system architecture.

Wikipedia article corpus as the open-domain source of data for iLFQA. Our classification module uses a BERT model, namely RoBERTa [15], to select Wikipedia articles most relevant to the given user question. Without the classification module, iLFQA would be forced to calculate *TF-IDF* for every paragraph in the knowledge corpus. If the knowledge corpus becomes sufficiently large, the calculation of *TF-IDF* for each passage becomes prohibitively expensive. Once a question has been classified, the appropriate documents are retrieved and sent to the passage selection module along with the user's question. It can be seen in Table 1 that by implementing a classification module we are able to decrease the response time and increase the overall BERTScore.

5.2 Passage Selection Module

The passage selection module accepts a series of tokenized documents and the user's question as input. *TF-IDF* is calculated for every paragraph in the input document. We select the most relevant paragraphs (5 paragraphs by default) using *TF-IDF*. Once we have selected the best candidates, those paragraphs are sent through a BERT model [15] that calculates a confidence score for each paragraph. In addition to a confidence score, we also determine the span of text within each paragraph that may contain the answer. We concatenate the best spans and the paragraph(s) with the highest confidence score together and pass them as an input to text generation module.

5.3 Text Generation Module

The text generation module takes the user's question, a series of possible answer spans, and the input paragraph(s) as input. We use a BART [14] model pretrained on the ELI5 data set [9] to produce our text. It takes the text we have generated as input along with the given question, and attempts to generate an answer. The text generation model abstracts the information from the best answer spans and the selected paragraph(s) into a long-form answer.

6   Experimentation

6.1   Experimental Methodology

The results for iTA and iLFQA were obtained on a Google Cloud Platform Virtual Machine instance. Our VM instance uses an Intel Broadwell CPU, one NVIDIA TESLA K80 GPU with 12GB of memory, and a 64GB persistent SSD. The operating system was Ubuntu 18.04. iTA's passage selection model was used from a shared-norm LSTM checkpoint [5] that trained with Adadelta optimizer with a batch size of 60 and mainly used TriviaQA-unfiltered data. iTA utilizes TF-IDF to select 5 paragraphs which are most closely related to a given question. An LSTM model selects the paragraph to use as input to the sequence to sequence model. The highest confidence value will be sent to the BART [14] that was trained on ELI5 [9] generative model to get the answer. Four students from various majors posed questions based on what they learned in the textbook. We've compiled a list of 75 important questions. The experiments and conditions are shown in the paper [8].

In iLFQA, once a user inputs a question, that question is classified, the appropriate documents are selected based on that questions topic classification. The documents are searched for the paragraphs that most likely contain the answer to the question using TF-IDF. Once TF-IDF is obtained, a BERT model named RoBERTa is used to find the span of text within the each selected paragraph that may contain the answer. All the spans of text and the paragraph that has the span of text with the highest confidence score are sent to the same BART model used for sequence to sequence generation in iTA.

For iLFQA three different Transformer based model architectures are used. The Zero-Shot classification model [24] was used from a checkpoint trained on the Multi-Genre Natural Language Inference corpus (MultiNLI) [29]. MultiNLI consists of a large number of sentence genre pairs. We used a model named RoBERTa [15] to extract the spans of text which contain possible answers after selecting five paragraphs using TF-IDF. The RoBERTa model we used was trained on the Stanford Question Answering Dataset 2.0 (SQuAD 2.0) [23]. SQuAD 2.0 is a dataset that contains a series of questions with an associated paragraph containing the answer. Some questions in the SQuAD 2.0 dataset have associated paragraphs with no answer. iLFQA uses the same BART model [14] as iTA, it was trained on the ELI5 dataset [9].

The questions for iLFQA were selected by reading through the source material, which includes five textbooks and slightly more than 300 Wikipedia articles. Human volunteers read through the source material, and wrote questions that should have answers in the material. The accuracy measures that were collected are based on comparing the answer generated by iLFQA to the passage that contains the answer.

Our results show the results of iLFQA that was implemented with two different versions. One version uses a selection of five textbooks as the source document. The other version of iLFQA used for evaluation includes a selection of Wikipedia articles as well. Articles were obtained by downloading the entire English Wikipedia article base[2] that was presented on July, 20 2021. Articles that were present in some of the 1000 most viewed articles were filtered out of the entire Wikipedia article base and compiled into

---

[2] dumps.wikimedia.org.

a text document that was made available as a source document for iLFQA. The purpose of evaluating two different versions of iLFQA with the same hardware was to establish that we could produce, accurate reasonable results from using a source document that was truly open domain. The difference in performance between the two versions will be explored in the Experimental Results Section.

We have evaluated iLFQA by two metrics named BERTScore [31] and BLEU [17]. BLEU stands for Bilingual Evaluation Understudy. These scores were obtained by submitting a total of 115 questions to iLFQA. BLEU evaluates a machine generated text by matching n-grams between the generated text and the reference text. An n-gram is a sequence of text that appears contiguously. BLEU can be calculated by setting n to different values. Typically, the higher the value of n, the more rigorous the evaluation.

BLEU is an excellent method for evaluating direct translation, however BLEU fails to capture semantic meaning. There are currently few methods for evaluating the semantic correctness of generative question answering platforms. We have chosen evaluate our responses using a metric BERTScore in addition to BLEU because BERTScore was designed to go beyond n-gram matching.

BERTScore uses a pre-trained BERT model to evaluate the cosine similarity between the generated text and the reference text. The embeddings of the generated text and reference text are compared to determine the similarity in meaning between the two texts. This method of evaluation is more rigorous and appropriate for evaluating the semantic correctness of the generated response than BLEU. BERTScore works by taking the gold standard-sequence and the generated sequence. It embeds both sequences in the BERT model's embedding space. Each token in the gold sequence, and each token in the generated sequence are paired with the most similar token in the embedding space. There now exist two sequences of vectors that represent the most similar tokens in the embedding space, one sequence from the gold standard answer, and one sequence from the generated text. Precision, Recall, and F1 are calculated based on these two sequences of vectors. These are called BERT_P, BERT_R, and BERT_F1 respectively. The similarities of the two texts are compared token by token. Precision, recall, and F1 are calculated based on the matches by similarity of tokens in the languages embedding space. This is better than BLEU but not perfect. BERTScore works best for machine translation and summarizing. It does not necessarily take into account the given context when generating an answer. Despite this, BERTScore is one of the best existing metrics to evaluate semantic correctness.

6.2   Experimental Results

We have evaluated iLFQA by two metrics named BERTScore [31] and BLEU. BLEU stands for Bilingual Evaluation Understudy. These scores were obtained by submitting a total of 115 questions to iLFQA and 76 questions to iTA. In addition, we have also evaluated a version of iLFQA which uses the same design and architecture as iLFQA but which uses a selection of several hundred Wikipedia articles as it's knowledge corpus. BLEU evaluates a machine generated text by matching n-grams between the generated text and the reference text. An n-gram is a sequence of text that appears contiguously. BLEU can be calculated by setting n to different values. Typically, the higher the value of n, the more rigorous the evaluation. The results as measured by BLEU for all architectures can be seen in Fig. 3. iLFQA_Wiki actually performed the best out of all the architectures, matching the most words in the context given for the question to the generated text. This is a promising result since it indicates that using a wide knowledge domain doesn't adversely affect the accuracy of the produced results, at least as far as n-gram matching is concerned.

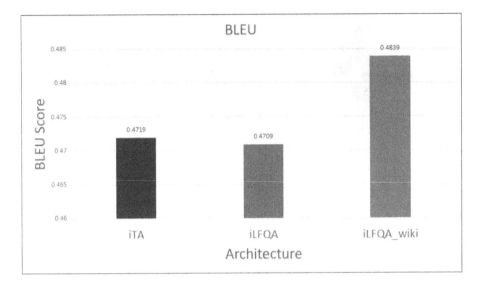

Fig. 3. Represents average BLEU score for all questions on each architecture.

BLEU is an excellent method for evaluating direct translation, however BLEU fails to capture semantic meaning. There are few methods for evaluating the semantic correctness of generative question answering platforms. We have chosen BERTScore because it was designed to go beyond n-gram matching. An explanation of how BERTScore is calculated was given in the Experimental Methodology section. BERTScore is a measure of accuracy, scores are on a scale from zero to one. One representing a perfect match, and 0 representing an inappropriate match. The BERTScore accuracy of each architecture can be seen in Figs. 5, 6, and 4. BERT_P represents BERTScore precision, BERT_R and BERT_F1 represent BERTScore recall and BERTScore_F1 respectively. In Fig. 4 it can be seen that BERT_R is highest for iLFQA without the wiki questions and lowest of iTA.

The version of iLFQA_Wiki, the one that contained Wikipedia articles had the lowest BERT_R score, likely due to the fact that it contained the largest amount of raw data. Despite this all scores are within 0.06 of each other on a scale from 0 to 1, indicating the performance was similar across all architectures.

Results for the other measures that BERTScore produces can be seen in Figs. 5 and 6. Now, when the results of BERT_F1 and BERT_P it can be seen that the results were even closer. iLFQA performed the best on BERT_F1 and iLFQA_Wiki performed the best on BERT_P, indicating that the iLFQA and iTA architectures are capable of returning reasonable correct semantic responses. This highlights our system's capability of producing efficient and accurate results using commonly available resources.

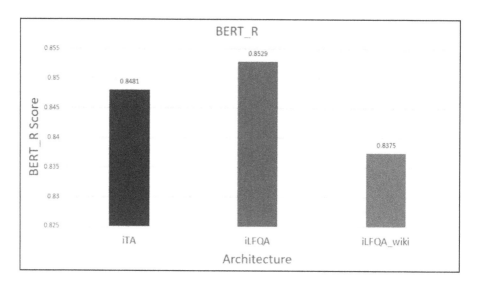

Fig. 4. Represents average BERT_R score for all questions submitted to each architecture.

Perplexity can be thought of as how well a model predicts a sample. In this case, perplexity is measured from the generated text against the golden answer. A lower perplexity indicates a better score. It can be seen from Fig. 7 that iTA actually scores the highest out of all three architectures. This is likely because iTA uses the smallest dataset and the possible selections of answers from said dataset are the smallest.

The results of the duration of each module are given in the Figs. 8, 9, 10, and 11. Each measure of duration represents the average duration taken in seconds for each module to complete it's work. Confidence duration shows the average amount of time taken in seconds for each different architecture to produce it's text responses. iTA is the longest, due to the method by which it determines it's confidence score. Initially an LSTM model was used to calculate the confidence scores during paragraph selection. Using an LSTM model as part of our passage selection module resulted in total response times close to 30 s on average. In iLFQA and iLFQA_Wiki we employed a BERT model to determine the confidence score which reduced this duration drastically in both iLFQA versions. It can be seen that time to generate text remained close to the same for all models, and was actually the shortest for iLFQA_Wiki. As expected the duration it takes to calculate TF-IDF for iLFQA_Wiki is the highest. This is due to the fact that TF-IDF is currently calculated for every paragraph in the classified documents. Due to the extra data present in iLFQA_Wiki's data corpus, this significantly increases the time taken to calculate TF-IDF in the iLFQA_Wiki architecture. The amount of time taken to classify each question into an appropriate category is indicated in the Fig. 11. Only iLFQA and iLFQA_Wiki had a classification module. iLFQA_Wiki actually performs slightly faster than iLFQA on average but the difference is almost negligible.

The classification module was added because it increased the accuracy and response time of the entire system. The results of iLFQA with and without the classification module are shown in Table 1. Overall accuracy and response times were increased. It should be noted that the response time was decreased even though there is extra com-

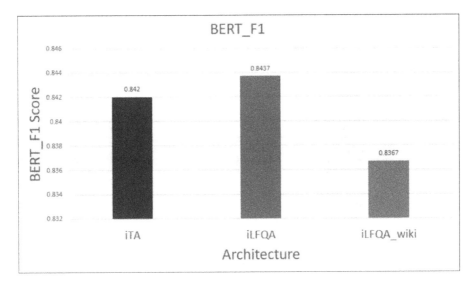

Fig. 5. Represents average BERT_F1 score for all questions submitted to each architecture.

putation involved in the classification module. This overall improvement is due to the fact that classifying questions allows us to process fewer total tokens when selecting a passage. This indicates that document classification during retrieval could significantly increase the amount of data that could be drawn upon to generate text without vastly decreasing the overall time taken to produce any given answer (Fig. 12).

Table 1. iLFQA results with different heuristics.

Classification comparison					
Version	BLEU	BERTScore P	BERTScore R	BERTScore F1	Response Time
iLFQA	0.476107211	0.836836127	0.8337135	0.834965846	10.20230839
Without classification	0.481451919	0.837526725	0.831856726	0.83424287	10.78334703

Many heuristic approaches to improving the accuracy and performance were attempted during the implementation of iLFQA. Several different combinations of best span and best paragraph concatenation were tried. These methods included using only best spans, using all the paragraphs retrieved at the TF-IDF stage, using two of the best paragraphs, and several other methods. The results of those methods can be seen in Table 2. Each version indicates a different method for generating the input for the text generation module. Five best answers sends the four best answer spans, and the highest scored paragraph to the text generation module. Four best context sends the four highest scoring paragraphs to the text generation module. Two answer two para sends two spans and two paragraphs to the text generation module. Two answer one para sends two spans and the highest scored paragraph to the text generation module. The configuration we have settled on uses the method described for five best answers.

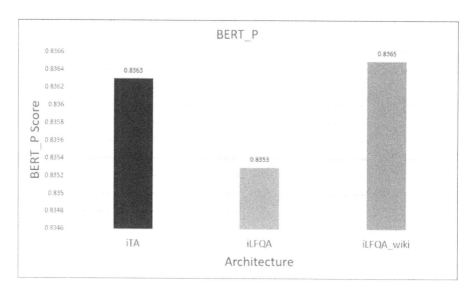

Fig. 6. Represents average BERT_P score for all questions submitted to each architecture.

Table 2. iLFQA implemented with different context selection methods.

Context generation comparison			
Version	BLEU	Perplexity	Response time
Five best answers	0.476107211	72.9835029	10.20230839
Four best context	0.478781827	73.62670021	10.21829904
Two answer two para	0.47847055	72.8883316	10.1626531
Two answer one para	0.480523763	73.39233174	10.08110274

In addition to different heuristic approaches to constructing the input for the text generation module, different methods of text generation were experimented with. T5 [22], GPT-2, and an alternate BART model were all experimented with. For each different experiment, we removed the current iLFQA text generation model and replaced it. The metrics for the results of those experiments can be seen in Table 3. One abstractive summarization model based on the T-5 architecture [22] sometimes produced results that were acceptable and semantically correct very quickly. This behavior was not consistent. Only the model used in the final iLFQA implementation has been trained on the ELI5 dataset. Training the models we did not select on a proper long form dataset would yield much better results.

Table 3. Experiments with different text generation architectures.

Different text generation model architectures					
Version	BLEU	BERTScore P	BERTScore R	BERTScore F1	Response time
iLFQA	0.476107211	0.836836127	0.8337135	0.834965846	10.20230839
T-5	0.697096606	0.843155287	0.805740035	0.82375022	5.252131491
GPT-2	0.558380775	0.810880085	0.810336336	0.81031971	5.399227395
DistilBART	0.539802913	0.820264539	0.817357293	0.818603523	6.840499799

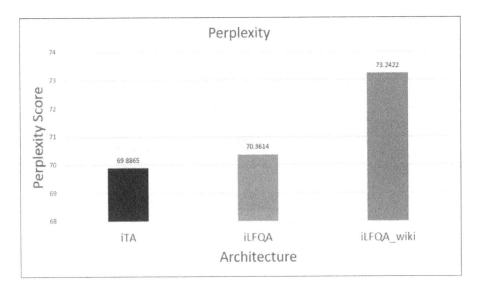

Fig. 7. Average perplexity for all questions submitted to each architecture.

Fig. 8. Average time taken to determine confidence score for each paragraph that makes it through the TF-IDF module.

Finally, we present some examples of iLFQA output. See Figs. 13 and 14 for an example of an iLFQA response. Note that for Figs. 13 and 14 the answers are contained within iLFQA's knowledge corpus.

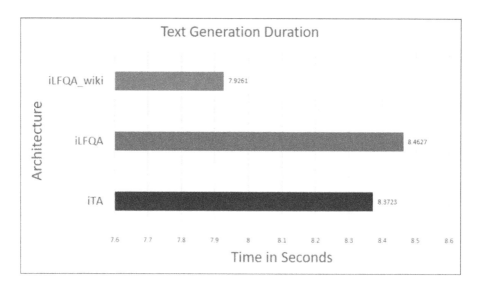

Fig. 9. Average time taken for each architecture to generate the text used for the response.

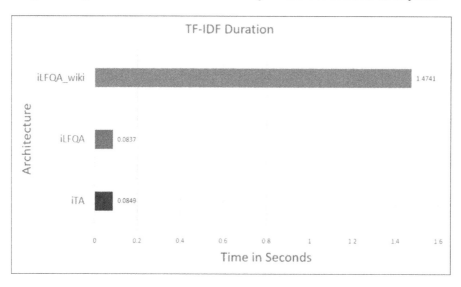

Fig. 10. Average time taken to calculate TF-IDF for each paragraph in an entire source document.

iLFQA does not perform well when a key phrase from the question is missing in the knowledge corpus, see Figs. 15 and 16. In these cases, iLFQA will return a response that is grammatically correct, but semantically incorrect. That is to say, the question may read as if it was human generated, but it will lack meaning with regard to the question. If the specific information that is asked for does not exist in the iLFQA reference documents, the text generation model will attempt to synthesize a response based on the question. Ultimately iLFQA returns an answer that briefly mentions the key topic of the

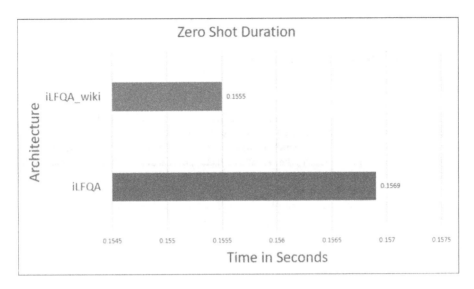

Fig. 11. Average time taken for question classification to occur, iLFQA_Wiki represents a version of iLFQA that incorporates selected Wikipedia articles as a source document.

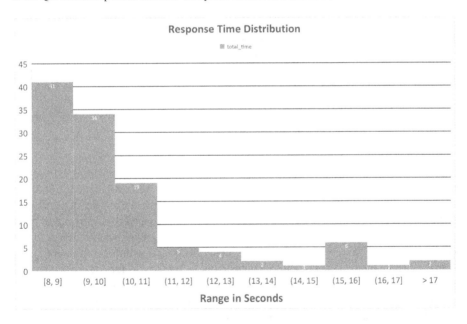

Fig. 12. Histogram representing distribution of response times for 115 questions.

question, but does not answer the question. At times when the key phrase or topic presented by the question is not contained in the reference documents, the text generation model will pull some information from the data that it was pre trained on. In this case

Question:

How does Data Science use statistical techniques?

Selected context:

make robust conclusions based on incomplete information Applications to real data sets choosing between hypotheses based on data in random samples careful reasoning attempt to choose between two views of the worldFor whatever aspect of the world we wish to study – whether it's the Earth's weather, the world's markets, political polls, or the human mind – data we collect typically offer an incomplete description of the subject at hand. A central challenge of data science is to make reliable conclusions using this partial information. In this endeavor, we will combine two essential tools: computation and randomization. For example, we may want to understand climate change trends using temperature observations. Computers will allow us to use all available information to draw conclusions. Rather than focusing only on

Fig. 13. Question.

Generated answer:

Data Science doesn't use statistical techniques, it uses data analysis techniques. Data analysis is the study of data and how it relates to other data. For example, if you want to know the average temperature of a region, you can look at the average temperatures of a bunch of people in that region. If you look at that data you can make a pretty good guess as to what the average is. You can also look at how the data relates to the other data you're looking at, and use that to make a more nuanced analysis.

Fig. 14. Response.

it is obvious that the ELI5 dataset contains some questions and explanations about the NBA. Even though it supplies this information in the response, the answer does not provide an adequate explanation for the question. We have managed to deploy a prototype version of our system to Google Assistant. Currently google assistant requires that it receive a response from any application associated back-end within 10 s. It can be seen from our results that we are not able to consistently hit that mark. We could significantly improve our response times by upgrading the GPU used in our implementation. However, this is a proof of concept, and should show that with reasonable resources, systems like ours can be deployed. This opens up opportunities for more rigorous testing and increased human evaluation.

Question:

How old was Tim Duncan when he won his first championship?

Selected context:

3.2 SDs above average 100 years chance 18 100 " years Sometimes , adjustments have to be made in the first or last bin , to ensure that the smallest and largest values of the variable are included . You saw an example of such an adjustment in the Census data studied earlier , where an age of " 100 " years actually meant " 100 years old or older . " We can see that there are 10 bins (some bars are so low that they are hard to see) , and that they all have the same width . We can also see that none of the movies grossed fewer than 300 million dollars ; that is because we are considering only the top grossing movies of all time . It is a little harder to see exactly where the ends of the bins are situated . For example , it is not easy to pinpoint exactly where the value 500 lies on the horizontal axis . So it is hard to judge exactly where one bar ends and

---

Fig. 15. Hard question.

Generated answer:

Depends on how old he was when he won his first championship. If he was 16, then he would have been 18 years old at the time of the championship. But if he was 23, then it would be a little over a year older than the NBA's maximum age limit, which is 25 years old. So Tim Duncan would have turned 23 on the day he won the championship, which would have made him 23 years old when the NBA changed the age limit to 25 years.

Fig. 16. Hard response.

7   Conclusions and Future Work

The field of question answering has been advancing rapidly in recent years. Some of the most promising advances, such as Transformer architecture, have improved our ability to perform generative question answering. We have implemented a question answer platform, named iLFQA: intelligent Long-Form Question Answering, that seeks to take advantage of these and other advances in natural language processing. iLFQA seeks to provide accurate, meaningful, and explanatory answers to user generated questions in a reasonable amount of time. iLFQA works by accepting a question as input, classifying the question by topic, and deducing the correct answer from an associated document. The information for the correct answer is synthesized into a response by and returned to the user. The architectures and systems that iLFQA leverages have been presented along with a detailed overview of each module. iLFQA has been evaluated with a meaningful metric that most accurately captures its semantic performance. There are very few long-form generative question answering systems that currently exist. It is our hope that this work will contribute to progress in that field.

It seems that a promising area of future work on this project or related subjects could be to fine tune an abstractive summarization model on a long-form documents. Many summarization models produce single sentence outputs and may repeat themselves when longer outputs are required. However very few examples of research on long-form summarization exist. Future work on iLFQA should consider training a summarization module on long-form documents and using that as a portion of the text-generation module. This might require actually creating a suitable database. Some summarization databases used for pre-training and fine tuning provide single sentence explanations of whole articles. This is not ideal for our work. Also, as mentioned earlier. BERTScore [31] is best used as an evaluation of machine translation. We can use it in our case because the reference context and generated text can be compared for evaluation in a manner consistent with how our system was designed. However, this is not an ideal metric to capture the correctness of the generated answer as it relates to the original question. A recently proposed metric named KPQA [13] may capture the relationship between then question and response more accurately. KPQA determines key phrases in the question and attempts to determine if machine generated answers capture the meaning of the key phrases from the question. Future implementations of iLFQA should incorporate this metric or any similar future metrics that are designed to evaluate the semantic accuracy of a generated response.

References

1. Adhikari, A., John DeNero, B.: The foundations of data science, September 2019. https://www.inferentialthinking.com/chapters/intro.html
2. Chandra, Y.W., Suyanto, S.: Indonesian chatbot of university admission using a question answering system based on sequence-to-sequence model. Procedia Comput. Sci. **157**, 367–374 (2019). https://doi.org/10.1016/j.procs.2019.08.179. https://www.sciencedirect.com/science/article/pii/S187705091931097X. The 4th International Conference on Computer Science and Computational Intelligence (ICCSCI 2019): Enabling Collaboration to Escalate Impact of Research Results for Society
3. Cho, K., et al.: Learning phrase representations using RNN encoder-decoder for statistical machine translation (2014)
4. Chowanda, A., Chowanda, A.D.: Generative Indonesian conversation model using recurrent neural network with attention mechanism. Procedia Comput. Sci. **135**, 433–440 (2018). https://doi.org/10.1016/j.procs.2018.08.194. https://www.sciencedirect.com/science/article/pii/S1877050918314844. The 3rd International Conference on Computer Science and Computational Intelligence (ICCSCI 2018): Empowering Smart Technology in Digital Era for a Better Life
5. Clark, C., Gardner, M.: Simple and effective multi-paragraph reading comprehension (2017)
6. Devlin, J., Chang, M.W., Lee, K., Toutanova, K.: Bert: pre-training of deep bidirectional transformers for language understanding (2019)
7. Donya Rooein, P.P.: Adaptive conversations for adaptive learning: sustainable development of educational chatbots (2020)
8. Duggirala, V., Butler, R., Banaei-Kashani, F.: iTA: a digital teaching assistant. In: Proceedings of the 13th International Conference on Computer Supported Education - Volume 2: CSEDU, pp. 274–281. INSTICC, SciTePress (2021). https://doi.org/10.5220/0010461002740281

9. Fan, A., Jernite, Y., Perez, E., Grangier, D., Weston, J., Auli, M.: ELI5: long form question answering (2019)
10. Guo, M., Dai, Z., Vrandečić, D., Al-Rfou, R.: Wiki-40B: multilingual language model dataset. In: Proceedings of the 12th Language Resources and Evaluation Conference, Marseille, France, pp. 2440–2452. European Language Resources Association, May 2020. https://www.aclweb.org/anthology/2020.lrec-1.297
11. Kim, S., Kwon, O.W., Kim, H.: Knowledge-grounded chatbot based on dual wasserstein generative adversarial networks with effective attention mechanisms. Appl. Sci. **10**(9), 3335 (2020). https://doi.org/10.3390/app10093335. https://www.mdpi.com/2076-3417/10/9/3335
12. Krishna, K., Roy, A., Iyyer, M.: Hurdles to progress in long-form question answering (2021)
13. Lee, H., et al.: KPQA: a metric for generative question answering using keyphrase weights (2020)
14. Lewis, M., et al.: Bart: denoising sequence-to-sequence pre-training for natural language generation, translation, and comprehension (2019)
15. Liu, Y., et al.: Roberta: a robustly optimized BERT pretraining approach (2019). http://arxiv.org/abs/1907.11692
16. Min, S., Michael, J., Hajishirzi, H., Zettlemoyer, L.: AmbigQA: answering ambiguous open-domain questions. In: Proceedings of the 2020 Conference on Empirical Methods in Natural Language Processing (EMNLP), pp. 5783–5797. Association for Computational Linguistics, November 2020. https://doi.org/10.18653/v1/2020.emnlp-main.466. https://www.aclweb.org/anthology/2020.emnlp-main.466
17. Papineni, K., Roukos, S., Ward, T., Zhu, W.J.: Bleu: a method for automatic evaluation of machine translation. In: Proceedings of the 40th Annual Meeting of the Association for Computational Linguistics, Philadelphia, Pennsylvania, USA, pp. 311–318. Association for Computational Linguistics, July 2002. https://doi.org/10.3115/1073083.1073135. https://www.aclweb.org/anthology/P02-1040
18. Pennington, J., Socher, R., Manning, C.D.: Glove: global vectors for word representation. In: In EMNLP (2014)
19. Petroni, F., et al.: Kilt: a benchmark for knowledge intensive language tasks. arXiv:2009.02252 (2020)
20. Radford, A.: Improving language understanding by generative pre-training (2018)
21. Rae, J.W., Potapenko, A., Jayakumar, S.M., Hillier, C., Lillicrap, T.P.: Compressive transformers for long-range sequence modelling. arXiv preprint (2019). https://arxiv.org/abs/1911.05507
22. Raffel, C., et al.: Exploring the limits of transfer learning with a unified text-to-text transformer. CoRR abs/1910.10683 (2019). http://arxiv.org/abs/1910.10683
23. Rajpurkar, P., Jia, R., Liang, P.: Know what you don't know: unanswerable questions for squad (2018)
24. Romera-Paredes B., T.: An embarrassingly simple approach to zero-shot learning (2017)
25. Tu, Z., Jiang, Y., Liu, X., Shu, L., Shi, S.: Generative stock question answering (2018)
26. Vaswani, A., et al.: Attention is all you need. In: Guyon, I., et al. (eds.) Advances in Neural Information Processing Systems, vol. 30. Curran Associates, Inc. (2017). https://proceedings.neurips.cc/paper/2017/file/3f5ee243547dee91fbd053c1c4a845aa-Paper.pdf
27. Vaswani, A., et al.: Attention is all you need (2017)
28. Wang, W., Yang, N., Wei, F., Chang, B., Zhou, M.: Gated self-matching networks for reading comprehension and question answering. In: Proceedings of the 55th Annual Meeting of the Association for Computational Linguistics (Volume 1: Long Papers), Vancouver, Canada, pp. 189–198. Association for Computational Linguistics, July 2017. https://doi.org/10.18653/v1/P17-1018. https://www.aclweb.org/anthology/P17-1018

29. Williams, A., Nangia, N., Bowman, S.: A broad-coverage challenge corpus for sentence understanding through inference. In: Proceedings of the 2018 Conference of the North American Chapter of the Association for Computational Linguistics: Human Language Technologies, Volume 1 (Long Papers), pp. 1112–1122. Association for Computational Linguistics (2018). http://aclweb.org/anthology/N18-1101

30. Wolf, T., et al.: Huggingface's transformers: state-of-the-art natural language processing (2020)

31. Zhang, T., Kishore, V., Wu, F., Weinberger, K.Q., Artzi, Y.: Bertscore: evaluating text generation with BERT. CoRR abs/1904.09675 (2019). http://arxiv.org/abs/1904.09675

Creation of a Teacher Support System for Technology-Enhanced Accelerated Learning of Math in Schools

Aija Cunska[✉] [ID]

Vidzeme University of Applied Science, Valmiera 4501, Latvia
`aija.cunska@va.lv`

Abstract. In this age of "high" technology and "cold" touch, the role of teachers is growing. Teachers develop and change students' lives, paving the way for life-long learning and career success. Remembering the knowledge and skills acquired at school, students continue to draw strength from the support and love provided by teachers. The influence of teachers is usually deep and lasting. Creative and purposeful teachers engage and influence students, families, other teachers, school leaders and local communities. Even when most students studied at home during the Covid-19 pandemic, the issue of the quality of education remained the responsibility of teachers. All teachers have the same intention to help their students achieve their goals and succeed. The work of teachers is responsible and full of great challenges, as their needs and conditions are dictated by the students, the school, the local government, the state and even emergencies. Surveys of students conducted in 2021 and interviews with Latvian teachers of mathematics in focus group interviews indicated critical problems and revealed a worrying picture in the acquisition of mathematics in general education schools during the last two years in connection with distance online learning. In focus group interviews, teachers indicated that, given the background of the Covid-19 pandemic, they needed significant support from both local governments and policy makers to maintain emotional balance affected by technological progress, and from academics and scientists to understand the conditions and modalities how to learn math smarter, faster and more efficiently in the future. This article seeks answers to the question: *How can we support those who help shape the future?* The article presents the results of student surveys and teacher focus group interviews. As a solution to the problem, a framework model developed in cooperation with math teachers is proposed, using design thinking approaches and techniques. The model will further help to create a support system for technology-enhanced accelerated learning of mathematics, as well as provide innovative character and promote strategic use for STEM industries.

Keywords: AI4Math · Teacher support system · Accelerated learning of mathematics · Interdisciplinary approaches · Technology-enhanced learning

1 Introduction

With the advent of the Fourth Industrial Revolution, the role of education is changing and it must respond to the new needs of society. Teachers need to learn throughout their lives,

© Springer Nature Switzerland AG 2022
B. Csapó and J. Uhomoibhi (Eds.): CSEDU 2021, CCIS 1624, pp. 189–211, 2022.
https://doi.org/10.1007/978-3-031-14756-2_10

collaborate with other teachers and professionals, lead change in line with technological progress and develop personal qualities [1]. If teachers used to have enough knowledge of the content and pedagogy of their subject, now teachers have to know the possibilities of technology at a fairly excellent level, have to be interdisciplinary [2], have to feel the emotional wellbeing of students [3], have to take into account neuroscience research [4] to accelerate learning. They also need to follow sustainability strategies [3] to adapt curricula to the needs of the circular economy, green thinking and other recovery needs of society.

Although schools have made extensive use of technology in recent years, the sudden and complete transition to distance learning online in 2020 and 2021 due to the effects of the Covid-19 pandemic has been an additional challenge and a huge stress for most schools. For example, in Latvia, the new competence approach with new teaching rules and materials developed within the "School2030" project was still in the implementation stage, and teachers had not yet managed to master the new standards when the pandemic forced everyone to learn about technology.

The subject of mathematics, which is taught at all levels of education, faced particular challenges, as mathematics is seen as the basis for life, society and many other sciences. It has become a key area in the development of technology, the circular economy and environmental modeling. Knowledge of mathematics offers many advantages when working in areas related to sustainability. But, according to a compilation of several studies [5], it was the subject of mathematics that suffered the most during the pandemic, and in particular: 1) students had the worst performance in mathematics compared to other subjects; 2) during remote online learning, students learned about half of the intended the content of mathematics, 3) parents are not knowledgeable enough to help students learn mathematics; 4) learning mathematics further increased the already existing feelings of stress and anxiety; 5) teachers of mathematics have the least knowledge of effective teaching strategies.

Thus, the support of teachers and the long-term effect of this support is one of the most powerful factors in shaping the future of the country [6]. The field of education is too important to experiment with and neglect in the future.

The central question in this article and study is "How can we support those who help shape the future?" Undoubtedly, the key is to show respect and recognition. Compensation for the hard and heroic work of teachers is also important. But teachers also need access to high-quality teaching resources, time and support so that they can develop and engage in continuing professional development [6]. Teachers need advice from other educators, emotional support and ideas for effective teaching strategies.

During the online distance learning in the spring of 2020, the Ministry of Education and Science of Latvia also conducted a survey [7], which was completed by 4662 teachers and indicated problems that require state support: 1) lack of digital skills, 2) lack of a common methodology for the use of technology, 3) lack of a unified system and digital platform for communication and solving various tasks, 4) in the effort to ensure quality, the learning material remains unlearned, 5) inability to plan time, anxiety and stress, 6) lack of motivation, encouragement and support, 7) lack of socialization, 8) it takes too long for teachers to provide feedback and correcting student works, 9) teacher overloading in preparation digital online lessons, for finding more creative tasks, for creating more than one type of tasks and for correcting students' work, 10) lack of interdisciplinary approaches and activities in nature, 11) a sedentary lifestyle and lack of sport activities, 12) lack of cooperation between education and IT professionals to create innovative solutions and reduce the tensions created by digital technologies.

The paper is part of the research project "Artificial Intelligence (AI) support for accelerated math acquisition approach (AI4Math), No: 1.1.1.2/VIAA/3/19/564". It is an interdisciplinary postdoctoral research project (1/1/2020 - 31/3/2023), where as a result of the cooperation of specialists in the fields of mathematics, pedagogy and ICT, the development of human resources is promoted by increasing the diversity, accessibility, motivation, involvement and compliance with the needs of the STEM sectors. The scientific goal of the research is to develop an interdisciplinary math support strategy for general education schools using Artificial Intelligence, Big Data and stimulating learning approaches, thus promoting Area 5 "ICT" and the 5th growth priority "We need to create a modern education system that meets the requirements of the future labor market and promotes economic transformation" set within the framework of the Latvian Smart Specialization Strategy.

This paper also includes preliminary results that have already been described in the author's publications [2, 8]. Publication [8] develops and proposes a definition and taxonomy for accelerated of mathematics Learning in mainstream schools. Publication [2] identifies interdisciplinary digital technology-based approaches that stimulate students' interest in learning, motivate and improve attitudes in the long run. This paper includes more up-to-date data, more detailed research findings from interviews of focus group of math teachers, and more refined structures and a more focused move towards a basic scientific goal: In line with the design thinking approach and methods, we need to identify, in collaboration with mathematics teachers, the basic elements to help create a support system for technology-enhanced accelerated learning of mathematics in schools, as well as to ensure innovation and strategic use for STEM industries.

2 Theoretical Framework

2.1 Accelerated Learning of Mathematics

To date, there is no clear and uniform definition of the concept of accelerated learning (AL). Researchers and educational professionals in various fields are constantly discussing the concept of AL. And some important components are also discussed in the literature. Initially, AL was defined as "faster acquisition of skills and knowledge" [9]. Other definitions also focus on the time factor, such as "any learning system that seeks to optimize learning time in relation to the content acquired" [10]. Qomario [11] have pointed out that AL is an approach that is used to improve students' learning abilities so that students can learn faster and more efficiently, and that the learning atmosphere is designed as a fun and active interaction between students and teachers. The involvement of "whole body, mind and human experience" in the learning process was initiated by Meier [12]. The Center for Accelerated Learning (Alcenter) has pointed out that AL is today the most advanced of the methods that make up a complete system for accelerating the learning process based on the latest brain research. AL is the way we learn using all human talents: physical, creative, musical, artistic, etc. AL is an activity-based and student-centered process with the following basic principles [12, 13]: 1) learning encompasses the whole human mind and body with all senses and emotions, 2) learning takes place on many levels simultaneously, 3) learning is the creation of new knowledge and skills rather than consumption, 4) learning is collaborative, 5) learning takes place through practice, 6) positive emotions greatly improve learning, 7) visual images make it easier to perceive and store information.

According to Dave Meier, the AL approach helps students to develop a positive attitude towards mathematics [12]. The National Institute for Excellence in Teaching (NIET) emphasizes that accelerated learning of mathematics (ALOM) is usually associated with children gifted with special programs or, conversely, with students with special needs. It is often controversial, but understanding it in general encourages greater involvement of teachers in the planning and regular evaluation of the learning process. Pupils thrive in an environment where their needs are taken into account and their readiness levels are determined [14].

Our brains are adaptable, and when students learn or change their approach to learning, incredible developmental results can be created. In recent years, a new science has developed, neuroplasticity, which specifically studies improvements in brain function and emphasizes that brain function can and should be improved at any age [4]. Based on research [4, 15], seven basic principles can be identified that significantly improve and accelerate the learning of mathematics:

- For growth to take place, the learning process must be regular and continuous;
- Mistakes and error correction improve the long-term sustainability of mathematical skills;
- Positive communication from parents and teachers, which inspires faith in the student's strengths, is especially important for promoting growth;
- Applying an interdisciplinary approach activates neural pathways and learning in general;
- Open and creative math problems are important, which promote deeper learning and retain attention;
- Meaningful collaboration and exchange of ideas accelerates neuronal flow and improves learning. That's why group and project work is so important in math lessons;
- There is a strong correlation between memory and time, and by repeating the subject, students are able to learn mathematics much faster.

Neuroscientists [4, 15] point out that the knowledge we currently have about brain function is so important that it should change the way we teach students and run schools. Our brains have a tremendous ability to grow and change at any stage of life. This is demonstrated by a study of black cab drivers in London, which showed that extensive and complex spatial training has led to a significant increase in the amount of brain hippocampus in humans, which is important for all spatial and mathematical perceptions [16]. Taxi drivers in London must undergo extensive training, learning how to navigate between thousands of places in the city. This training is colloquially known as "being on The Knowledge" and takes about 2 years to acquire on average. To be licensed to operate, it is necessary to pass a very stringent set of police examinations [16]. Similarly, neuroscience researchers [15, 17, 18] confirm that the best time for brain growth is when people work on challenges, make mistakes, correct them, fight, and move on. But teachers usually do their best to make students' work easier by breaking down problems into smaller tasks, or by giving answers in front of them, or by facilitating the content of teaching.

Based on the explanations of accelerated learning found in scientific research [8, 15–18], the following definition of ALOM is proposed: ALOM is a smart approach to the learning process to create a motivating environment for student-teacher collaboration, resulting in high-quality mathematical competence. The goal of ALOM is to provide quality learning opportunities for each student to promote the development and realization of their potential throughout their lives. The tasks of ALOM are: 1) a positive and creative atmosphere that promotes the well-being, competence and success of students; 2) productive interaction of students and teachers, which supports positive emotions; 3) a flexible learning process that responds to the needs of each student; 4) effective teaching strategies that are effective, interdisciplinary and based on scientific research; 5) Analytics and evaluation to track change, adapt to needs and promote growth.

2.2 Technology-Enhanced Learning

Since the 1980s, smart learning based on smart technologies has emerged as a new paradigm of education [19]. Smart technologies (Cloud Computing, Big Data and Analytics, IoT, Virtual Reality and Artificial Intelligence solutions) allow students to learn comfortably in both traditional and technology-based ways. As a result, Technology-Enhanced Learning (TEL) has gained increasing attention in educational research. TEL is used to provide a flexible learning process. Technological progress has affected the way we acquire knowledge and learn. It provides quick access to information, facilitates collaboration, affects learning efficiency and saves our time. In the field of education, technology enables access to digital content, creative expression and evaluation. With the development of mobile technology, learning is possible anywhere and anytime. This is particularly important for the Z generation, which was born in the age of technology, acquires knowledge mostly through technology and focuses on finding information quickly. Researchers [20] have found that Generation Z students 1) use technology more often than traditional teaching methods, 2) prefer mobile applications to video content available on the Internet, 3) prefer interactive online forms, 4) follow the example of teachers in the use of technology, 5) they observe others before they start doing the same.

There has been much discussion on how to maintain a balance between traditional learning models and the use of technology, which has led to the notion of blended learning. "Blended Learning is an approach that provides innovative educational solutions through an effective mix of traditional classroom teaching with mobile learning and online activities for teachers, trainers and students. It is the TEL to extend beyond the classroom walls and facilitates better access to learning resources" [21]. The main idea of this study is also based on smart schools, which use a combination of technology and teaching strategies to provide each student with basic math skills, adapt individual learning trajectories and improve the classroom atmosphere as a whole. Smart schools also use appropriate learning strategies online to make the learning process more engaging, motivating and stimulating [22].

2.3 Artificial Intelligence

A study [23] indicates that AI is a cross-disciplinary discipline that encompasses several disciplines, such as neuroscience, psychology, mathematics (including statistics), information science, and computer science. Today, AI, represented by deep data-based learning, is developing rapidly and is widely used in many fields. It is believed [24] that AI algorithms have existed since the late 1970s, but their widespread use with the available computing power in the world and modern AI chips began only 5–7 years ago. AI is a term used to describe a set of computer systems and computer programs that use human-like thinking features to perform tasks. AI systems are able to analyze images and videos, listen to sounds, understand and synthesize language, predict exchange rate fluctuations, predict electricity consumption and perform many other tasks that only humans have been able to do so far. If more and more new AI solutions appear in production (for example, in Latvia company "Balticovo" the quality of eggs is analyzed with the support of AI, before the eggs reach the final tests), then in school education AI is still at an early stage of development.

2.4 Interdisciplinary Approach

In the field of education, the word "interdisciplinary" has been around for years. Several studies [25–27] indicate that interdisciplinary learning is innovative, attractive and exciting, as well as driving 21st century education reforms. Interdisciplinary learning is a way of learning and thinking that is based on several disciplines in order to acquire new knowledge and skills. For interdisciplinary training to be effective, researchers [27, 28] emphasize the following framework conditions: 1) interactivity of different fields, 2) teamwork and cooperation between people, 3) internship in relevant fields in the workplace, 4) breadth of knowledge and skills, 5) creativity and curiosity, 6) constant access to knowledge and lifelong learning, 7) changes in school curricula, 8) support for teacher education and development, 9) digital and social skills.

3 Methodology

In order to find an answer to the research questions and achieve the set goal, the research design (see Fig. 1) was created with the following research plan: 1) research of theoretical literature in the fields of pedagogy, application of technology and acquisition of mathematics; 2) determination of the ecosystem of accelerated mathematics acquisition by making observations in Latvian educational institutions with applications of Phenomenology, Ethnography and Narrative approach; 3) analysis of ALOM experience data using qualitative methods (student surveys, interviews with specialists in the field and structured interviews with teachers' focus groups); 4) triangulation of research results and determination of correlations to increase the reliability of research data; 5) identification of key elements of a teacher support system for ALOM in schools, using a design thinking approach; 6) testing and summarizing conclusions.

Given that the concept of ALOM is new and little research has been done on it, the Qualitative approach was used. According to researcher Creswell [29], qualitative

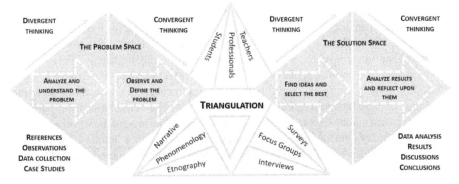

Fig. 1. Research design scheme, researcher's concept.

research is exploratory and useful in cases where the most important and testable values are not known. And Patton [30] has argued that research is qualitative if it aims to explore what people do, know, and think using a variety of data mining techniques, such as observation, interviews, surveys, or document analysis. The strategy of the qualitative approach allowed the study to be more creative and innovative in order to identify the key elements of a support system to school audiences that could affect a large part of the public in the future.

In order to fully explain the significance of the study results, their impact was studied from several perspectives using a triangulation approach [31]. By looking at different data sources (students, teachers, industry professionals), methods (student surveys, industry interviews, teacher focus group structured interviews) and types of analysis (ethnographic, phenomenological, narrative), the triangulation approach increased the likelihood of drawing the right conclusions and making better decisions, as well as increased the reliability and validity of the results.

As the main strategy of the qualitative approach to data analytics, phenomenology was singled out in order to understand the processes taking place in the human consciousness and in the human world during the acquisition of mathematics. "Phenomenology involves a change in the "sense of the world": everything acquires its sense and value only when it becomes the content of the lived experience of the subject correlated to his intentional acts. This is the main thesis of the phenomenological method aiming at overcoming the traditional opposition between rationalism and empiricism" [32].

An ethnographic approach was used to study in depth the daily habits of students, teachers and relationships in the natural school environment. "Educational ethnographic research is a research design that involves observing teaching and learning methods and how these affect classroom behaviors. This research model pays attention to pedagogy, its effects on learning outcomes and overall engagements by stakeholders within the classroom environment" [33]. As a result, observations were made and documented over three years (2019–2021) in several different schools in Latvia: one primary school, 5 primary schools, 2 secondary schools and one technical school.

A narrative approach was used for data analysis to document, summarize and structure experiences in social accounts and media spaces, especially during the Covid-19 pandemic, when there was no opportunity to meet in person. "Like many research approaches, narrative research is taking new directions in the digital age" [34].

The study is a compilation of current information and many years of experience with an interdisciplinary approach, and the methodology includes: 1) A selection and brief description of eight technology-based interdisciplinary approaches based on experience, 30 years of mathematics lessons, teacher success stories from around the world and expert advice; 2) Primary data, so as to reveal the problems of distance learning, the Ministry of Education and Science of Latvia (IZM) in cooperation with the company "Edurio" surveyed 4662 teachers, 8352 parents, 10177 students in May and June, 2020 [7].

The design thinking approach widely used today was used to identify the elements of the teacher support system, which is best described by David Kelley. "The main tenet of design thinking is empathy for the people you're trying to design for. Leadership is exactly the same thing – building empathy for the people that you're entrusted to help." and "Deep empathy for people makes our observations powerful sources of inspiration. We aim to understand why people do what they currently do, with the goal of understanding what they might do in the future" [35]. The design thinking approach is human-oriented, which helped to organize active involvement, successful teacher collaboration and generate new ideas. The principles of design thinking helped to identify problems in learning mathematics and to adapt to the needs of the target audience of mathematics teachers. The 53 teachers interviewed were between the ages of 30 and 62, all with tertiary education, and 70% were female. Thus, the diversity of teams also helped to create a diversity of ideas.

4   Results

4.1   Results of the Student Survey

Data were collected in several ways and over a longer period of time. First, electronically, by interviewing students during distance learning in April, May and September 2021. The electronic questionnaire was created with Google Forms and was filled in by 557 students from 1st to 12th grade from different levels of Latvian and Maltese schools. The questionnaire contained 26 questions, which were evaluated according to the Likert scale. The complexity of the questions and the length of the questionnaire were adapted to the age of the students.

In order to verify the usefulness of the electronic questionnaire, discussions of industry experts were organized, as a result of which two important permits were obtained for the research: 1) from The Academic Ethics Commission of Vidzeme University of Applied Sciences, 2) from the Government of Malta for Education Directorate for Research, Lifelong learning and Employability.

Second, several structured focus group interviews were conducted, in which 53 teachers were interviewed in face-to-face conversations, telephone conversations, and online ZOOM conversations. They also participated in the evaluation of the student survey data.

Figure 2 clearly shows that during distance learning more than half of the students (60%) had stress in learning mathematics, 68% of students lacked patience to read and understand the conditions of the tasks correctly, 63% of students felt additional workload, 80% of students lacked teacher explanation, 84% of students did not have the ability to plan time and learning process independently, 50% of students needed to learn additional digital skills despite Generation Z abilities and existing skills. Teachers also pointed out these as the most disturbing things in the focus group interviews, marking them with sad emotion (☹). The following students' answers were mentioned as positive emotion (☺) from viewpoints of educators: "the mathematics teacher helps me enough" (69%), "I also have individual consultations available" (68%), "the math teacher corrects the work fast enough and answers my questions" (63%), "my workplace at home is comfortable and appropriate" (76%), "I have the right technology" (83%), "I have a correspondingly fast data transfer" (61%), "I am happy with the technologies and programs we use" (79%). By the teachers' neutral emotions (😐) were given and by the students the answer "agree" were given to the following statements: "I feel support from my school in learning math" (42%), "The explanation of the math teacher is pleasant and understandable" (53%), "I have the opportunity to collaborate with classmates while learning math" (47%), "My parents help me to learn math" (26%). On the other hand, three statements caused the most discussion for teachers (?), for which the answer "agree" has been chosen by relatively few students: "More creative and problem-solving tasks contribute to deeper learning of mathematics" (37%), "Mathematics can be learned in collaboration with other subjects" (27%), "Mathematics can also be acquired qualitatively through distance learning" (33%).

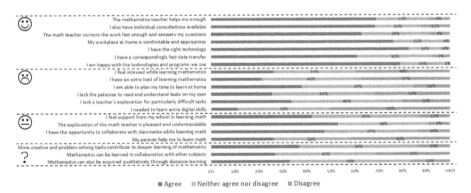

Fig. 2. Survey of students on the acquisition of mathematics during distance learning in 2021.

Second, teaching higher mathematics to students of the Faculty of Engineering of Vidzeme University of Applied Sciences in a 3-year period (2019–2021), the author surveyed 271 students to find what their main interests are, which occupy a large part of their lives and make their daily lives happier and more creative. Students were able to provide answers in a free-form form, and the summary provided a daily student profile (see Fig. 3) that accurately demonstrated the necessity of interdisciplinary approaches to learn math with interest and pleasure as a creative and multidimensional subject.

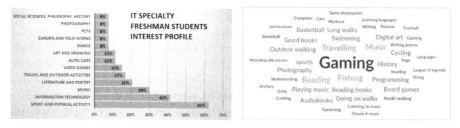

Fig. 3. IT specialty freshman students interest profile

As shown in Fig. 3, 62% of students are interested in sports, 42% of students are interested in Information Technology, 30% of students are interested in music, 21% of students are interested in literature and poetry, 17% of students are interested in traveling and outdoor recreation, 15% of students are interested in video games, 11% of students are interested in cars. Students are also interested in art and drawing (11%), dance (8%), garden and field work (8%), dogs and cats (8%), photography (8%), social sciences and history (8%). Other student's interests include: handicrafts (6%), logic puzzles (4%), films (4%), cooking (2%), languages (2%), theaters (2%), self-improvement (2%).

4.2   Results of Teachers Focus Group Interviews

The main problems of learning mathematics during online distance learning in 2020 and 2021, which were mentioned by Latvian teachers in their interviews, are summarized in the form of quotations:

Teacher No 1: "Not all mathematics topics were taught during online distance learning. There are huge knowledge gaps left for students."

Teacher No 2: "I've totally over-worked, constantly looking for more creative tasks, creating presentations and correcting students' work."

Teacher No 3: "I am not familiar with the new standards for teaching mathematics that were created as a part of the project School2030."

Teacher No 4: "The municipality has not explained to me the goals of the latest development strategy in the field of education in the region."

Teacher No 5: "I am not familiar with national documents describing guidelines for the subject of mathematics."

Teacher No 6: "I do not know how to take surveys of my students to determine their class desires and emotional state."

Teacher No 7: "No one has taught us the tools to collect and visualize data properly and clearly."

Teacher No 8: "Can student survey data really work for us?"

Teacher No 9: "No one has told us about effective teaching strategies that could make a significant contribution to learning math."

Teacher No 10: "It is very difficult to keep students' attention and motivation during distance online learning."

Teacher No 11: "I would like to learn about interdisciplinary approaches so that I can collaborate more with other teachers."

Teacher No 12: "We - teachers still lack the digital skills to work with the latest technology tools."

Teacher No 13: "I would love to understand how Artificial Intelligence can help with learning math."

Teacher No 14: "It is difficult to run online lessons when students have such different power technologies. Some students even don't have microphones and cameras."

Teacher No 15: "I was used to working in a classroom, where many posters with formulas and visual images were put on the walls. I don't see how I can provide supportive environment on my computer."

Teacher No 16: "It seems to me that home works and tests in mathematics are done by older or smarter friends instead of students themselves."

Teacher No 17: "The students confessed me that they use the possibilities of the Photomath app to solve math problems at home."

Teacher No 18: "I would really like to collaborate more with my students and colleagues through the computer, but I don't know how."

Teacher No 19: "Students are so stressed to hear the word "mathematics" that I've also been stressed over the last year."

Teacher No 20: "There is a serious imbalance between the use of technology and the acquisition of learning content."

Teacher No 21: "I would like students to help me more in teaching other students. When we were in class, they talked and helped. Now students do only their own things and do not help each other."

Teacher No 22: "I just don't have time for professional and self-development."

Teacher No 23: "The offer of courses and seminars is so big that I am confused. The school management is forcing me to take courses that I think are unnecessary."

Teacher No 24: "I would like to tell students about the relationship between circular economics and mathematics, but I don't know how."

Teacher No 25: "I would like to know how to calculate the mathematical gaps caused by the Covid-19 pandemic."

Teacher No 26: "The Covid generation has become lazier, but more demanding of higher quality content, presentations and digital materials that we do not have much yet."

Teacher No 27: "Students have symptoms of depression and I find it difficult to understand how to support their social and emotional needs."

Teacher No 28: "Parents can be more helpful in learning math for junior students than for senior students."

Teacher No 29: "I tried very hard to work with the individual needs of each student, both the talented and the hard, but it took too much time".

In focus group interviews, teachers indicated that, given the background created by the Covid-19 pandemic, they needed significant support from school leaders, policy makers, academics and scientists to understand the conditions and ways to learn mathematics in a more innovative, faster and efficient way in the future.

In summarizing, structuring and analyzing the set of problems identified, the mathematics teachers made a careful selection and suggested that the following eleven key elements should be included in the future model of support system:

- Understanding of the framework of normative and strategic documents;
- Research and data analytics;
- Choice of learning strategies;
- Identification of technology support;
- Creating a supportive learning environment;
- Promotion of cooperation;
- Ensuring balance;
- Bridging social gaps;
- Use of science-based research;
- Professional development;
- Identifying and anticipating future trends.

Following the recommendations of mathematics teachers, the key elements of the ALOM support system are designed as a daily guide and a guide for mathematics teachers, practitioners, school leaders and policy makers in order to help to take into account all circumstances and create a supportive environment for accelerated learning. This is just a suggestion that any user can supplement and develop according to one's vision.

Understanding of the Framework of Normative and Strategic Documents. Teacher focus group interviews have shown that many teachers have difficulty teaching mathematics content and developing practical tasks in line with program objectives. Teachers also have incomplete knowledge about mathematics curricula, about the development of lesson plans, about the possibilities of solving pedagogical problems. It is therefore important to support teachers by educating and raising awareness of national regulations and strategic documents that contribute to the achievement of educational goals not only at the classroom level, but at the national level. For example, the development of the Latvian mathematics curriculum takes into account the statement of the Sustainable Development Strategy "Latvia 2030": education must be high-quality, lifelong, creative-oriented in order to respond to the challenges of global competition and demography [36]. For example, In September 2018, President Xi Jinping remarked at the National Education Conference that China's focus should shift from 'capacity' to 'quality', and that the modernization of education should support the modernization of China. In 2019, the Chinese State Council published two significant plans: 1) China's Education Modernization 2035 Plan and 2) the Implementation Plan for Accelerating Education Modernization (2018–2022), are aimed at substantially modernizing China's education system by 2035, the year the country is determined to realize socialist modernization and become an education powerhouse [37].

Research and Data Analytics. Big data can show big solutions. It is especially important for teachers to learn Data Analytics to understand how data can work for us. If information could once be obtained through conversations, libraries and observations, now technology helps. Data is stored on the Internet, mobile devices, servers and elsewhere. There is a lot of data and fragmentation, but you need to be able to select and use it wisely. How and why we access information and what we do with it is more important than the information itself to ensure personalization, knowledge persistence and transfer. How and why students learn and what they do with the resources they teach is more important than the content itself! Researchers point out that it is important to analyze existing data, standards and information in order to determine the level of each student and develop individual developmental trajectories in accordance with the most current learning standards [14].

Choice of Learning Strategies. Properly chosen learning strategies usually provoke students to think actively and engage in the learning process. Researchers [38] indicate that deep involvement or immersion in activities and tasks creates positive emotions, enjoyment and well-being. Pupils' concentration increases if learning challenges and approaches match their skill levels and individual differences. Successfully chosen strategies are an important support for teachers to make mathematics a more accessible and meaningful subject [39]. Modern math lessons need to be interactive, student-centered, collaborative, on-demand, enriched with relevant technologies that generate interest and curiosity. For example, inspirational stories can activate brain activity, allowing you to engage emotionally in the learning process and better understand the concepts of mathematics. Telling stories about historical events in mathematics or the use of numbers in current environmental or social problems can facilitate content perception and increase student engagement.

Based on the identified problems during distance learning, neuroscience research and pedagogical success stories, the following technology-based interdisciplinary approaches to ALOM can be suggested (Table 1):

Table 1. Technology-based interdisciplinary approaches to ALOM.

No	Name	Description
1	Contextualizing	Mathematics through the prism of history
2	Problem-centering approach	Applying math and science to solving real-world problems
3	Visualization	Combinations of mathematical numbers and images
4	Active approach	Mathematics in step with sports
5	Parallel approach	Mathematics in combination with a foreign language
6	Modeling	Math applications
7	Coding and programming	Math through logic and algorithms
8	Artificial Intelligence	Synergy of mathematics and neuroscience

Contextualizing – Mathematics Through the Prism of History. Research [40] indicates that contextualization is a review of mathematics and science in history. It shows the connections between the theories of science and mathematics and their historical and cultural roots. The historical foundations of a particular mathematics topic can serve as a core for a better understanding and more effective acquisition of the topic. In the lessons, it is valuable to create a link between the natural sciences and the humanities by creating an interdisciplinary link. For several years now, in the course of studying applied mathematics with students, studying specific topics, we look at the history of the origin of this topic. A good example of this approach can be seen by studying the topic of solving systems of linear equations with the Gaussian method, when students find the answers to many mathematically mediated questions: Why is the ingenious German scientist Karl Gauss called the King of Mathematicians and the Innovator of Science? How can I count the numbers from 1 to 100 the fastest? How to divide a circle into equal parts using a circle and a ruler? How did number theory become science? How can astronomy solve complex mathematical problems? What are trigonometric or Gaussian sums? How did the surface theory for the development of geodesy come about? How did the unit of measurement "gauss" in physics come about? The example shows that in one lesson, mathematics can successfully meet history, science, art, astronomy, astrology, geography and physics to stimulate students' interest in real applications of mathematics. The strength of this approach is the creation of a cultural-historical reference, which forms a strong basis and support for the development of students' personal knowledge of mathematics.

Problem-Centering Approach – Applying Math and Science to Solving Real-World Problems. Research [41] indicates that problem-focused approaches give students the opportunity to explore mathematics for themselves and offer sensible solutions. All attention in the classroom is focused on the problem to be solved, as a result of which there is an active discussion both among the students and between the teacher and the students. In addition, these types of classes can take place in nature, outside classrooms, using digital technologies and other aids (maps, tape measures, rulers, compasses, measuring instruments, thermometers, etc.). The strength of this approach is personal involvement, which increases motivation and interest. Leading a math group for 3rd to 7th grade students during the Covid-19 pandemic, when distance learning was identified and students mostly learned from home, we developed 10 sets of tasks that encouraged students to go to nature to find the information they needed and make calculations independently. For example, one of the tasks was as follows: A number is called symmetric if it can be read equally from the right and the left. There has been one important symmetrical year in the development of the "Kalnamuiža" park territory. Find this year on the information board of the park territory. Name the next symmetric year. Calculate the difference between the two symmetric years. Is the difference a symmetric number? If not, can the digits make a symmetric number and which one?

Visualization - Combinations of Mathematical Numbers and Images. It is said that a picture is worth a thousand words. This is especially true in mathematics, where an image or some other type of visual model can be useful in describing a mathematical idea [42]. Over the last twenty years, with the development of computer visual imaging capabilities, a new field of mathematics called visual mathematics has emerged. Especially for younger students, it is important to work with constructions and colors. In this way, students can better see how planes and three-dimensional shapes are formed and how formulas work. Today, many students have a visual perception. They learn the substance best when they can see what is happening, and a non-visual approach can even hinder their efforts to solve the problem. A good example of a visualization strategy for learning mathematics is the famous Königsberg "bridge walking" problem, which can be easily explained by creating a graph diagram as a visual representation of the situation.

Active Approach - Mathematics in Step with Sports. Parents, educators and health professionals have pointed out that today's children do too little exercise, which is further exacerbated by distance learning at home. It is well known that without movement a child cannot grow up healthy. The more and different movements a child has, the more active the brain is and the more intense intellectual development takes place. The movement creates emotional well-being and improves perception. Pre-school and primary school teachers in particular have given a lot of thought to developing an interdisciplinary approach between sports and math lessons. But the most innovative solution has been created by the Canadian technology company LÜ (https://play-lu.com/). It has developed an interactive area that uses light, sound and video effects to transform any gym into an engaging and comprehensive learning environment. LÜ is a smart space that understands the behavior and interaction of the people in it in real time. Using information from ceiling-mounted 3D cameras, students can learn math from wall-mounted tasks in sports.

Parallel Approach - Mathematics in Combination with a Foreign Language. In Latvian schools, English is mostly studied as the first foreign language and German or Russian is chosen as the second foreign language. Taking into account that teachers often lack digital mathematics tasks and worksheets in Latvian, they search for them on the Internet in foreign languages. In this way, an interdisciplinary approach is promoted while learning mathematics and a foreign language. Open tasks can be mastered in this way especially productively. A study [43] indicates that tasks are considered open if their starting or target situation is not specified. Open problems usually have more than one solution and can be accomplished in more than one way. This leaves students free to tackle the task and allows them to use different ways of thinking. The approach works very well in mixed classrooms, where the level of knowledge of the students is from the lowest to the highest, because everyone can provide answers according to their knowledge and skills. The teacher has the opportunity to find tasks, the simplest of which are possible for all students in the class, but higher levels of difficulty challenge the abilities of the most talented students. A good example of an open problem strategy in mathematics lessons is the task: 10 participants come to the seminar, and each shakes hand with each other exactly once. How many handshakes have occurred? There is only one answer - and it is 45, but there are at least seven methods of solving: looking at all possibilities, with the help of graphs, with the help of visualization, with the help of a table, with the method of mathematical induction, with combinatorial formulas. Finally, students can be asked to find a general case for the task.

Modeling - Math Applications. Not all students love math at first glance. To some it may seem tedious and complicated, especially if only theory, formulas and methods are taught. Math lessons can be made more interesting and digitally engaging with apps that can be found on the Internet for free and for different ages. The following applications, for example, will provide an interdisciplinary and visually appealing approach between mathematics and technology lessons: Mathspace (https://mathspace.co/us), CK 12 (www.ck12.org), Shapes 3D (www.mathsisfun.com), Khan Academy (www.khanacademy.org), Photo Math (photomath.app), GeoGebra (www.geogebra. org), etc. In turn, the application of mathematics and modeling will be helped by more professional programs, such as MATLAB (www.mathworks.com) or Wolfram|Alpha (www.wolframalpha.com).

Coding and Programming - Math Through Logic and Algorithms. Coding and programming are great helpers and powerful methods for students to better understand mathematics. For example, first-year students of the Faculty of Engineering learn applied mathematics much more quickly and train algorithmic thinking if, in parallel with theoretical lessons, the functions of the mathematics math module of the Python programming language are mastered. Primary school students, on the other hand, become much more interested in mathematics and develop logical thinking more deeply if they attend robotics lessons and are able to write working codes for robot movements without errors.

Artificial Intelligence - Synergy of Mathematics and Neuroscience. The highest point of the interdisciplinary approach is AI solutions that promote socialization, quick feedback, interactivity, involvement, multiple repetition, generation of different tasks according to each student's individual abilities. Research in the new field of neuro-education [44, 45] emphasizes that, in collaboration with AI, individual brain activity data can be used in the future to understand each student's strengths and weaknesses and make math learning much faster, deeper and more personalized. AI and mathematics have the strongest synergies, as AI is based on mathematics (linear algebra, statistics, probabilities, logic, etc.) and AI-based solutions can be successfully used to learn math faster. Researchers [46] describe an example where an online adaptive learning solution called "Mathematical Adaptive Platform" (PAM) has been developed in Uruguay, the content of which is adapted to the national mathematics curriculum. PAM provides personalized feedback according to each student's skill level, based on an analysis of the student's experience. PAM provides assistance to students through more than 25,000 differentiated tasks and 2,800 feedbacks to explain the solution to each task.

Identification of Technology Support. In order to be able to take an active part in society, many cross-cutting competencies are needed, of which digital skills are essential. In households, the digital transformation has largely taken place, but the situation in the field of education is ambiguous. If today's children are to enter the labor market freely in the future, they must acquire digital skills from the start of school. It is no longer a matter of showing the use of digital devices at school, students learning to create and program smart tools and robots. This requires a digital transformation of education and the use of appropriate technologies. In this case, researchers recommend the use of (1) The Technology Integration Matrix (TIM) as a good tool, which provides a framework for describing the use of targeted technologies for a quality learning process. A total of five meaningful learning environment characteristics (active, collaborative, constructive, authentic, goal-directed) and five levels of technology integration (entry, adoption, adaptation, infusion, transformation) form a productive 25-cell evaluation matrix [47], (2) SAMR (substitution, augmentation, modification, redefinition) model as a valuable tool for determining the impact of technology on the learning process, especially in the online learning process [48], (3) TPACK (technological, pedagogical, content, knowledge) framework, which describes knowledge necessary for teachers to effectively integrate technology into the teaching process [49].

Creating a Supportive Learning Environment. Math classroom promotes immersion and involvement in the learning process. As a result, the learning process becomes more meaningful, cognitive, creative, safer, more equal for all and more relevant to intellectual, mental and physical development. In this way, the learning environment is oriented towards mutual trust, cooperation and support, as well as promotes students' motivation and independence. A high-quality learning environment for a mathematics subject can be created not only in the classroom, but also in a library, museum, company, park or forest, by adjusting the duration of studies accordingly.

Promotion of Cooperation. Collaborative learning usually affects all partners. In mathematics, collaboration is especially important to influence the thinking, perseverance and understanding of at least one (at best each) student. According to researchers [50], when students work together on a task, they develop a coordinated approach to meaning making and a goal-oriented approach. Sharing and solving tasks allows students to communicate and talk to make joint decisions. Technology can be seen as a positive tool for equal interaction in the learning process to support social interactions and thus change the nature and effectiveness of these interactions [50].

Ensuring Balance. Fear of math is for most people, and the problem of math anxiety has been one of the most studied for many years. Researcher Ashcraft has pointed out that math anxiety is a feeling of tension and fear that hinders the full mastering of mathematics. It can appear at any age. The higher the student's stress, the lower his/her math skills and motivation [51]. A team of researchers [52] has shown that emotional well-being is inextricably linked to motivation. Therefore, in the process of learning mathematics, it is very important to strike a balance between motivation and achievement, so that students work with pleasure and joy [52]. It is equally important to take into account the pace of development in technology and education. Where education was once a driver of technological progress, technology now dominates educational capacity, which reduces prosperity and creates social inequalities. Given the rapid pace of technological development, meaningful and fundamental changes in education are needed to achieve more inclusive and sustainable development for all [53, 54]. This is clearly shown in Fig. 4 - if motivation has a higher priority than results in mathematics lessons, then there is joy and well-being, otherwise students develop fear and stress. If technology is used in a balanced and thoughtful way in mathematics lessons, students become motivated, otherwise there is discomfort and social tension. In those periods when a balance is achieved between motivation and results, as well as between technology and the content of education, a "feeling of happiness" emerges, which is also our greatest goal.

Bridging Social Gaps. As one of the most innovative pedagogical strategies of 2020, The Open University named Social Justice Pedagogy in its annual publication Innovating Pedagogy, which aims to educate and enable students to become active participants who understand social discomfort and can help build equality in the classroom. Researchers [55] point out that teachers need to identify students' interests, unique experiences and needs in order to apply an appropriate and equitable distribution of learning resources and technologies that allow everyone to integrate productively into the learning process.

Use of Science-Based Research. Ensuring a successful mathematics learning process requires a research-based and interdisciplinary approach. Recent research in the fields

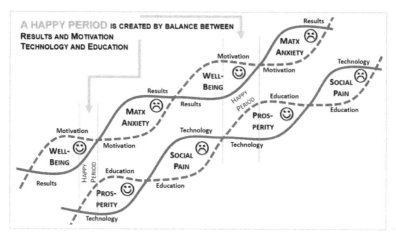

Fig. 4. A happy period is created by balance [8].

of education, psychology and neuroscience can enrich teachers' understanding of more effective teaching approaches. There have been many neuroscience studies in recent years that explain the link between mathematical thinking and brain activity. Neuroscience research has been able to uncover information that is not visible to the eye at the level of student behavior. Researchers have found that each part of the brain is subject to specific perceptual and cognitive processes that successfully process visual and audible information, perceive numerical information faster, or hold attention for longer [56]. Understanding the neurocognitive processes that underpin the development of mathematical competence can provide important insights into the optimization of mathematical education [57].

Professional Development. Professional development must become an integral part of every school's culture, as the education sector is evolving rapidly and teachers must not afford not to learn. Teachers usually improve during their free time when using course, seminar and conference offers. There should be no teachers who complain about classroom behavior but do not attend workshops on relevant topics themselves. There should be no teachers who do not understand the new teaching standards but do not attend long-term planning courses. Also, an athlete cannot become a winner if he participates only in competitions, but does not train every day [58]. Professions change so fast that you have to find time for continuous self-improvement.

Identifying and Anticipating Future Trends. The ability to research and anticipate oneself must be acquired through successive processes of reflection and action. It is the ability to take a critical position, to take responsibility, to decide, to choose and to act in a way that is sometimes already known. It is the ability to look at situations from different and different perspectives. According to OECD research, foresight mobilizes cognitive skills (analytical and critical thinking) to predict what may be needed in the future in a particular area or how actions taken today may affect the future. Both reflection and anticipation are harbingers of responsible action [3].

5   Conclusions and Further Studies

In this article, we presented the partial results of the postdoctoral research project "Artificial Intelligence (AI) support for accelerated math acquisition approach (AI4Math), No. 1.1.1.2/VIAA/3/19/564", the main goal of which is to develop an interdisciplinary support strategy of mathematics for general education schools thus promoting the 5th area "ICT" and the 5th growth priority "We need to create a modern education system that meets the requirements of the future labor market and promotes economic transformation" set within the framework of the Latvia Strategy of Smart Specialization.

Several student surveys and focus group interviews with Latvia teachers of mathematics revealed the main problems encountered in learning mathematics during the distance online training in 2020 and 2021. Teachers indicated that taking into account the background of the Covid-19 pandemic, they needed significant support from local governments and policy makers to maintain the emotional balance affected by technological progress, and from academics and scientists to understand the conditions and ways to learn math smarter, faster and more efficiently in the future. Teachers acknowledged that they felt lonely and vulnerable, that their horizons were not as broad as they should be, that schools had not worked hard to support teachers, and that there was a lack of full collaboration between teachers to share experiences and develop interdisciplinary approaches.

To answer to the study's key question "How can we support those who help shape the future?" 53 Latvia teachers of mathematics worked together to select and recommend the following 11 basic elements for the future ALOM model of support system: 1) Understanding of the framework of normative and strategic documents; 2) Research and data analytics; 3) Choice of learning strategies; 4) Identification of technology support; 5) Creating a supportive learning environment; 6) Promotion of cooperation; 7) Ensuring balance; 8) Bridging social gaps; 9) Use of science-based research; 10) Professional development; 11) Identifying and anticipating future trends.

100% of all teachers indicated that the ALOM framework developed in the study will be an important support for the future development of school curricula and the environment. It will help teachers and provide quality mathematics opportunities for every student to develop and fulfill their potential throughout their lives.

The study has suggested that educators need to raise awareness of the benefits of the ALOM principles. An important contribution of the research is the collection and systematization of existing but fragmented knowledge in a transparent system, which provides a valuable basis for planning and promoting learning in the future. ALOM is a new and evolving field, so the proposed model is innovative but with a recommendatory nature. It is designed as a daily guide and helper for math teachers, practitioners, school leaders and policy makers to help take all circumstances into account and create a supportive environment for accelerated learning. It can also be used successfully to support teachers in other subjects. And anyone interested can open it with the help of a QR code (see Fig. 5) and add new actual and future-oriented cases.

Our research shows how many more challenges will need to be addressed in the context of fast technological development. Impacts will vary at the level of individuals, schools and also at national level. No matter how large and fast each development has been, there is an urgent need to develop skills and new approaches to learning. Fast

Fig. 5. ALOM definition and Support System [8].

and targeted intervention by policy makers and industry professionals will be needed. Educational leaders in turn should ask themselves the following questions: Do we use design thinking and stimulating strategies to improve learning processes and decision-making? Are we encouraging our teachers to broaden their horizons? Are we taking into account advances in neuroscience and the potential of Artificial Intelligence to create deeper, faster, and more personalized learning?

Researchers in the field of education need to conduct additional research to clarify the following issues: "Why are students still so low motivated to learn math?", "Why do teachers use interdisciplinary approaches so little in the learning of math?", "Why do students think that it is not possible to learn mathematics through distance learning?", "How to relieve the work of math teachers with smart approaches and technology support?".

Acknowledgments. The research is carried out within the framework of the postdoctoral project "Artificial Intelligence (AI) Support for Approach of Accelerated Learning of Mathematics (AI4Math) (1.1.1.2/VIAA/3/19/564)" at Vidzeme University of Applied Sciences with the support of ERAF.

References

1. Panagiotopolos, G.A., Karanikola, Z.A.: Education 4.0 and teachers: challenges, risks and benefits. Eur. Sci. J. ESJ **16**(34), 114–120 (2020). https://doi.org/10.19044/esj.2020.v16n34p114

2. Cunska, A.: Technology-based interdisciplinary approaches to accelerated learning of mathematics. In: Proceedings of the 13th International Conference on Computer Supported Education - Volume 2: CSEDU, pp. 114–121 (2021). https://doi.org/10.5220/0010473901140121

3. OECD: OECD Future of Education and Skills 2030. OECD Learning Compass 2030. A series of concept notes. Well-being 2030 (2019)

4. Boaler, J.: Limitless Mind. Learn, Lead and Live Without Barriers. CI Group (UK) Ltd., Croydon (2019). 248 pp

5. Sawchuk, S., Sparks, S.D.: Kids Are Behind in Math Because of COVID-19. Here's What Research Says Could Help, Education Week Homepage. https://www.edweek.org. Accessed 21 Oct 2021

6. Elliott, K., Juliuson, J., Katz, N., Parris, O.J.: In Support of Educators: Strategies that Work. EDC, Education Development Center, Waltham (2021)
7. IZM: Mācību gada noslēguma aptaujas IZM un Edurio aptauju rezultāti (2020)
8. Cunska, A.: Taxonomy of technology-based support for accelerated learning of school math. In: Proceedings of ICERI 2021 (The 14th Annual International Conference of Education, Research and Innovation Conference 8th–9th November 2021), pp. 3647–3656 (2021)
9. Andrews, D.H., Fitzgerald, P.C.: Accelerating Learning of Competence and Increasing Long-term Learning Retention, U.S. Air Force Research Laboratory. Warfighter Readiness Research Division, Arizona (2010)
10. Sottilare, R., Goldberg, B.: Designing adaptive computer-based tutoring systems to accelerate learning and facilitate retention. Cogn. Technol. **17**(1), 19–33 (2012)
11. Qomario, Q.: Pengaruh Pendekatan accelerated learning Terhadap Kemampuan Pemecahan Masalah Matematis. J. Elem. Sch. (JOES) **1**(2) (2018). https://doi.org/10.31539/joes.v1i2.364
12. Meier, D.: The Accelerated Learning Handbook: A Creative Guide to Designing and Delivering Faster. More Effective Training Programs. McGraw-Hill Education – Europe, New York, NY, United States (2000)
13. Alcenter Homepage. www.alcenter.com. Accessed 11 Oct 2021
14. NIET Homepage. https://www.niet.org/. Accessed 07 Oct 2021
15. Duval, A.: Everyone Can Learn Mathematics to High Levels: The Evidence from Neuroscience that Should Change our Teaching. AMS Blogs. On Teaching and Learning Mathematics (2019)
16. Maguire, E.A., et al.: Navigation-related structural change in the hippocampi of taxi drivers. Proc. Natl. Acad. Sci. **97**(8), 4398–4403 (2000). https://doi.org/10.1073/pnas.070039597
17. Coyle, D.: The Talent Code: Greatness Isn't Born, It's Grown, Here's How. Bantam Books, New York (2009)
18. Moser, J., Schroder, H.S., Heeter, C., Moran, T.P., Lee, Y.H.: Mind your errors: evidence for a neural mechanism linking growth mindset to adaptive post error adjustments. Psychol. Sci. **22**, 1484–1489 (2011)
19. Zhu, Z.-T., Yu, M.-H., Riezebos, P.: A research framework of smart education. Smart Learn. Environ. **3**(1), 1–17 (2016). https://doi.org/10.1186/s40561-016-0026-2
20. Szymkowiak, A., Melovic, B., Dabic, M., Jeganathan, K., Kundi, G.S.: Information technology and Gen Z: the role of teachers, the internet, and technology in the education of young people. Technol. Soc. **65**(C) (2021). https://doi.org/10.1016/j.techsoc.2021.101565
21. Chandra, S.R.V.: Blended learning: a new hybrid teaching methodology. J. Res. Sch. Prof. Engl. Lang. Teach. JRSP-ELT **13**(3) (2019)
22. Kalantarnia, Z., Rostamy, M., Shahvarani, A., Behzadi, M.: The study of application of algebrator software for mathematical problems solving. Math. Educ. Trends Res. 1–7 (2012). https://doi.org/10.5899/2012/metr-00003
23. Guo, T., Han, C.: Artificial intelligence mechanism and mathematics implementation methods. Scientia Sinica Mathematica **50**(11), 1541–1578 (2020). https://doi.org/10.1360/SSM-2020-0031
24. Southgate, E., Blackmore, K., Pieschl, S., Grimes, S., McGuire, J., Smithers, K.: Artificial Intelligence and Emerging Technologies in Schools. Research Report. Commissioned by the Australian Government Department of Education (2019)
25. Kim, E.: The case for interdisciplinary education: a student's perspective. NJCSS J. Soc. Stud. (2020)
26. AldertKampAdvies Homepage. https://www.aldertkamp.nl/post/interdisciplinary-education-a-wave-of-the-future. Accessed 09 Oct 2021
27. RSE: Embedding interdisciplinary learning in Scottish schools. Advice paper. Roy. Soc. Edinburgh **20**(2) (2020)

28. McPhee, C., Bliemel, M., Bijl-Brouwer, M.: Transdisciplinary innovation. Technol. Innov. Manag. Rev. **8**(8), 3–6 (2018). http://hdl.handle.net/10453/127522
29. Creswell, J.W.: Research Design. Qualitative, Quantitative, and Mixed Methods Approaches, 3rd edn. Sage Publications, California (2009)
30. Patton, M.Q.: Qualitative Research & Evaluation Methods, 3rd edn. Sage Publications, California (2008)
31. Hoffman, J.: Triangulation: Raise Your Probability of Making Good Decisions. https://jay mehoffman.medium.com/triangulation-raise-your-probability-of-making-good-decisions-687e92176222. Accessed 07 Oct 2021
32. Farina, G.: Some reflections on the phenomenological method. Dialogues Philos. Ment. Neuro Sci. **7**(2), 50–62 (2014)
33. FORMPLUS Homepage. https://www.formpl.us/blog/ethnographic-research. Accessed 01 Oct 2021
34. METHODSPACE Homepage. https://www.methodspace.com/what-is-narrative-research-methodology/. Accessed 05 Oct 2021
35. Center for Building a Culture of Empathy Homepage. http://cultureofempathy.com/Refere nces/Experts/David-Kelley.htm. Accessed 25 Oct 2021
36. Skola2030 Homepage. https://www.skola2030.lv/lv/jaunumi/3/valdiba-pienemts-valsts-pam atizglitibas-standarts. Accessed 09 Sept 2021
37. Australian Government: China's Education Modernisation Plan Towards 2035. Australian Government, Department of Education, Skills and Employment (2020)
38. Kukulska-Hulme, A., Bossu, C., Coughlan, T., et al: Innovating Pedagogy 2021: exploring new forms of teaching, learning and assessment, to guide educators and policy makers. Open University Innovation Report 9. Institute of Educational Technology. The Open University (2021)
39. Persico, A.: What Math Teaching Strategies Work Best? 16 Math Education Experts Share Their Suggestions (2019)
40. Nikitina, S., Mansilla, V.B.: Three Strategies for Interdisciplinary Math and Science Teaching: A Case of the Illinois Mathematics and Science Academy. GoodWork@ Project Series, Number 21 (2003)
41. Biccard, P., Wessels, D.C.J.: Problem-centred teaching and modelling as bridges to the 21st century in primary school mathematics classrooms (2011). https://directorymathsed.net/dow nload/Biccard.pdf. Accessed 01 Oct 2021
42. Tenannt, R.: Visualizing Mathematics: Imagery Techniques for Learning Abstract Concepts (2006)
43. Pehkonen, E.: Open-Ended Problems: A method for an educational change (1999)
44. Bidshahri, R.: Neuroeducation Will Lead to Big Breakthroughs in Learning (2017)
45. Wilson, D., Conyers, M.: Five Big Ideas for Effective Teaching: Connecting Mind, Brain, and Education Research to Classroom Practice. Teachers College Press, New York (2013)
46. Perera, M., Aboal, D.: The Impact of a Mathematics Computer-Assisted Learning Platform on Students' Mathematics Test Scores. Maastricht Economic and Social Research Institute on Innovation and Technology (UNU-MERIT) (2019)
47. Ruman, M., Prakasha, G.S.: Application of Technology Integration Matrix (TIM) in teaching and learning of Secondary School Science subjects. IOSR J. Humanit. Soc. Sci. **22**(12), 24–26 (2017)
48. Terada, Y.: A Powerful Model for Understanding Good Tech Integration. Technology Integration. Edutopia, George Lucas Educational Foundation (2020)
49. Koehler, M.J., Mishra, P., Kereluik, K., Shin, T.S., Graham, C.R.: The technological peda-gogical content knowledge framework. In: Spector, J.M., Merrill, M.D., Elen, J., Bishop, M.J. (eds.) Handbook of Research on Educational Communications and Technology, pp. 101–111. Springer, New York (2014). https://doi.org/10.1007/978-1-4614-3185-5_9

50. Zurita, G., Nussbaum, M.: Computer supported collaborative learning using wirelessly inter-connected handheld computers. Comput. Educ. **42**, 289–314 (2004). https://doi.org/10.1016/j.compedu.2003.08.005

51. Ashcraft, M.H.: Math anxiety: personal, educational, and cognitive consequences. Curr. Dir. Psychol. Sci. **11**(5), 181–185 (2002). https://doi.org/10.1111/1467-8721.00196

52. Schweinle, A., Turner, J.C., Meyer, D.K.: Striking the right balance: students' motivation and affect in elementary mathematics. J. Educ. Res. **99**(5), 271–293 (2006). https://doi.org/10.3200/JOER.99.5.271-294

53. OECD: The Future of Education and Skills. Education 2030. The Future We Want (2018)

54. Goldin, C., Lawrence, F.K.: The Race Between Education and Technology. The Belknap Press of Harvard University Press, Cambridge (2010)

55. Kukulska-Hulme, A., Beirne, E., Costello, E., et al: Innovating Pedagogy 2020: exploring new forms of teaching, learning and assessment, to guide educators and policy makers. Open University Innovation Report 8. Institute of Educational Technology. The Open University (2020)

56. Menon, V., Chang, H.: Emerging neurodevelopmental perspectives on mathematical learning. Dev. Rev. **60** (2021). https://doi.org/10.1016/j.dr.2021.100964

57. Hyde, D.C., Ansari, D.: Advances in understanding the development of the mathematical brain. Dev. Cogn. Neurosci. **30**, 236–238 (2018). https://doi.org/10.1016/j.dcn.2018.04.006

58. Pillow, A.: We All Hate Professional Development, But Some of Y'all Need It. Education Post Homepage. https://educationpost.org/we-all-hate-professional-development-but-some-of-yall-need-it/. Accessed 07 Oct 2021

Investigating STEM Courses Performance in Brazilians Higher Education

Laci Mary Barbosa Manhães[1]([envelope]) [iD], Jorge Zavaleta[2] [iD], Renato Cercear[3,4] [iD], Raimundo José Macário Costa[5] [iD], and Sergio Manuel Serra da Cruz[2,5] [iD]

[1] Department of Science (PEB), Fluminense Federal University, Estr. João Jasbick s/n, Santo Antônio de Pádua, Brazil
`mary_manhaes@id.uff.br`
[2] Federal University of Rio de Janeiro, PPGI/UFRJ, Rio de Janeiro, Brazil
`{jorge.zavaleta,serra}@ppgi.ufrj.br`
[3] Department of Education and Research, INC: Instituto Nacional de Cardiologia, Rio de Janeiro, Brazil
[4] Telehealth Center, Rio de Janeiro State University, Rio de Janeiro, Brazil
[5] Department of Computer Science, Federal Rural University of Rio de Janeiro, Seropédica, Rio de Janeiro, Brazil
`macario@ufrrj.br`

Abstract. STEM (Science, Technology, Engineering, and Mathematics) has become a popular term in Education worldwide and has received increasing attention in recent years in Brazil. However, developing a reliable assessment of interdisciplinary learning in STEM courses has been challenging. This chapter performed an extended investigation on all the Brazilian STEM courses using the National Assessment of Student Achievement (ENADE) datasets and improved our previous innovative methodology. ENADE datasets are maintained by National Institute for Educational Studies and Research Anísio Teixeira and correspond to data about all final year STEM students who performed the exams in 2005, 2008, 2011, 2014, and 2017. Our dataset contains 527,058 events; it was used to compare groups of female and male students' academic performances, the grades obtained through the years, the mean ages, and compare the performance between the types of Brazilian universities.

Keywords: Data science · Data analysis · Educational data mining · ENADE · STEM · Tertiary education

1 Introduction

Scientific and technological development is based on science, technology, engineering, and mathematics (STEM) professionals who generate the flow of scientific discoveries and technological innovations that allow the economy to move, grow, and consequently generate well-paid jobs, and social welfare improves. Achieving professional maturity is a consequence of cognitive development and skills learned in the student phase that allows one to face a variety of occupations and develop as a citizen. However, students

© Springer Nature Switzerland AG 2022
B. Csapó and J. Uhomoibhi (Eds.): CSEDU 2021, CCIS 1624, pp. 212–231, 2022.
https://doi.org/10.1007/978-3-031-14756-2_11

who cannot specialize in STEM sometimes move to another field or drop out of higher Education. Altogether, in part due to deficiencies in teaching, learning, and STEM student support [1].

In the North American undergraduate system, the improvement in STEM is continuous. Both teaching and learning opportunities contribute significantly to the growth of the country's economy, national security, health, and well-being of the people [1].

Like the US, Brazil must examine agreement mechanisms with the various STEM actors involved to encourage or increase investments or efforts to maximize to reduce dropout rates from STEM higher education degrees. Besides that, the country must support STEM teachers, inspire students to engage in STEM fields, and increase students from underrepresented groups in STEM courses.

Our contribution aims to support the use of Educational Data Education (EDS) to Brazilian National Assessment of Student Achievement datasets (ENADE) to benefit the educational institutions' academic staff and academic managers to improve the quality of higher Education. EDS is known as a specific application of Data Science in the field of Education [2]. For Cao [3], EDS uses relevant concepts related to Data Science and presents new opportunities in exploring new domains of data, such as economy and Education.

This chapter represents an extended version of the work presented at the 13th International Conference on Computer Supported Education - CSEDU [4]. We presented an expanded investigation of the characteristics of students of the final year enrolled in the STEM tertiary degrees in Brazil who performed the ENADE exams in 2005, 2008, 2011, 2014, and 2017, including a deeper analysis of previous elements and a new section related to the analysis of grades and types of university. It is organized as follows: Sect. 2 introduces the National Higher Education Exams, describes the context and performance measurement tools used in Brazil, Sect. 3 presents the ENADE Dataset, which contains the discussion of related works, the description of the dataset, and the analytical procedures, Sect. 4 presents the discussion and analysis of the distribution by sex, Sect. 5 presents the discussion and results about age distribution, Sect. 6 shows the discussion and results of the analysis of the notes and their different perspectives, Sect. 7 presents the analysis of the different categories of higher education institutions. Finally, the conclusions are described in the final section.

2   Brazilian National Higher Education Exams

The Brazilian educational system is complex and dependent on various national policies and administrative levels. The system generates abundant amounts of semi-structured data every year, and it must analyze in this research to seek relevant information for STEM related actors.

Brazilian Ministry of Education (MEC) coordinates a federal research agency devoted to obtaining information from primary education to higher education [5]. This agency is called as National Institute for Educational Studies and Research Anísio Teixeira (INEP). It gathers and maintains data and detailed information about private and public educational institutions.

INEP has the attribution to organize the events and apply the Brazilian National Assessment of Student Achievement (ENADE). The annual examination is part of the

National Higher Education Assessment System (SINAES). Those datasets are essential to understanding the Brazilian educational scenario.

ENADE exam was created to assess the quality of the higher education system, to cover all higher education institutions throughout Brazil, such as university centers, universities, and non-university institutions, from public or private sectors [5]. It aims to measure and evaluate the performance of distance education or classroom students based on the information about the skills, competencies acquired during their courses, and the content of the course program [6]. In short, it was designed to evaluate the tertiary education system using questions related to three main components: undergraduate course evaluation, performance evaluation of students, and institutional evaluation.

Each year-end, three major groups of courses, are requested to invite students to take the ENADE exams. The educational institutions must enroll students in a set of subject areas classified as (1) STEM, pedagogical, literature, and related fields; (2) Health, agriculture, natural resources, and related fields; and (3) Social sciences, humanities, culture, and design fields. A three-year cycle is scheduled in each area for both educational institutions and students [5].

The evaluation results of each cycle are published annually by INEP, attributing a final grade ranging from zero to five for each higher course in each educational institution. The higher the grade obtained by the students, the better the grade. Finally, the ENADE datasets are formally published on the INEP website [5].

3 Exploring the ENADE Dataset

Academic managers must analyze the datasets to assess their respective institutions or courses and compare incoming and graduating students [8, 9]. The ENADE datasets can help to estimate the performance and the efficiency of Brazilian higher education institutions. Another critical task is about supporting teachers' work.

A detailed analysis of ENADE datasets can reveal information and perceptions about the quality and competencies of the training of new Brazilian professionals. However, few educational institutions can afford sophisticated analysis or learn from the ready-made data, despite the vast quantity of educational datasets available. Time constraints and production goals force many academic managers to fill out administrative forms as simple data collectors, not decision-makers.

3.1 Related Works

There are several works in this area. Some authors explored minor parts of the datasets (Table 1) [10–15]. Initial studies using the ENADE 2014 dataset investigated students' performance [10–12], applying different statistical techniques and variables, and focusing on different areas. Later, a novel evolution was performed, expanding the students' performance studies to analyze multidisciplinary [13] and medical courses [14], using ENADE 2010 dataset.

Another approach integrates datasets in time, producing a composed dataset using ENADE 2004, 2007, 2010, and 2013 [15]. The authors used simple statistical methods to analyze undergraduate courses of dentistry (oral medicine).

The work considered state of the art in this area corresponds to a study analyzing all Brazilian STEM degree courses [4]. It integrates ENADE datasets 2005, 2008, 2011, 2014, and 2017. They considered all types of Brazilian academic organizations (university centers, universities, and non-university institutions), using public or private sectors data, and about in-class and distance learning. They focused on the ones who performed the ENADE exam as final-year students.

Table 1. Related works using ENADE datasets.

Authors	Dataset	Aims
Crepalde and Silveira (2016)	ENADE 2014	Investigate students' performance (originated from public and private institutions) considering sex, race, and financial income
Silva et al. (2017)	ENADE 2014	Analyze Math and Science students' performance; the variables were analyzed using multiple linear regression techniques and the Stepwise method
Vista, Figueiró and Mozzaquatro (2017)	ENADE 2014	Verify by statistical analysis the performance of undergraduate students in the Computer Science degree of the state of Rio Grande do Sul, Brazil
Santos and Noro (2017)	ENADE 2010	Compare the students' performance that participated in a specific multidisciplinary project called 'PET-Saúde'
Neto et al. (2018)	ENADE 2010	Investigate the factor that affects student performance in the Brazilian undergraduate medical programs; they considered seven variables associated with results and applied a multivariate analysis model of binary logistics regression
Moimaz, Amaral, and Garbin (2017)	ENADE 2004, 2007, 2010, and 2013	Use simple statistical methods to analyse undergraduate course of dentistry (oral medicine)
Manhães et al. (2021)	ENADE 2005, 2008, 2011, 2014, and 2017	Investigate all types of Brazilian academic organizations to study variables of STEM degree courses

3.2 Dataset Description

INEP started to apply ENADE exam in 2004. Since that time, it has offered its results as public files and microdata using the CSV format. Each file corresponds to a specific exam [5] published yearly on the INEP website following open data principles. As this becomes available, the information set about all ENADE exams and the data for each student who performed the exam is published as public records with anonymized information.

The register holds in columns the data attributes collected or generated during the registration or execution of the exam. Multiple students' attributes are presented in records, such as age, sex, grades, the tertiary degree in which the student is enrolled, and other attributes [5].

In short, it consists of two parts the register of students' performance. The initial part corresponds to a General knowledge (GK) evaluation which contain ten questions. A second part, referring to domain knowledge, with 30 multiple choice questions and the grades of three additional essay-type questions.

A final grade (FG) is calculated considering 25% of the GK grade and 75% of the DK grade. A data dictionary is presented in the Portuguese language, and it accompanies each microdata and describes all attributes.

It is essential to mention that many students have boycotted the first three cycles of the ENADE exam due to a political movement. Because of it, some of them gave no answers to the questions, which can justify the high number of null values. After that first cycle, the ENADE shows good adhesion. For example, a consolidation of ENADE datasets from 2005, 2008, 2011, 2014, and 2017 corresponds to 527,058 exams, corresponding to a significant adhesion. In 2020, due to the COVID-19 pandemic, the ENADE was not applied.

3.3 Analytical Procedures

Some steps were necessary to execute the analysis. Briefly, the steps consist of a sequence of execution that:

1. Downloads microdata from the INEP site.
2. Defines attributes to be analyzed.

   - Focused on degree code attribute, select the STEM final year students.
   - Executes data cleansing process and data checks (missing values and to identify invalid data).

3. Remove outliers and null data.
4. Executes data analysis, including statistics.

– Produce as graphics the data visualization.

As study variables, we used sex (tp_sexo), degree code (co_grupo), age of the student (nu_idade), and type of universities (such as private or public institutions). The grades variables were also included to analyze the information across private and public tertiary universities, using grades in General Knowledge (GK) (nt_fg), grades in the Domain Knowledge (DK) (nt_ce), and final grades (FG) (nt_ger).

We developed the analytical task using the R programming language (R Core Team, 2020), RStudio (RStudio Team, 2020), and Microsoft Excel (2018). Tables and figures were produced to enhance the explanations, and Boxplot was used to visualize outliers.

All ENADE public datasets (microdata) for the years 2005, 2008, 2011, 2014, and 2017 were obtained, and a filter was applied to select all students enrolled in the STEM tertiary degrees present in the microdata. After that, data analysis, including statistics, was performed the final year students enrolled in STEM tertiary degrees using four core features:

- Distribution by sex.
- Distribution by age.
- Analysis of grades, specifying the GK, DK, and, consequently, the exam's final grade FG.
- Analysis of grades, specifying the GK, DK, and the exam's final grade FG, between the private and public tertiary universities.

4   Analyses of the Distribution by Sex

Initially, to demonstrate the distribution of final year students enrolled in STEM tertiary degrees, we can tabulate the percentage of males and females (Table 2). A histogram helps to understand the information over the years (Fig. 1).

Table 2. The number of male and female final year students in STEM degrees by ENADE exams.

ENADE exams	Male STEM	Female STEM	Total of students
2005	24,612	16,468	41,080
2008	29,469	17,178	46,647
2011	48,351	27,725	76,076
2014	102,098	56,080	158,178
2017	133,532	71,545	205,077

A piece of important evidence was detected about the distribution by sex. A reduction year by year in the STEM field was observed in the percentage of females, and it means that it is predominately composed of males the workforce of the STEM field (see Fig. 1).

Some authors corroborate those observations [16–18]. They propagate that female's participation in the STEM field are decreasing in the past years [16], in the field of computer science was demonstrated a continuous under-representation of women [17], and similar results were identified when comparing degree programs in China, India, Russia, and the United States [18].

The Organization for Economic Co-operation and Development (OECD) produced the important document "Education at a Glance report" [19]. The 2020 edition provides a possible explanation for the under-representation of women in some fields. They suggest that women fear they will not have equal career opportunities in those fields after completing their education. Our research using Brazilian datasets emphasizes that women are being underrepresented in technology.

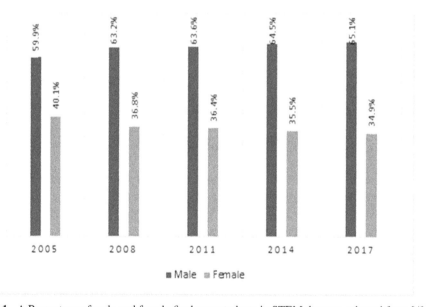

Fig. 1. A Percentage of male and female final year students in STEM degrees, adapted from [4].

5 Analyses of the Distribution by Age

This section was designed to identify the distribution by ages of the final year students to determine the students' average age in STEM degrees by sex.

We can verify that, over the years, females have been younger than males in all years (Fig. 2) in Brazilians STEM courses. As a reference, we can use the OECD indicators about the age of students in many countries [19].

We can see that mean the ages of males tend to increase. We can identify the evolution in 2005 (average 26.4 years old), 2008 (average 27.5 years old), 2011 (average 27.7 years old), and 2017 (average 27.4 years old).

It could be noted a lower mean age of the female students occurs in the same period. We identify the females' mean age evolution in 2005 (average 25.7 years old), 2008 (average 24.9 years old), 2011 (average 26.0 years old), 2014 (average 26.4 years old), and 2017 (average 25.8 years old).

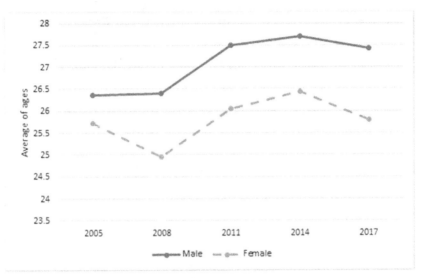

Fig. 2. Average age of final year students enrolled in STEM degrees by sex, adapted from [4].

Exploratory analysis shows that many male students over 30 years old are finishing STEM degrees (Fig. 3). The third quartile of the years (2011, 2014, and 2017) demonstrates that information. The median age is increasing year by year. Based on outliers' analysis, many males finishing over 40 years were noted (see Fig. 3).

Study the female boxplot show us that 25% of student females are finished the STEM degree over 26 years old (Fig. 4). The third quartiles from 2005 to 2017 are more regularity ranging from 26 to 28 years old.

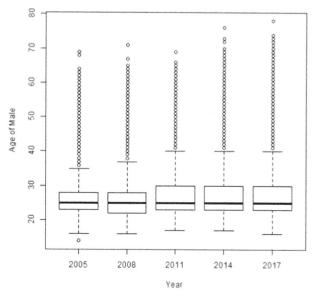

Fig. 3. A Boxplot of the age of male final year students in STEM degrees, adapted from [4].

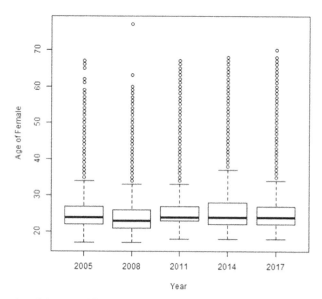

Fig. 4. A Boxplot of the age of female final year students in STEM degrees, adapted from [4].

6   Analysis of Grades, Specifying the GK, DK and Final Grade

We aim to estimate the level of knowledge acquired during the undergraduate course in STEM fields. After analysis, we obtained the trends of students' grades to foresee the future labor force performance in the STEM field.

ENADE registers have grades in general knowledge (GK), domain knowledge (DK), and final grades (FG), where the range of the grades is [0,100].

6.1   General Knowledge (GK) Grades

These grades refer to students' competencies and skills. They capture the level of knowledge about broad themes. It constitutes a proxy to education quality and corresponds to a part of the overall level of professional excellence.

Figure 5 demonstrates the decreasing tendency to the average of general knowledge (GK) grades that occur to males and females.

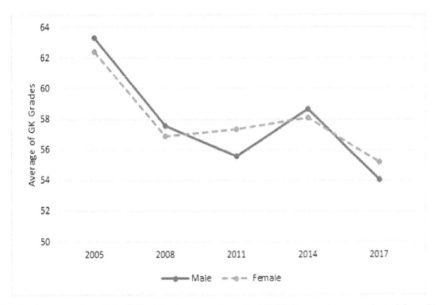

Fig. 5.   Average of grade in GK of final year students in STEM degrees by sex, adapted from [4].

Exploratory analysis from male and female boxplots about performance in GK (Fig. 6 and 7) demonstrate that the average is around 60% in both cases. As mentioned before, note that the number of outliers could be explained by the number of students that boycotted the first ENADE exams.

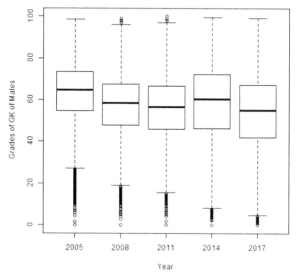

Fig. 6. Boxplot of male final year students in STEM degrees by GK grades, adapted from [4].

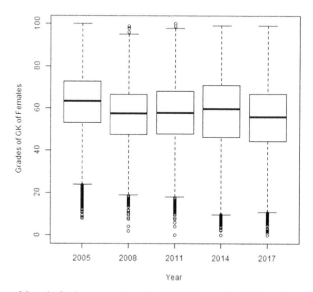

Fig. 7. Boxplot of female final year students in STEM degrees by GK grades, adapted from [4].

6.2 Domain Knowledge (DK) Grades

The grade intends to analyze the Competency-Based STEM Curriculum learned by the student to indicate the professional competencies and education qualities in the STEM field. In short, it intends to capture the level of knowledge of students in the STEM disciplines.

Notably, the grades obtained by males are more significant than that of females over time (Fig. 8). We can identify those lower initial grades explained by the number of students who boycotted the first ENADE exams.

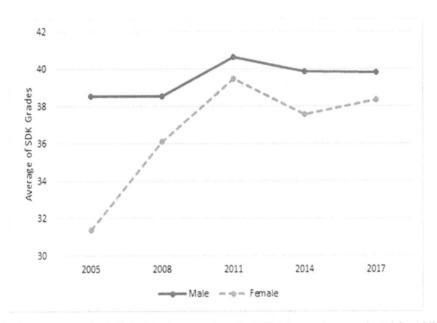

Fig. 8. Average grades in DK of final year students in STEM degrees by sex, adapted from [4].

Exploratory analysis was made using males and female boxplots about performance in DK (Fig. 9 and 10).

Male performance in DK shows a significant incidence of outliers (Fig. 9). As it indicates the number of students who obtained a grade over 80 in DK it shows a high level of future STEM professionals in the Brazilian market, implying a labor force improvement. However, the graphic shows that students are under 40 in around 50%, demonstrating an insufficient level of knowledge.

The female performance in the DK is lower than the male's performance (Fig. 10).

It is important to note that it is impossible to determine the reason for those results based exclusively on these results. These results reinforce the need to understand better and study the consequences for the labor force in STEM fields and female performance.

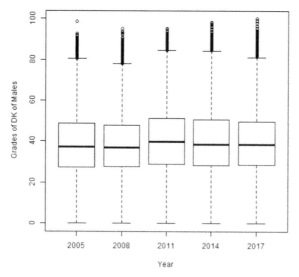

Fig. 9. Boxplot of male final year students in STEM degrees by DK grades, adapted from [4].

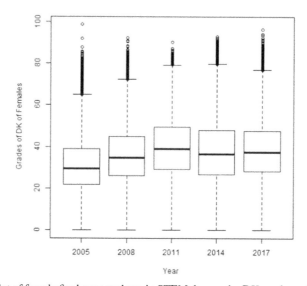

Fig. 10. Boxplot of female final year students in STEM degrees by DK grades, adapted from [4].

6.3 Final Grades (FG)

The final grade represents the overall result of the ENADE exam, it results from the weighted composition of 25% of the GK and 75% of the DK.

Remarkably, the average of grades for both sexes is under 45 (Fig. 11). We can note the difference between males' and females' averages.

The final grades also express the boycotted the first ENADE exams as a composition of GK and DK notes. Many registers contain null values or zero grades.

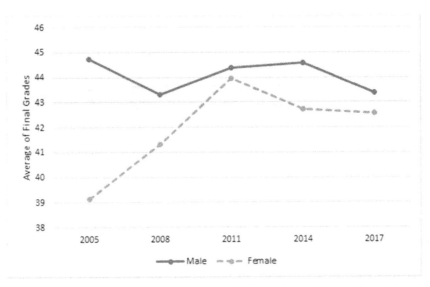

Fig. 11. Final grade average of final year students in STEM degrees by sex, adapted from [4].

Exploratory analysis from males and female boxplots about performance in FG demonstrate some exciting elements (Fig. 12 and 13).

Male performance in upper quartile outliers, influenced by DK's measures, demonstrate that students obtained grades over 80 in FG (Fig. 12). It demonstrates high-level knowledge and corresponds to an outstanding grade. It is essential to the point that most male students have unsatisfactory grades, with low average performance.

Female students' performance also has unsatisfactory grades, but we can note a tiny increase in their performance in the last exams (Fig. 13).

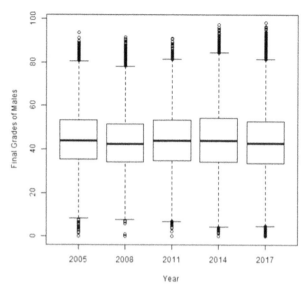

Fig. 12. Boxplot of male final year students in STEM degrees by FG grades in the ENADE exams, adapted from [4].

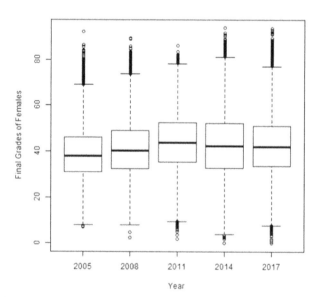

Fig. 13. Boxplot of female final year students in STEM degrees by FG grades in the ENADE exams, adapted from [4].

7   Analysis of Type of University

In the Brazilian context, a lot of educational institutions contribute to tertiary education. The Brazilian education system in higher education institutions comprises universities, university centers, and non-university institutions from public or private sectors. Public supports can be provided by the Federal, State, or Municipal government.

In this section, we study the students' performance from different financial sources using their grades. These sources are represented by private units (from the private sector) or Federal, State, or Municipal units (from the public sector). ENADE terms used to call this like 'type of university'.

We will perform an analysis of grades, specifying the GK, DK, and the exams final grade FG, between types of university (in other words, between the private and all public tertiary universities).

7.1   General Knowledge (GK) Grades

Figure 14 shows the level of knowledge greater in the Federal Public Universities group. Over time, all types of universities demonstrate the same level of achievement, maintaining the group trend.

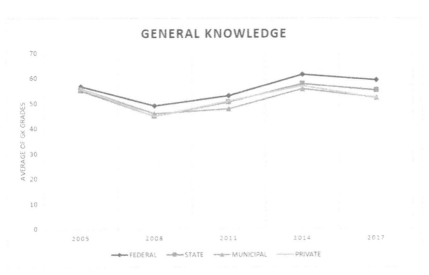

Fig. 14. Average of GK of final year students in STEM degrees in different types of universities.

7.2   Domain Knowledge (DK) Grades

Figure 15 shows the level of domain knowledge greater in the Federal Public Universities group. We can verify that Municipal Public Universities has the worst performance among all types evaluated.

We emphasize that Private Universities also have lower performance than Federal and State Public Universities.

Once again, all types of universities follow the same group performance over time.

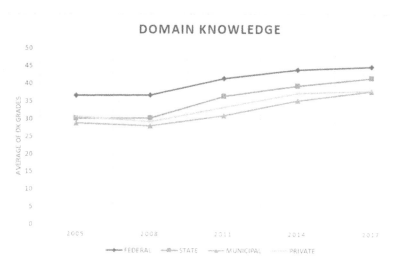

Fig. 15. Average of DK of final year students in STEM degrees in different types of universities.

7.3 Final Grades (FG)

Figure 16 shows the FG's average of final year students in STEM degrees in different types of universities. It intends to represent the overall result of the ENADE exam between the public or private sectors.

Visual analysis is sufficient to demonstrate that the average Final Grade is below 50%, but students' performance is more significant in Federal Public Universities. This result indicates that Brazilian educational institutions need to review the curricula of undergraduate STEM courses.

FINAL GRADE

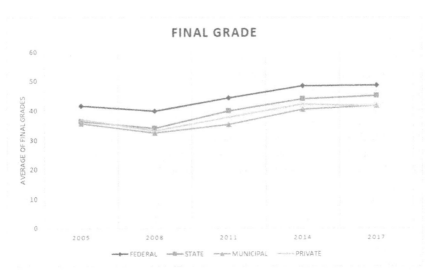

Fig. 16. Final Grade average of final year students in STEM degrees in different types of universities.

8 Conclusions

The National Assessment of Student Achievement (ENADE) public datasets are adequate to explore records of students of the tertiary education level. We select the records of students in STEM fields to obtain information about degree courses in this field.

Our study consolidates the archives from exams applied from 2005, 2008, 2011, 2014, and 2017. We execute experiments to analyze the characteristics of final year students in STEM degrees from those who took the exams in those years.

The analyses were focused on the understanding of distribution by sex, distribution by age, analysis of grades and types of university.

The distribution by sex in the STEM field shows a majority male adhesion and a reduction year by year in the percentage of females. Our research emphasizes that women are being underrepresented in technology.

The age factors demonstrate that, over the years, females have been younger than males in all years. Our study brought evidence that the mean ages of males tend to increase in the STEM field.

Our results about the analysis of grades and types of university expand the knowledge about the STEM field, especially about female representation.

Students' competencies and skills, measured by general knowledge (GK) grades, demonstrate a decreasing tendency to the average grades in males and females. The average is around 60% in both cases.

The measured domain knowledge (DK) grades, which is referent to the professional competencies and education qualities in the STEM field, show that grades obtained by males were greater than that of females over time. Our results demonstrate an insufficient level of knowledge, with a significant incidence of outliers in Male performance measures. Unfortunately, it is impossible to determine the reason for those results based exclusively on these results.

The weighted composition from both measures cited before defines the overall result of the ENADE exam as the final grade. We demonstrate that the average of grades for both sexes is under 45 which means that, in general, the students have unsatisfactory grades, with low average performance. Female students' performance has a tiny increase in their performance in the last exams.

Our study demonstrates that it is possible to realize analyzes using their grades and the information about the financial sources (or the type of university information). We observed the relations between the private and all public tertiary universities groups.

Is remarkable that the Federal Public Universities group in Brazil shows better results in the level of knowledge and domain knowledge. The curriculum analysis from Private Universities also has lower performance and Municipal Public Universities had the worst performance among all types evaluated. This result indicates that Brazilian educational institutions need to review the curricula of undergraduate STEM courses.

Our results show pieces of evidence about women in the STEM labor market. We demonstrate that without a STEM degree, women are less likely to occupy certain positions in this increasing market.

References

1. National Academies of Sciences, Engineering, and Medicine: Indicators for Monitoring Undergraduate STEM Education. The National Academies Press, Washington, DC (2018). https://doi.org/10.17226/24943
2. Romero, C., Ventura, S.: Educational data science in massive open online courses. Wiley Interdiscip. Rev.: Data Min. Knowl. Discov. 7(1), e1187 (2017)
3. Cao, L.: Data science: a comprehensive overview. ACM Comput. Surv. (CSUR) 50(3), 1–42 (2017)
4. Manhães, L., Zavaleta, J., Cerceau, R., Costa, R., Cruz, S.M.S.: Investigating undergraduate Brazilians students' performance in STEM courses. In: Proceedings of the 13th International Conference on Computer Supported Education - Volume 2: CSEDU, pp. 122–130 (2021). https://doi.org/10.5220/0010495201220130. ISBN 978-989-758-502-9, ISSN 2184-5026
5. INEP: Microdados. INEP - Instituto Nacional de Estudos e Pesquisas Educacionais Anísio Teixeira. Ministério da Educação (2021). http://inep.gov.br/web/guest/microdados. Accessed 10 Jan 2021
6. OECD: Assessing higher education learning outcomes in Brazil. High. Educ. Manag. Policy 24/2 (2013). https://doi.org/10.1787/hemp-24-5k3w5pdwk6br
7. INEP: Higher Education Assessments (2020). http://portal.inep.gov.br/higher-education-ass essments. Accessed 21 Dec 2020
8. de Brito, M.R.F.: O SINAES e o ENADE: da concepção à implantação. Avaliação: Revista da Avaliação da Educação Superior (Campinas) 13(3), 841–850 (2008). https://doi.org/10.1590/S1414-40772008000300014
9. Zoghbi, A.C., Rocha, F., Mattos, E.: Education production efficiency: evidence from Brazilian universities. Econ. Model. 31, 94–103 (2013). https://doi.org/10.1016/j.econmod.2012.11.018. ISSN 0264-9993
10. Crepalde, N.J.B.F., Silveira, L.S.: Desempenho Universitário No Brasil: Estudo Sobre desigualdade educacional com dados do Enade 2014. Revista Brasileira de Sociologia - Rbs, [s.l.] 4(7), 211–238 (2016)

11. Silva, L.F., da Rocha, M.E.P.S., de Araujo Fagundes, R.A.: ENADE: math and science students' performance analysis. IEEE Lat. Am. Trans. **15**(9), 1742–1746 (2017)
12. Vista, N.P.B., Figueiró, M.F., Mozzaquatro, P.M.: Técnicas de mineração de dados aplicadas aos microdados do ENADE para avaliar o desempenho dos acadêmicos do curso de Ciência da Computação no Rio Grande do Sul utilizando o software R. In: I Seminário de Pesquisa Científica e Tecnológica, s. l., vol. 1, pp. 1–11 (2017)
13. Santos, B.C.S., Noro, L.R.A.: PET-Saúde como indutor da formação profissi-onal para o Sistema Único de Saúde. Ciência Saúde Coletiva, [s.l] **22**(3), 997–1004 (2017)
14. Neto, T.A., Pereira, P.D.S.F., Nogueira, M.L., de Gody, J.M.P., Moscardini, A.C.: Factors that affect the national student performance examination grades of Brazilian undergraduate medical programs. GMS J. Med. Educ. **35**(1), 1–17 (2018)
15. Moimaz, S.A.S., Amaral, M.A., Garbin, C.A.S.: Enade: uma análise quanti-qualitativa dos exames nacionais de Odontologia. ABENO, [s.l.] **17**, 97–108 (2017)
16. Christie, M., O'Neill, M., Rutter, K., Young, G., Medland, A.: Understanding why women are under-represented in science, technology, engineering and mathematics (STEM) within higher education: a regional case study. Production **27**(SPE) (2017)
17. Papadakis, S., Tousia, C., Polychronaki, K.: Women in computer science. The case study of the Computer Science Department of the University of Crete, Greece. Int. J. Teach. Case Stud. **9**(2), 142–151 (2018)
18. Loyalka, P., et al.: Computer science skills across China, India, Russia, and the United States. Proc. Natl. Acad. Sci. **116**(14), 6732–6736 (2019)
19. OECD: Education at a Glance 2020: OECD Indicators. OECD Publishing, Paris (2020). https://doi.org/10.1787/69096873-en

Towards the Use of Augmented Reality for Physics Education

Morcos Adly[1], Nada Nasser[1], and Nada Sharaf[2(✉)]

[1] The German University in Cairo, New Cairo, Egypt
morcos.bishay@student.guc.edu.eg, nada.nasser@guc.edu.eg
[2] The German International University, Cairo, Egypt
nada.hamed@giu-uni.de

Abstract. Physics is around us everywhere. Young students should be taught the physics concepts from n early age. The goal of the work presented in the paper is to make Physics education more interactive and engaging through using an AR serious game. The game introduces the student to some physics concepts.

Keywords: Augmented reality · Physics education · Education

1 Introduction

Augmented Reality (AR) is gaining more popularity with the widespread of new mobile cell phones and tablets. It offers a different way of seeing and understanding [22]. It affected various fields including training, medical imaging, ... etc.

AR has also been used for learning [10].

Serious games are also increasingly being used in different areas including education [13].

AR is argued to increase concentration for learning since more senses of the students are involved [7].

AR makes items clearer and more interactive [16]. Studies also show that students using AR have higher scores [7, 16, 17].

AR was used in different applications related to Physics concepts such as the work presented in [5].

The proposed platform *PHYAR* provides a game with educational purposes. It is specifically designed for the user to gain or strengthen his/her knowledge about three main important Physics concepts which are: States of Matter, Light and Gravity [2]. PhyAR combines AR and serious games to provide an attractive and efficient learning platform.

The main purpose of the study is to investigate the effect of using PHYAR app on students' academic scores in Physics. This is done in comparison to the scores of the students when using traditional methods of teaching and learning.

2 AR in Education

AR makes use of the affordability of this real-time reality by giving extra and relevant data that enlarge students' understanding of the real world. AR may be founded on

© Springer Nature Switzerland AG 2022
B. Csapó and J. Uhomoibhi (Eds.): CSEDU 2021, CCIS 1624, pp. 232–259, 2022.
https://doi.org/10.1007/978-3-031-14756-2_12

and go with innovation [20]. Given the energizing improvements of events and the show of AR usefulness as an improved UI innovation, scientists accept that AR has huge potential ramifications and various advantages for the expansion of educating and learning conditions. For instance, AR can possibly:

1. connect with, animate, and persuade understudies to investigate class materials from various edges
2. help instruct subjects where understudies couldn't attainably increase real direct understanding (for example space science and topography)
3. upgrade joint effort among understudies and educators and among understudies
4. cultivate students' innovation and creative mind
5. assist understudies with assuming responsibility for their learning at their own pace and on their own way, and
6. make a real learning condition reasonable to different learning styles

Analysts have investigated the utilization of AR applications inside an assortment of fields and teaches, a considerable a lot of which are now legitimately or in a roundabout way identified with training. Also, others have taken a gander at the utilization of AR for clinical preparing reenactments. Likewise, experts have checked on the 1990 s s AR clinical showcase writing widely. Different analysts have analyzed the utilization of AR as a device for building where Web3D and AR advances permit understudies to investigate mechanical designing ideas. Kaufmann and his group have concentrated their AR look into applications in science and geometry by making an AR framework to encourage learning among teachers and understudies. Furthermore, specialists have explored the practicality of different AR applications for use inside the field of web based business, inside structure, and science instruction [21].

AR as a blended and upgraded reality has convincing features for instructive purposes; its latent capacity and affordances can be additionally broadened when an AR framework is planned by interfacing various kinds of advancements. In this area, we distinguish highlights and affordances of AR frameworks in five viewpoints dependent to investigate that fully AR usage is for instructive purposes. As per the examination, AR could empower (1) learning content in 3D viewpoints, (2) global, community and arranged learning, (3) students' faculties of essence, promptness, and submersion, (4) picturing the imperceptible, and (5) connecting formal and casual learning. We talk about every viewpoint as follows.

In the first place, AR can improve learning encounters by utilizing 3D engineered objects for understudies to interface with. AR empowers understudies to utilize 3D manufactured items to enlarge the visual impression of the objective framework or condition. Understudies can examine the 3D object from a wide range of points of view to improve their getting. A researcher indicated a case of utilizing 3D AR in educating cosmology. The investigation included two kinds of meetings: AR and customary instructing meetings. In the AR meeting, instructors and primary school students utilized a blend of innovations including a whiteboard, projector, web camera, AR tile, and virtual 3D displaying bundle to watch and turn a virtual 3D turning earth to find out about earth and sun, and day and night. The customary training meeting included perusing of a print book, address about the nearby planetary group, and a showing utilizing 3D physical items (e.g., a tennis ball, a string, and a light) to become familiar with similar points. Researchers investigated the inquiries educators posed in the two meetings and instructors' meetings after the meetings. They found that educators understood the advantages of utilizing 3D symbolism and accepted that AR can make distant topic accessible to understudies. Be that as it may did not look at whether the 3D learning experience made by AR was fundamentally more useful to understudies than the control of true 3D physical models. Early research on science training may give understanding into this inquiry. An analyst found that understudies in the gathering of utilizing both computerized and physical models performed essentially better than the gatherings utilizing both of the models. Other researchers likewise contended that both computerized and physical models ought to be given through class guidance on the grounds that various understudies have inclinations for various sorts of models and image frameworks. As far as research in AR, more proof is required to help for the utilization of the AR-based 3D virtual models over this present reality 3D models [20].

The second part of affordances identifies with the utilization of handheld mobile devices in AR. With cell phones, remote association, and location registered innovation, the unavoidable or mobile AR framework could empower ubiquitous, collective and arranged learning. The affordances of such a framework could incorporate versatility, social intuitiveness, setting affectability, network, and distinction. For instance, a few versatile AR games, for example, Environmental Detectives and Mad City Mystery were created to help learning outside of homerooms. In Environmental Detectives, learners utilized mobile devices to lead examinations, assembled information remarkable to the area, investigated and deciphered the information, and proposed arrangements touchy to the specific circumstance. It is demonstrated that connecting with understudies in playing virtual games in real spaces may raise understudies' setting affectability, and result in settling on progressively educated choices considering all ecological related elements. Besides, utilizing handheld gadgets in portable conditions could make understudies be occupied and increment task interferences. Since an AR system could distinguish students' areas and working status, give task updates, and offer choices to pull together understudies' consideration, these inserted consideration mindful highlights may decrease assistance with these entrusting interferences and deal with students' consideration. Moreover, social intuitiveness could be improved when understudies work together through arranged cell phones just as eye to eye collaborations, and system redid to various ways of examinations could be given to advance distinction [20].

Thirdly, it is guaranteed that AR and other vivid media for learning, for example, genuine games and virtual universes offer affordances of quality, promptness, and submersion. AR could give an intervened space that gives students a feeling of being in a spot with others. Such feeling of quality may improve understudies' acknowledgment of network of students. Furthermore, an AR system could incorporate constant criticism and give verbal and nonverbal signals to cultivate understudies' feeling of quickness. Given that instantaneousness is imperative to encourage the emotional side of learning, AR that unites students, virtual articles or data, and characters in a genuine situation can possibly increase immediacy. At last, vivid media like AR could give students a feeling of submersion, which is the abstract impression that one is taking part in a complete, reasonable encounter. Immersion could make conceivable the learning arranged in genuine problems, issues and conditions. An ongoing mobile AR study expands on the AR affordances of upgrading students' faculties of essence, quickness and inundation to help understudies' learning of a socioscientific issue on atomic vitality use and radiation contamination with regards to the atomic mishaps at the Fukushima Daiichi Nuclear Power Plant after the 3.11 seismic tremor in Japan. Ninth graders utilized Android tablets to gather information of mimicked radiation esteems on their grounds, which was speculatively a grounds around 12 km away from the Nuclear Power Plant and theoretically on the primary day after the hydrogen gas blast at the force plant. The investigation found a noteworthy connection between understudies' impression of the AR action and change in atomic perspectives, giving proof that AR can influence students' full of feeling spaces toward certifiable issues [20].

The fourth part of affordances is that AR superimposing virtual articles or data onto physical items or situations empowers representation of imperceptible ideas or occasions. AR system could bolster students in envisioning theoretical science ideas or undetectable marvels, for example, wind stream or attractive fields, by utilizing virtual items including particles, vectors, and images. For instance, Augmented Chemistry permitted understudies to choose synthetic components, make into 3D atomic models, and pivot the models. Some other researchers expanded a paper-based shading book with 3D content and gave youngsters a spring up book understanding of envisioning the book content. These enlarged genuine articles make new representations that can possibly upgrade understudies' comprehension of theoretical and imperceptible ideas or marvels [20].

The fifth part of affordances distinguished from the writing is that AR can possibly overcome any barrier between learning in formal and casual settings. For instance, the CONNECT venture utilized AR and different advances to build up a virtual science topical park condition. The earth had two modes: the exhibition hall mode and school mode. Situations created in the earth incorporate both virtual and traditional field excursions to science exhibition halls, pre-and post-visit curricular exercises, and investigation and displaying exercises. In this undertaking, hence, science learning at school was associated with learning encounters of the virtual and ordinary exhibition hall visits, with the utilization of AR to enlarge understudies' representation, examinations and models. An underlying assessment of the CONNECT venture showed that nature decidedly impacted understudies' inherent inspiration for learning science and applied comprehension of the grinding idea. However, the convincing highlights and affordances recognized in this area may not be extraordinary to AR, since some of them could be found in different frameworks or conditions (e.g., pervasive and versatile learning situations) with comparable advancements or ideas. To exploit the affordances of AR, in this manner, it is critical to investigate how the utilization of AR could be lined up with various instructional methodologies so as to accomplish proposed instructive destinations [20].

As a result, AR could have a paramount role specifically in educational field, for that it could easily show explicitly what could somehow be imaginary for students to understand. In addition to its interactive advantageous feature and how it could magically unveil mystery, which facilitates students' understanding.

2.1 AR in Physics

Educational field is of very wide branches, however researchers have found out that physics education could be clearly taught and shown to students with the use of AR technology for its countless advantages of two-way communication and the easiness of display of whatever real life models in front of users.

Augmented Reality has just shown a few zones where it has positively affected training. There are various fields in which AR has just demonstrated the possibility to realize clearing enhancements. For a certain something, the blend of AR reenactments and preparing works out, joined with material input interfaces, have been appeared to improve people's presentation in learning an assortment of physical abilities [21].

AR that plans to improve real life physics tests for students with innovations, for example, AR so as to make them progressively unmistakable and conceivable with new types of interaction. With the assistance of intelligent trials, physical associations are to be made more obvious for students of science, designing and mathematics training. Physical standards of mechanics and thermodynamics, are made intelligently researchable continuously [14].

Physics gives us an amazing asset that permits us to communicate inventiveness to see the world in new manners and to change. Physics is the reason for present day innovation and for the apparatus and the hardware utilized for an innovative work in science, building and clinical sciences (Cornell University, 2011), yet it is notable that Physics is viewed as the most troublesome subjects at school. Material science Education Research shows that there is a hole between what understudies realize and what instructors expect essentially (Zuza and Guisasola, 2014). For examining physics, understudies need to comprehend the ideas and standards of the physical world which in some cases can't be seen. Computer recreations as learning and encouraging instruments to assist understudies with seeing more substance of material science (Jimoyiannis and Komis, 2001). Understudies who have incredible theoretical thinking capacities will profit by reproduction based learning [18].

For teachers and planners, characterizing AR from a wide perspective would be progressively beneficial in light of the fact that such a definition proposes, that AR could be made and executed by changed advances, for example, computer devices, handheld gadgets, head-mounted shows, etc. That is, the thought of AR isn't restricted to innovation and could be reexamined from an expansive view these days [20].

2.2 Advantages of Using Augmented Reality in Education

AR can improve the mentality towards science teaching. A few scientists brought up that AR upgrades consideration and concentration for learning as more senses of the students are involved [7]. AR innovation can be seen as enchantment by learners as it mirrors the presence of items on a bit of paper. The objects' animation in AR applications is astounding for learners and draws their consideration [16]. Studies show that students who used AR technology have significantly achieved higher scores in comparison with the ones who only followed basic teachers' guidance in classroom. Those clear results are because of the better comprehension of issue situations using AR, improved abilities of analyzing and relating the subjects. Moreover, 3D models are effectively to be stuck in students' memories as of the feature of zooming and review the objects from various angles, and of course AR is far way increasingly practical, realistic and fascinating than conventional learning, thus learners do not get bored rapidly [7]. Another study showed that AR-supported science teaching makes positive commitments to learners' accomplishment and perspectives towards the course. Positive results can likewise be accomplished by means of AR innovation in various courses. For instance, the attitude of the trial bunch utilizing AR innovation had increasingly inspirational mentalities towards the course contrasted with the individuals who learned by means of traditional strategies. Students were eager to utilize AR once more and demonstrated no tension while utilizing AR applications. Students who learned science ideas by means of AR held better mentalities towards the course, and in this way, made more prominent scholarly progress in the course [16]. AR explanations in this manner are introduced in direct closeness to the real objects to improve learning impacts and encourage retention by the students [17].

Realizing impacts What and how do students learn in AR learning situations? Research has shown that AR frameworks and conditions could assist students with creating aptitudes and information that can be learned in other innovation improved learning situations yet in an increasingly attractive manner [6]. Computers and Education control an assortment of learning items and handle the data in an intelligent manner. AR situations can likewise encourage expertise procurement. In [11] AR versatile games permitted students to arrange, look and assess information and data; accordingly, students' aptitudes in exploring essential and auxiliary information could be created through these games. Another arrangement of aptitudes that are significant and basic in a data based economy can likewise be advanced in AR learning situations (Mathews, 2010; Rosenbaum et al., 2007). For instance, Rosenbaum et al. (2007) demonstrated that the feeling of realness offered by an AR learning condition advanced students' comprehension of dynamic models and complex causality. Besides, AR situations could build understudies' inspiration, which thusly may assist them with growing better examination abilities and increase information precision on the subject (Sotiriou and Bogner, 2008). In particular, understudies' spatial capacities can be improved in the wake of utilizing vivid and cooperative AR applications (Kaufmann and Schmalstieg, 2003; Kaufmann et al., 2005; Martin-Gutierrez et al., 2010). Different teacher-learner communication situations could likewise be upheld by AR frameworks, along these lines amplifying move of learning (Dede, 2009; Kaufmann and Schmalstieg, 2003). Another new arrangement of aptitudes that could most likely be advanced in AR are psychomotor-psychological abilities since AR could utilize viewable signals just as haptic prompts to upgrade users' encounters (Feng et al., 2008). Kotranza et al. (2009) demonstrated an AR framework in clinical medication that inserted touch-sensors in a physical domain, gathered sensor information to quantify students' exhibitions, and afterward changed the presentation information into visual criticism. By utilizing this AR framework, students could get ongoing, instant reactions that may help improve their exhibitions and upgrade their psychomotor aptitudes in an intellectual errand.

Furthermore, AR frameworks give answers for learning challenges that have been distinguished in past research. For instance, understudies as a rule experience challenges envisioning imperceptible marvels, for example, spinning of the earth (Kerawalla et al., 2006). AR permits students to control virtual articles or watch marvels that may not be effectively found in a regular habitat (e.g., biological systems of wetland or life patterns of wetland animals). These learning encounters thus could advance students' reasoning aptitudes and applied understandings about imperceptible marvels (Liu et al., 2009) and correct their misguided judgments (Sotiriou and Bogner, 2008). Albeit so far a larger part of AR frameworks have been created for showing science and arithmetic since learning these subjects require perception of conceptual ideas, there were additionally a couple of frameworks intended for understudies with extraordinary necessities and language learning. For example, Liu (2009) developed an AR learning condition with a setting mindful learning game to assist understudies with beating learning hindrances and adequately improve students' English talking and listening abilities. Besides, AR situations advance significant practices and skill levels that may not be created and sanctioned in other technology enhanced learning conditions (Squire and Jan, 2007; Squire and Klopfer, 2007). The AR game in Squire and Jan (2007) furnished understudies with chances to encounter how researchers think and do, and to apply their logical understandings to determine ebb and flow issues occurring in their nearby network.

AR instructive exercises can be basic in providing nowadays students with a forward-thinking, 21st century training which sets them up for the difficulties and exercises they will look in our current, quickly changing and innovation improved world. For instance, present day students need to figure out how to take care of issues as a component of an intelligent and circulated group, in anticipation of confronting difficulties in their future professions which really are too enormous to be unraveled, or maybe even conceptualized, by people acting alone [21].

AR shows better approaches for cooperating with this present reality and can make encounters that would not be conceivable in either a totally real or virtual world. AR has the exceptional capacity to make vivid half and half learning situations that consolidate real and virtual items. AR advancements empower clients to encounter logical wonders that are impractical in reality, for example, certain compound responses, making out of reach topic accessible to some understudies. The control of virtual articles and perception of wonders that are hard to see in reality can be encouraged through AR. This type Advances in Human-Computer Interaction of learning experience can support thinking aptitudes and increment calculated comprehension of wonders that are either imperceptible or hard to see just as right any misguided judgments. AR tends to learning challenges that are regularly experienced with envisioning in observable marvels. The abilities and information that understudies create through innovation improved learning situations might be grown all the more adequately through AR innovation. The subjective remaining task at hand might be diminished by coordinating different wellsprings of data. The submersion and connection features offered by AR may urge understudies to take part in learning exercises and may improve understudy inspiration to learn. AR gives profoundly intelligent encounters and can create valid student action, intuitiveness, and an elevated level of authenticity. Association with the world is significant in the learning procedure, and, aside from the real world, AR is probably the most ideal methods of encouraging this connection.

3   Related Work

3.1   Experimenting AR in Education for Primary Schools

Their difference between educators' utilization of AR with their use of conventional teaching materials has delineated that AR can be utilized to help 10-year-old kids see how the earth and sun connect in 3D space to offer ascent to day and night. However, they likewise found that youngsters instructed using their AR programming were less connected with than those associated with role-play; educators were bound to ask the kids to watch an AR activity and portray it, contrasted with the role-play meetings wherein kids were urged to make and control the jobs of the on-screen characters. The talked with educators additionally perceived the capability of AR innovation, yet said that they might want it to be progressively adaptable and controllable, so they could include and evacuate separate components, and delayed or stop the animations. Their feedback recommends that these highlights would empower them to include their youngsters more in, for instance, finding out the connections between components by adjusting parameters."Making it real": exploring the potential of augmented reality for teaching primary school science [9].

The points of their examination were to distinguish the impact of AR innovation on center school understudies' scholastic accomplishment and mentalities towards their science course, and to decide their perspectives towards AR innovation. Moreover, another point of their work was to research the connections between these students' scholarly accomplishment, mentalities towards the course and perspectives towards AR applications. It was seen that the understudies in the test bunch had more significant levels of scholastic accomplishment and inspirational mentalities towards the course than those in the benchmark group. This critical group distinction can be taken as proof of the constructive outcome of AR innovation. The effect of Augmented Reality Technology on middle school students' achievements and attitudes towards science education [16].

3.2   Applying AR for Physics Education

Motion Understanding Using an Augmented Airtable. An air table trial utilized for the comprehension of laws behind object motion and collisions was the objective of that work. Two distinctive AR executions were introduced, one planned for single user experimentation with a HMD, and a subsequent one utilizing projective AR focusing on multi-client shared outcomes. In the projective AR a substantially more fulfilling final product was achieved contrasted with the HMD variant which experienced restrictions caused fundamentally by equipment issues, for example, the field of perspective on the gadget. Later on, these methodologies will be assessed at their areas of deployment (college and innovation historical center) so as to evaluate the user experience, the connection perspectives and educational estimation of these system [14].

Light Experiments in AR. AR learning application is planned as a strengthening learning tool, which is proposed to be utilized with existing printed learning materials. For example explicit illustrations in a material science study book could be utilized as markers, connecting to relating content in our AR application. An understudy contemplating this book could begin the AR application whenever, point the cell phone towards the delineation and immediately have the option to get extra intelligent content and data with respect to the learning subject. The AR application content never ought to replace or basically copy the book but help to understand the content. The application could give little intelligent activities which depend on the real laboratory or book exercises. The AR application is to reenact a basic optical bench with one focal lens, one light source and screen. The focal point type and qualities can be modified interactively on the screen [19].

Another study [12] for example shows the utilization of an ARToolKit based versatile AR application, utilizing various cards with printed markers to reproduce virtual apparatuses. Those apparatuses can be utilized to recreate physical impacts. Among others it is conceivable to point a virtual laser pointer and in this way the laser beam on a virtual focal point that refracts the virtual light. The researchers approach requires for explicitly printed cards with markers on it, as those are being moved by the user in space to allow the apparatuses to collaborate. The distribution alludes to "intuitive AR" as "interesting and entertaining" and sees an "extraordinary potential for edutainment".

Another direction [5] is using a virtual lens experiment in a material science course. Their arrangement comprised of markers set on a rail that reassembles a genuine optical bench. Understudies can utilize the test arrangement like a conventional lens test arrangement, for example move the focal point on the rail and watch the difference in the picture on the virtual screen. The creators utilize a normal computer display (data projector) to make the result of the AR noticeable. They directed the experiments with two eighth grader classes, one with the AR form and the other with a normal arrangement. Their experiments demonstrated that, the gathering utilizing the AR exploratory arrangement didn't score a fundamentally better outcome in after tests however had a positive memory on the experience itself.

Another study [8] used an AR application for directing different analyses in the field of mechanical physics. It is conceivable to encounter virtual bodies behave physically right. The application utilizes a material science motor to give a practical simulations of masses and forces. The user needs to wear a head-mounted VSD(Video-See-Through Device) and is followed remotely by an expert motion tracking system. While pointing toward supported learning, no assessment about its adequacy for learning had been directed. Because of its expert tracking framework this sort of AR applications these days are restricted to specially prepared labs.

4   Methodology

ARphy game is designed to teach youngsters approximately within ages 9–12 in primary level some crucial physics concepts of paramount importance, specifically three topics, which will be discussed in the upcoming section. Its main goal not only the teaching, but also sticking their learnt information to their minds. That is because the most difficult part is not to instantly understand the taught concept, but is that the memory never misses it as long as possible.

The inclusion of AR technology is the reason for the previous discussed points. Young students got used to reading from syllabus books and sometimes see some illustrative figures related. However, it could be more interesting, expressing and engaging to view real-time animated figures and models with interactive user-interface, which gives the user the opportunity to be in control of what he/she wants to experience. Moreover, physics being the center of the app is the best-friend of AR than any other sciences as AR technology has various capabilities to show detailed physics concepts in terms of movements, sequencing and aspect ratios in a professional unforgettable manner.

4.1 Topics Included

ARphy game app consists of three main scenes:

- **Matter** section focuses on states transformation (Solid, Liquid, Gas) and how these transformations affect the particles behaviour and arrangement, along with illustration of the type of change whether heating or cooling.
- **Light** section consists of two main Light concepts which are 1- Reflection: The ARphy app describes Reflection part by the concept of seeing anything in a mirror including in which form light propagates and it shows explicitly the sequence of light rays journey starting from the light source itself till finally seeing that thing by the eyes. 2- Shadow: This part shows the idea behind the shadow formation and how it is affected by the distance between the light source and the object, as well as the light intensity.
- **Gravity** section is mainly to differentiate between the behaviour of free fall of any object in space against its behaviour on earth. Furthermore, the effect of air resistance presence is simply cleared against its absence and how masses affect the arrival time of the object to reach the ground, if both of the objects were thrown at the same time.

The above topics were selected out of many others related physics topics for numerous reasons. First of all, there is a common gap among most of related papers which is the target age. The target age for users was excluding the age criteria target for ARphy app that ranges between 9 and 12. Second reason is that Matter and Gravity topics specifically were never mentioned in any AR app of the referenced papers. Light topics is mentioned in [16]. However, it was designed for older age group with different concepts. Thirdly, the chosen topics Matter, Light and Gravity are the top of schools' physics curriculum for age group 9–12. Last but not least, The above mentioned topics' concepts are so AR-friendly in terms of explicitly unveiling every tiny detail about the structures as well as behaviours and reactions which make learning more fun and boosts user's understanding and achievements.

4.2 Markers

Marker is the target for the AR camera to be detected to display the models related to each marker, which will be discussed in details in ARphy structure section.

Markers for ARphy app are unique for the specified age group not only because they contain colorful illustrating shapes, but also they contain picky bullet points for memory located at the back of each marker. Moreover, the chosen markers are so easy to handle, replace and play with. That's why **cards** are used as markers for all of the previous mentioned reasons.

Marker cards' structure for the topics included in ARphy are shown in Figs. 1, 2.

Fig. 1. Cards [2].

Fig. 2. Cards part II [2].

4.3 ARphy Structure

This section is a walk-through experience for each scene including transition scenes in ARphy app. The walk-through will be as follows. Starting with the main menu, followed by Matter, then Light and finally Gravity.

Main Menu. The Main Menu page contains three main buttons for the three main topics mentioned before. In addition to an Exit button and Sound On/Off button as shown in Figs. 3 and 4.

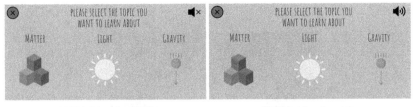

(a) Main Menu Sound Off   **(b)** Main Menu Sound On

Fig. 3. Sound menu [2].

Fig. 4. Exit warning.

Matter. When Matter button is clicked in the main menu, the Matter Menu appears with those two buttons: 1-Transformation Cycle 2-Random, as shown in Fig. 5. Those two buttons will be discussed next.

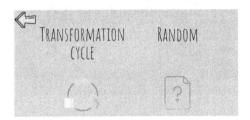

Fig. 5. Matter menu [2].

All Transformations. This scene is reached, as shown in Fig. 6, when Transformation Cycle button is pressed. First, the solid card must be detected in order to begin the change of states journey.

The ice cube appears first showing the atoms arrangement focusing on their vibrating behaviour by a surprising mobile vibration gesture as shown in Fig. 7.

A melt button of a fire icon can be noticed located at the top right, when clicked the ice cube starts to melt along with the particles moving faster and sliding over each other till the ice cube is transformed to water completely as shown in Fig. 7.

Fig. 6. Request for solid card detection [2].

(a) Ice Cube with its Particle

(b) Melting Ice

Fig. 7. Ice states.

An extra icon of a Snowflake appears at the top left corner which indicates freezing, as well as the fire icon, but this time it indicates evaporation. Giving the choice to the user whether to freeze the water back to ice or evaporate it as shown in Fig. 8.

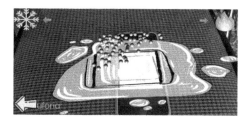

Fig. 8. Melted Ice (water) with its particles [2].

If the Snowflake icon is pressed, then the particles will start moving slower and closer to each other forming the ice cube once again with the mobile vibration gesture.

If the Fire icon is pressed, then the atoms starts to move faster and randomly away from each other till all water vanishes completely forming cloud as shown in Fig. 9.

Only the Snowflake reappears indicating condensation of the vapour. Once the Snowflake icon is clicked, the atoms slow down and start to slide over each other reforming water.

Random. Once the Random button is clicked the user will be asked to detect one of the solid, liquid or gas cards. When each card is detected separately its 3D models appears relatively as show in Fig. 10.

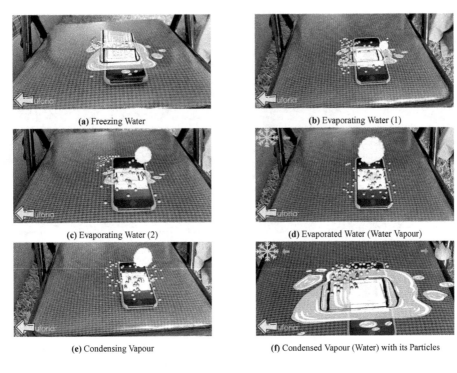

(a) Freezing Water

(b) Evaporating Water (1)

(c) Evaporating Water (2)

(d) Evaporated Water (Water Vapour)

(e) Condensing Vapour

(f) Condensed Vapour (Water) with its Particles

Fig. 9. Water and gas states [2].

Light. As soon as Light button is chosen, the user will have to choose which light concept to learn about Reflection or Shadow as shown in Fig. 11.

Customization. ARphy app is customized for both boys and girls as Light section offers Reflection and Shadow fitting boys/girls interests when any of the two concepts is chosen as shown in Fig. 12.

Reflection. First, the user must detect the Reflection card, instantly a scene consisting of Messi, Ballon D'or, mirror and moon are seen, if boy is chosen and similarly if girl is chosen, the user will see Elsa, Olaf, mirror and moon. In both choices, the user can see a Sun icon at the top right corner of the screen and is asked to turn on the light, as initially the moon is up and sun is down as shown in Fig. 13.

Once the light is on, light rays will be seen travelling from the sun to Ballon D'or/Olaf hitting the mirror then finally reaching Messi/Elsa's eyes. At the same time light reaches Messi/Elsa's eyes Ballon D'or/Olaf can be seen in the mirror. Moreover, Messi/Elsa displays a message and makes a known character gesture indicating that they are happy and the user can turn off the light again by finding the moon icon as shown in Figs. 14.

(a) Request for Solid/Liquid/Gas Card detection

(b) Ice Cube (Solid Card)

Water (Liquid Card)

Vapour (Gas Card)

Fig. 10. Random detection [2].

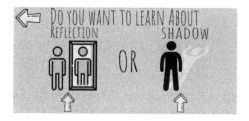

Fig. 11. Light menu [2].

Fig. 12. Gender menu [2].

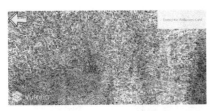

(a) Request for Reflection Card detection

(b) Messi after detecting Reflection Card if the Boy is chosen

(c) Elsa after detecting Reflection Card if the Girl is chosen

Fig. 13. Reflection [2].

(a) Light Rays journey to Messi's eyes

(b) Ballon D'or appears in the Mirror

(c) Light Rays journey to Elsa's eyes

(d) Olaf appears in the Mirror

Fig. 14. Messi and olaf [2].

Shadow. The user must detect the Shadow card, in order to see Ballon D'or/Olaf, white board and a light source which is a torch. There are buttons for light intensity control and a button to switch on/off light totally as shown in Fig. 15.

As long as light is not off, the user can swipe up to move Ballon D'or/Olaf closer to the torch. It is obvious that the shadow size displayed on the white board increases as shown in Fig. 16.

Or swipe down to move Ballon D'or/Olaf away from the torch. It is obvious that the shadow size displayed on the white board decreases as shown in Figs. 17.

Gravity. When the Gravity button is chosen, the user is asked to detect the gravity card as shown in Fig. 18.

(a) Request for Shadow Card detection   (b) After detecting Shadow Card if the Boy is chosen

(c) After detecting Shadow Card if the Girl is chosen

Fig. 15. Shadow [2].

(a) Olaf is far from Torch   (b) Ballon D'or is close to Torch
Olaf is close to Torch

Fig. 16. Shadow for olaf and ballon D'or case 1.

(a) Olaf is far from Torch   (b) Ballon D'or is far from Torch
Olaf is far from Torch

Fig. 17. Shadow for olaf and ballon D'or case 2.

Fig. 18. Request for gravity card detection.

Once the gravity card is detected space scene appears along with an apple and a paper kite. The paper kite and apple positions almost stay the same and they do not fall, as Space misses gravity. An extra button for mode switching from Space to Earth appears at the top right corner of the screen as shown in Fig. 19.

Fig. 19. After gravity card detection space appears [2].

If the earth button is pressed grass appears instead of space as well as a sound is heard saying Gravity on, which indicates that Gravity's effect is on. In addition to two extra buttons are shown along with the Space/Earth mode switching button. The first button to specify whether to turn on/off Air Resistance. The second button (Play icon) is to free fall the apple and paper kite, to analyze their behaviour as shown in Fig. 20.

Fig. 20. After switching to earth mode.

Assuming that Air Resistance button is Off and the Play button is clicked. Not only the play button disappears and a reset button is shown instead, but also both the apple and the paper kite hit the ground at the same time, as there is no opposing force which is the Air Resistance as shown in Fig. 21.

Fig. 21. Air resistance off (Both reaches the grass at the same time).

Reset button is to reset the positions of the apple and paper kite for retrial as shown in Fig. 22.

Assuming that the Air Resistance button is On this time, a sound is heard saying Air Resistance On. Then, if the play button is clicked. The user will observe that the apple reaches the ground faster than the paper kite, as the apple is heavier and the Air resistance acting on the paper is larger because the paper has larger surface area as shown in Fig. 23.

Fig. 22. Reset positions and turn on air resistance.

Fig. 23. Air resistance on (Apple reaches the grass first) [2].

For the gravity scene, all buttons are featured with voice-over to alert the user whether that button's related feature is On or Off.

4.4 New Features

We have also added new category for the game which is electric circuits. It allows the students to interact with circuits to learn the concepts of parallel and series circuit connections.

The student has to first choose the concepts as shown in Fig. 24.

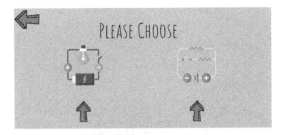

Fig. 24. Caption.

The student can then start to do the connections to see the effect on a light bulb and a buzzer as shown in the Fig. 25 and Fig. 26.

As shown from the figure, the bulb's light is stronger when connected in parallel.

5   Experimental Design

This section tracks the procedure followed to examine the effect of AR serious game (ARphy) on students' academic achievements compared to utilizing traditional learning methods introduced in [2].

It was proven by [15] that the learning outcome is not correlated with students' engagement. As a result, each of them were tested as two independent factors.

(a) Light and Buzzer Off (Series Case)

(b) Light and Buzzer On (Series Case)

Fig. 25. Series circuit.

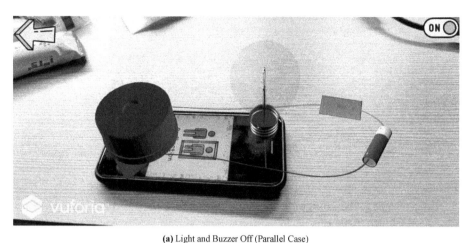

(a) Light and Buzzer Off (Parallel Case)

(b) Light and Buzzer On (Parallel Case)

Fig. 26. Parallel circuit.

The Experimental Design strategy is composed of the following steps which will be discussed next.

1. Test Setup
2. Model Criterion Selection
3. Evaluation Material Preparing
4. Testing Process

5.1 Test Setup

– **Primary Dependent Variable:** The learning gain of the child.
– **Secondary Dependent Variable:** The engagement level reported during the learning process.

– **Independent Variables:** The only independent variable to be observed is the method of learning, which is AR serious game versus traditional learning.

Hypothesis. AR aided Physics learning boosts students' academic achievements compared to traditional learning methods.

Children were divided into two groups. The **first group** was refreshed with the Physics concepts, then students of this group had to answer a test about past physics concepts they have already learnt in school which are included in ARphy game, then experience same content but using ARphy app, and finally solving the same previous test. On the other hand, the **second group** solved the Physics test first, then refreshed with the same Physics concepts that they have already learnt before. Then, answering same previous test again. Experiencing ARphy game comes next followed by last solving phase of the same Physics test that was answered two times before.

5.2 Model Criterion Selection

One of two well-known models was to be chosen. The first one is Between Group Design which begins with answering a physics test by each student. Students separation comes next, half of the sample is exposed to traditional learning methods and the other half experiences same content using ARphy app. Finally, each student of the two groups is then asked to re-answer the physics test.

The second model choice is Within-Subject Design. This is done by exposing each student to both traditional learning method followed by experiencing the AR game ARphy.

The first model was the one intended to be used for its higher accuracy results. However, due to pandemic of COVID-19 if the first choice was utilized, the students testing number would be divided by two in order to test each medium separately, and this involves huge number of students which we could not gather because of the virus. Thus, the second choice was utilized for its moderate students number involvement and its fair results accuracy as well.

There are many advantages for opting the Within-Subject Design, one of them is equating the conditions prior to experiment by using the same participants in each condition which removes individual differences and makes it easier to detect differences across levels of the independent variable. Another pro, is that Within-Subject Design save vast amounts of time. On the other side of the coin, there are some disadvantages associated with the Within-Subject Design, all of which were avoided as much as possible in this study. One major con is known as the carryover effect. The carryover effect is an effect that "carries over" from one experimental condition to another. Despite the fact that it was gruelling to get over this disadvantage, a sequence of steps were adjusted to overcome this negative aspect, as at first they receive the Physics content via traditional learning then, they were exposed to ARphy serious game. Thus, if there would be a carryover effect, it would be due to classic learning which would be revealed by the test answered by students before (the test after traditional learning) and after experiencing ARphy app.

5.3   Pre and Post Tests

There is no difference between the Pre and Post Tests of Traditional Learning and AR, so that the learning gain could be measured correctly. The test simply contains eleven multiple choice questions presented in hard copy (listed in Appendix A). The questions were divided on the three topics (Matter, Light, Gravity) as follows, Matter section consists of five questions, light three questions and gravity also contain three questions.

There are three types of questions. The first type is basic multiple choice question, including one correct answer and choose-all-correct-answers type of questions. The second one is arranging sequence question and the last type is narrative question.

All the questions have multiple choices in range four to seven choices, including options like "All of the above" and "Nothing from the above". This options make it more challenging for the students to solve the test. These choices were added to avoid having inaccurate data if the student chooses an answer haphazardly.

The questions were designed to test the students' knowledge and also were opted carefully to examine the crucial parts that ARphy game covers from the three discussed physics concepts. These two tests' results will be used to carry out the learning gain test, which will be covered in the upcoming subsection.

5.4   Learning Gain Test

The Learning Gain Test is the evaluation method used to assess youngsters' academic levels by solving the pre and post test of Traditional Learning method and ARphy game, which are exactly the same to check the results difference for each child of his/her tests. This difference is how the Learning gain is obtained in order to prove/disprove the research hypothesis.

The Learning Gain of Classical Learning method results are exactly calculated by subtracting the pre Classical Learning test from the post Classical Learning test. Similarly, The Learning Gain of ARphy results are calculated by subtracting the pre ARphy test, which is the same result of the post Classical Learning test, from the post ARphy test. For the previous calculated learning gains, if the results are positive this indicates that there is an increase in academic scores of the child whether due to the Classical learning or AR dependably. However, if the results were negative, this indicates that there is a decrease in the academic level. Finally, if the learning gain of the Classical Learning is zero then, Classical Learning has no effect in boosting child's academic level, and the same for the learning gain of the AR.

Finally, the learning gain of both Classical Learning and AR are compared to notice which method has a greater impact on youngsters academic level.

5.5   Engagement Test

This study is analysing the engagement level of the child while receiving the educational content using ARphy serious game app. In order to achieve this goal, a test that aims at measuring the engagement level of the students was utilized using a Likert Scale survey. The Likert Scale is a five (or seven) point scale that is used to rate the agreement or disagreement of the participant with a specific statement, or question. This approach is widely used in subjective research topics. [3] The Likert Scale survey used in this study is a standardised questionnaire that is generally used to test the engagement level that the user experiences while being in the experiment. It relies on asking the user various questions that point out many aspects regarding how involved, in control, and entertained the user was during the experiment. The questionnaire is composed of 11 measured aspects (listed in Appendix A) that evaluate the overall flow of any activity through assessing three major factors: control, enjoyment, and engagement. The questionnaire was answered by the children directly after completing all the Physics topics of ARphy game. In a nutshell, the results of the measures included in this questionnaire can be summed up to represent the overall engagement, in control, enjoyment levels of using ARphy app.

5.6 System Usability Scale

The SUS(System Usability Scale) is a basic, reliable survey that assesses the usability of a system. It is formed of 10 Likert scale items (listed in Appendix A) [4]. These standardised questions evaluates the app's impact, efficiency, and user satisfaction. This tool is very helpful for the study as it is important to collect the users' feedback with respect to the usability measures of ARphy game. It was mainly used to assess the usability of the AR serious game "ARphy".

5.7 Testing Process

To begin with, some Ethical Aspects were taken into consideration:

– Having children's parents or teachers permission.
– Giving a clear introduction of the study's main goal before the experiment.

A target population of ten children, in the age group of 9–12 years were considered as control and experimental group of population n = 10.

Five of the ten participants took part in the following activities: Answering Pre Classic Learning test followed by Refreshment for the targeted physics concepts which was delivered to each child followed by the post Classical Learning which is similarly the pre ARphy test, ARphy serious game experience, post ARphy test. The remaining five participants participated in the same activities except the Pre Classic Learning test. Lastly, the 10 students answered the engagement test and system usability scale. The whole session lasted for a duration of twenty-five to thirty minutes.

Prior to playing ARphy game, the participants were given the pre-test precisely described in Subsect. 4.3.1 and were asked to answer the questions without any help unless a clarification was needed. Afterwards, the participants were asked to play ARphy.

Upon successful covering of all topics in the game, the post-test link was provided to the participants in order to evaluate their learning gain. Finally, a questionnaire survey link was provided to the participants so that they can fill the questionnaires corresponding to the engagement test and system usability scale.

6   Results

6.1   Learning Gain

The Learning Gain is to show the effect of Traditional Learning methods versus AR Serious Game. Learning Gain is calculated by subtracting the pre test results from the post test results in each case.

First Group. The children got an average pre classical learning test score of 3.8 out of 18 points (n = 5, M = 3.800, SD = 0.748), an average post classical learning test/pre ARphy test score of 7.8 out of 18 points (n = 5, M = 7.800, SD = 1.327), and an average post ARphy test score 17.2 out of 18 (n = 5, M = 17.2, SD = 0.748). This has indicated a learning gain for the classical method by an average of 4 points (n = 5, M = 4.000, SD = 1.155), and a learning gain for ARphy serious game by an average of 9.4 points (n = 5, M = 9.400, SD = 0.800).

Figure 27 shows explicitly the comparison between the classical learning method gain and ARphy gain in group one which shows the p-value significance (p <0.05).

Fig. 27. Classical method gain vs ARphy gain (Group one).

Second Group. The children got an average pre ARphy test/post classical learning test score of 9.6 out of 18 points (n = 5, M = 9.600, SD = 2.059), and an average post ARphy test score of 17.6 out of 18 points (n = 5, M = 17.600, SD = 0.800). This has indicated a learning gain by an average of 8 points (n = 5, M = 8.000, SD = 2.449).

Figure 28 shows explicitly the comparison between the classical learning method gain in group one and ARphy gain in group two which shows the p-value significance (p <0.05).

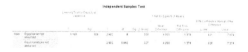

Fig. 28. Classical method gain (Group one) vs ARphy gain (Group two).

There is a significant difference between both tests' outcome. It is obvious that AR aided learning has boosted children's academic scores after using ARphy app.

6.2   Engagement Level Test Results

The Engagement test results were collected from each student after experiencing ARphy game, in order to assess its extent for user involvement, entertainment and engagement level while playing. The end result was calculated by getting the average answer of all questions of each user's test, then averaging the previous obtained results. Finally, getting an average answer for all tests which is 4.5 representing 90%.

6.3   System Usability Scale Results

The usability of ARphy game test was answered by each participant after using ARphy app. Its results were obtained by the following steps:

1. The even numbered questions were calculated by subtracting one from the score.
2. The odd numbered questions were calculated by subtracting the score from 5.
3. The new obtained values were summed up then multiplied by 2.5 for the overall score to be out of 100.

The end average result obtained was approximately 94 which is obviously above the average of 68 according to [1]. Thus, the overall usability of ARphy is considered to be totally above the average that is over-satisfactory for the game.

7   Conclusions and Future Work

Engaging AR technology in Physics concepts' illustration is one of the focused on topics these days. The attention was drawn in this study to analyse the compact of AR teaching using ARphy app versus traditional learning on students' academic levels.

The main goal was to prove or disprove the research hypothesis that was AR aided Physics learning boosts students' academic achievements compared to traditional learning methods.

For testing the above hypothesis, the following measures were followed. ARphy app was implemented and tested for the Physics topics Matter, Light and Gravity. Then, an experiment was carried out having two centered goals. The first objective was to obtain the learning gain of each student and the second objective was to monitor engagement levels of students while utilizing ARphy game. Youngsters were divided into two groups, both were Within-Subject Design, each group contains 5 children of age between 9–12 years.

To evaluate the learning gain, each of the youngsters of the second group was given a pre test before the ARphy experience, but after some refreshing for the included topics that they have already learnt traditionally in School. In addition to the post test which was given to them after utilizing ARphy. On the other hand, the first group took an extra test but before the refreshment. The Classical learning method gain of the first group was compared to the Arphy gain of the first group and second group.

To assess the engagement level, a standardised five Likert scale survey was filled out by the students, which include many questions evaluating the involvement, enjoyment, and engagement levels during an activity. Upon conducting the experiment, gathering data, and analysing it, it was concluded that the average percentage is 90%.

One last test used to measure the usability of ARphy app is the System Usability Scale. It is a standardised 10 Likert Scale survey, and the youngsters who took part in experiment were asked to give their opinion on these items to get their feedback. The collected results revealed that the usability of ARphy exceeded the standardised average score of this test.

To sum up, the results revealed the significant positive impact of using ARphy app, the AR aided Physics learning game, on students' academic achievement. Consequently, ARphy app has a noticeable effect in Physics learning compared to classical learning method which could be taken advantage of for promoting students' academic level.

8   Limitations

Some limitations were encountered while applying the experiment. To begin with, is the small number of participants of the tested groups. There are two main reasons behind this. The first reason is that COVID-19 spread was during testing time so parents rejected the idea of their child's participation in the experiment fearing from infection. The second reason is that the Augmented Reality technology is not supported by all mobile devices, and we were only capable of providing one device for the experiment; thus, the experiments could not be conducted in parallel or not even sent to be tested at homes and receiving survey results.

The problem of having a small sample was overcome by using the t-tests in evaluating the learning gain as well as the engagement levels because t-tests are useful for analysing data of small samples. However, Having a larger number of participants would have assisted in obtaining more indicative results.

8.1   Future Work

The future work could be summarized in two main points:

1. **Extending Physics Topics on ARphy App:** As currently ARphy app has three main topics which are Matter, Light and Gravity, there could be a great opportunity for more topics extension. The extended topics could even be suitable for broader age range with varying complexities.
2. **Improving Features and Sound Effects:** Some features like the chosen models could be modified along with the sound effects that could vary with the current scene to be more expressible instead of having uni-toned background music.
3. Testing the new circuits extension of Arphy.

References

1. SUS Score. https://usabilitygeek.com. Accessed 13 July 2020
2. Adly, M., Nasser, N., Sharaf, N.: PHYAR: introducing a mixed/augmented reality platform for physics concepts. In: Csapó, B., Uhomoibhi, J. (eds.) Proceedings of the 13th International Conference on Computer Supported Education, CSEDU 2021, Online Streaming, 23–25 April 2021, vol. 2, pp. 338–345. SCITEPRESS (2021). https://doi.org/10.5220/0010496103380345

3. Allen, I.E., Seaman, C.A.: Likert scales and data analyses. Qual. Prog. **40**(7), 64–65 (2007)
4. Brooke, J.: SUS: a quick and dirty'usability. Usability Eval. Ind., p. 189 (1996)
5. Cai, S., Chiang, F.K., Wang, X.: Using the augmented reality 3d technique for a convex imaging experiment in a physics course. Int. J. Eng. Educ. **29**(4), 856–865 (2013)
6. El Sayed, N.A., Zayed, H.H., Sharawy, M.I.: ARSC: augmented reality student card. In: 2010 International Computer Engineering Conference (ICENCO), pp. 113–120. IEEE (2010)
7. Fidan, M., Tuncel, M.: Integrating augmented reality into problem based learning: The effects on learning achievement and attitude in physics education. Comput. Educ. **142**, 103635 (2019)
8. Kaufmann, H., Meyer, B.: Simulating educational physical experiments in augmented reality. In: ACM SIGGRAPH Asia 2008 Educators Programme, pp. 1–8 (2008)
9. Kerawalla, L., Luckin, R., Seljeflot, S., Woolard, A.: "Making it real": exploring the potential of augmented reality for teaching primary school science. Virtual Reality **10**(3–4), 163–174 (2006). https://doi.org/10.1007/s10055-006-0036-4
10. Khan, T., Johnston, K., Ophoff, J.: The impact of an augmented reality application on learning motivation of students. Advances in Human-Computer Interaction, **2019** (2019)
11. Klopfer, E., et al.: Augmented Learning: Research and Design of Mobile Educational Games. MIT press, Cambridge (2008)
12. Lai, C.L., Wang, C.L.: Mobile edutainment with interactive augmented reality using adaptive marker tracking. In: 2012 IEEE 18th International Conference on Parallel and Distributed Systems, pp. 124–131. IEEE (2012)
13. Laine, T.H.: Mobile educational augmented reality games: a systematic literature review and two case studies. Computers **7**(1), 19 (2018)
14. Minaskan, N., Rambach, J., Pagani, A., Stricker, D.: Augmented reality in physics education: motion understanding using an augmented airtable. In: Bourdot, P., Interrante, V., Nedel, L., Magnenat-Thalmann, N., Zachmann, G. (eds.) EuroVR 2019. LNCS, vol. 11883, pp. 116–125. Springer, Cham (2019). https://doi.org/10.1007/978-3-030-31908-3_8
15. Repenning, A., Lewis, C.: Playing a game: the ecology of designing, building and testing games as educational activities. In: EdMedia+ Innovate Learning, pp. 4901–4905. Association for the Advancement of Computing in Education (AACE) (2005)
16. Sahin, D., Yilmaz, R.M.: The effect of augmented reality technology on middle school students' achievements and attitudes towards science education. Comput. Educ. **144**, 103710 (2020)
17. Strzys, M.P., et al.: Physics holo. lab learning experience: using smartglasses for augmented reality labwork to foster the concepts of heat conduction. Eur. J. Phys. **39**(3), 035703 (2018)
18. Techakosit, S., Nilsook, P.: The learning process of scientific imagineering through ar in order to enhance stem literacy. Int. J. Emerg. Technol. Learn. (IJET) **11**(07), 57–63 (2016)
19. Wozniak, P., Vauderwange, O., Curticapean, D., Javahiraly, N., Israel, K.: Perform light and optic experiments in augmented reality. In: Education and Training in Optics and Photonics, p. DTE05. Optical Society of America (2015)
20. Wu, H.K., Lee, S.W.Y., Chang, H.Y., Liang, J.C.: Current status, opportunities and challenges of augmented reality in education. Comput. Educ. **62**, 41–49 (2013)
21. Yuen, S.C.Y., Yaoyuneyong, G., Johnson, E.: Augmented reality: an overview and five directions for AR in education. J. Educ. Technol. Dev. Exch. (JETDE) **4**(1), 11 (2011)
22. Zarzuela, M.M., Pernas, F.J.D., Martínez, L.B., Ortega, D.G., Rodríguez, M.A.: Mobile serious game using augmented reality for supporting children's learning about animals. Procedia Comput. Sci. **25**(Supplement C), 375–381 (2013)

Design of an International Scrum-Based Collaborative Experience for Computer Engineering Students

Alberto Fernández-Bravo⬡, Ignacio García Juliá⬡, and Olga Peñalba⁽⌧⁾⬡

Escuela Politécnica Superior, Universidad Francisco de Vitoria, Carretera Pozuelo-Majadahonda Km 1.800, 28223 Pozuelo de Alarcón, Madrid, Spain
{a.fernandezbravo,i.garcia.prof,o.penalba}@ufv.es

Abstract. Globalization is a phenomenon that continues to permeate more and more domains in our lives. Students and educators cannot remain oblivious to this reality, particularly in the Information Technologies (IT) realm where, more likely than not, the former will have to cooperate and work together with people from other countries with different languages and cultures.

Developing English fluency is essential, but effective communication and collaboration require acquiring international and intercultural skills. With this aim, we -at Universidad Francisco de Vitoria- are promoting and implementing new formative activities and approaches within our official curricula. In this contribution, we describe one of them: A web application development project carried out by a team composed of Dutch and Spanish students who work together according to the Scrum framework using a DevOps approach.

The project described here has been run for three consecutive years in partnership with The Hague University of Applied Sciences and the company Delta-N. This work elaborates and extends the experience described in the communication "Computer Science and Engineering: Learning to Work in International and Multicultural Teams" presented at CSEDU 2021 [1].

Keywords: International skills · Higher education · Project-based learning · Scrum · DevOps

1 Introduction

Over the last decades, there has been a growing trend towards globalization. It is becoming commonplace for large companies and even small and medium enterprises (SMEs) to run their projects in an international setup where customers, providers, and even development teams are likely to belong to different cultures.

This reality has been facilitated by the emergence of increasingly abundant and sophisticated online collaboration tools and environments, which remove many previously insuperable barriers. However, we should not simplistically assume that, for the students to cope in such a scenario, it is enough to help them develop a good level of English or become competent in using the previously mentioned tools. As the OECD

© Springer Nature Switzerland AG 2022
B. Csapó and J. Uhomoibhi (Eds.): CSEDU 2021, CCIS 1624, pp. 260–273, 2022.
https://doi.org/10.1007/978-3-031-14756-2_13

Science, Technology and Industry Scoreboard 2017 on Digital Transformation reports: "management and communication skills seem to be especially important in digital-intensive industries" as "workers in those industries operate in a more independent and decentralized fashion (e.g. through teleworking), perform relatively more non-routine tasks, or have to deal with continuously changing settings" [2].

Acknowledging this reality, we at the Polytechnic School of the Universidad Francisco de Vitoria (SP) (from now on UFV) started to look on possibilities to accommodate in our programme ways to equip our students with intercultural and conflict management competencies that complement the quantitative, ICT, and STEM skills traditionally associated to engineering training, so that we help them to perform efficiently in such international environments.

We explored some of the various initiatives that have emerged in the area of online training, such as Erasmus + Virtual Exchange [3], E-Tandem [4], Teletandem [5], Cultura [6], or Evaluate [7]. Conceding the value of each of these proposals, we wanted to offer an experience closer to the one you get in the professional realm, where interaction goes beyond a friendly game, and your "crust" is at stake. We wanted that our students had an academically graded experience of close cooperation with people from other cultures, with shared interests but who simultaneously may have different values and codes of behaviour.

The following sections describe how The Hague University of Applied Sciences (NL) (from now on THUAS) and UFV have collaborated to provide our respective students such a learning experience, the challenges we have met, and comment on some of the observed results. We would like to thank professors Gerda V. Geld, Loess Tromp, Job Habraken and Sonya Spry from THUAS for leading this project together with us and for the opportunity to carry out this experience for three consecutive years. We would like as well to thank Mr Fokko Veegens and his company Delta-N for the excellent support and coaching provided with the DevOps infrastructure.

2 Project-Based Learning in an International Online Environment

THUAS and UFV started this collaboration project in 2019 as a pilot with the aim of providing an international and intercultural experience to third-year students enrolled in their respective Software Engineering and Computer Engineering degrees. We have already run three consecutive editions of it.

Despite the similarities between both degrees, it was decided since the beginning of the project to take a pragmatic approach that allowed each institution to enrich its academic offering while avoiding the complexities associated with a tighter alignment of theoretical subjects. We agreed that it was more sensible to set a realistic goal, defining a practical challenge, which were long enough to make sense to be organized as a project and provide meaningful growth, and short enough to easily fit within a term, where each university has different starting and ending days, holiday periods, etc.

The devised project thus spans ten weeks and focuses on developing a web application according to a set of requirements that allow the students to integrate knowledge on various topics, such as relational databases, web development, human-computer interaction, and others. Collaboration is organized according to the Scrum framework [8], and the development follows a DevOps philosophy [9] where changes can be easily put into production thanks to the combination of configuration management tools and repos, extensive use of automated testing and the definition of pipelines that are triggered each time the developers commit verified changes.

Students are granted access to a Microsoft Azure DevOps environment that integrates the previously mentioned functions with backlog management and Scrum or Kanban boards. These artefacts facilitate the transparency among stakeholders and inspection of the ongoing product. In addition, they get accounts in the Azure cloud, which have associated a moderate (100 €/student) credit so that they can experiment deploying the resulting increments in different kinds of virtual machines, databases, etc.

2.1 Opportunity Analysis

As usual, it is not that easy to turn an attractive ambition into a concrete reality, like the one described in the previous paragraphs.

In our case, firstly, we needed to find a suitable partner -preferably in the higher education community- that shared a similar interest. Secondly, there was a need to find candidate courses whose contents allowed an integrated and meaningful experience for the students, which would not distort or complicate their assessment. Thirdly, we needed to pay attention to the time dimension, as courses -obviously- take place at specific periods. Last but not least, we needed to assess what was the best academic year for the students to have this experience.

We were lucky enough to find THUAS, an institution with which we had an excellent relationship, having already hosted some of our Erasmus students for several years. Our colleagues there were devising a project integrating their theoretical and practical courses on IT Operations and Process and Project Management with another on Global Cooperation. After examining our syllabi, we found that we were substantially aligned, and we could try the collaboration with what appeared to be a low risk of failure.

When it comes to course selection, it is essential to be flexible and identify at least two meaningful courses, aware that there will not be a perfect match from the contents point of view. Also, bear in mind the calendarization of the courses and select at least one in each semester so that you can accommodate the needs of your potential partners. In addition, consider each institution's bank and mid-term holidays, a culturally-rooted factor that will likely differ and require adjustments to the plan to put all participants at ease. In our case, the first two years, we coordinated this project with an advanced course on Software Engineering. Last year (2021), we had to align it with a Project Development course due to a shift in our academic calendar.

Finally, we decided that junior students (third year) would be the optimal candidates, as they have developed a substantial competence that allows them to take the challenge. We believe that first-year or sophomores, in general, are not mature enough. On the other hand, fourth-year students have their idiosyncrasy, focused on completing their studies and exploring or already initiating their professional careers, which might prevent them from focusing on the project.

2.2   Assessing the Number and Profile of Students

To maintain a low level of risk and ensure an optimal level of support, we decided to start by offering participation in the project as an optional activity to a small number of students who showed both motivation and an adequate level of English.

In the three editions held so far, ten, eight and sixteen students have participated respectively, representing 27%, 22% and 30% of the total number of students enrolled in the course each year.

For the time being, we have decided to continue with this approach with the ambition of establishing it as part of compulsory education in a couple of years, when our first class of the bilingual section (starting in the 2021–2022 school year) will be in their third year.

In an international collaboration project such as this, attention must be paid to the students' profiles and not base your screening just on the academic record. Of course, this is important, especially for knowledge and competencies relevant to the scope, but you should look too at the student's motivation and critically judge their English level. Keep in mind that the wrong choice will have an impact not only on your students but also on the rest of the team.

In regards to English, B2 is a strict minimum. However, the students will take full advantage of the experience if they have a C1 level or higher, particularly if we consider that the participants need to deal with cultural nuances frequently expressed verbally. If they are not fluent, there is a high possibility that they will become more introverted, and the overall team performance will be affected. At this point, it is worth mentioning that in our second and third editions, THUAS has made a valuable contribution, organizing conferences on multicultural collaboration as part of the project kick-off. They have helped the students to acquire consciousness on the one hand of some facts and, on the other, of stereotypes much needed to develop a multicultural sensitivity.

Another aspect to consider is the motivation of the candidates. Taking Susan Fowler's motivational outlooks model as a reference [10], we observe that, frequently, students want to participate either by an "inherent" or "external" motivation.

The first group includes dynamic students who basically enjoy getting involved in proposed activities and -quite often- do not correctly estimate the workload of more serious pursuits, like the one we are describing. By not measuring the effort needed, this group can easily have work management issues affecting their performance, both in the project itself and other courses. This problem has shown to be particularly true for students who have part-time jobs and need to be more conscious of their tight schedules. The second one would gather those who either want to add "an extra line in their CV" or those who mainly value the opportunity to visit The Hague during the kick-off.

We would not exclude those students right away but prefer to guide them to have a more reflective approach and help them identify how this activity connects and integrates with their values and purpose in becoming engineers. In doing so, we believe that we contribute to their having a richer experience and developing a sense of responsibility towards themselves and their Dutch peers.

Finally, it is advisable to have completed the screening process a few weeks before the project starts. On the one hand, both sides need to know whether there will be enough students to create balanced groups and devise alternatives if that were not the case. On the other, there is a need to coordinate administrative arrangements (i.e., setting up accounts and permissions) usually impacted by some inevitable bureaucracy.

2.3 Designing the Experience

Defining the Scope. As mentioned earlier, both THUAS and UFV wanted to develop an educational experience that helped our students to put into practice skills and knowledge relevant to their engineering training while enriching it with an intercultural dimension.

We both agreed that the project should reinforce software engineering and IT operation concepts taught at both institutions. In this respect, we concluded that building a web application according to the following set of requirements would be a challenge adequate to their academic development:

- The application must be architected according to the Model-View-Controller (MVC) Web pattern [11].
- Data persistence will be provided using a SQL relational database, containing at least one instance of one2many and many2many relationships and supporting CRUD operations.
- For the sake of maintainability, all the associated artefacts must be subject to version control. The application shall be documented so that it is in a shape that allows a different team to evolve it.
- The application must follow responsive design principles so that it is accessible from different devices.
- It must support open authentication (OAUTH) through at least an external identity provider.
- It must conform with General Data Protection Regulation (GPDR) requirements [12].
- Automated test suites and user acceptance tests are required to verify and validate the quality of the application.

Regarding team collaboration, and in the spirit of experiencing Agile practices, we also agreed to use Scrum as a valuable framework, which has proven its effectiveness to structure non-hierarchical collaboration as the one expected for this project. Together with commitment, focus and courage, Scrum promotes the values of openness and respect, which fit perfectly with our purpose and helped us strengthen the foundations for proper multicultural interaction.

Assessing the Roles. Another aspect to think about is what is the challenge that you want to pose to your students. It must align with the contents of the courses you plan to integrate with the project-based international experience. No matter how much we believe in the value of the proposed activity, if we link it to a course, students will rightly demand assessment criteria consistent with its content.

As mentioned earlier, at UFV, we were integrating an advanced software engineering course, with a substantial focus on project management topics (where the students learn to reflect on product value, backlog management, planning – both traditional and Agile-) , but with no specific code developing practices. Considering the scope of this collaboration project, where a significant effort goes on coding, testing, developing pipelines, and so on, this posed to us a problem from an evaluation perspective as we could not assess and grade our students on areas that do not belong to the scope set in the Teaching Guide.

We opted for an approach that we deemed consistent with our syllabus and valuable enough for our partners, suggesting that our students participated as product owners (POs) within each of the Scrum teams formed. On the one hand, the role requires significant interaction with the development team. Among other responsibilities, the POs need to clarify requirements, negotiate Sprint goals, and put into action the cultural and communication skills we want to promote. On the other, it is a role that can be played effectively without requiring so many hours on the part of the student, an aspect that we needed to consider as the ECTSs involved are significantly lower than for the Dutch students.

However, this option posed some problems. When the Scrum Guide describes the various roles in the framework, it dictates that the PO role must be a single person. Although the participants are no longer teenagers, putting just one student amidst 6 or 7 peers could be somewhat intimidating, especially when the group has not met before. Besides, Agile practitioners often acknowledge that this type of collaboration is better suited for professionals with significant experience, a condition rarely met in undergraduates. Therefore -without questioning the need to adhere to this rule in real organizations- we agreed, after some discussion, to relax this requirement for pedagogical purposes. We decided that two students would play the role but ensuring that they interacted with a "single voice" with the rest of the Scrum team. According to the observed results, we can confirm that the change did not negatively impact the project, neither on the ability to deliver the product nor in the Scrum framework learning experience.

2.4 Running the Project

As previously mentioned, the project spans over ten weeks of the term corresponding to the affected courses.

The week right before the kick-off is mainly devoted to honing the final administrative arrangements so that students can start their work smoothly once it concludes. It is also the time for UFV students to fine-tune their product presentations and rehearse the pitch they will deliver to their Dutch peers, ensuring that they are concise, well-articulated and convey the associated value and impact. In our first edition, we asked our students to prepare a one-minute introductory clip. The intention was to facilitate the team-building phase during the kick-off. However, in subsequent years, we discarded this option as we did not perceive a significant benefit.

The Kick-Off. It is a momentous event in any project but particularly in collaboration experiences involving students such as this. Therefore, it needs to be carefully prepared and executed. We have run this as a one day and a half activity, excluding travelling time. Considering that the number of Dutch students involved in the project is significantly higher, UFV participants and teachers have travelled to The Hague, while our peers at THUAS have kindly taken care of the local aspects such as tools, venues, and amenities.

The meeting starts in the morning of the first day with the introduction of the participating teachers. It is an occasion to provide additional details on both institutions to the students, reminding the specific goal of the project they are about to start and share the agenda for both days.

This part usually doesn't require more than 30–45 min. Right away, students acting as POS get around fifteen minutes to pitch their ideas and answer the questions posed by the audience. For a total of ten or twelve groups, this requires about three hours, including a short coffee break. After the pitches, developers and PO have a short negotiation where the former tries to convince the latter about teaming up with them. After 20–30 min, groups get formed, and a lunch break is taken, which offers a first opportunity for the team members to socialize and get to know each other.

After lunch, the students receive a lecture on multicultural collaboration provided by an expert in the field, which for most of them represents a first serious exposure to the subject. These sessions have proven to be most valuable, as it is crucial that the students go beyond a "buddy rapport" and understand cultural differences and nuances on aspects such as authority, punctuality, commitment and get away from stereotypes. An honest reflection on what they have learned prepares them to face the challenges of the coming weeks more objectively and efficiently.

Pursuant to the lecture, the students are given two to three hours to start high-level discussions on their application and next all participants are invited to join a visit to the city which ends with a welcome dinner. Once again, we cannot fail to see the potential that these activities have in terms of facilitating group cohesion.

Experience, again, has shown us how important it is to run this event as a face-to-face meeting. Despite the convenience and breadth of collaboration tools available today, the opportunities arising from on-site discussions are irreplaceable. Interaction is faster and easier; non-verbal cues are more easily understood, and, in general, the chances to generate a friendly environment are much higher.

The covid-19 pandemic prevented us from celebrating our 2021 edition onsite. As usual, everything was carefully prepared but the event itself was more challenging to manage. Even though we are talking about countries with excellent communication services, handling an interactive activity such as this turns out to be quite hard. When everybody is connected to the same space, such as in a webinar, there are no particular issues beyond someone losing shortly their connection and quickly reconnecting. However, when communication goes from broadcast mode to smaller group discussions and input is required from several tenths of people that have to switch virtual spaces, things get much more complex. Turning microphones and cameras on and off does take some time. There are as well slight delays that are not disruptive when communication is unidirectional or involves just a few people, but that in this situation hinder the exchanges and end up with less openness in the participants. Acknowledging that for most of the students this was not an impediment, we noted that in this edition some of the teams needed more time to reach a performing stage and in one of them -for the first time-cooperation did not work at all.

On the second day, Delta-N provides lectures on DevOps as a working paradigm, as well as on the Microsoft Azure and Azure DevOps and the infrastructure that the team is going to use to develop their project. In addition, time is reserved so that the newly formed teams can establish the schedule and the background rules that will norm their cooperation for the coming weeks and start building a first draft of the product backlog that will drive their efforts.

The event is thus closed allowing both the teams and the teachers to have the opportunity to share a final lunch, and then UFV participants will return home.

Executing the Project within a Scrum Framework. Once the students have learned the basics of the project, agreed on a high-level view of their product, and established a basic set of rules for collaboration, the actual development starts.

As previously mentioned, we agreed to run the project using the Scrum framework. Dutch students hold the Scrum Master (SM) and developer roles and Spanish students the product owner (PO) role. The allocation of roles relates to the integrated courses and the total amount of ECTS involved. THUAS runs this project integrating several subjects, which total up to fifteen ECTS. Our practical sessions on advanced Software Engineering accounts for four and a half ECTS.

This imbalance in the academic load prevents our students from opting for a developer profile that requires a higher dedication and more frequent coordination with the rest of the team members. We explored the possibility that they held the Scrum Master role. Considering that most of the team and the infrastructure resources are in The Hague, we finally decided to allocate it to a Dutch student.

In such a situation, our students must help their peers to set the right expectations on their involvement as they cannot devote the same hours to the project. They need to communicate the actual number they can put on the project and their distribution over the week. Not understanding this need has been a source of stress and conflict on several occasions during the first weeks.

It is also worth stressing that they must respect and use efficiently the time allocated to the events where their participation is required. Though, in the beginning, this situation is challenging, it is also good for the students to flesh out some of the theoretical notions they learn in their courses. For instance, you cannot always choose the people with whom you collaborate. Moreover, they see models such as Tuckman's in practice and funny concepts as storming, norming, and so on, acquire, all of a sudden, a real meaning.

The length of the Sprints, one week, is determined by the teachers, which help to reduce the risk of product drifting by more frequently reviewing both the increments and the team performance. The students are free to define the specific timing of the daily Scrums, the Sprint planning, review and retrospective, as they need to accommodate different needs involving courses and part-time jobs schedules. In our case, classes take place in the evening, and THUAS students have them in the morning.

As with the dual PO role, you can be flexible to a certain extent for pedagogical purposes but stretching too much the flexibility could finally lead to a way of working that has little to do with Scrum, thus spoiling one of the goals of the collaboration. For instance, we believe that it might be worth relaxing some of the rules, like the one indicating that the PO must not participate in the daily Scrum. We advise our POs to attend this event at least a few times over the first weeks. In doing so, they get a better understanding of the framework dynamics. On the contrary, despite an obvious need to conciliate the academic and job dimensions, you must remind some students that they cannot collapse all the Scrum events in one or two days. Otherwise, the framework principles of transparency, inspection and adaptation get seriously compromised.

All in all, we have striven to follow the framework as close as possible to the original intent of its creators. They emphasize not changing Scrum's roles, events, artefacts, or rules, as the outcome would not be Scrum. Acknowledging this, we all believe that you can still provide significant Scrum learning to the students while making the previously mentioned trade-offs. To our minds, the essential point is that the changes are justified and explained to the students and compared to the original setup.

Apps Presentation and Closure. The last week of the project is devoted to share the resulting products, where the whole team shall work on a slide deck, presenting the initial vision, the resulting outcome and describing how they would proceed if the project were to continue by means of a roadmap. Although we both THUAS and UFV hope to hold this event face-to-face in the future, so far we have always run it online.

Beyond the technical, collaboration, and management skills acquired during the project, this event provides an excellent opportunity for the students to enhance and complete their communication skills by showing their ability to defend and pitch their product.

The academic perspective is represented by THUAS and UFV teachers, while Delta-N representatives provide the market view. The defense is public, and all the teams can watch their contenders.

Bearing in mind that being exposed to a larger audience could be intimidating for some students, we have striven to find the right balance between academic rigour and a friendly atmosphere, where the students should expect objective and constructive feedback from all participants. One way we found to contribute to this goal is by allowing all the audience to vote their preferred application, which helps to close the experience on a high note.

2.5 Student's Support

Teachers' Involvement. For the teachers' part, we mostly rely on asynchronous collaboration, using tools such as Slack. The main point is to be attentive to issues that may pop up at each site and share them quickly enough to apply corrective actions. We would advise defining a light protocol identifying a simple standard regarding contents and minimum frequency to report progress.

Time demands to support the students should be carefully dimensioned. Although third-year students are largely autonomous, they need guidance related to their role and on conflict management. Sprint reviews and retrospectives offer an excellent opportunity to coach the teams. The first event focuses on how value is incorporated into the product, while the second examines team performance.

Nevertheless, if you consider short Sprints and eight or more teams as we do, attending these events for each group and Sprint becomes impossible unless significant staff is involved, conditions that pitifully are rare. Even for the parameters considered, if you want to provide minimal support, you need no less than three teachers from each site focusing on just two or three milestones of the project.

Project Infrastructure. In a project such as this, infrastructure selection is essential for success. Advised by our colleagues at THUAS and Delta-N, we have used Microsoft

Azure to provide cloud computing services and Microsoft Azure DevOps as a comprehensive environment, enabling teams to plan, deliver, deploy, and even operate their applications in an agile and transparent way. The latter is a feature-rich environment that allows the students to familiarize themselves with Continuous Integration practices. They can define and maintain backlogs, track work with Scrum or Kanban boards, keep development efforts synchronized with Github repositories and have automated tests and builds run seamlessly.

Apart from providing a rich set of features for developers, Azure DevOps offers to our students as POs a wealth of built-in and third-party components that help them to manage their product backlogs easily. The environment allows them to hierarchically structure user stories, assign values and priorities, and collect estimations with the development team using poker planning, etc.

2.6 Performance Assessment

One important point to take into consideration while designing a project such as this is how to evaluate the participants. The criteria need to be relevant both to the project itself and -even more importantly- to the courses integrated with it.

Finding a perfect match between the competences associated to your courses and those you can actually measure in the project can be challenging. You need to start from the former, so that you guarantee fairness towards your students and to make sure that internal and external standards concerning national and international higher education qualification requirements relevant to your degree are preserved.

Table 1 shows the competences in our Advanced Software Engineering course that we deemed relevant to the scope of the collaboration project.

Table 1. Competences belonging to the UFV SW Engineering course relevant to the project.

Type of competence	Description
Basic	Students can convey information and ideas on problems and solutions to both specialized and non-specialized audiences
Generic	Knowledge to carry out measurements, calculations, valuations, appraisals, surveys, studies, reports, task planning and other similar computer works
Specific	Ability to assess customer needs and specify software requirements to meet these needs, reconciling conflicting objectives by finding acceptable compromises within the constraints of cost, time, the existence of already developed systems and the organizations themselves
Specific	Ability to identify, evaluate and manage the potential associated risks that may arise
Specific	Ability to identify and analyze problems and to design, develop, implement, verify, and document software solutions based on an adequate knowledge of current theories, models and techniques

On the other hand, while it is important to respect the autonomy of each institution, performance, criteria need to be coordinated to ensure that students at each university have common interests at stake. In the end we proposed to focus on three areas so that we can assess:

- Attitudes and collaboration
- The adequacy in performing the product owner role, and
- The students' ability to communicate the goal, context, benefits, and impact of their product, together with a critical assessment of the collaboration effort.

Attitudes and Collaboration. Teamwork, collaboration, motivation, and conflict management are topics that we cover in our course. In addition, at UFV we promote a culture that keeps our students away from individualism and always bears others in mind. We try to make this objective part of our students' core values, so that the creation of a pleasant working environment and the idea of the common good arises spontaneously in them.

To preserve the need for autonomy of the students, during the kick-off, the teams define their own set of background rules, which usually are associated to:

- Events active attendance and punctuality
- Showing respect to inputs and comments proposed by other members
- Show commitment with the set goals and be open to support other members as required
- The need to be transparent in communicating issues so that impact of materialized is more easily managed by the team

Sprint retrospectives offer a recurring opportunity to review together with the participants how well they honor their "pledge". In the early weeks, attention must be given to false consensus, particularly in those groups gathering shyer or less assertive members. Not dealing on time with issues related with background rules can be quite damaging. Trust among participants starts to dissolve and with it the lack of confidence on the agreements made in the meetings. This will result in a lack of accountability from the team members and eventually in not reaching their goal, consistent with team dysfunctions as described by Patrick Lencioni [13].

Product Owner Performance. This is the dimension that more clearly maps with the competences that we want to develop in our course. We want our students to grow their knowledge and skills in product management, understanding the value their product provides to different stakeholders and reflecting on how they can bring to reality their proposal by means of meaningful increments. We therefore measure:

- Their ability to define and articulate a product idea. We ask them to identify the most important stakeholders, to describe the context required to understand the value proposition and to explicitly mention the impact they expect they will provide. They are encouraged to go beyond vague statements and provide some measure of the impact.

- Their ability to breakdown the epic representing the web application into features and actionable user stories that can be managed in the context of Sprints. Here we also value that the students arrange their priorities in a way that increments are meaningful.
- Their ability to collaborate with the development team in the grooming of the product backlog and creating Sprint backlogs.

Communication Skills. As communication among team members is covered in the first area, here we focus on our students' ability to isolate essential and accidental information about the product. They must examine the different stakeholders' perspectives to show that they can adapt their pitches depending on the audience.

Once more, our peer at THUAS professor Sonya Spry has added great value to our collaboration by sharing with us their approach and offering to those of our students that could accommodate their schedule to participate in specific presentation skills sessions together with their Dutch peers. This has allowed us to extend our assessment criteria to consider:

- How effective are the presentations' openings
- The clarity to state the purpose and the outline of their pitch
- The effectiveness to use "sign-posts" that allow the audience to remain connected to the speech
- How persuasive is their language
- Use of rhetorical resources

3 Results

As we have previously mentioned, we have successfully run this project three times, starting in 2019. This reference is worth highlighting because the covid-19 pandemic hit us in this collaboration project's second and last edition.

Since the beginning, we have aimed to capture the results of the students' experience in a formal record that allows us a more objective analysis. However, due to confinements and general restrictions, like many other institutions, we had to modify our plans and focus our efforts on adapting our infrastructure and methods to that situation. Therefore, we are still working on the definition of practical instruments to gather students' feedback according to relevant items to measure this experience's impact on them.

Notwithstanding, we have got plenty of informal feedback from the participants. Except for two of them, this last year (who had some issues with the language and did not correctly manage conflict), they consistently value the experience as one that has helped them grow as engineers and especially be better prepared to cope with cultural differences in a future job.

Even if they do not participate as developers, they are all enthusiastic about the possibility they got to work in an environment such as Azure DevOps. Defining, building, and monitoring a product from a single environment is certainly formative and exciting, as it helps them integrate theoretical and practical knowledge received in other courses.

They also feel that they have learned to structure collaboration through Scrum, and somehow, they come out in better shape for their future professional performance. The participants could experience in a controlled environment many of the circumstances they will find once they graduate. For instance, in the first year, some groups were affected by different events, which reduced their initial capacity, forcing them to adapt to the new conditions quickly. In some cases, this meant that they had to give up their original ambition. Learning to identify the critical-to-quality aspects of the product, identify the proper criteria for prioritization, and preserve team cohesion seems an excellent life lesson from the project.

They also appreciate that they end this project with a higher consciousness of the need to be attentive to cultural differences when collaborating in an international group. Even in countries like The Netherlands or Spain, which are both European and have a substantial set of shared values, both parties need to detect the differences and be willing to accommodate each other for the benefit of the project.

4   Conclusions

The experience described focuses on the design and implementation of an international collaboration experience between software engineering and computer engineering students from two different countries. However, we hope that, by having described it along the stages of a project, it could be valuable for other institutions, even considering more distant disciplines.

The shared founding premise is that being proficient at the technical level is no longer enough. There is a need to develop new learning initiatives to develop practical intercultural skills, which professionals operating in a globalized world will need more than ever. If someone intends to become an engineer, the need to learn and become proficient working with technical tools and environments is undeniable. However, many projects and organizations fail because they do not pay suitable attention to the human factor.

Awakening and promoting this sensitivity in the students can be achieved through a short-term collaboration using a structured but straightforward framework as Scrum that promotes transparency and adaptation and offers plenty of opportunities for inspecting the project outcome and the team performance. This process is further promoted through coaching and activities that explicitly focus on cultural differences affecting values and behaviors.

References

1. Fernández-Bravo, A., Peñalba, O., García Juliá, I.: Computer science and engineering: learning to work in international and multicultural teams. In: Proceedings of the 13th International Conference on Computer Supported Education - Volume 2: CSEDU, pp. 346–352 (2021)
2. OECD: OECD Science, Technology and Industry Scoreboard 2017: The Digital Transformation, p. 41. OECD Publishing, Paris (2017). https://doi.org/10.1787/9789264268821-en
3. Erasmus+ Virtual Exchange Homepage. https://europa.eu/youth/erasmusvirtual. Accessed 26 Oct 2021
4. N.D.: ETandemLearning autonomous language learning with a partner. https://www.languages.dk/methods/tandem/eTandem_syllabus_en.pdf. Accessed 26 Oct 2021

5. Teletandem Homepage. http://www.teletandembrasil.org. Accessed 26 Oct 2021
6. Cultura Homepage. https://cultura.mit.edu. Accessed 26 Oct 2021
7. Evaluate Homepage. https://sites.google.com/unileon.es/evaluate2019/. Accessed 26 Oct 2021
8. Schwaber, K., Sutherland, J.: The 2020 Scrum Guide™. https://www.scrumguides.org/scrum-guide.html. Accessed 29 Nov 2020
9. Kim, G., Humble, J., Debois, P., Willis, J.: The DevOps Handbook. IT Revolution Press, Portland (2016)
10. Fowler, S.: Why Motivating People Doesn't Work… and What Does, 1st edn. Berrett-Koehler Publishers Inc., San Francisco (2014)
11. Fowler, M.: Model View Controller. https://martinfowler.com/eaaCatalog/modelViewController.html. Accessed 26 Oct 2021
12. General Data Protection Regulation (GDPR) Compliance Guidelines. https://gdpr.eu. Accessed 26 Oct 2021
13. Lencioni, P.: The Five Dysfunctions of a Team, 1st edn. Jossey-Bass, San Francisco (2002)

Comparative Analysis on Features Supporting Students' Self-regulation in Three Different Online Learning Platforms

Tuija Alasalmi(✉)

Haaga-Helia University of Applied Sciences, Ratapihantie 13, 00320 Helsinki, Finland
`tuija.alasalmi@haaga-helia.fi`

Abstract. The paper discusses results of a survey focusing on student expectations and experiences on learning analytics and tools used for supporting self-regulation. The students participating in the survey were from three different organisations and they were also users of three different learning platforms. The results show differences between the students' self-efficacy levels and identified tools and pedagogical solutions in each platform. The paper also discusses students' preferences and need for learning analytics dashboards and peer comparisons as well as their potential for functioning as supporting measures for self-regulation.

Keywords: Learning analytics · Self-regulation · Learning platforms

1 Introduction

Learning analytics is typically used for designing learning from the point of view teachers, academic advisors or other administrators of an educational organization. It often focuses on identifying and monitoring possible drop-outs through certain indicators, or it is used to map the study paths to find the challenging points which call for redesign of the online course. The perspective for applying tools of learning analytics has been mainly administrative [1]. To make it clear, in this article learning analytics is viewed as the process of gathering data about student activities and efforts in a digital learning environment, interpreting the collected data and producing reports and analysis based on the data to the users of the digital learning platforms [2, 3].

A lot of In-built learning analytics tools have been developed in the digital learning platforms for the purpose of reducing teachers' manual work, to collect acts of learning into countable units (such as clicks, views and time used) which can be reported to the platform users. Data mining connected with learning analytics usually utilizes system logs (such as time and number of logins), the amount and duration of views per document or links offered as study materials, the amount of forum posts, assignment submissions and test attempts. The tools used for analyzing learning vary in digital learning platforms. In general, student dashboards, progression tracking tools, test and questionnaire tools, gradebooks or grading views and course overview repots offer organized information

© Springer Nature Switzerland AG 2022
B. Csapó and J. Uhomoibhi (Eds.): CSEDU 2021, CCIS 1624, pp. 274–289, 2022.
https://doi.org/10.1007/978-3-031-14756-2_14

about the learner's actions. For teachers the platforms usually offer a wider range of reports based on logs.

Learning analytics may help teachers design better online courses, but the collected data must be relevant. It has been suggested that in order to collect meaningful data in an online course, teachers need to incorporate certain pedagogical elements which produce digital footprints in the learning process [4]. These elements can be constructed using the different tools or activities in the digital learning platforms. In this article a learning analytics tool means any kind of activity or method for collecting the users' data and presenting it to them. These tools may be automated (the platform collects and organizes the data) or manual (the teacher collects and organizes the data), or both. A combination of manual and automated learning analytics is for example a grading table in which a teacher adds a manually graded assignment.

A well-made online course design that consistently collects digital footprints of the learners' actions may help teachers in offering timely support and feedback, but does the collected data really benefit students directly? Do the designed pedagogical elements and measures of collecting data enhance student self-regulation or motivation? What kind of data about the learning process would be useful or meaningful to the students? Learning analytics may play a role on a metacognitive level, for example by directing students into using better learning strategies, developing self-regulatory skills or improving their emotional and cognitive awareness of themselves as learners.

These issues were explored in the MOPPA project (Motivation och självreglering på inlärningsplatta med hjälp av inlärningsanalytik) funded by the Swedish Cultural Foundation in Finland. Three educational institutes in Finland participated in a comparative survey focusing on mapping the students' needs and experiences on using digital learning platforms during online or blended learning courses. The participating institutes were Haaga-Helia University of Applied Sciences, Prakticum, and Axxell Utbildning. Haaga-Helia is a Finnish-speaking higher educational institute while the latter two are Swedish-speaking vocational upper secondary schools. All three organizations have different digital learning platforms: Haaga-Helia uses Moodle, Axxell has ItsLearning and Prakticum operates on Google Classroom (G Suite for Education, henceforth also GC). Therefore, the responses from each organisation were viewed separately in comparison with each other to find out if there are any differences concerning student experiences with the platforms. Some categorizations were made within all respondents to be able to highlight overall trends but mostly this paper concentrates on the comparative discussion of the survey responses to the structured questions. The views which emerged in the open-ended questions and interviews have been discussed further in the proceedings of the 13th International Conference on Computer Supported Education and they are addressed only briefly here. This paper extends the analysis with a closer look at students' self-efficacy and the identified learning analytics tools on three different online learning platforms.

2 The Survey Framework and Participants

The survey was conducted between November 2020 and February 2021. It was followed by semi-structured student interviews with seven volunteer respondents. The amount of survey respondents was 93, of which 47 were from Haaga-Helia, 34 from Axxell and only 12 from Practicum. The interviewed focus group participants were only from Haaga-Helia.

The questionnaire contained 25 questions, consisting of multiple-choice questions based on ready-made alternatives, value scale questions, yes/no questions, and open text answers. The first set of 8 questions was concerned with background information, preferences on the use of learning platform and general knowledge about learning analytics. The next 12 questions were structured, mapping the students' experiences on online courses in respect to course design and identifying which aspects were considered important in online learning. Finally, there were some open-ended questions focusing on the student motivation, expectations and experienced barriers to learning. The fields of study of the students who responded to the survey were not defined in advance, but those who studied in blended learning implementations were specifically selected as the respondent groups. Due to the COVID-19 situation, however, the respondents were asked to assess the proportion of online studies in their study program. The share of online studies was estimated at 5% at the lowest and 100% at the highest. There were only minor differences with the three organizations: in Haaga-Helia the average share of online studies was 80,2%, in Axxell 91,8% and in Practicum 87,1%.

The respondents of Axxell were all studying in the program of early childhood education and all the respondents od Prakticum were students of ICT. In Haaga-Helia, there were more study programs involved although the majority of the respondents studied ICT and digital services or business and entrepreneurship. Nevertheless, the results are not completely comparable, as the students from Haaga-Helia had a variety of courses available taught by different teachers. In vocational basic education programs, student groups usually have a tutor teacher and most of the vocational qualification units are taught by the same teachers.

3 Self-regulation: The Theoretical Framework of the Survey

The question of self-regulation has been much addressed in educational science. It is widely recognized that self-regulation skills have a fundamental impact on learning [5], and they may be particularly important when studying independently in an online environment. According to a study by Roll and Winne, support for self-regulation is what students actually expect from learning analytics [5]. There's only minor evidence on the use of learning analytics having any effect on learning results [6, 15], and measuring competence development through learning analytics reliably may be difficult. However, learning analytics may be used as a tool for supporting metacognitive processes during learning [7, 8].

Self-regulatory processes include for example ability to plan and schedule one's actions, personal goal-setting, use of different learning strategies for different purposes, reflective self-evaluation, identification of personal beliefs supporting self-efficacy and ability to seek help and additional information [9]. Barry Zimmerman stresses that self-regulation is not a measurable skill as such: "it is the self-directive process by which learners transform their mental abilities into academic skills" [9]. Zimmerman distinguishes three phases of self-regulation in a learning process. The preparatory (forethought) phase engages students in setting personal learning goals and planning suitable learning strategies. Students also need to become aware of their self-motivational beliefs such as their interests and orientation on the topic and how competent they feel. The performance phase activates students to observe their learning habits and chosen strategies and change them if needed. Adaptation happens through monitoring of errors and successes. In the performance phase students also use various methods of self-control (such as visualizations). The self-reflection phase pushes students to evaluate and compare their performance and efforts with the expected learning outcomes, standards and results achieved by others. In the final phase, self-regulated learners try to find reasons for succeeding or failing and observe their level of satisfaction [9].

In any learning process, students' self-regulation can be enhanced simply by offering ways to initiate and implement these processes. In an online learning context this would mean for instance providing students opportunities to set their own goals with respect to the expected learning outcomes, suggesting learning paths and schedules for studying, offering tips on alternative learning strategies, offering tests and other means to check one's competence level, and helping students explore their interest and motivation through assignments which are meant for orientation.

In the performance phase students may benefit from well-timed notifications or reminders (to focus their attention), visualizations explaining the study process or the concepts used, and other motivational support. To help student observe their learning habits, it may be useful to plan checkpoints in which students are able to report their feelings, level of commitment and even express their opinion on the workload. Any kind of dashboard collecting numerical data on the students' actions in the learning platform may help them understand and possibly change their behavior. For instance, it may be useful to see how many minutes or hours students have spent reading or writing or going through the course material.

For self-regulation support in the final self-reflective phase, teachers may build in self-assessment practices such as questionnaires or simple self-evaluation questions in connection with assignment submission. There may be sessions for collaborative reflection, peer review practices, feedback given in various ways and oral or written summaries available in the platform.

The link between self-regulation and learning analytics was discussed in a study by Schumacher and Ifenthaler, mapping students' expectations towards learning analytics. Schumacher and Ifenthaler used Zimmerman's cyclical view on self-regulation as the framework and identified some key expectations in each phase. In the forethought phase students hope for planning tools, motivational kicks, to-do lists or comprehensive views on deadlines, clear goals for learning and personalized recommendations. In the performance phase students expect to have continuous update on their performance and skills development compared to the required competences and learning outcomes. Students also asked for additional material suitable for their skills level, possibilities for social learning and recognition of their learning efforts offline. In the self-reflective phase students look for self-reflective practices and assignments. They also wish for personal feedback given at the right moment [10].

The MOPPA project survey aimed at investigating if any of these recognized expectations are incorporated in the course design and furthermore, if they exist, how does the pedagogical design of online courses with certain functionalities contribute to student self-regulation and motivation?

4   Results

The questions of the MOPPA project survey did not directly map the use or usefulness of certain kinds of platform-based tools such as progress tracking, learning analytics dashboard, or reports for a number of reasons. First, the learning platforms of the participating organizations are different and the learning environment tools used by different teachers in their own courses vary. Secondly, at the time of drafting the questionnaire, it was not fully known if there were any analytical tools available and activated in the three learning platforms. It was also not known how well the students were able to use their organization's learning platform and its functionalities, including the analytics tools that may be available to them.

The questions were drafted in a way that the students did not need to know which functionalities of the learning platform collect data or what kind of student activities can accumulate information about their learning in the system. The starting assumption in compiling the survey was that students may not be familiar with the concept of learning analytics at all. Therefore, in the first part of the questionnaire, respondents were asked to explain the concept of learning analytics. Out of the 93 respondents, 44 (47,3%) reported that they do not know what learning analytics means or that they heard the concept for the first time. 31 students (33,3%) tried to explain the concept but failed essentially. Only 18 respondents (19,4%) were able to describe the concept correctly – and those were students from ICT and digital services in higher education. It has been previously suggested that students do not really understand learning analytics nor the tools used for collecting data [11], and these results point at similar conclusion.

The respondents were also asked if learning analytics had been discussed in their studies. Only 5,8% of all respondents reported that learning analytics had been addressed. 51,7% answered that learning analytics had never been discussed, and 42,5% were not sure if it was ever mentioned.

4.1   Self-efficacy Levels

In the second question set of the survey, the respondents were first asked to name the course which they based their answers on. Subsequently, it was asked how well the students felt they were able to operate on the chosen course area. The purpose of the question was to map the level of self-efficacy with respect to the use of learning platforms. The results are described in Fig. 1.

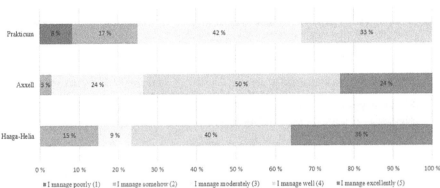

Fig. 1. The respondents' assessment of their self-efficacy level.

It is noticeable that none of Prakticum's students thought they managed excellently, and moreover, none of Haaga-Helia's or Axxell's students managed poorly. Furthermore, Axxell's students seem to be the most confident in general, whereas Haaga-Helia students' perceptions are more divided: the majority feels they manage well or even better, but more than 20% did not feel as competent. The results indicate that some students may have problems with self-regulation, even though it was not directly asked. The lower figures may also refer to other barriers to learning, lack of digital skills or to some problematic issues in the online course design, and therefore respondents were able to briefly explain their answer if they wanted. The answers mostly highlighted the following issues: the structure of the platform was too complicated, the course layout and structure were disorganized and there was too much information or some material was missing, the pedagogical solutions, guidance and communication were confusing, the assignments were difficult, schedules and deadlines were hidden and making personal plans and schedules was hard. The only positive remark referred to the clear weekly plan being used, offering a comprehensive view on how and when to proceed with the course material.

4.2 Pedagogical Solutions Supporting Self-regulation as Tools in Learning Platforms

Next, the respondents were asked to analyze if they had identified certain pedagogical practices in the online courses which they participated in. The pedagogical practices listed in the question were recognized, as discussed previously, as ways to support self-regulation processes in the forethought phase, performance phase and self-reflection phase. They were also identified as functionalities which may contribute to learning analytics. Therefore, it was interesting to see if students had had any experience on them in different online learning platforms. The percentages of positive answers are presented in Table 1.

Table 1. The amount of positive identification of certain pedagogical practices.

Phase	Did the course include the following pedagogical practices (Yes/No)? A course functionality supporting self-regulation	Haaga-Helia	Axxell	Prakticum
		Yes	Yes	Yes
Forethought	Initial competence assessment (in the beginning of the course)	27,7 %	84,4 %	16,3 %
	Setting personal competence goals	23,4 %	75,8 %	25 %
	Strategies or tools to plan your progression in the course	25,5 %	78,1 %	0 %
	Strategies or tools to schedule your studies in the course	21,7 %	65,6 %	8,3 %
Performance	Study process monitoring (a comprehensive view on all course assignment submissions and grading)	58,7 %	69,7 %	25 %
	Follow-up on your emotions during the course	14,9 %	43,8 %	16,7 %
Self-reflection	Self-assessment on personal skills development during different phases of the learning process	19,1 %	72,7 %	16,7 %
	Self-reflection on your learning	38,3 %	75 %	16,7 %

The first four statements are connected with the forethought phase, the next two with the performance phase and the final two statements with the self-regulation phase. The table shows Axxell's students highly agree on the existence of pedagogical solutions and/or platform tools supporting self-regulation in all the phases. Haaga-Helia's students seem to have experienced measures for self-regulation to some extent in the performance phase. Even Prakticum's students reported proportionally more instances of personal goal setting in the forethought phase, although at large the responses indicate that students in Prakticum are hardly offered any supportive measures for self-regulation.

The experienced differences may be explained with the tools available in the organizations' learning platform. On the other hand, these solutions may be incorporated with very simple tools, if only teachers design these elements to be part of their course. For example, initial competence assessment can be done with simple questionnaire document or with the platform's built-in test or questionnaire tool. Similarly, while study progress monitoring may be organized through a dashboard tool, it can be carried out as a simple spreadsheet or to-do list just as well.

The positive experience rates of Axxell's students may be due to the fact that ItsLearning has a student dashboard (360 reports) for monitoring study progression and the feature is being used in Axxell. ItsLearning also has a built-in learning outcome tool through which students are able to follow which assignments add up to certain learning outcomes and there is also a comprehensive view collecting all the required learning outcomes on a course. A view including all course assignments and grading is available to students as a separate tab. The teachers have also invested on the overall pedagogical design of their online course. These partly explain why the amount of positive experiences of these elements is significantly higher in Axxell.

In Haaga-Helia, measures supporting self-regulation in the forethought phase seem to be rarely offered. Haaga-Helia's students are Moodle users and on a course level they may monitor their assignment submissions and grades from the gradebook which is visible by default. However, the gradebook view may not be very informative and it may be disorganized; it depends on how well the teachers have adjusted the gradebook settings and if they have turned on the assignment settings correctly. Another tool available to Moodle users is the course progression block, but in Haaga-Helia it is not necessarily activated in all courses. It remains unclear if the positive answers with study process monitoring refer to the gradebook or to the course progression block (which also collects all the assignments and materials to be completed in a course area), or both. However, in the open-ended responses, the progress bar was distinguished by several students as one of the most useful tools of the course. Overall, the survey and the interviews revealed a strong need for progression tracking tools [12].

In Haaga-Helia there seems to be very little opportunities available for following personal skills development on a course level. Moodle does have tools for this: the competence block and the learning outcomes tool. Neither of these is used in Haaga-Helia's Moodle, and therefore the self-assessment is probably carried out as questionnaires or as questions in connection to assignments. Practices for self-reflection are more common, but still the support for self-regulation in the final, self-reflective phase seems to be rather rare.

In Google Classroom there is "View my work" link where students may check their assignment grades and all the graded assignments in a list view. Most likely the positive responses on course progress monitoring are connected with this grading view, since GC does not seem to offer any other methods to follow one's progression or learning outcomes in a course. If this is the case, either the grading view alone is not enough, or students do not know how to use it or alternatively the teachers do not use GC tools for handing out assignments, as the percentage of positive responses of study progress monitoring in Prakticum is so low. Finally, setting personal goals and initial competence assessment in Google Classroom have most likely been done through an assignment created by the teacher, because there is no learning outcome or competence tool available within the platform.

All three platforms have a calendar tool and teachers may set deadlines for assignments. It seems unclear if any tools are offered or suggested for making personal schedules, but at least in Moodle students are able to add their own calendar events into the general platform calendar which collects all deadlines from all of their courses. On a course level, the calendar tool is not available by default and thus usually not visible. In GC, students may also add calendar events and view their own calendar simultaneously with the course schedule. It is possible that the Google calendar is not used in Prakticum, based on the low amount of positive answers. On the basis of Table 1 percentages, ItsLearning seems to support personal scheduling, but it was not clear if it was done with a platform tool.

Interestingly enough, none of the platforms seem to highlight gathering data of the students' emotions. This could be done with a simple tool utilizing emoticons, for example. Alternatively, teachers may add questions or assignments where students report their emotional status or describe their feelings in relation to the learning material and assignments. It would be interesting to study how monitoring personal feelings and moods could help students with measures of self-control in the performance phase. In a study by Silvola and colleagues concerning learning analytics and student engagement, students raised a need for tools that would support emotional engagement in learning environments [13]. Monitoring the cumulative data on one's emotions could help students reflect on their performance and competence development, and thus it could potentially help in self-regulation as well.

4.3 Self-efficacy Levels and Pedagogical Practices Supporting Self-regulation in Online Learning

To explore the link between the students' perspectives of their own self-efficacy in online courses and their experiences of certain pedagogical practices in the respective courses, the responses were categorized according to their self-efficacy estimates and the alternatives presented in Table 1. In Table 2, the responses are divided in two columns: the left column represents the share of students who reported being able to operate less than moderately in their online courses. The right column represents the share of students who felt they managed at least moderately. The figures show the percentage of students in these two categories who had identified the listed pedagogical practices in their course.

Table 2. Percentages of yes answers on experienced measures of self-regulation reported by students with lower or higher self-efficacy levels.

Did the course include the following pedagogical practices (Yes/No)	All respondents (n=93)	
	Self-efficacy under moderate (percentage of positive answers)	Self-efficacy moderate or better (percentage of positive answers)
Initial competence assessment (in the beginning of the course)	45,5	45,1
Setting personal competence goals	18,2	45,1
Strategies or tools to plan your progression in the course	9,0	44,0
Strategies or tools to schedule your studies in the course	9,0	39,0
Study process monitoring (a comprehensive view on all course assignment submissions and grading)	63,4	57,3
Follow-up on your emotions during the course	18,2	25,6
Self-assessment on personal skills development during different phases of the learning process	18,2	51,2
Self-reflection on your learning	18,2	39,0

Typically, students who were confident about their ability to operate on the course more often reported experiencing measures for supporting self-regulation. In particular, offering tools for scheduling one's studies are often identified by those with high level of self-efficacy in all the organizations. Similarly, providing opportunities for self-reflection are more often experienced by students with higher self-efficacy beliefs. Study progress monitoring is a bit more controversial: even though monitoring options are reported to exist, they are more often identified by those students who struggle in their online course. The same phenomenon is visible with initial competence assessments: students with lower self-efficacy levels more often identified having initial competence assessments in their courses. It is perhaps incorrect to claim that study progress monitoring and initial competence assessments reduce students' self-efficacy; rather, it may just as well be the lack of other measures supporting self-regulation that influence the self-efficacy beliefs. Or alternatively, the results also echo the findings of Park & Jo in 2015, reporting that students are unable to correctly interpret and utilize the data dashboards in learning environments [14]. Another possible interpretation is that learning analytics dashboards are in some ways demotivating for students with lower achievement level [16]. Nevertheless, having study progress monitoring available may lead to a situation where students are continuously informed about their failures to deliver the scheduled tasks. This combined with the reported problems of the courses (incoherent structure, lack of guidance and communication, and unclear scheduling) creates confusion and dissatisfaction, which may also result in lower levels of self-efficacy and motivation.

4.4 Monitoring Learning with Dashboards

Study progress monitoring in online learning platforms is surely more than just checking final grades; it is about keeping track of all the course material, observing what is completed and what more there is to do. Monitoring one's behavior and keeping track of the different acts of learning is what matters for self-regulation in the performance phase. Therefore, the survey respondents were asked if a dashboard view available in the learning environment would help them in their studies (Fig. 2).

Axxell students' appreciated dashboard views the most in general, although Haaga-Helia peaked with the most extreme opinion. The differences in the percentages between very beneficial and extremely beneficial may be explained with the already available 360 reports view in ItsLearning. The site-level dashboard view in Haaga-Helia Moodle is quite limited: it only shows the overall percentage of completion of the courses which utilize completion tracking. On a course level, no dashboard view is available. This may explain why Haaga-Helia's students feel a dashboard view would benefit them, whereas in Axxell the need has already been satisfied. In Prakticum the need for a dashboard view was slightly lower; but nevertheless, in all the organizations the need for a comprehensive dashboard in the learning platform clearly emerged.

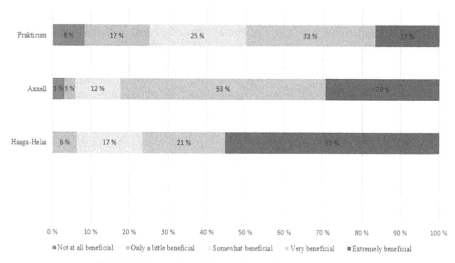

How beneficial do you consider having a comprehensive dashboard view of the current course status (including all deadlines, weekly workload, study time required etc.?)

Fig. 2. Estimated usefulness of a monitoring dashboard.

Respondents were also asked if they get any visual information on their actions within the learning platform (Fig. 3). The responses from Prakticum indicate that Google Classroom offers a bit less graphics than the other platforms. The opinions of Haaga-Helia Moodle users are divided, but on the whole, there are some visualizations evidently available. Most of Axxell's students reported having information in visual form to some extent or quite a lot. As pointed in Table 1, ItsLearning users reported higher positive answers (69,7%) with the availability of study progress monitoring than Moodle users (58,7%). Moreover, as most ItsLearning users feel that there is quite a lot visual information available, it is fair to say that ItsLearning seems to display the data in a more graphic way as compared to Moodle or GC. Interestingly enough, in the follow-up question nearly all Axxell's respondents (96%) reported that the amount of visualizations was enough for them, whereas with Haaga-Helia's students the number was only 60%, and about 38% responded that there weren't enough visualizations available. In Prakticum, the percentage of students satisfied with the amount of visualizations was 82% and only 18% wished for more graphs.

Why is it that Moodle users crave for more visualizations even though they are at least to some extent available? It may well be that nearly all those who responded having only a few or no visualizations (41,3% of the Haaga-Helia respondents) belong to the 38% who hoped for more graphs. One explanation for this could be that the visual tool selection is differentiated in Haaga-Helia Moodle courses as it depends on whether the teacher activates the visual progression block or not. The use of available blocks and activities is not regulated nor forced through administration, leaving teachers with the responsibility to apply the tools in their courses. It is also interesting to note that while

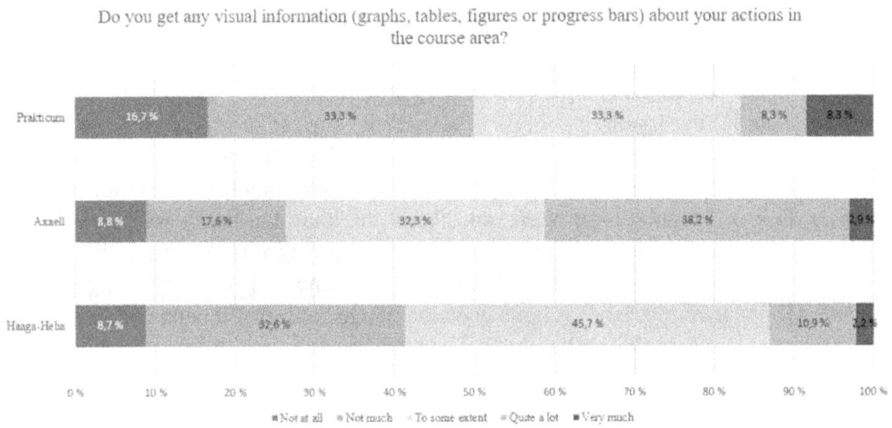

Do you get any visual information (graphs, tables, figures or progress bars) about your actions in the course area?

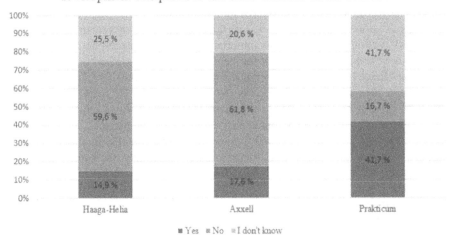

Fig. 3. Estimated amount of visualizations in the learning platforms.

Prakticum's Google Classroom seems to offer less visualizations than the other learning platforms, the students were quite happy with the situation.

4.5 Peer Comparison and Self-regulation

The respondents were asked if they would benefit from having information on how they succeed and progress on the course compared to their fellow students. The respondents' opinions are summarized in Fig. 4.

Is it useful to have information on your progression and development compared to the other students of the course?

Fig. 4. Students' opinions on the usefulness of peer comparison.

Most of the students in Axxell and Haaga-Helia did not consider peer comparisons useful while students in Prakticum were more positive towards it. The difference may be explained by the respondents' age. The survey participants in Prakticum were all young, between 18 and 20 years of age, while in Axxell the age range was from 19 to 61 and in Haaga-Helia from 21 to 56. Zimmerman has pointed out that novices are more likely to value peer comparisons than experts. Novices pay less attention to the forethought phase in their self-regulation process, and therefore "they fail to set specific goals or to self-monitor systematically, and as a result, they tend to rely on comparisons with performance on others to judge their learning effectiveness" [9]. Others suggests that the desire for peer comparisons is not a question of age but rather it depends on the level of motivation or goal-setting skills, and that peer comparisons could help students to self-regulate when the motivation or the ability to set personal goals is low [8]. Interestingly enough, one of the respondents who was interviewed mentioned how peer comparisons could potentially reduce motivation by increasing the students' sense of inferiority and by adding extra pressure to succeed on the same level as others, thus also affecting self-regulation negatively [12]. Auvinen and colleagues tested predictive visualizations based on comparative peer data, and their results suggest that such comparative visualizations only help students to self-regulate if they already have some level of interest to do so and who also perform on a higher level [15].

The median age of the respondents in Haaga-Helia (34) and Axxell (38) suggest that in these two organizations, students are more experienced and do not have the need to compare their skills or progression with other students. This view was strengthened by the open-ended answers in which many students commented how their prior work experience defined their learning path, progression and study efforts a lot and therefore peer comparisons were in a way useless. Also Zimmerman argues that for experts, self-evaluation against their own learning objectives is more important, as it affects their learning efforts and produces a higher level of self-efficacy [9].

Nevertheless, there is an interesting link between the preferences on peer comparisons and the factors which are considered important in online learning. The respondents were offered various statements on the positive aspects of online learning, and they were asked to choose the three most important ones for them personally. The chosen preferences were quite similar except for the first choice, as presented in Table 3.

The top choice in each organization is marked with green, the second popular choice is marked in blue and the third place is indicated in yellow. It is interesting to see that studying with peers was clearly the most important aspect of online learning in Prakticum, where students also appreciated having peer comparisons. In Axxell and Haaga-Helia, studying with peers had little or no significance. Being able to study at one's own pace is highly rated in all organizations, and finally, the ease of accessing material and submitting assignments in online learning is somewhat important to all respondents.

The results may suggest that peer interaction may play a role in self-regulation of younger students. The importance of peer collaboration may also be due to the pedagogical design of the course: if the study process is designed to include lots of group work, peer collaboration evidently becomes important also in online studying.

Table 3. The positive aspects of online learning perceived as important by the respondents.

	Haaga-Helia	Axxell	Prakticum
I can study together with my peers	4 %	0 %	92 %
I can do assignments at my own pace whenever I want	85 %	89 %	75 %
I can progress according to my own skills level	23 %	32 %	8 %
I get more guidance than in classroom	4 %	4 %	0 %
I get feedback on the assignments from the teacher	21 %	11 %	17 %
I get feedback on the assignment from other students	2 %	0 %	8 %
I can go through the course material as many times as I like	40 %	61 %	33 %
I can access and store the study material in one place	36 %	43 %	17 %
I can download the material from the platform to be able to access them offline	19 %	0 %	8 %
I can check the study schedule in the platform	23 %	11 %	8 %
Submitting assignments is easy	34 %	50 %	33 %
I can access all the important services and links to other systems and applications in the platform	4 %	0 %	0 %
Something else	2 %	0 %	0 %

5 Conclusion

The comparative analysis of the responses indicates that in Axxell, students have the strongest self-efficacy beliefs concerning their ability to study in their learning platform. They also had the most experiences on certain pedagogical practices which are proven to support self-regulation. The students in Prakticum had the lowest self-efficacy when it comes to studying in the online platform; in addition, they reported having the least experience on the pedagogical practices supporting self-regulation. In Haaga-Helia, the self-efficacy levels were moderate even though the incidences of pedagogical practices supporting self-regulation were quite uncommon in the learning platform. On the basis of the survey it is impossible to define whether the differences are due to the learning platforms being used. Some explanations were suggested based on habitual observations made of the tools available in Moodle, ItsLearning and Google Classroom. Ultimately, the differences may depend either on the tools available on the learning platform or pedagogical choices made by the teachers. It is highly likely that both aspects are reflected in the prevalence of certain measures supporting self-reflection.

On the basis of the survey responses, some issues may be highlighted with respect to learning analytics. Student dashboards collecting and presenting learning data are much appreciated and quite useful for self-regulation in the performance phase. Peer comparisons in the form of comparative progression or competency reports and ranking lists may be useful and engaging for younger students, but generally adult students did not have a need for comparative analytics in dashboards. However, the influence of comparative peer visualizations on student motivation and self-regulation remains unclear, and the ambiguity in previous research findings call for more studies on the topic.

It is worth asking if the information provided by descriptive learning analytics dashboards truly help students to self-regulate their learning. In her research, Hooli noted that students had mixed experiences on visualizations: they weren't considered useful for planning studies and the information was not meaningful [17]. This survey suggests that student consider learning analytics dashboards beneficial, but it remains unclear if monitoring learning through dashboards is enough to support or improve self-regulation. To ensure full support for self-regulation, learning analytics dashboards should collect data on activities during all the phases of self-regulation [10, 17] and therefore attention should also be paid to the pedagogical design of online courses. As the results show that students do not understand learning analytics properly, they should be familiarized with the use of learning analytics tools. In particular, students should be provided with workshops where they can learn concrete ways in which they can follow their own study data and make plans based on it.

References

1. Leitner, P., Khalil, M., Ebner, M.: Learning analytics in higher education—A literature review. In: Peña-Ayala, A. (ed.) Learning Analytics: Fundaments, Applications, and Trends. SSDC, vol. 94, pp. 1–23. Springer, Cham (2017). https://doi.org/10.1007/978-3-319-52977-6_1
2. Society for Learning Analytics Research (SoLAR): What is Learning Analytics? https://www.solaresearch.org/about/what-is-learning-analytics/. Accessed 8 Nov 2021
3. Siemens, G.: Learning analytics: the emergence of a discipline. Am. Behav. Sci. **57**(10), 1380–1400 (2013)
4. Aksovaara, S., Koskinen, M.: Lähtökohtia oppimisanalytiikalle osaamisen kehittämisen tukena. In: Hartikainen, S., Koskinen, M., Aksovaara, S. (eds.) Kohti oppimista tukevaa oppimisanalytiikkaa ammattikorkeakouluissa. Jyväskylä University of Applied Sciences Publications 274. Jyväskylä: Jyväskylä University of Applied Sciences (2020)
5. Roll, I., Winne, P.H.: Understanding, evaluating and supporting self-regulated learning using learning analytics. J. Learn. Anal. **2**(1), 7–12 (2015)
6. Viberg, O., Hatakka, M., Bälter, O., Mavroudi, A.: The current landscape of learning analytics in higher education. Comput. Hum. Behav. **89**, 98–110 (2018)
7. Kleimola, R., Leppisaari, I.: Kohti uudistuvaa arviointia oppimisanalytiikan avulla. In: Hartikainen, S., Koskinen, M., Aksovaara, S. (eds.) Kohti oppimista tukevaa oppimisanalytiikkaa ammattikorkeakouluissa. Jyväskylä University of Applied Sciences Publications 274. Jyväskylä: Jyväskylä University of Applied Sciences (2020)
8. Sedrakyan, G., Malmberg, J., Verbert, S., Järvelä, S., Kirschner, P.A.: Linking learning behavior analytics and learning science concepts: designing a learning analytics dashboard for feedback to support learning regulation. Comput. Hum. Behav. **107** (2018). https://doi.org/10.1016/j.chb.2018.05.004

9. Zimmerman, B.: Becoming a self-regulated learner: an overview. Theory Pract. **2**, 63–144 (2002)
10. Schumacher, C., Ifenthaler, D.: Features students really expect from learning analytics. Comput. Hum. Behav. **78**, 397–407 (2018)
11. Slade, S., Prinsloo, P.: Learning analytics: ethical issues and dilemmas. Am. Behav. Sci. **57**(10), 1510–1529 (2013). https://doi.org/10.1177/0002764213479366
12. Alasalmi, T.: Students' expectations on learning analytics – learning platform features supporting self-regulated learning. In: Csapó, B., Uhomoibhi, J. (eds.) Proceedings of the 13th International Conference on Computer Supported Education, Volume 2. SciTePress, Portugal (2021)
13. Silvola, A., Näykki, P., Kaveri, A., Muukkonen, H.: Expectations for supporting student engagement with learning analytics: an academic path perspective. Comput. Educ. **168** (2021)
14. Park, Y., Jo, H.: Development of the learning analytics dashboard to support students' learning performance. J. Univ. Comput. Sci. **21**(1), 110–133 (2015)
15. Auvinen, T., Hakulinen, L., Malmi, L.: Increasing students' awareness of their behavior in online learning environments with visualizations and achievement badges. IEEE Trans. Learn. Technol. **8**(3), 261–273 (2015)
16. Russell, J.E., Smith, A., Larsen, R.: Elements of success: supporting at-risk student resilience through learning analytics. Comput. Educ. **152** (2020). https://doi.org/10.1016/j.compedu.2020.103890
17. Hooli, H.: Students' experiences of learning analytics in academic advising for supporting self-regulated learning. A master's thesis. Oulu: University of Oulu (2020)

Learning/Teaching Methodologies
and Assessment

MakeTests: A Flexible Generator and Corrector for Hardcopy Exams

Fernando Teubl[✉][iD], Valério Ramos Batista[iD], and Francisco de Assis Zampirolli[iD]

Centro de Matemática, Computação e Cognição,
Universidade Federal do ABC (UFABC), Santo André, SP 09210-580, Brazil
{fernando.teubl,valerio.batista,fzampirolli}@ufabc.edu.br
http://www.ufabc.edu.br

Abstract. Customarily, exams are still in a venue with a printout for each candidate. The challenge is to correct them all for big classes. Some tools generate and correct them automatically in case of multiple-choice tests. Here we expand this capability with **MakeTests**, an automatic generator and corrector of exams that is free of charge and open source. Moreover, **MakeTests** enables: (1) highly parametrized questions drawn from database; (2) inclusion of many question styles besides multiple-choice (true/false, matching, numerical and written response, only this one with manual correction); and (3) real-time correction with webcam upon handing in. We demonstrate how to prepare exams with **MakeTests** particularly for several question styles with parametrization, and how to proceed with the automatic correction. Our technique is illustrated by two experiments: the first focused on feedback immediately upon handing in a test given to 78 students, and the second managed by a professor without any programming knowledge. We conclude that **MakeTests** enables quick elaboration of parametric questions and fast correction, even for users that lack technical knowledge.

Keywords: Automatic item generation · Multiple-choice questions · Parametrized quizzes

1 Introduction

Information and Communication Technologies (ICT) are increasingly widespread in evaluating students with several computational resources. Their employment begins already in the first year of the primary school. Many tests are described in [4], with the conclusion that online activities show more reliability at evaluating students. This is because they feel that unlike a person machines perform unbiased evaluations, even if teachers tend to give higher marks than the ones from ICT automatic evaluation.

Activities in virtual laboratories also adopt ICT. Their differences from hands-on laboratories are systematically revised in [3], both in undergraduate and secondary education. This work indicates that the virtual tools represent a valuable complement to activities in hands-on laboratories. Regarding Computer-Based Assessment (CBA) or e-assessment, in [6] the authors study some works focused on the students' anxiety

© Springer Nature Switzerland AG 2022
B. Csapó and J. Uhomoibhi (Eds.): CSEDU 2021, CCIS 1624, pp. 293–315, 2022.
https://doi.org/10.1007/978-3-031-14756-2_15

when these are entirely evaluated by computer, including the final mark. This work shows that both acceptance and implementation of CBA present problems that are not quite understood yet and therefore demand further research.

For an introductory programming course in Computer Science, the question of whether more exams change students' final grades is surveyed in [1]. They considered the Spring and Fall classes with a total of five and three exams, respectively. From a statistical point-of-view there was no difference but the students feel that more exams reduce anxiety and increase both motivation and self-confidence to learn. This work cites others indicating that more evaluations for a class generally improve its performance compared with other classes'.

According to [5], "[a]ssessments are increasingly carried out by means of computers enabling the automatic evaluation of responses, and more efficient (i.e., adaptive) testing". This work presents two experiments for generating heterogeneous items (variations of items/questions). In the authors' study 983 individuals were evaluated by Heterogeneous Computer-based Assessment Items (HCAI). Their experiment counted on students from 34 German schools, who were 14 to 16 years old. The authors discuss the complexity of creating heterogeneous items. As an example, changing the word order in a sentence or a single letter in a word may turn a statement unclear. For evaluations they selected 40 out of 70 questions, of which 10 and 30 had hard and easy levels of difficulty, respectively.

Another resource is the use of means like tablets and smartphones in evaluations, called Mobile-Based Assessment (MBA). Its acceptance from students was gauged in [11] with 145 pupils of a European secondary school evaluated in this way for three weeks. This work shows that they do prefer to answer multiple-choice and true/false tests on smartphones.

In [9] the authors compared CBA with traditional methods on 74 undergraduate modules and their 72,377 students. The modules belonged to a variety of disciplines (25% in Science & Technology, 22% in Arts & Social Sciences, 14% in Business & Law, 9% in Education & Languages, and 30% in others). This work indicates that the time spent on evaluation activities had a significant relation to the passing rates. Another conclusion is that the balance between weekly evaluations and other activities by CBA has a positive influence on the passing rates.

An analysis of 123,916 undergraduate students' behaviour during the preparatory weeks for their final exam was performed in [10]. The authors considered students from 205 modules of the Open University, and concluded that the more engagement and participation in CBA, the better performance in the evaluations.

Some of the aforementioned works suggest that students achieve a better performance with more evaluations [1,9,10]. For this purpose we need ICT that make teachers' tasks easier by means of generating and correcting tests automatically. Preferably they ought to adopt heterogeneous items [5] as well as the paper-and-pencil modality, according to [6].

2 Related Works

At the Introduction we cited works indicating that frequent assessments improve learning. But in order to simplify the process of creating and correcting parametric (or heterogeneous) items in various styles, the corresponding ICT are discussed in this section. They are more related to the purpose of this article.

A visual modelling to teach science and mathematics was presented in [13], including solution of scientific problems. Teachers, pupils, educational projects and educational studies are grouped in a uniform relational database employed by their ICT, which was implemented with PHP programming language and MySQL. The ICT was tested with more than 1,000 pupils from Russian secondary schools. For instance, groups of 5–6 participants had to solve exercises by means of Newton's Second Law. They had to visualize graphics, write down values on tables and make analytical decisions. The authors focused on methods to create activities rather than on graphical interfaces, hence it remained inconclusive how their ICT contribute to students' performance in tests.

In [2,7] the authors introduce ADLES, an open-source language devoted to formal specification of hands-on exercises on virtual computing, networking and cybersecurity. This language allows for design, specification and semi-automatic deployment of a virtual machine (VM) for classes, tutorials or competitions, so that users will access it in many activities.

Moodle (moodle.org) is a Virtual Learning Environment (VLE) endowable with the plugin Virtual Programming Lab (VPL) available at vpl.dis.ulpgc.es. Introduced in [12], VPL enables students to submit codes programmed in C/C++, R, Java, Python and others, as solution to exercises. The Moodle server remains protected by redirecting code execution to several VMs. Multiple-choice parametrized questions can be prepared directly on Moodle, however without VPL, and this VLE also lacks printing and automatic correction of hardcopies. Therefore, Moodle is inadequate for large class paper-and-pencil tests.

The platform MCTest was presented in [17]. Accessible at vision.ufabc.edu.br and developed in Django with MySQL, MCTest's source code is open and available on GitHub. Therefore, any institution can install MCTest and assign a system administrator (SA) to register departments, courses, disciplines and professors. Any course must be associated to a coordinator, who will link it to Topics, Classes, Questions, Exams, Professors and Students. These capitalized words indicate *classes* of Object Oriented Programming (OOP). The related to professors may also create Classes, Questions and Exams. All these entities are created on web browser windows. Classes and Questions can also be imported from CSV files, in which the students' Id, name and email are specified in the case of a Class, but Questions follow another CSV formatting. Specially [17] explains how to prepare parametric questions, either multiple-choice or written-response ones, with some Python encoding that MCTest uses to produce individualized exams, one for each student, but all released as a single PDF file per class. In fact, through MCTest a professor with several classes can set up a unified exam for them all. In this case MCTest will generate a ZIP-file that includes a PDF for each class. The correction is automatic for multiple-choice questions providing one digitizes the answer cards into another PDF to be uploaded by the system, which will then email to the professor a detailed CSV spreadsheet containing each student's name, Id, their scored chosen alternatives and total mark.

MCTest has been constantly improved and now some of its most recent versions allow for the students to submit programmed codes to Moodle via VPL plugin. Program codes are then corrected automatically. In [15] the authors developed a method in which VPL configuration files were adapted for the student to submit Java codes together with a TXT file that specifies the question model drawn for that student, e.g. "model: F". That work relied on test cases elaborated manually for each question model. More recent versions of MCTest (see [16]) identify the student's question model through their Name/Surname and therefore dismiss that TXT file. Moreover, test cases are now generated automatically. We shall resume this method again soon but in the next sections the reader will see that **MakeTests** includes more different question styles than MCTest.

In [8] the author explains the importance of creating personalized multiple-choice questions to reduce plagiarism. For this purpose he also shows that just shuffling questions and the order of their linked alternatives are both insufficient. His strategy includes the elaboration of efficient *distractors*, namely wrong alternatives, otherwise students can simply discard them to guess the right answer. There the author deals with the following questions: "Would it be possible to architect a generic framework to enable personalized multiple-choice examinations?" and "What are the technical, pedagogical, and administrative challenges posed by personalization, and how these might be overcome?"

Personalized evaluations count on three approaches as reported by [8]: 1) parametrization, in which some parameters take random values; 2) databank, from which questions are selected at random; 3) macro, which is a program fragment (inside a question) that is replaced by a new phrase whenever executed. Their multiple-choice tests are responded on *optical answer sheet*, also called *bubble/Scantron sheet*, commercialized by Scantron Corporation®. The student writes her/his Id and also the script Id on this sheet. The two Ids determine the exam variation assigned to the respective student.

Still in [8] an automatic generation of heterogeneous multiple-choice items of XYZ-type is acquainted. This type means that any of the Boolean variables X, Y and Z take either value, which produces a truth table with eight rows (see [8, Fig. 2]). One must write items in HTML with macros identified by CSS markers, e.g.

```
<span class="cws_code_q">$0</span>
```

in which $0 is replaced by the macro with some value. XYZ-items are prepared with an XML file which must include right and wrong alternatives (see [8, Figs. 5–6]). Unfortunately the author omits how specifically some HTML, XML and CSS contents produce items, but here we detail the way **MakeTests** does it for every type: standard multiple-choice, XYZ, True/False, matching, and written response questions. Moreover, our source code is available at github.com/fernandoteubl/MakeTests.

MakeTests was originally published in [14]. The present work differs from that paper in the following sense: here we include technical specifications regarding the process of elaborating and correcting exams, particularly in Sects. 3.1 and 3.4. Moreover, an additional example is included in Sect. 3.2 besides supplementary information all along the text.

The main distinctions between MakeTests and the aforementioned works are: (1) with MakeTests the professor develops questions by means of various Python scientific libraries such as graphics (`matplotlib`), algebraic analysis (`sympy`), and many others. (2) Activities are individualized and generated in hard copy, and they make use of various styles of questions. (3) Corrections can be carried out automatically through a mobile device. (4) The professor can also resort to answer keys generated by MakeTests in order to make corrections easier. Details will be given in the next sections.

MakeTests is ideal for examiners that prefer written response tests because, in the case of a medium difficulty test, the automatic correction will already show the student's right and wrong answers. Hence the examiner can concentrate their attention only on what avails of the student's solutions that led to wrong answers. Roughly speaking, the examiner's manual correction will be just half of the whole work. Of course, in the case of large classes MCTest can spare even the whole manual correction, useful when the examiner just wants to give a preparatory exam.

MakeTests' greatest advantage is that it enables configuration and elaboration of exams in a highly flexible way. MakeTests profits the user that is familiar with Python language, shell commands and JSON format. Otherwise one can resort to templates that are easy to adapt, since question elaboration is what mostly needs teachers' and professors' endeavour.

Each question prepared with MakeTests is represented by an abstract Class (in OOP) that defines a model, namely the user must implement it according to its aims. The question consists of a text, which may include parametrization, and the answer format, which defines its type and also its procedure of correction. Thus MakeTests enables an unlimited diversity of questions and answer types, which just rely on their implementation. The next section explains the details of our methodology.

3   Method

In this section we explain how to prepare and issue exams with MakeTests step-by-step. Firstly one must obtain a Database of Questions (DQ) classified according to group categories such as subject, difficulty or any other characteristic. See more on that in Subsect. 3.1. Subsection 3.2 shows a general and typical example of a question in MakeTests.

Secondly the user must configure exams conforming to the explanations in Subsect. 3.3. The configuration includes the exam layout (header, fonts, logotype, etc.) as well as the groups of questions to be drawn with their respective weights. Here one must give both the DQ and the class roll paths. After generating the tests they can be printed for the examiner. Finally Subsect. 3.4 presents the three types of automatic correction available through MakeTests:

Manual: The user prints an answer key from MakeTests in order to correct the exams manually;

Scanned: The user scans all the exams into a single PDF that MakeTests will process and render a spreadsheet with marks and feedback;

Real-time: The user corrects the exams instantly with a (web)cam by pointing it to the answer sheet. This also generates the aforementioned spreadsheet.

The spreadsheet comes in CSV-format and the feedback lists each question number with the respective student's answer and score. We also have developed the library SendMail.py to email this student's information to her/his private address at the user's discretion. In this way each student promptly gets an individual summary of her/his performance. Subsection 3.4 describes how each student receives that email.

All these resources are organized in a structure of files and folders generated by MakeTests.py, a library that is available on github.com/fernandoteubl/MakeTests. Figure 1 illustrates the directory structure, and besides MakeTests.py (only 2,609 lines including examples and comments) the reader will also find another three files on GitHub: SendMail.py (only 270 lines), README.md and convertPdf Text2PdfImage.sh. MakeTests' code was all developed in Python 3.8.

Fig. 1. File and folder structure of **MakeTests**. The master file is MakeTests.py, which can generate template questions indicated by the dotted rectangle. It can also produce a student list for an exam (Students.csv), all the exam issues in a single file (Tests.pdf), answer key (AnswerKeys.pdf), and so on. These resources and their employment are all detailed in README.md (from [14]).

Linux and macOS are the recommended operating systems for you to run **MakeTests**. Compatibility with Windows has not been fully checked yet but this option will be available in the near future. Besides Python you must have LATEX installed on your computer, more specifically the command `pdflatex` to generate PDF files, together with the Python package managers `conda` (www.anaconda.com) and `pip`. All MakeTests' dependencies are free of charge and open source, among which the following packages available by `conda` and `pip`: `cv2`, `qrcode`, `PyPDF2`, `pybaz`, `barcode`, `tesseract` and `pytesseract`.

The next subsections explain each of the foregoing steps.

3.1 Elaborating Questions

The DQ is organized pursuant to groups, namely directories indicated by a dotted rectangle in Fig. 1. For a large DQ one can also add subdirectories that determine subgroups, which simplify the disposal of all questions. As an example, you can arrange questions by theme in directories, together with their respective level of difficulty in subdirectories. A question of an exam issue corresponds to a unique Python file, and reciprocally. For instance, if you want an exam with three questions amounting to all the difficulty levels in Fig. 1, by running MakeTests it will draw one from each folder inside the directory `Questions`. An error is reported if a folder lacks the minimum number for the required specification.

Questions in MakeTests are structured according to the Class Tree Diagram shown in Fig. 2. Therefore, whenever the users programme a question they must make it inherit the characteristics from one of the five Classes indicated by the leaf nodes, and therefore implement its corresponding methods. New styles of question are created by adding their respective leaf Class to the tree. For example, supposing we used Artificial Intelligence to evaluate written response questions, then a new child leaf Class named `QuestionAI` would derive from `Question`, and there we should implement methods to read and interpret scanned handwriting. In this new Class some algorithms in Convolutional Neural Network would help correct answers automatically. Each object question acquires specific data of a student contained in the class roll `Students.csv` like name, Id and email, this file depicted in Fig. 1. As a Python Class the user can parametrize any of its parts.

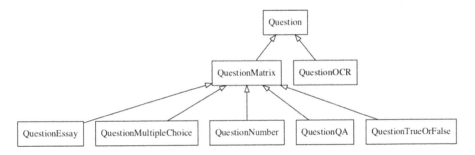

Fig. 2. Hierarchy of question Classes.

Question is an abstract Class with the following methods and attributes:

```
Class  Question :
    info                                        dict ()
    def  makeVariables ( self ):            -> None
    def  getQuestionTex ( self ):           -> str
    def  answerAreaAspectRate ( self ):     -> float
    def  drawAnswerArea ( self , img ):     -> img
    def  doCorrection ( self , img ):       -> str
    def  getAnswerText ( self , LaTeX ):    -> str
```

The attribute info contains all the student's data: Id, email, name and any other relevant piece of information added by the user in Students.csv. Methods are all abstract and described below:

makeVariables: the first to be invoked and works as a Class constructor: it generates all the variables employed in the elaboration of the question. These variables are applied in the parametrization of both statements and alternatives, draw of questions, and shuffling of items and answers, among others. MakeTests executes the method random.seed immediately before calling this function, which generates seed in such a way that to each student a unique constant value is assigned. Namely, if you need to re-run MakeTests it will generate the *same* exam issue previously assigned to that student.

getQuestionTex: renders the question statement (and linked alternatives) in LaTeX format. Here you can include figures, diagrams and any other element supported by LaTeX.

answerAreaAspectRate: shows the image aspect devoted to the space of the answer card ($width/height$). This information is necessary because MakeTests generates a blank image that is forwarded to the method drawAnswerArea described right next.

drawAnswerArea: depicts an image of the upcoming answer card. For instance, this function can draw two circles for the student to choose one and fill it out as an answer. Here we utilize drawing methods from the library CV2 (OpenCV). Figure 3(a) exemplifies the space of an answer card generated by this method.

doCorrection: receives the digitized image of the previously mentioned answer card already filled out by the student. This method interprets that digitized image and gives the corresponding question score. MakeTests automatically segments the image of the answer card with the proportions, orientation and element positions according to the format generated by drawAnswerArea. Figure 3(b) illustrates an image received by this method.

getAnswerText: returns a text that faithfully corresponds to the student's answer. Optionally the professor can print this text as part of the answer keys.

In order to segment the space of an answer card as in Fig. 3(b) MakeTests needs to identify the exact position of this space in the whole image of the digitized exam. For that we resort to special markers printed together with the exam. They will help the computer vision algorithms to segment the space of the answer card as shown in Fig. 4. These markers are both generated and detected by MakeTests, a process that is unknown to the Class Question.

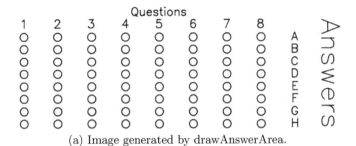

(a) Image generated by drawAnswerArea.

(b) Image received by doCorrection.

Fig. 3. Generation and correction of the image of the answer card.

Firstly all markers of the digitized exam are identified. Each question is delimited by four markers, of which the glyph is always two concentric nested circles.

The image undergoes an initial treatment that removes noise, then an edge detection algorithm is applied followed by another for contour detection. This one produces a Contour Hierarchy, which contains all image contours and the counting of nested contours. This is precisely what makes the glyph of a marker easily identifiable, as part of MakeTests's strategy (see Fig. 4). Such a technique is also employed in the detection of QRCode patterns, which always contain three markers whose glyph is two nested squares. For merely aesthetic reasons MakeTests adopted circles instead of squares but it added an extra contour hierarchy by letting the inner circle be a circumference. This trick results in a better robustness of marker detection since it discards false positive cases. The inclusion of the extra contour hierarchy was only possible because MakeTests adopts markers that are relatively bigger than the ones from QRCode.

Right after detecting all markers MakeTests combines all possible pairs and draws a line for every single one. At this step MakeTests looks for a barcode between markers, and whenever one is found the respective pair will represent an upper or lower bound of the space for an answer card (see Fig. 4). The barcode utilizes the code128-symbology and contains both question and student's Id, together with a cipher, which indicates if it is the upper or lower bound of the answer card. MakeTests prefers barcode rather than QRCode because the height of the former can be very short since its information comes horizontally. Namely, QRCodes would altogether take much more vertical space, which would require more pages to print the exam. Though it is hard to identify a barcode along arbitrary orientations of an image MakeTests circumvents this problem because any of our barcodes will lie between a pair of markers, hence its orientation is always known.

This process turned out to be robust even after changing the image orientation and its perspective. The space of the answer card is segmented and adjusted to the rectangular format with a quite good precision. Therefore our method `doCorrection` succeeds in segmenting the space of answer cards without any further processing. Section 3.4 presents the practical results of this process.

Fig. 4. `AnswerArea` identification and segmentation.

Of course, image processing methods are by far non-trivial. Except for a specialist in digital image processing it will be unfeasible to elaborate items only with the abstract Class `Question`. For **MakeTests** we have developed the Class `QuestionMatrix` inherited from `Question`, in which computer vision algorithms were already implemented. This Class supports questions that use matrices as answer cards whose entries can be chosen for the student to inscribe an answer. It has the following attributes and methods:

```
Class  QuestionMatrix ( Question ):
    rows                                    -> [ str ]
    cols                                    -> [ str ]
    hlabel                                  -> str
    vlabel                                  -> str
    def  getScore ( self ,  matrix_answer ) -> str
    def  getTemplate ( self )               -> matrix [ bool ][ bool ]
```

`QuestionMatrix` is abstract and yet one must implement three methods to use it. However its use does not require any knowledge of computer vision. Now we describe the methods that must be implemented:

makeVariables. This function is identical to the one from the Class `question` except for four additional attributes that must be initialized:
  - **rows** and **cols**, which contain a list of each row and column labels, respectively. In Fig. 3(a) the row labels are represented by letters 'A' to 'H', and the column labels by characters from '1' to '8'. The matrix size is determined by the size of these lists.
  - **hlabel** and optionally **vlabel** represent a horizontal and a vertical label, respectively. In Fig. 3(a) the horizontal label is 'Questions', whereas the vertical one is 'Answers'.

getScore. Gets a `boolean` matrix as input, which stands for the alternatives chosen by the student. This function returns the question score.

getTemplate. Returns the correct answers in a `boolean` matrix, which is applied as visual feedback conforming to the explanations in Sect. 3.4.

getQuestionText & getAnswerText. These are the same ones previously described in the Class `Question`.

Though the abstract Class `QuestionMatrix` does not require any knowledge of computer vision one still needs advanced Python programming to handle it.

MakeTests furnishes pre-implemented codes that simplify the elaboration of new questions whenever the user creates new corresponding PY-files. In the present day MakeTests comprises the following Classes for question styles:

Essay: Written response to be corrected manually, and the score recorded in a spreadsheet of students' marks;

MultipleChoices: An Array Class (similar to ArrayList in Java), in which any element is a typical multiple-choice question. The user defines the number of questions to be drawn;

TrueOrFalse: An Array Class in which any element is a group of questions, and the answer key of any group is a Boolean sequence. The user defines groups with their respective questions, and afterwards the number of groups to be drawn;

QuestionAnswer (QA): A Class that generates matching questions. The user defines a list of questions and another of answers;

Number: The student chooses digits in a matrix to form a number. The user defines the expected number of digits and the tolerance (zero for exact and weighted for approximate answer).

3.2 Example

Here we detail an example of parametrized multiple-choice question named `MYchoices.py` in Fig. 1. In order to see its layout the user can run the following on the shell:

```
./MakeTests.py -e choices >
                Questions/Easy/Mychoices.py
```

The Class `myQuestionMultipleChoice` was implemented in `MYchoices.py`, and its code is reproduced in Fig. 5. There one sees the method `makeSetup`, which creates two floating-point variables `x` and `y` on lines 10–11. These take random values between 1 and 48 with two decimals. On line 8 we see the variable `op` taking only the addition operation, namely `'+'`. With the command `op=random.choice(['+','-','*','/','**'])` we rewrite line 8 in order to enable all basic arithmetic operations. Moreover, lines 9–16 must be encapsulated by the conditional `if op == '+':` and adapted for the other operations. In fact MakeTests will shuffle the entries of `vetAnswers` for each exam issue, no matter what operation was drawn to define this vector. On lines 9–12 notice that `round` is superfluous for `'+'` and `'-'` but necessary for the others. Hence we kept it because `MYchoices.py` can serve as a template that will require little customization from future users. By the way, this is one of the mindsets that enable good employment of MakeTests.

```
 1  from MakeTests import QuestionMultipleChoice
 2  class myQuestionMultipleChoice(QuestionMultipleChoice):
 3      def makeSetup(self):
 4          import random
 5          self.questionDescription = "Check the correct alternative:"
 6          self.questions = [self.mySubQuestion(random.choice(['+']))]
 7          self.correctionCriteriaDescription = ""
 8      def mySubQuestion(self, op='+'):
 9          import random
10          x = round(random.uniform(1, 49), 2)
11          y = round(random.uniform(1, 49), 2)
12          vetAnswers = [[round(x + y, 2), True],
13                        [round(x + 1.2 * y, 2), False],
14                        [round(1.2 * x + y, 2), False],
15                        [round(1.2 * x + 1.2 * y, 2), False]
16                       ]
17          return {"statement": '''If x={x} and y={y}, what is
                x{op}y?'''.format(x=x, y=y, op=op),
18                  "alternatives": sorted(vetAnswers, key=lambda k:
                       random.random()), "itensPerRow": 4}
19      def calculateScore(self, correct, wrong, blank):
20          if   correct == 1: return 100
21          else:              return   0
```

Fig. 5. Code of a question implemented in the Class `myQuestionMultipleChoice` (from [14]).

With MakeTests one can utilize, modify or even create question styles, which in their turn will allow for many question types. As an example, the aforementioned Class QA for the matching style enables us to work with the types one-to-one, one-to-many and many-to-many. Hence for each style the user can aggregate new types either by resorting to an already implemented Class or by creating a new one from scratch.

Parametrization allows the question in Fig. 5 for many possible renderizations, one of them depicted in Fig. 6, which will be discussed in the next subsection.

Figure 7 shows another example of a different answer style for a question. Here it inherits the attribute `QuestionQA.full_list_question_answer_text` from the Class `QuestionQA`. This attribute consists of a vector where each entry is a tuple with a noun and its respective definition. The question also inherits the method `QuestionQA.auxShuffle`, which shuffles nouns inserted in `QuestionQA.full_list_question_answer_text` that amount to n (equals to 8 in this example), so that each student gets a different variation. The method also assigns the indexes selected in `QuestionQA.list_question_answer_index` to the respective students. In this way the Class `QuestionQA` generates an exam with selected questions and the student must choose an association between numbers and letters (see Fig. 8). We remark that in this example a gradual scoring criteria was established by means of the method `QuestionQA.calculateScore`.

3.3   Configuring the Exam

All the information regarding the generation of an exam to a specific class must be configured through a JSON file. MakeTests outputs a standard JSON file that can be

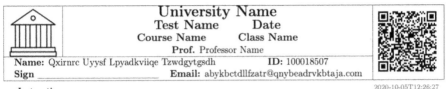

Instructions:

2020-10-05T12:26:27

- **Remark:** Fill up the corresponding circle without smudging.
- There are 4 pages and 4 questions. Make sure you have received all the sheets and the questions are legible.
- The score is calculated by the weighted arithmetic mean of all questions (there is no rounding).
- Conversion criteria:

Score	[0%,50%[	[50%,65%[	[65%,75%[	[75%,85%[	[85%,100%]
Grade	F	D	C	B	A

Question 1 of 4 (Weight 2): Check the correct alternative:

1.1. If x=43.62 and y=10.77, what is x+y?

A) 60.85   B) 54.39   C) 63.11   D) 69.58

Fig. 6. Cutout of an exam that includes the question in Fig. 5. Here we see the header and the answer card for the automatic correction. A single answer card can be configured to comprise many such questions at once (from [14]).

changed by the user. For instance, in order to get the file `config.json` in Fig. 1 we run the shell command

```
./MakeTests.py -e config > config.json
```

The JSON format was adopted in **MakeTests** because of its both simple representation and easy portability to web systems. As a matter of fact, **MakeTests** handles an *extended* JSON syntax that includes comments and line breaks. In order to generate the exam in PDF from JSON this file is endowed with fundamental fields handled by **MakeTests**, namely

includeJSON: Allows for merging another JSON file to modify or complement the original one;

questions: Contains both the DQ path and the configuration to select questions, as depicted in Fig. 9, which shows the parameters `db_path` and `select`, respectively.

```
from MakeTests import QuestionQA
class MyQuestionQAnswer(QuestionQA):
    def makeSetup(self):
        self.full_list_question_answer_text = [
            [''' Triangle ''', '''Three sides '''],
            [''' Quadrilateral ''', '''Four sides '''],
            [''' Pentagon ''', '''Five sides '''],
            [''' Hexagon ''', '''Six sides '''],
            [''' Heptagon ''', '''Seven sides '''],
            [''' Octagon ''', '''Eight sides '''],
            [''' Enneagon ''', '''Nine sides '''],
            [''' Decagon ''', '''Ten sides '''],
            [''' Hendecagon ''', '''Eleven sides '''],
            [''' Dodecagon ''', '''Twelve sides '''],
            [''' Tridecagon ''', '''Thirteen sides '''],
            [''' Tetradecagon ''', '''Fourteen sides '''],
            [''' Pentadecagon ''', '''Fifteen sides '''],
            [''' Hexadecagon ''', '''Sixteen sides '''],
            [''' Heptadecagon ''', '''Seventeen sides '''],
            [''' Octadecagon ''', '''Eighteen sides '''],
            [''' Enneadecagon ''', '''Nineteen sides '''],
            [''' Icosagon ''', '''Twenty sides '''],
        ]
        self.list_question_answer_index = QuestionQA.auxShuffle(self.full_list_question_answer_text, 8)
        self.question_description = "Associate the Questions with the Answers."
        self.labels       = {"questions": "Questions", "answers": "Answers"}
        self.correction_criteria_description = '''
\\textbf{IMPORTANT:} Each question can only be associated with one answer and vice versa.
A correct association will be disregarded if there is another association of the same question or answer.

\\begin{center}
\\begin{tabular}{| c | c | c | c | c | c | c | c |} \\hline
\\textbf{Correct qty:} & 0-3 & 4 & 5 & 6 & 7 & 8 \\\\ \\hline
\\textbf{Score:}       & 0\\%% & 20\\%% & 40\\%% & 60\\%% & 80\\%% & 100\\%% \\\\ \\hline
\\end{tabular}
\\end{center}
'''

    def calculateScore(self, correct, total):
        if   correct >= 8: return 100
        elif correct >= 7: return 80
        elif correct >= 6: return 60
        elif correct >= 5: return 40
        elif correct >= 4: return 20
        else:              return 0
```

Fig. 7. Code of a question implemented in the Class myQuestionQAnswer.

Each question of an exam issue is drawn from a group, and this field also defines the weight of the corresponding question, together with a prefix as exemplified in Fig. 9;

input: Contains the class roll in CSV. For instance, in order to produce the header in Fig. 6 we used the default file Students.csv generated by MakeTests;

output: Sets path and file names of both the PDF that contains all exam issues and the PDF with the corresponding answer keys. MakeTests suggests creating the path in case it does not exist (for **questions** and **input** the user just gets an error message in this case). In Fig. 1 they appear as Tests.pdf and AnswerKeys.pdf, respectively;

correction: Sets all the criteria for the automatic correction. In this field we write a Python code to compute the total score, choose a name for the CSV file with the students' scores, and also the directory to store this file and the image of the corrected exams. In Fig. 1 it appears as Correction, and more details will be given in the next section;

tex: Contains all the information about the exam formatting, like LATEX preamble, headers and structuring of questions. Notice that MakeTests neither contains nor pro-

Name: Qxirnrc Uyysf Lpyadkviiqe Tzwdgytgsdh   ID: 100018507
Sign _____ Email: abykbctdllfzatr@qnybeadrvkbtaja.com

2021-09-17T18:11:07

Instructions:
- **Remark:** Fill up the corresponding circle without smudging.
- There are 2 pages and 1 questions. Make sure you have received all the sheets and the questions are legible.
- The score is calculated by the weighted arithmetic mean of all questions (there is no rounding).
- Conversion criteria:

Score	]0%,50%[	[50%,65%[	[65%,75%[	[75%,85%[	[85%,100%]
Grade	F	D	C	B	A

Question 1 of 1 (Weight 2): Associate the Questions with the Answers.

Questions		Answers	
1	Triangle	**A:**	Nine sides
2	Octagon	**B:**	Eleven sides
3	Pentagon	**C:**	Five sides
4	Hexadecagon	**D:**	Sixteen sides
5	Pentadecagon	**E:**	Fifteen sides
6	Enneadecagon	**F:**	Eight sides
7	Hendecagon	**G:**	Nineteen sides
8	Enneagon	**H:**	Three sides

IMPORTANT: Each question can only be associated with one answer and vice versa. A correct association will be disregarded if there is another association of the same question or answer.

Correct qty:	0-3	4	5	6	7	8
Score:	0%	20%	40%	60%	80%	100%

Answer to question 1:

Fig. 8. Cutout of an exam that includes the question in Fig. 7.

duces anything in LATEX, but JSON instead (e.g. `config.json` in Fig. 1). Therefore, we can alter the JSON file to generate the exams in HTML (the default is PDF from LATEX).

We have chosen to configure exams in JSON because this gives the user total flexibility to design them. The users can create their own models with customized images and headers, and so relegate the JSON file to specific details like date of the exam, class roll and DQ.

```
"questions": {
    "salt": "",
    "db_path": "Questions",
    "select" : [
        {"path": "Easy",    "weight": 3, "replaces": {"%PREFIX%": "Weight 3"}},
        {"path": "Medium",  "weight": 2, "replaces": {"%PREFIX%": "Weight 2"}},
        {"path": "Medium",  "weight": 2, "replaces": {"%PREFIX%": "Weight 2"}},
        {"path": "Hard",    "weight": 3, "replaces": {"%PREFIX%": "Weight 3"}}
    ]
},
```

Fig. 9. Cutout of a typical config.json to create an exam with four questions (one easy, two medium and one difficult with their respective weights). Each path must be as depicted in Fig. 1 (from [14]).

3.4 Correcting the Exam

There are three means to correct exams with MakeTests, as explained below.

I Manual Correction. One prints the aforementioned file AnswerKeys.pdf as a guide to speed up the manual correction, and also do it in parallel with Teaching Assistants (TAs) without a computer. They can take notes on the hardcopy whenever they find mistakes in the answer keys, which might happen in case a question statement gives rise to an interpretation other that made by the professor. Figure 10 shows part of such a file that begins with the student "Qxirnrc" from Fig. 6.

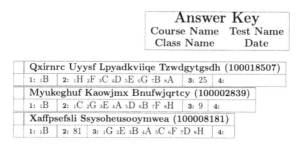

Fig. 10. Cutout of AnswerKeys.pdf generated by MakeTests (from [14]).

The two other means of correction resort to Computer Vision, and they generate automatic reports with the students' scores and also feedback. As a matter of fact, they complement the manual correction in the case of written response tests. For instance, the automatic correction shows the student's right and wrong answers and therefore the manual correctors may restrict their work on what avails of the student's solutions that led to wrong answers. Roughly speaking, their whole work will be halved for a medium difficulty test.

Anyway, we shall see that the automatic corrections allow for emailing a feedback to each student even in real time, namely at the moment they hand the test in. Afterwards the professor can digitize the corrected exams and email them to the students, so that each one will receive only their corresponding corrected test.

In the case of a mere preparatory exam just to evaluate the students' performance, the professor can opt for a traditional multiple-choice test and profit MakeTests' DQ. In this case each student can get their scores immediately upon handing in the exam.

II Digitizing the Solved Exams. This consists of piling up the solved exams and scanning them all at once through a document feeder. The user gets a single PDF file, or even separate ones at will, to be processed by MakeTests. For the automatic correction MakeTests needs to access the DQ, the class roll and the exam configuration. Each student's name, Id and scores are saved as an individual row of a CSV-file that the professor can open with a spreadsheet program. For each student MakeTests also creates a folder containing their scanned exam with the corresponding answer key, and also the image of each solved question. With MakeTests the professor can get these folders separately compressed and sent to the respective students. Hence, in case of a distance learning course the professor does not have to schedule a meeting with the whole class to discuss the answer key. Therefore, any student who takes objection can furnish arguments with the separate image(s) through email.

Figure 11 shows some items of such a student's folder with the five question styles already available in MakeTests. These images were generated with the shell command

```
./MakeTests.py -p scannerFile.pdf
```

Notice in Fig. 11 that a red cross indicates a wrong choice, whereas the missed right one is shown with a red question mark in a blue background. Right answers are indicated by a checked green circle. Exception is made for Fig. 11(e), in which MakeTests promptly chooses the computed mark. This one can be changed by the professor in order to include the correction of written response questions.

III Feedback. The individual corrections are emailed to the students by the `SendMail` tool available in MakeTests. Though `SendMail` has been developed to email individual corrections to the students it is in fact an independent program that can be employed with purposes other than MakeTests' correction feedback. `SendMail` reads a JSON-type configuration file with the parameters

input: the name of the CSV-spreadsheet containing all the students' information;
sender: the sender's email (normally the professor's);
SMTP_server: SMTP email server address to be used;
SMTP_port: SMTP server port, which nowadays enforces STARTTLS;
SMTP_login: SMTP access login, normally the same of `sender`'s;
SMTP_password: SMTP server password, which will be required at the dispatch lest you leave it blank;
subject: the subject of the email;
message: the message of the email.
columns: defined as
  – **email.** The name of the spreadsheet column that contains the email addresses, which can even be more than one per line.

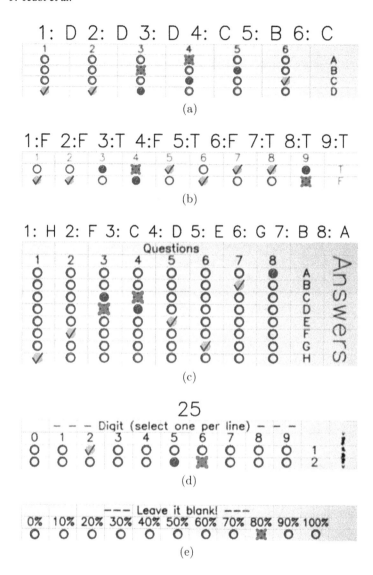

Fig. 11. Cutouts generated by MakeTests after submitting the PDF to the automatic correction: (a) Multiple-Choice; (b) TrueOrFalse; (c) QA; (d) Number (they got a wrong answer 26 instead of 25); (e) Essay (from [14]). (Color figure online)

- **attachment.** The name of the spreadsheet column that contains each file to be attached to the respective email. If it is instead a folder then the attachment will consist of all files therein.
- **filter:** Optional Python code that receives each student's information and decides whether the message will be sent. This filter is useful for omitting feedback in some

cases, e.g. of students that did not sit the exam. It can also filter the student's mark in order to send different messages to approved and failed ones.

Messages are sent to each student individually, hence both mark and email are kept private. In order to personalize messages with the student's mark each name in the column of the CSV spreadsheet works as an identifier that will replace its respective token in both title and scope of the message.

For example, if the total exam score is in the column `Final_Score`, for an email with title "Your final score is Final_Score" the student who got "B" will receive a message entitled "Your final score is B". The same works for the content of the message. Notice that `Final_Score` is a token that will be replaced by a value and in your message this will happen to *any* word that coincides with the name of a column. Therefore you must avoid adopting trivial column names, otherwise they will spoil your message. MakeTests generates a standard CSV file with column names encapsulated by the % character in order to avoid such improper replacements. See more details at github.com/fernandoteubl/MakeTests.

IV Real-Time Correction. The real-time correction is similar to the one explained in Sub-subsection II. MakeTests uses Computer Vision to read the answer cards of Fig. 11 and correct them. Student information is contained in the barcodes, as depicted in Fig. 6. The student's scores are automatically computed and then sent as feedback to their personal email.

This process is enabled by choosing either the built-in camera of the professor's computer, or a USB webcam attached thereto, namely with the shell command:

```
./MakeTests.py -w 0
```

where 0 stands for the former, to be replaced with 1 for the latter.

With the real-time correction each student gets a lower-bound estimate of their scores already at the time they hand in the exam. Of course, manual correction will later add whatever avails of the handwritten solutions, and also the scores of the dissertation questions. Figure 12 exemplifies a webcam image of an answer card processed by MakeTests.

Although the feedback will be generated again with the second correction method discussed in Sub-subsection II some students might want the real-time correction in order to organize their time for upcoming main exams. But a conventional webcam may generate images in poor quality, and therefore the second method will fine-tune the feedback sent to the students.

4   Results and Discussions

Now we present some preliminary results obtained with MakeTests which aim at assessing its potential use in new evaluation modalities. All the practical exams involving MakeTests happened at our institution for different courses of the programme Bachelor in Science and Technology (BST).

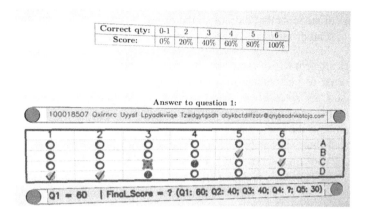

Fig. 12. Cutout of an image captured by webcam. Here we see an exam whose first question consists of six parts (1.1 to 1.6), and each one is a multiple-choice item with four alternatives (A to D). Part 1.4 was left blank but MakeTests indicates any missed right answer by a blue circle with a red question mark. (from [14]) (Color figure online)

4.1 First Experiment Report

We tested our third correction method in a class with 78 students of the course Object-Oriented Programming (OOP). Generated by MakeTests the main exam consisted of three parametric questions, namely

Q1) Multiple-Choice: The statement shows part of a parametrized source code for the student to deduce what it does, and only one alternative describes it correctly. Four parts of source codes were chosen from a DQ at random;

Q2) TrueOrFalse: Some sentences about OOP concepts are listed, and the student must decide which ones are true or false. With MakeTests we selected twelve parametric sentences in a DQ at random;

Q3) QA: The statement includes a two-column table, the first one containing several software Design Patterns (DP) and the second one the respective descriptions but in a shuffled order. The student must find which description matches the DP. Five items were randomly selected from a DQ.

Correction was performed according to Sub-subsection IV, and students could promptly check their scores by email. In that experiment we used a USB webcam Microsoft HD 720p 30fps connected to an Apple MacBook Dual Core 512GB SSD with 8GB RAM. As depicted in Fig. 12, the examiner just had to point the camera to the answer cards with barely good enframing and ambient light (no flash). MakeTests also sends the scores to the examiner's computer, so that students could also glance at its screen for a quick evaluation. Each exam was corrected in less than 20s, hence the queue was always short, but also because students were previously warned that they should not ask questions at that moment. Moreover, about 20% of the students turned down the automatic correction.

After the exam we proceeded to the method discussed in Sub-subsection II. As already explained, this second correction enables MakeTests to process answer cards without either distortion or blurring. Moreover, the PDF contains the image of each student's complete exam, which can be emailed to them for accurate checks. Indeed, that was done in the experiment, as described at the beginning of Sub-subsection II. See Fig. 11 for an example of what the student receives by email. Of course, this feedback is optional and many professors prefer to discuss exam corrections privately with students during the assistance.

On this experiment we remark two important facts:

– The professor detected no cheating in the test. Since questions are parame-trized and randomly drawn to each student, then cheating was significantly hampered;
– Far fewer students took objection to their marks compared with standard written response exams already given by that same professor. Though he has not analyzed this reduction statistically, he claimed that the usual amount of recorrection requirements dropped more than 50%. This is possibly because that exam did not include any written response question, besides the fact that the students had already received the correction by email.

A difficulty arises in the case of large classes if the professor wants nominal exams, as shown in Fig. 6 for the student "Qxirnrc". In the experiment it took almost 10min for the class to begin the test. The professor could have reduced this time by warning the students to look for their seats in alphabetical order.

Another strategy is to replace nominal exams with at least five versions of the same test. Hence any student that sits beside or behind a colleague will not have more than 20% chance of receiving the same version. Moreover, the professor can print extra copies lest some of the students smudge their answer cards. Such a strategy requires a manual association of each student with their corresponding version but is useful in the case of late matriculated ones that may turn up to sit the exam.

During that exam circa 2% of the answer cards were smudged, so the examiner had to use the answer keys to correct them manually (notice that each issue had three answer cards). Sometimes smudges can be erased by a correction pen/tape/liquid/etc., but Computer Vision cannot always decide which alternative the student really wanted to choose. In this case MakeTests' default is to give naught to the answer, and in Fig. 11 we would get a column with an *extra* red cross.

4.2 Second Experiment Report

MakeTests' present day version requires intermediate knowledge in Python programming. However, the professor can draft questions and resort to a technician that will implement them in Python. Here we comment on a maths professor who does not have programming knowledge but was willing to use MakeTests for a test. He gave an extra exam to 36 students of the course Foundations of Mathematics. His draft was implemented by a technician, and consisted of three questions:

Multiple Steps: similar to the one whose answer card is shown in Fig. 12 but with five steps 1.1 to 1.5 of a mathematical proof. In one of the versions the statement was "prove that $x > y$ and $y > z$ imply $x > z$". Each step had six alternatives, and only one justified that step correctly.

Multiple-choice with Justification: similar to a standard multiple-choice question but requiring a written response besides choosing an alternative. Naught is promptly given to any wrong choice but the professor additionally checks the student's solution if it led to the right one.

TrueOrFalse: On a list of mathematical proofs the student must find out which are wrong or right.

The technician used the professor's draft to prepare the exam with MakeTests. Afterwards the professor gave him all the solved exams digitized in a single PDF, so that MakeTests generated the CSV spreadsheet with each question number and the respective student's answer and score, as mentioned in Sect. 3. The professor just had the task of validating the second question manually in case of right choice. He expressed a positive opinion regarding the exam variations and the quick correction by MakeTests, but the only hindrance was his lack of programming knowledge.

5   Conclusion and Future Works

We have just introduced the open source system MakeTests, which generates and corrects various exams whose texts are individualized through Python random variables. The printouts allow for instantaneous correction, which can be passed on to each respective examinee by email. Furthermore, the user can either choose several pre-defined types and styles of questions already built in MakeTests, or even create new ones. These are the main innovations offered by MakeTests that we have not found in any other system yet.

As described in Sect. 3.1, MakeTests already counts on five question styles. Skilled Python programmers will benefit from MakeTests by themselves, but other users can request some help offered by technical staff.

Compared with [14] the present work detailed how questions are created specially regarding technical elements of object oriented programming, for which the implemented Class hierarchy of MakeTests is essential. Herewith we have presented more examples that can be adapted by the reader to create new questions. The complete material is available at github.com/fernandoteubl/MakeTests.

The real-time correction with automatically emailed feedback resulted in a very positive practice according to our experiments. For the time being, this option is only possible by means of a computer with either Linux or macOS. In MakeTests' future versions we shall add Windows and also smartphone with either Android or iOS.

MakeTests' further developments include turning it into a complete platform on which the Python programming will be a concealed facility. Hence the user will be able to write exams in either plain text, LaTeX or Markdown through an interface, behind which an automatic translator to Python will get the MakeTests' exams we have been working with.

For the time being MakeTests produces exams only in PDF but it is going to incorporate online formats like XML, which is exportable to Moodle. Last but not least, we plan to resettle MakeTests' DQ into a Databank of Questions maintained by various professors who share lists of exercises. Hence, examiners will be able to pick questions therefrom and customize tests without having to create exercises from scratch. Also, MakeTests' DQ and its entire code can be settled into a WebService for professors to maintain all of their resources in cloud and synchronized instead of on local machines.

References

1. Adkins, J.K., Linville, D.: Testing frequency in an introductory computer programming course. Inf. Syst. Educ. J. **15**(3), 22 (2017)
2. Allen, J., de Leon, D.C., Goes, Haney, M.: ADLES v2.0: managing rapid reconfiguration of complex virtual machine environments. In: The Colloquium for Information System Security Education (2019)
3. Burkett, V.C., Smith, C.: Simulated vs. hands-on laboratory position paper. Electron. J. Sci. Educ. **20**(9), 8–24 (2016)
4. Csapó, B., Molnár, G., Nagy, J.: Computer-based assessment of school readiness and early reasoning. J. Educ. Psychol. **106**(3), 639 (2014)
5. Engelhardt, L., Goldhammer, F., Naumann, J., Frey, A.: Experimental validation strategies for heterogeneous computer-based assessment items. Comput. Hum. Behav. **76**, 683–692 (2017)
6. Hakami, Y.A.A., Hussei, B., Ab Razak, C., Adenuga, K.I.: Preliminary model for computer based assessment acceptance in developing countries. J. Theor. Appl. Inf. Technol. **85**(2) (2016)
7. de Leon, D.C., Goes, C.E., Haney, M.A., Krings, A.W.: ADLES: specifying, deploying, and sharing hands-on cyber-exercises. Comput. Secur. **74**, 12–40 (2018)
8. Manoharan, S.: Cheat-resistant multiple-choice examinations using personalization. Comput. Educ. **130**, 139–151 (2019)
9. Nguyen, Q., Rienties, B., Toetenel, L., Ferguson, R., Whitelock, D.: Examining the designs of computer-based assessment and its impact on student engagement, satisfaction, and pass rates. Comput. Hum. Behav. **76**, 703–714 (2017)
10. Nguyen, Q., Thorne, S., Rienties, B.: How do students engage with computer-based assessments: impact of study breaks on intertemporal engagement and pass rates. Behaviormetrika **45**(2), 597–614 (2018). https://doi.org/10.1007/s41237-018-0060-1
11. Nikou, S.A., Economides, A.A.: Mobile-based assessment: investigating the factors that influence behavioral intention to use. Comput. Educ. **109**, 56–73 (2017)
12. Rodríguez del Pino, J.C., Rubio Royo, E., Hernández Figueroa, Z.J.: VPL: laboratorio virtual de programación para moodle. In: XVI Jornadas de Enseñanza Universitaria de la Informática, pp. 429–435. Universidade de Santiago de Compostela. Escola Técnica Superior d'Enxeñaría (2010)
13. Smirnov, E., Bogun, V.: Science learning with information technologies as a tool for "Scientific Thinking" in engineering education (2011, online submission)
14. Teubl, F., Batista, V.R., Zampirolli, F.A.: MakeTests: generate and correct individualized questions with many styles. In: Csapó, B., Uhomoibhi, J. (eds.) Proceedings of the 13th International Conference on Computer Supported Education, CSEDU 2021, Online Streaming, 23–25 April 2021, vol. 1, pp. 245–253. SCITEPRESS (2021). https://doi.org/10.5220/0010337902450253
15. Zampirolli, F.A., et al.: Parameterized and automated assessment on an introductory programming course. In: Anais do XXXI Simpósio Brasileiro de Informática na Educação, pp. 1573–1582. SBC (2020)
16. Zampirolli, F.A., Sato, C., Savegnago, H.R., Batista, V.R., Kobayashi, G.: Automated assessment of parametric programming in a large-scale course. In: Latin American Conference on Learning Technologies (LACLO) (2021)
17. Zampirolli, F.A., Teubl, F., Batista, V.R.: Online generator and corrector of parametric questions in hard copy useful for the elaboration of thousands of individualized exams. In: 11th International Conference on Computer Supported Education, pp. 352–359 (2019)

Evaluating Academic Reading Support Tools: Developing the aRSX-Questionnaire

Nanna Inie[(⊠)][iD] and Bjørn Hjorth Westh

Center for Computing Education Research, IT University of Copenhagen,
Copenhagen, Denmark
{nans,bjwe}@itu.dk
http://www.itu.dk

Abstract. This paper presents and evaluates a new survey metric, the active Reading Support indeX (aRSX), which was created to help researchers and designers evaluate whether a specific software or hardware tool supports active reading of academic texts. The aRSX is comprised of questions in five categories: *The Text, Cognitive Workload, Physical Workload, Perceived Learning, User Experience and Aesthetics,* and *Flow,* as well as an open-ended question for additional comments. In this paper, we present our initial development of two beta-versions of the questionnaire in two studies of $n = 100$ and $n = 53$ deployments, evaluating paper, laptops, iPads, and reMarkable tablets as reading support tools. These studies led to the current version of the aRSX, and the initial results suggest that the metric is a reliable and valid indicator of a tool's ability to support reading of academic texts.

Keywords: Reading support tools · Digital reading · Evaluation metrics · User experience design · User experience methodology

1 Introduction

Reading academic literature with digital tools is becoming more and more of a normalcy for students, yet the diversity of digital reading support tools is surprisingly low [24,32]. Digital textbooks or research papers are rarely designed to look different from their physical instances, and they often distributed as PDFs, a format which was developed primarily for printing, and which has the purpose of making a document look *stable* on all devices and in all printing conditions – not for accommodating different reading platforms or preferences. The hardware used to consume digital texts is largely confined to personal laptops and, less often, tablets [24,32,35]. As Pearson, Buchanan and Thimbleby claimed in their work on developing lightweight interaction tools, existing digital document formats are "far from ideal", and both the software and hardware used for reading often supports casual reading much better than attentive, close interpretation of the text [32].

From an educational perspective, this is at best an under-utilization of the great potential of digital tools that we could take advantage of. At worst, this may be a causal factor of a declining trajectory in reading abilities and motivation of students from elementary school through college [15,30,40]. In one meta-study of research on digital

© Springer Nature Switzerland AG 2022
B. Csapó and J. Uhomoibhi (Eds.): CSEDU 2021, CCIS 1624, pp. 316–335, 2022.
https://doi.org/10.1007/978-3-031-14756-2_16

versus physical reading, Delgado and colleagues [8] found that students seem to have become worse at reading in digital formats, and suggested that one of the causal factors may be *the shallowing hypothesis*, a rationale that states that because the use of most digital media consists of quick interactions driven by immediate rewards (e.g. number of "likes" of a post), readers using digital devices may find it difficult to engage in challenging tasks, such as reading complex text, which requires sustained attention [2,8].

One promising avenue for supporting better digital reading is to study the particularities of the *user experience design* (UX) of digital tools (both hardware and software) – exploring the way these tools afford and support what is necessary for students to engage in *active reading* [1,29,32]. The interaction design of digital tools is particularly interesting to explore, because out of the factors identified as central for reading experience and reading preferences when choosing between analog and digital formats (for instance; ability to concentrate, ability to remember what was read, convenience and expenses, and technological limitations/possibilities [29]), *interaction design* (both possibilities and affordances of the technological environment) is the main factor that researchers in reading support software have a an essential opportunity to improve [32].

With this paper, we suggest that rather than looking at differences between physical and digital platforms in a broad sense, we should investigate the user experience of different tools in more detail. A tablet is not just a tablet and a computer is not only a computer, just as a book is not the same as a loose sheet of paper, and a highlight pen is not the same as a blunt pencil. Different digital document readers can be used on the same tablet or computer, rendering broad comparisons between "digital" and "physical" somewhat meaningless. Rather, we should try to evaluate which interface features and interaction formats work well for different users and their learning preferences.

This paper presents a novel evaluation metric for reading support tools. We present the beta versions of the active Reading Support indeX, aRSX, a first step towards a standardized evaluation scheme that can be used by researchers and developers to, relatively quickly, identify how a reading tool supports user experience and learning preferences when reading academic texts. In addition, such an evaluation tool can be used to indicate "robust moderating factors" that influence reading performance in different media [8].

In the paper, we evaluate two beta versions of the aRSX and present the current version of the questionnaire. Although the form is in continuous development, we find it useful to share our early experiences with utilizing such a survey metric. This paper explores the question: *How might we evaluate academic reading support tools with a focus on user experience?*, and provides initial findings that the aRSX can give reliable, robust evaluations of the user experience of different digital reading support tools.

2   Background and Related Work

Surveys of attitudes and preferences continue to conclude that college students slightly prefer physical formats for focused academic reading, generally stating they feel like paper-based reading let them concentrate and remember better [25,29]. Research in Human-Computer Interaction (HCI) has suggested that especially annotation and note-taking on a computer are activities that compete with the reading itself, due to the lack of direct manipulation [19]. The cognitive workload required to interact with digital devices may be higher than that of interaction with physical media, and annotating introduces another cognitive workload [32,33]. Indeed, studies have indicated that the simplest textual and interactive environment is associated with the highest comprehension outcomes and better user experience [3,10,33].

It is a noteworthy challenge for the development of good reading support software and hardware that few existing studies of digital reading specify the reading environment provided to participants, and even fewer evaluate its interaction design in detail. This means that it is difficult to identify the UX designs that influence the reading experience and performance in a positive way; and thus, difficult to improve the design of novel digital tools for academic reading in a rigorous, systematic manner [18].

2.1 Reading Support Tools

A reading support tool can be defined as *any tool that can be used by people to read or support the reading of documents that primarily consist of written text.* A reading support tool can be analog or digital, and it can be software or hardware. Hardware platforms, of course, need software to display a text.

Digital reading software – programs for displaying ePub and PDF formats – is often not recognized as a specialized tool, because it mainly displays content, rather than support the reader actively. With the advent and spread of literature in digital formats, however, active reading support tools are in great demand [32]. Reading on an iPad with GoodReader may yield different reading performance and user experience than using Adobe Acrobat Reader on a Samsung Galaxy Tab, even though these could both be categorized as reading on 'tablets'. There is a difference in evaluating the iPad versus the Samsung Galaxy, or evaluating GoodReader versus Adobe Acrobat. Providing clarity and distinctions between these tools is necessary for research to be comparable and findings to be widely applicable.

2.2 User Experience of Reading Tools

There are numerous ways of evaluating *usability* of products and tools (such as walk-throughs, think-aloud tests, contextual inquiry, etc.), but a tool's ability to support reading of academic texts is more complex than its usability. The UX goals of reading tools are not efficiency or performance metrics, but rather, that the user feels cognitively enabled to focus on a text content for as little or as much time as necessary [32], that the user feels enabled to process the text in any way they might need, as well as whether the tool is a good "fit" for the student's preferences and reading context.

User experience as a research agenda is concerned with studying the experience and use of technology in context. The UX of a product is considered a consequence of

> "a user's **internal state** (predispositions, expectations, needs, motivation, mood, etc.), the **characteristics of the designed system** (e.g. complexity, purpose, usability, functionality, etc.) and the **context** (or the environment) within which the interaction occurs (e.g. organisational/social setting, meaningfulness of the activity, voluntariness of use, etc.)" [14], our emphases.

Models of UX usually separate a product's pragmatic from its hedonic qualities, where pragmatic attributes advance the user toward a specific goal and depend on whether the user sees a product as simple, predictable, and practical. Hedonic attributes, on the other hand, are related to whether users identify with a product or find it appealing or exhilarating [16]. Pragmatic attributes are often found to exert a stronger influence on the evaluation of a product than hedonic attributes.

Although text is presented linearly, learning by reading is not a linear process. Reading, and particularly academic reading, is open-ended. An academic reader depends on constant self-evaluation of whether the material is understood and internalized or not, rather than defined and well-known external objectives.

Generally, metric-based research investigating students' opinions and experiences of reading tools has consistently found positive correlation between interaction design and reading performance [10,11,21,22,43], and between user attitudes and learning outcomes [20,36,41]. However, few studies investigate which features in particular foster a good learning experience, although with some exceptions, e.g. [3,5,32,33].

One survey identified some of the most important themes for academic students when choosing between digital and paper as the following: Flexibility, ability to concentrate, ability to remember what was read, organizing, approachability and volume of the material, expenses, making notes, scribbling and highlighting, and technological advancement [29]. Out of these, the flexibility of the reading support tool, the approachability of the content, the affordances of annotation, and the technological advancement are factors which are (to at least some degree) influenced by the design of the reading support tool.

2.3 Questionnaire-Based UX Evaluation

While qualitative research such as detailed interviews and observations are traditional methods for conducting UX evaluations, these methods are time-consuming and not easy to implement on a large scale. The goal of the aRSX is develop a quantitative, questionnaire-based evaluation with a foundation in UX research. Quantitative questionnaires are non-costly and time-efficient to execute, and they have been used for decades as a valuable indicator of tool specifications and requirements [12,13,27]. They also have the advantage that most HCI researchers and designers are already familiar with the format and the reporting of its results.

The NASA Task Load indeX (TLX) [13] has been used for over 30 years as an evaluation method to obtain workload estimates from 'one or more operators', either while they perform a task or immediately afterwards. The TLX consists of six subscales: Mental, Physical, and Temporal Demands, Frustration, Effort, and Performance. All questions are on a 21-point scale of "Very Low" to "Very High." The assumption is that a combination of these dimensions represents the workload experienced by most people performing most tasks. Each subscale is furthermore "weighted" by the individual performing the task, so that the final score is given based on how important each subscale is to the individual. The NASA TLX has been translated into more than a dozen languages, and modified in a variety of ways [12]. It is being used as a benchmark of evaluation, and has proved its value in a wide range of fields from nuclear power plant control rooms to website design.

Other fields have had great success appropriating the TLX to evaluate task-specific tools, for instance *creativity support* [4,6]. The Creativity Support Index (CSI) is based on the NASA TLX, and is a psychometric survey designed to assess the ability of a digital creativity support tool to support the creative process of its users. Its structure is very similar to the NASA Task Load Index, but its theoretical foundation is based on concepts from creativity and cognition support tools, such as creative exploration, theories of play, Csiksentmihalyi's theory of flow, and design principles for creativity support tools. The CSI includes six subscales or "factors": Collaboration, Enjoyment, Exploration, Expressiveness, Immersion, and Results Worth Effort.

While both the TLX and the CSI are incredibly valuable tools, the surveys do not address the particularity of learning from reading. Inspired by the CSI, we decided to create an evaluation form tailored to uncover the ability of a tool to support textual knowledge acquisition based on theory of learning and UX research.

3   Methodology

3.1   Criteria for a Usable Evaluation Form

In order to evaluate the usefulness of the aRSX, we specified the following criteria as ideals for the questionnaire:

1) *Theoretical foundation*: The evaluation should be grounded in prior research on reading, learning, and user experience design.
2) *Operationlizability*: The evaluation should be operational and useful for researchers and designers developing and evaluating reading support tools. It should be clear and usable for both participants and those who administer the evaluation.
3) *Generalizability*: The evaluation must enable researchers to analyze different kinds of reading support tools with different types of populations in different types of settings.
4) *Validity*: The survey should accurately measure the factors that it intends to measure.
5) *Reliability*: The survey should produce reliable results, aiming for a Cronbach's alpha above .70.
6) *Empirical grounding*: The framework must be thoroughly tested in practice.

In the study presented in this paper, we focused especially on developing the *theoretical foundation*, *validity*, *reliability*, and *empirical grounding*. The operationalizability and generalizability are somewhat inherent in the original NASA TLX survey, and we borrow some credibility from this thoroughly tested metric.

Further theoretical and empirical grounding must be developed through applying and evolving the evaluation form in different studies and communities. Through this paper we share the aRSX with other researchers and invite them to use, evaluate, and modify the evaluation form.

3.2   Experimental Setups

The first beta-version of the aRSX was created and tested in *Study 1: Laptop and paper reading*. The findings from this study led to the second beta-version, which was much longer than the first. It is common practice in psychometrics to create longer, temporary versions of a survey when developing a new metric. This allowed us to conduct an exploratory factor analyses of the responses, in the interest of identifying the items or questions that performed the best. The second beta-version was tested in *Study 2: iPad and reMarkable reading*. Following this section, we will describe the studies and findings chronologically.

Both studies were designed as controlled within-group studies, where we invited a group of students to read half a text on one medium, asked them to evaluate it, then switched to a different medium for the second half of the text, and asked the participants to evaluate the second medium. That means that 77 students filled out a total of 153 evaluations (one student only completed a reading in one medium), 100 of the first beta-version, and 53 of the second beta-version. The within-group comparison of two reading tools per participant allowed us to explore the aRSX with four different media, as well as to conduct four different reliability analyses of Cronbach's alpha (one per reading tool). Although thorough development of psychometric evaluation forms require hundreds and sometimes thousands of participant numbers for statistically sound analyses to be conducted, we find it valuable to share our initial findings at an early state of development, both to document the development process of the aRSX openly, as well as to share the beta-versions of the questionnaire with the research community for feedback and comments.

The beta-versions of the aRSX were designed as "Raw TLX", eliminating the part of the original TLX which is concerned with pairwise ranking of the subscales to reflect personal importance attributed to each subscale or factor. The raw-TLX approach is simpler to employ, and does not appear to yield less useful results [12], but more importantly, we wished to thoroughly develop and evaluate subscales to explore which questions and factors carried higher loading. When the question wordings and factors are more resolved, it may make sense to add a pairwise factor rating to the survey, as in the TLX and the CSI [6, 13]. Consequently, no cumulative or final score for each tool was calculated for the beta versions of the aRSX, as the final score is traditionally dependent on each participant's ranking of the different factors.

4 Study 1: Laptop and Paper Reading

4.1 First Beta-Version Subscales: Cognitive Workload, Physical Workload, Perceived Learning, User Experience and Aesthetics, and Flow

The first beta-version of the aRSX is shown in Fig. 1. It had 10 basic questions.

Cognitive and Physical Workload. The first three questions were copied from the TLX-questions concerning mental and physical demand. In the first beta-version, the cognitive and physical workload were collected under one subscale, as we did not anticipate the physical workload to play a significant role to the evaluation, other than as a potentially distracting factor to the cognitive workload.

One of the most basic issues in the study of cognitive workload is the problem of how to actually measure it, but since we are concerned with the user *experience* of cognitive workload, it is appropriate to let the user self-evaluate this aspect. The

COGNITIVE & PHYSICAL WORKLOAD

1 **How mentally demanding was the task?**   Very low I__I__I__I__I__I__I__I Very high

2 **How physically demanding was the task?**   Very low I__I__I__I__I__I__I__I Very high

3 **How hard did you have to work to complete the task?** Very little I__I__I__I__I__I__I__I Very hard

PERCEIVED LEARNING

4 **I felt like I was learning something from reading the text** Highly disagree I__I__I__I__I__I__I__I Highly agree

USER EXPERIENCE AND AESTHETICS

5 **I enjoyed using this system or tool to read the text** Highly disagree I__I__I__I__I__I__I__I Highly agree

6 **The system or tool allowed me to annotate the text in a way that was helpful to me** Highly disagree I__I__I__I__I__I__I__I Highly agree

7 **The interface of the tool was pleasant to look at** Highly disagree I__I__I__I__I__I__I__I Highly agree

FLOW

8 **While I was reading, I forgot about the tool I was using and became immersed in the text** Highly disagree I__I__I__I__I__I__I__I Highly agree

9 **I could imagine using this tool to read texts on a regular basis** Highly disagree I__I__I__I__I__I__I__I Highly agree

10 **Any additional comments about the task or the tool?** Open answer

Fig. 1. The first beta-version of the aRSX, from [17].

cognitive attention required to read and annotate on paper is minimal for most people, which often means that people can do it without thinking [34]. One of the main goals of a reading support tool is to minimize the cognitive effort required of the reader to interact with the tool itself, so they can focus completely on the content [9,32].

Perceived Learning. The questionnaire should evaluate whether the tool allows for learning from reading the text. This aspect is an evaluation of the tool's *pragmatic* qualities, i.e. whether it advances a user toward their specific goal [16]. An academic reader depends on constant self-evaluation of whether the material is understood and internalized or not [7,42]. Self-evaluation is often used in learning research, and has been proven reliable [28,35]. The fourth question of the survey simply asked the reader whether they believed they learned from the reading.

User Experience and Aesthetics. As described in Sect. 2.2, good user experience and interaction design have a positive correlation with learning outcomes. Although we expected readers to have higher *pragmatic* expectations of reading support tools than *hedonic* expectations, user enjoyment and aesthetics of the reading tool are extremely important to the overall user experience and technology adaptation [16,29]. The fifth and seventh question asked whether the reader enjoyed using the tool and whether the tool was pleasant to look at. Since the questionnaire was designed for evaluating *active* reading [1], it should evaluate how the tool supports annotation of the text. When engrossed in academic reading, many users will accompany reading with annotating; for example, highlighting, commenting and underlining. Annotations can be considered a by-product of the active reading process, and they should be supported by digital reading tools [31]. Therefore, the sixth question asked whether the tool allowed annotation in a form that was helpful to the user.

Flow. According to Csikszentmihalyi's concept of *flow* in a learning context, student engagement is a consequence of simultaneous occurence of *concentration, enjoyment,* and *interest* - of high challenge in combination with high skill [38]. While theory of learning is often concerned with the *content* of a given text, the aRSX focuses on the capacity of the *tool* to allow the reader to process and engage with the text. The experience of flow can happen to individuals who are deeply engrossed in activity which is intrinsically enjoyable, and the activity is perceived as worth doing for its one sake, even if no further goal is achieved [26]. The experience of flow while reading academic texts can occur as the result of a well-written, interesting or challenging text, but it can be enhanced or disrupted by *contextual factors* such as the tool used to consume the text [32,39]. Question eight and nine of the aRSX addressed whether the tool fosters immersion while reading, and whether the tool would fit into the reader's regular practices.

Finally, the questionnaire finished with an open-ended question, so we could learn about other factors that may have been relevant to the reader during the first evaluation of the questionnaire.

4.2 Participants and Execution

Subjects and Treatments. 50 students from the IT University of Copenhagen, Denmark, were recruited during fall 2019. The test setup was a controlled, *within-group* setup, where each student was subjected to both treatments; a paper reading treatment, and a digital reading treatment. The study is described in further detail in Inie and Barkhuus [17].

In the **paper reading** treatment, students were provided with the text on printed A4 paper, two highlighter pens, one pencil, one ball pen, and sticky notes.

In the **digital reading** treatment, students were provided with the text on a laptop. The text was a PDF file and formatted exactly as the paper reading for comparability. The text was provided in the software Lix, a reading support software for PDF readings.

After reading the first half of the text, each student filled out the aRSX survey on paper for that treatment. They then read the other half of the text in the opposite medium, and filled out the aRSX on paper for that medium. 26 of the students read the first half of the text on paper and the second half of the text on computer (condition A) and 24 students read the first half of the text on computer and the second half of the text on paper (condition B)[1]. The students in condition A and B were not in the same room. They were not informed about the focus on the evaluation form. The students were instructed to read the text "as if you were preparing for class or exam, making sure to understand the major points of the text".

---

[1] This slight unevenness in distribution of condition A and B was due to student availability at the time of the experiment.

4.3 Findings from Study 1

The experiment was run as an explorative experiment, where we were interested in discovering if the evaluation was generally meaningful for participants, and whether it yielded significant and useful results. The first type of data we gathered was **experimental observations**, primarily questions from participants about the wording of the survey or how to complete it. Those data did not need thorough analysis, as the questions we received were quite straightforward.

The second type of data from the study were the **quantitative results** of the evaluations. We performed Cronbach's alpha tests on the responses for reliability.

The third type of data was the **qualitative responses** to the final open-ended question. The question was optional, and we received 25 comments regarding the paper reading and 37 comments regarding the digital reading. The length of the comments varied from one to eight sentences. We clustered the comments into themes according to their relevance to the questionnaire, rather than their distinct evaluation of the tool.

Experimental Observations

Finding 1: "Annotating" Is Not an Obvious Concept. Several participants asked what was meant by 'annotating' in question 6: "The system or tool allowed me to annotate the text in a way that was helpful to me". This could be exacerbated by the fact that only few of the students were native English speakers, and the survey was conducted in English. In addition, we observed this theme in the open-ended survey responses (9 out of 37 participants commented on highlighting features), e.g.: *"It is very easy to highlight, but a little more confusing to make comments to the text"* (Participant 190802, digital reading). According to reading research, annotating a text consists of, for instance, highlighting, underlining, doodling, scribbling, and creating marginalia and notes [23,32]. Construction of knowledge and meaning during reading happens through activities such as these, making the possibility of annotation extremely important when supporting active reading. The question of annotation should be clarified.

Finding 2: "Interface" Is a Concept That Works Best for Evaluation of Digital Tools. The word 'interface' in question 7: "The interface of the tool was pleasant to look at" prompted some questions in the paper treatment. An interface seems to be interpreted as a feature of a digital product, and this was not a useful term when evaluating an analog medium. In the interest of allowing the aRSX to be used in the evaluation of both digital and analog tools, this question should designate a more general description of the aesthetics of the tool.

Quantitative Results

Finding 3: The Survey Appears to Be Internally Reliable. We performed an ANOVA two-factor analysis without replication, and calculated a Cronbach's alpha of .732 for paper, and .851 for laptop, which indicates satisfying reliability. Question one, two and three (pertaining to cognitive workload) ask the user to rate their mental and physical strain or challenge from 1 (Very low) to 7 (Very high). In these questions a high score corresponds to a negative experience, and the scores therefore had to be reversed to calculate sum score and Cronbach's alpha. Further tests are needed to investigate whether positively/negatively worded statements produce different results.

Finding 4: "Physical Demand" Should Be Specified. The average scores for question two (physical demand) were very low for both paper and laptop. The question is copied directly from the NASA TLX, and was deemed relevant because eye strain from digital reading has often been mentioned as a negative factor of screen reading in previous research (e.g. [37]). 'Physical demand', however, may be associated with hard, physical labor, and should be specified further to gain useful knowledge from the score. This was exacerbated by some of the comments from the open-ended question: *"I think it would be better to do the test on my own computer. The computer was noisy and the screen was small"* (Participant 190211, digital reading), and *"Reading on a pc is not pleasant when the paper is white. Use some sort of solarized"* (Participant 170303, digital reading). These comments demonstrate that types of experienced physical strain can very a lot, and the question of physical demand does not, in itself, yield useful insights.

Qualitative Responses. The open-ended question responses generally showed that participants were aware of the evaluation setup, and that they were focused on evaluating the usability and experience of the software tool. The responses also showed that many of the students were willing to reflect on and compare the different tools in a meaningful way during the same setup or session. The responses were extremely valuable in elaborating the measured experience reflected in the quantitative measures, and we would recommend to keep this question in future iterations of the survey.

Finding 5: The Content May Influence the Evaluation of the Tool. A theme in the comments which was not addressed by the questions in the aRSX was the content of the specific text which was read. Nine participants commented on the text e.g.: *"Really interesting text"*. (Participant 170001, digital reading) and *"The text was more of a refresher than new learning"* (Participant 150302, paper reading). Although it seemed from the comments like the students were able to distinguish the text from the tool, and some of these effect would be mitigated by the fact that the students read from the same text in both treatments, we believe it to be a relevant observation that the text which is being read may influence the experience of using a tool. Furthermore, the type of text may also require different tools for annotating, cf. *"When I tried to highlight mathematical formulas it would sometimes try to highlight additional text that I couldn't remove from the little highlight box"*. (Participant 180403, digital reading).

5   Study 2: iPad and reMarkable Reading

5.1   Second Beta-Version Subscales: The Text, Cognitive Workload, and Physical Workload

The second beta-version of the aRSX is shown in Fig. 2. Based on the findings of Study 1, we rephrased some of the questions of the aRSX, as well as added several questions.

For this beta-version, we wanted to conduct a factor analysis to determine optimal statement wording, and we therefore split many of the questions into several options.

THE TEXT

| 0 | How would you rate the difficulty level of the text? | Very easy I__I__I__I__I__I__I Very difficult |

COGNITIVE WORKLOAD

1a	It was mentally effortless for me to complete the task	Highly disagree I__I__I__I__I__I__I Highly agree
1b	Reading the text took a lot of mental exertion	Highly disagree I__I__I__I__I__I__I Highly agree
1c	My brain did a lot of work during this task	Highly disagree I__I__I__I__I__I__I Highly agree
1d	I experienced the work load for this task as low	Highly disagree I__I__I__I__I__I__I Highly agree

PHYSICAL WORKLOAD

2	It was physically effortless for me to complete the task	Highly disagree I__I__I__I__I__I__I Highly agree
2a	If you experienced physical strain, describe which kind	Open answer
2b	I felt physically tired/uncomfortable during reading	Highly disagree I__I__I__I__I__I__I Highly agree

PERCEIVED LEARNING

3a	The content of the text was easy for me to understand	Highly disagree I__I__I__I__I__I__I Highly agree
3b	I felt like I was learning something from reading the text	Highly disagree I__I__I__I__I__I__I Highly agree
3c	I did not acquire a lot of information from this text	Highly disagree I__I__I__I__I__I__I Highly agree

USER EXPERIENCE AND AESTHETICS

4a	I enjoyed using this system or tool to read the text	Highly disagree I__I__I__I__I__I__I Highly agree
4b	I found it fun to use this tool or system to read the text	Highly disagree I__I__I__I__I__I__I Highly agree
4c	This tool was annoying to use for reading	Highly disagree I__I__I__I__I__I__I Highly agree
5a	It was easy to interact with the tool or system	Highly disagree I__I__I__I__I__I__I Highly agree
5b	The functions of this tool or system were difficult to use	Highly disagree I__I__I__I__I__I__I Highly agree
5c	It was straightforward to use the tool or system in the way I wanted to	Highly disagree I__I__I__I__I__I__I Highly agree
6a	I liked the way the tool or system looked	Highly disagree I__I__I__I__I__I__I Highly agree
6b	The tool or system was attractive or beautiful	Highly disagree I__I__I__I__I__I__I Highly agree
6c	The tool or system was ugly or boring	Highly disagree I__I__I__I__I__I__I Highly agree
Int.a	The system or tool allowed me to highlight the text in a way that was helpful to me	Highly disagree I__I__I__I__I__I__I Highly agree
Int.b	The system or tool allowed me to take notes in a way that was helpful to me	Highly disagree I__I__I__I__I__I__I Highly agree

FLOW

Flow	While I was reading, I forgot about the tool I was using and became immersed in the text	Highly disagree I__I__I__I__I__I__I Highly agree
7a	I could imagine using this tool to read texts on a regular basis	Highly disagree I__I__I__I__I__I__I Highly agree
7b	I would like to own this tool to use for reading more often	Highly disagree I__I__I__I__I__I__I Highly agree
7c	I think this tool fits my reading requirements	Highly disagree I__I__I__I__I__I__I Highly agree
8	Write your additional comments about the task or tool	Open answer

Fig. 2. The second beta-version of the aRSX.

We added an initial question pertaining the general difficulty level the text response to Finding 5: 'The difficulty of the text may influence the evaluation of the tool'. If participants experience the text as very difficult, this may impact their perception of the tool. While the aRSX is not designed to evaluate the quality of the text, this question acknowledges that there is an difference between evaluating a text as difficult, and experiencing difficulty reading it.

Cognitive and Physical Workload was split into two subscales, and we added an open question to understand the Physical Workload further: "If you experienced physical strain, describe which kind". This was added in response to Finding 4: "'Physical demand" should be specified'.

Furthermore, we split the question about annotations into two questions, one about the tool's ability to support highlights and one about the support of creating notes cf. Finding 1: "'Annotating" is not an obvious concept'. We recognize that not all students may need to highlight or annotate the texts they read, in which case we anticipated the scores would be neutral.

Finally, the word "interface" was removed from the questionnaire in the interest of making it usable for evaluation of analog tools as well as digital, cf. Finding 2: "'Interface" is a concept that works best for evaluation of digital tools'.

In the second beta-version, we primarily focused on analyzing quantitative data from the responses. The qualitative responses to the open-ended questions fell in two categories; they elaborated either a participant's individual state, i.e. "I felt very tired due to lack of sleep", or the answers to the evaluation, i.e. "I normally don't highlight, so that was not relevant to me". The answers indicated that the task of assessing the tool's reading support ability was straightforward to understand to the participants.

5.2 Participants and Execution

Subjects and Treatments. 27 students at the IT University of Copenhagen, Denmark, were recruited to participate in the second study. It took place in fall 2021. Like Study 1, the setup for Study 2 was a controlled, within-group setup, where each student were subjected to both treatments; an iPad reading treatment, and a reMarkable[2] reading treatment. In the **iPad reading** treatment, students were provided with the text in PDF Expert, as well as an Apple pencil for annotations. In the **reMarkable** reading treatment, students were provided with the text in reMarkable's own PDF reading software and the reMarkable pen. The text was a PDF file and separated into two halves. After reading the first half of the text, each student filled out the aRSX survey on paper for that treatment. They then read the other half of the text in the opposite medium, and filled out the aRSX on paper for that medium. 16 students read on the iPad first and the reMarkable second, and 11 students read on the reMarkable first, and the iPad second. The students were not informed about the focus on the evaluation form, but were told to evaluate the tool they were using. The students were instructed to read the text "as if you were preparing for class or exam, making sure to understand the major points of the text".

5.3 Findings from Study 2

The focus on Study 2 was to investigate optimal wording for the individual questions. We opted to identify two to three items per subscale, so that each subscale or fac-

---

[2] https://remarkable.com/.

tor can be tested for reliability in future deployments. This beta-version of the aRSX contained 26 individual items, clustered around very similar questions to those in the first beta-version (although the questions were shuffled around randomly in the actual deployment, rather than organized in their respective subscales), as well as an open-ended question at the end. The open-ended question was rephrased in this version to an imperative "Write your additional comments ...", and this wording yielded much more detailed qualitative responses to the evaluations. 20 out of 26 (one student only read in on the reMarkable, and not the iPad) participants wrote comments for this question for the iPad, and 23 out of 27 participants wrote comments for the reMarkable, all comments of at least three sentences or more.

The overall averages of the evaluations are shown in Fig. 3. We conducted an exploratory factor analysis for the responses, and calculated Cronbach's alpha for the responses for iPad versus reMarkable, respectively. We also calculated Cronbach's alpha for each subscale.

Reliability. The overall Cronbach's alpha for the questionnaire was .919 for the iPad evaluation, and .904 for the reMarkable evaluation, indicating an extremely high internal reliability. This is likely a consequence of the many questions, and also indicates that we can safely remove some questions and likely still achieve a high reliability.

Factor Analysis. As we were interested in the factor loadings for each question, we present the rotated component matrix in Fig. 4 (KMO .790, $p < .001$). We can see that the questions cluster in 5 different components, which actually *almost* match the number of subscales (6), and that they are somewhat matched to their subscales: questions 7a, 7b, and 7c are all in cluster 1, question 1a, 1b, 1c, and 1d are all in cluster 2, and so forth.

With this analysis, we are interested in *eliminating* questions, so that we identify the two items that are most likely to represent the underlying factor. We see that question 7c and 2b load heavily (above .40) in more than one cluster, so we would like to eliminate those. Both 5a, 5b, and 5c also load in two clusters, but 5c loads lowest in both clusters, so we would also like to eliminate that. Using this process – identifying the questions that load highest in *one* category alone, we were left with questions 0, 1a, 1b, 2a, 3b, 3c, 5a, 5b, 6a, 6b, 7a, 7b, and the two questions about interaction. The questions about the text (0) and about flow are kept as they are, as they have not been divided into subquestions.

Post Item-Removal Reliability. After eliminating some questions, we re-calculated the reliability for the whole questionnaire, as well as for each individual factor. The results are shown in Table 1. We see that the survey is still highly internally reliable, both in total, and on each subscale level.

6   Discussion and Current Version of the aRSX

The current version of the aRSX is shown in Fig. 5. A main change is that the subscale User Experience and Aesthetics has been split into two: Interaction Design and

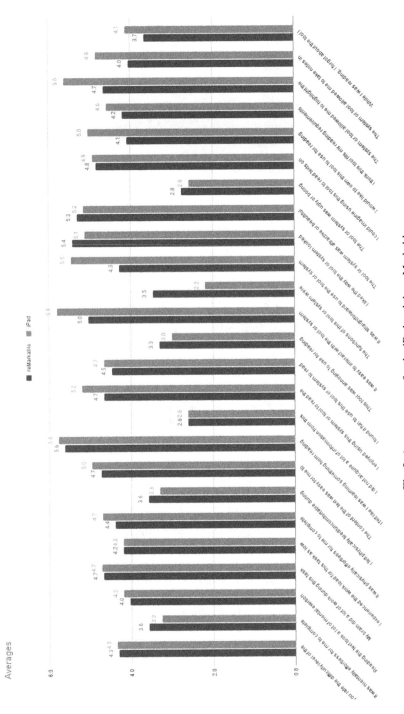

Fig. 3. Average scores for the iPad and the reMarkable.

Rotated Component Matrix[a]

	Component				
	1	2	3	4	5
4b	.908				
7b	.887				
7a	.839				
4a	.837				
7c	.798				.422
6a	.752				
4c	.751				
6b	.705				
Flow	.682			.528	
6c	.665				
1a		.828			
@0		-.826			
1b		.826			
1d		.714			
3a		.707			
1c		.697			
Interaction			.801		
5a	.497		.673		
5b	.454		.563		
5c	.417		.533		
Interaction			.511		
3b				.805	
3c				.784	
2b		.456			.627
2a					.621

Extraction Method: Principal Component Analysis.
Rotation Method: Varimax with Kaiser Normalization.

a. Rotation converged in 9 iterations.

Fig. 4. Rotated component metrix showing factor loadings for all questions.

Aesthetics, in an attempt to distinguish between pragmatic and hedonic qualities of the reading support tool. The 16 questions are based on findings from Study 1 and Study 2, and should provide a good basic evaluation of a given reading support tool.

Overall, it was simple to use the aRSX as an evaluation method. We identified some possibilities for improvement, which have been integrated into the current iteration. In our studies, the aRSX was distributed on paper, which was simple in terms of execution, and a little more cumbersome in terms of digitizing the data - especially transcribing the open-ended question responses might be problematic with large participant numbers. The survey may also be distributed digitally, which we hypothesize could have a positive effect on the open-ended question responses, both due to the possibility of making the question mandatory, and because of the ease of writing comments on the computer versus in hand. In its simple form, the survey should be straightforward to moderate for other studies. The evaluation does thus fulfill the criterion of *Operationalizability*, as per Sect. 3.1.

Table 1. Cronbach's alpha for the adjusted survey. The scales "The Text" and "Physical Workload" are not calculated, as there is only one question in these categories.

	iPad	reMarkable
The Text	N/A	N/A
Cognitive Workload	.837	.872
Physical Workload	N/A	N/A
Perceived Learning	.752	.670
UX & Aesthetics	.905	.790
Flow	.874	.874
Total	**.883**	**.861**

THE TEXT

1   **How would you rate the difficulty level of the text?**   Very easy |_|_|_|_|_|_|_| Very difficult

COGNITIVE WORKLOAD

2   **It was mentally effortless for me to complete the task**   Highly disagree |_|_|_|_|_|_|_| Highly agree

3   **Reading the text took a lot of mental exertion**   Highly disagree |_|_|_|_|_|_|_| Highly agree

PHYSICAL WORKLOAD

4   **It was physically effortless for me to complete the task**   Highly disagree |_|_|_|_|_|_|_| Highly agree

4a **If you experienced physical strain, describe which kind**   Open answer

PERCEIVED LEARNING

5   **I felt like I was learning something from reading the text**   Highly disagree |_|_|_|_|_|_|_| Highly agree

6   **I did not acquire a lot of information from this text**   Highly disagree |_|_|_|_|_|_|_| Highly agree

INTERACTION DESIGN

7   **It was easy to interact with the tool or system**   Highly disagree |_|_|_|_|_|_|_| Highly agree

8   **The functions of this tool or system were difficult to use**   Highly disagree |_|_|_|_|_|_|_| Highly agree

9   **The system or tool allowed me to highlight the text in a way that was helpful to me**   Highly disagree |_|_|_|_|_|_|_| Highly agree

10 **The system or tool allowed me to take notes in a way that was helpful to me**   Highly disagree |_|_|_|_|_|_|_| Highly agree

AESTHETICS

11 **I liked the way the tool or system looked**   Highly disagree |_|_|_|_|_|_|_| Highly agree

12 **The tool or system was attractive or beautiful**   Highly disagree |_|_|_|_|_|_|_| Highly agree

FLOW

13 **While I was reading, I forgot about the tool I was using and became immersed in the text**   Highly disagree |_|_|_|_|_|_|_| Highly agree

14 **I could imagine using this tool to read texts on a regular basis**   Highly disagree |_|_|_|_|_|_|_| Highly agree

15 **I would like to own this tool to use for reading more often**   Highly disagree |_|_|_|_|_|_|_| Highly agree

16 **Write your additional comments about the task or tool**   Open answer

Fig. 5. The current version of the aRSX.

We believe the current iteration of the aRSX is well founded in theory, described in the criterion *theoretical foundation*. Neither reading support nor user experience design are new fields, and there is a solid foundation of knowledge on which to build the selection of good evaluation questions. The novelty consists primarily in developing a consistent, reflective practice around such evaluation, so that both developers and researchers can best benefit from the work of colleagues and peers.

The aRSX is simple to deploy and analyze for any designer, developer or researcher, and the current iteration should present a question wording which is general enough to be applicable in a multitude of contexts. Building upon findings from well-established frameworks, such as the NASA TLX and the CSI, the current state of the survey is *generalizable*.

In this paper we presented an initial analysis of the survey's *validity*. We recognize that further studies and higher participant numbers are necessary to make rigorous claims about the survey's validity.

The survey responses had high reliability scores, which satisfy the criterion *Reliability*. All deployments of the survey had a general Cronbach's alpha above .70. This criterion will require more test-retest and split-half studies to confirm.

Finally, we have achieved better *empirical grounding* of the aRSX in conducting the first two studies and evaluating their outcomes.

7   Conclusion

In this paper, we presented two beta-versions of the aRSX, a novel survey metric designed to evaluate the ability of a tool to support active reading. We tested the two beta versions in two different studies, one focused on defining the questions to ask in such an evaluation, and one focused on defining the most salient factors to evaluate. Future work will include further deployment of the current version of the aRSX in different contexts and with different users. So far, the aRSX has shown to provide meaningful data with a relatively small sample, and we believe the iterations suggested in this paper will make it a stronger evaluation tool.

We believe that the aRSX is a very promising avenue for evaluating reading support tools based on user experience, and we invite the research community to apply and appropriate the survey.

Acknowledgments. We thank the students who participated in this experiment. This research was funded by the Innovation Fund Denmark, grant 9066-00006B: Supporting Academic Reading with Digital Tools.

References

1. Adler, M.J., Van Doren, C.: How to Read a Book: The Classic Guide to Intelligent Reading. Simon and Schuster, New York (2014)
2. Annisette, L.E., Lafreniere, K.D.: Social media, texting, and personality: a test of the shallowing hypothesis. Pers. Individ. Differ. **115**, 154–158 (2017)

3. Buchanan, G., Pearson, J.: Improving placeholders in digital documents. In: Christensen-Dalsgaard, B., Castelli, D., Ammitzbøll Jurik, B., Lippincott, J. (eds.) ECDL 2008. LNCS, vol. 5173, pp. 1–12. Springer, Heidelberg (2008). https://doi.org/10.1007/978-3-540-87599-4_1

4. Carroll, E.A., Latulipe, C., Fung, R., Terry, M.: Creativity factor evaluation: towards a standardized survey metric for creativity support. In: Proceedings of the Seventh ACM Conference on Creativity and Cognition, pp. 127–136 (2009)

5. Chen, N., Guimbretiere, F., Sellen, A.: Designing a multi-slate reading environment to support active reading activities. ACM Trans. Comput.-Hum. Interact. (TOCHI) **19**(3), 1–35 (2012)

6. Cherry, E., Latulipe, C.: Quantifying the creativity support of digital tools through the creativity support index. ACM Trans. Comput.-Hum. Interact. (TOCHI) **21**(4), 21 (2014)

7. Conway, M.A., Gardiner, J.M., Perfect, T.J., Anderson, S.J., Cohen, G.M.: Changes in memory awareness during learning: the acquisition of knowledge by psychology undergraduates. J. Exp. Psychol. Gen. **126**(4), 393 (1997)

8. Delgado, P., Vargas, C., Ackerman, R., Salmerón, L.: Don't throw away your printed books: a meta-analysis on the effects of reading media on reading comprehension. Educ. Res. Rev. **25**, 23–38 (2018)

9. DeStefano, D., LeFevre, J.A.: Cognitive load in hypertext reading: a review. Comput. Hum. Behav. **23**(3), 1616–1641 (2007)

10. Freund, L., Kopak, R., O'Brien, H.: The effects of textual environment on reading comprehension: implications for searching as learning. J. Inf. Sci. **42**(1), 79–93 (2016)

11. Haddock, G., Foad, C., Saul, V., Brown, W., Thompson, R.: The medium can influence the message: print-based versus digital reading influences how people process different types of written information. Br. J. Psychol. **111**, 443–459 (2019)

12. Hart, S.G.: NASA-task load index (NASA-TLX); 20 years later. In: Proceedings of the Human Factors and Ergonomics Society Annual Meeting, vol. 50, pp. 904–908. Sage Publications, Los Angeles (2006)

13. Hart, S.G., Staveland, L.E.: Development of NASA-TLX (task load index): results of empirical and theoretical research. In: Advances in Psychology, vol. 52, pp. 139–183. Elsevier (1988)

14. Hassenzahl, M., Tractinsky, N.: User experience-a research agenda. Behav. Inf. Technol. **25**(2), 91–97 (2006)

15. Hayles, N.K.: How We Think: Digital Media and Contemporary Technogenesis. University of Chicago Press (2012)

16. Hornbæk, K., Hertzum, M.: Technology acceptance and user experience: a review of the experiential component in HCI. ACM Trans. Comput.-Hum. Interact. (TOCHI) **24**(5), 1–30 (2017)

17. Inie, N., Barkhuus, L.: Developing evaluation metrics for active reading support. In: CSEDU (1), pp. 177–188 (2021)

18. Inie, N., Barkhuus, L., Brabrand, C.: Interacting with academic readings-a comparison of paper and laptop. Soc. Sci. Humanit. Open **4**(1), 100226 (2021)

19. Kawase, R., Herder, E., Nejdl, W.: A comparison of paper-based and online annotations in the workplace. In: Cress, U., Dimitrova, V., Specht, M. (eds.) EC-TEL 2009. LNCS, vol. 5794, pp. 240–253. Springer, Heidelberg (2009). https://doi.org/10.1007/978-3-642-04636-0_23

20. Kettanurak, V.N., Ramamurthy, K., Haseman, W.D.: User attitude as a mediator of learning performance improvement in an interactive multimedia environment: an empirical investigation of the degree of interactivity and learning styles. Int. J. Hum. Comput. Stud. **54**(4), 541–583 (2001)

21. Léger, P.M., An Nguyen, T., Charland, P., Sénécal, S., Lapierre, H.G., Fredette, M.: How learner experience and types of mobile applications influence performance: the case of digital annotation. Comput. Sch. **36**(2), 83–104 (2019)

22. Lim, E.L., Hew, K.F.: Students' perceptions of the usefulness of an E-book with annotative and sharing capabilities as a tool for learning: a case study. Innov. Educ. Teach. Int. **51**(1), 34–45 (2014)

23. Marshall, C.C.: Annotation: from paper books to the digital library. In: Proceedings of the Second ACM International Conference on Digital Libraries, pp. 131–140 (1997)

24. Mizrachi, D., Boustany, J., Kurbanoğlu, S., Doğan, G., Todorova, T., Vilar, P.: The academic reading format international study (ARFIS): investigating students around the world. In: Kurbanoğlu, S., et al. (eds.) ECIL 2016. CCIS, vol. 676, pp. 215–227. Springer, Cham (2016). https://doi.org/10.1007/978-3-319-52162-6_21

25. Mizrachi, D., Salaz, A.M., Kurbanoglu, S., Boustany, J., ARFIS Research Group: Academic reading format preferences and behaviors among university students worldwide: a comparative survey analysis. PLoS ONE **13**(5), e0197444 (2018)

26. Nakamura, J., Csikszentmihalyi, M.: Flow theory and research. Handb. Posit. Psychol. 195–206 (2009)

27. Nielsen, J.: How to conduct a heuristic evaluation. Nielsen Norman Group **1**, 1–8 (1995)

28. Paas, F., Tuovinen, J.E., Tabbers, H., Van Gerven, P.W.: Cognitive load measurement as a means to advance cognitive load theory. Educ. Psychol. **38**(1), 63–71 (2003)

29. Pálsdóttir, Á.: Advantages and disadvantages of printed and electronic study material: perspectives of university students. Inf. Res. **24**(2) (2019). http://InformationR.net/ir/24-2/paper828.html

30. Parsons, A.W., et al.: Upper elementary students' motivation to read fiction and nonfiction. Elem. Sch. J. **118**(3), 505–523 (2018)

31. Pearson, J., Buchanan, G., Thimbleby, H.: HCI design principles for eReaders. In: Proceedings of the Third Workshop on Research Advances in Large Digital Book Repositories and Complementary Media, pp. 15–24 (2010)

32. Pearson, J., Buchanan, G., Thimbleby, H.: Designing for digital reading. Synth. Lect. Inf. Concepts Retr. Serv. **5**(4), 1–135 (2013)

33. Pearson, J., Buchanan, G., Thimbleby, H., Jones, M.: The digital reading desk: a lightweight approach to digital note-taking. Interact. Comput. **24**(5), 327–338 (2012)

34. Pearson, J.S.: Investigating Lightweight Interaction for Active Reading in Digital Documents. Swansea University (United Kingdom) (2012)

35. Sage, K., Augustine, H., Shand, H., Bakner, K., Rayne, S.: Reading from print, computer, and tablet: equivalent learning in the digital age. Educ. Inf. Technol. **24**(4), 2477–2502 (2019). https://doi.org/10.1007/s10639-019-09887-2

36. Sage, K., Rausch, J., Quirk, A., Halladay, L.: Pacing, pixels, and paper: flexibility in learning words from flashcards. J. Inf. Technol. Educ. **15** (2016)

37. Sheppard, A.L., Wolffsohn, J.S.: Digital eye strain: prevalence, measurement and amelioration. BMJ Open Ophthalmol. **3**(1), e000146 (2018)

38. Shernoff, D.J., Csikszentmihalyi, M.: Cultivating engaged learners and optimal learning environments. Handb. Posit. Psychol. Sch. **131**, 145 (2009)

39. Shernoff, D.J., Csikszentmihalyi, M., Schneider, B., Shernoff, E.S.: Student engagement in high school classrooms from the perspective of flow theory. In: Csikszentmihalyi, M. (ed.) Applications of Flow in Human Development and Education, pp. 475–494. Springer, Dordrecht (2014). https://doi.org/10.1007/978-94-017-9094-9_24

40. Spichtig, A.N., Hiebert, E.H., Vorstius, C., Pascoe, J.P., David Pearson, P., Radach, R.: The decline of comprehension-based silent reading efficiency in the united states: a comparison of current data with performance in 1960. Read. Res. Q. **51**(2), 239–259 (2016)

41. Teo, H.H., Oh, L.B., Liu, C., Wei, K.K.: An empirical study of the effects of interactivity on web user attitude. Int. J. Hum. Comput. Stud. **58**(3), 281–305 (2003)
42. Tulving, E.: Memory and consciousness. Can. Psychol./Psychologie canadienne **26**(1), 1 (1985)
43. Zeng, Y., Bai, X., Xu, J., He, C.G.H.: The influence of e-book format and reading device on users' reading experience: a case study of graduate students. Publ. Res. Q. **32**(4), 319–330 (2016). https://doi.org/10.1007/s12109-016-9472-5

Enhancing the Quality and Student Attitude on Capstone Projects Through Activity-Based Learning and Reduced Social Loafing

Uthpala Samarakoon[1]([✉]), Kalpani Manathunga[1] [iD], and Asanthika Imbulpitiya[2] [iD]

[1] Faculty of Computing, Sri Lanka Institute of Information Technology, Malabe, Sri Lanka
{uthpala.s,kalpani.m}@sliit.lk
[2] School of Engineering, Computer and Mathematical Sciences, Auckland University of Technology, Auckland, New Zealand
asanthika.imbulpitiya@autuni.ac.nz

Abstract. Collaborative project work is inherently common in Undergraduate degree curricula. Yet many institutions face issues when implementing group work and assessing individuals to measure their skills gained due to diverse nature of groups, inconsistent knowledge levels and varied motivations. This chapter describes the integration of Activity-based learning into the development of an enhanced project module in an undergraduate degree in Information Technology. The learning design changed to incorporate several guidelines and mechanisms to improve the required skills in building a real-world solution for a problem at-hand. We report the results of applying this step-by-step approach when conducting the project module in consecutive two years of 2019 and 2020. Survey data collected from 120 students were analyzed to gain their perspective about the newly improved teaching-learning process. Further, the observations and experiences of the lecturers while conducting the session were also taken into consideration when analyzing the impact of the approach. The results indicated that introducing project-related in-class activities improves the student's enthusiasm towards the project module. Mechanisms for reducing social loafing in fact reduced the number of free-riding complaints from the students. When considering the overall qualitative feedback and quantitative measures it was evident that students preferred project-based learning compared to module-based learning and the newly enhanced approach managed to reduce many issues that occurred in conducting the project modules using a traditional approach.

Keywords: Project-based learning · Social loafing · Free riding · Student perspective · IT group projects · Activity-based learning

1 Introduction

Project-based learning (PBL) is a student-centered pedagogy where students acquire deeper knowledge through effective investigation of real-world challenges and problems. In PBL, learners are expected to attain essential expertise and abilities while

B. Csapó and J. Uhomoibhi (Eds.): CSEDU 2021, CCIS 1624, pp. 336–358, 2022.
https://doi.org/10.1007/978-3-031-14756-2_17

working on a project over a period. PBL is popular among IT undergraduates where students need to gain more technical skills through hands-on experience. Thus, most of the computing degree curricula include one or more PBL activities as group projects which help to enhance technical skills, collaborative learning, and soft skills such as teamwork, communication, leadership, and problem-solving [1–3].

Sri Lanka Institute of Information Technology (SLIIT) is a leading higher education institute which offers BSc in Information Technology degrees. In the curriculum of this degree program, "Information Technology Project" (ITP) is a module offered in the second year second semester, in which the learning outcomes are to strengthen technical skills and to inculcate soft skills like collaborative learning, communication among IT undergraduates. This module demands students to be engaged in a semester-long project in developing, managing a software product that may achieve industry goals to a certain extent by following appropriate software development methodologies. The complete process involves identifying the problem at-hand (a real-world problem given by an external client) and the expected project goals, gathering requirements from the client, preparation of project plans, design the solution and implementation while complying to a series of milestones during the process. Hence, the learning designing of the ITP module is entirely a project-based module in which students are assessed based on the course work and milestone achievements/completions throughout the entire semester. At the end of the semester, the student groups are expected to produce a fully functional software solution that meets the external client requirements. This ITP module intends to provide opportunities for students to apply knowledge and skills with a hands-on experience to improve their technical skills as well as real-world problem-solving skills.

Over the years, ITP module had been offered using diverse approaches and assessments which led the researchers to realize some inherent challenges in group projects-based modules and collaborative work. One problem is the lack of student passion and attention towards the lecture component of this module. Hence, the attendance for traditional face-to-face lecture sessions was reducing over the time during the semester leading only very few students to remain in the classrooms towards the latter part of the semester. Another major problem was social loafing or free riding in group projects. Social loafing is a behaviour where a particular student fails to contribute their fair share of effort when compared to the other students of a group [4]. This will lead to discouraging hardworking students in the group leading to unsuccessful group activities. Also, it was noticed that lack of student engagement in group activities due to disliking of traditional lecture delivery method for a hands-on module like ITP. Such problems were severely affecting the quality of the module as well as the students' performance.

Furthermore, due to Covid 19 pandemic, this module was shifted from a physical teaching-learning environment to a fully online environment in the year 2020 which introduced other challenges like no physical meetings between group members, internet, and other infrastructure issues during online synchronous meetings etc. Hence, the situation provided an opportunity for the researchers to investigate the appropriateness and successfulness of a project-based module in an online teaching-learning environment. Also wanted to identify which learning environment is most suitable (or preferred by the learners) for project-based learning modules.

Thus, the main intention of this research work is to study diverse mechanisms to enhance the quality of a project-based module by improving student enthusiasm during project-related lecture sessions and to reduce social loafing in group projects. Also, researchers investigated the most appropriate teaching-learning environment for project modules and the student perspective on project-based modules over the module-based learning to improve technical skills among IT undergraduates. Hence, this research holds four main objectives as given below:

- How students' passion could be enhanced during lecture sessions of project modules?
- How social loafing can be reduced in undergraduate group projects during an Information Technology undergraduate degree program?
- What is the most appropriate teaching-learning environment for group project-based modules as per student perspective?
- What is the impact of project-based learning over module-based learning to improve technical skills among IT undergraduates?

Reduction of social loafing and impact of project-based learning over the module-based learning was discussed in the previous research studies published by the authors [5, 6]. In this chapter, the authors present an overall learning design adopted to ITP module with a series of potential group activities that helped learners to trigger better collaborations and improve the learning experience. Moreover, this chapter reveals a framework utilized to reduce social loafing aspects and how students perceived these strategies and the overall satisfaction and learning experience students had over the cause of two consecutive academic years (2019 and 2020). The chapter is organized as below: authors present the background of the study, the methodology section revealing the details of the strategies adhered and the social loafing reducing framework followed by a section analyzing and discussing different results obtained (both qualitative and quantitative) and the concluding remarks.

2 Project Based Learning

Project-Based Learning (PBL) has become a broadly used method of teaching practical subjects. Most researchers agree that teaching with PBL has numerous advantages for students especially in the field of computer science such as applying their technical knowledge, acquiring practical skills in programming, getting involved in team processes, and understanding factors in project management [7]. Further, this paradigm has created a major shift in teaching and learning changing the traditional teacher-centered learning environment to a student-centric active one. Project-based learning is a comprehensive approach to teaching by involving students to find a solution for a prevalent real-world problem [8]. Many researchers across many disciplines highlight the importance of using project-based learning as an educational methodology to teach different subjects [8–12].

2.1 Project-Based Learning in CSEd

So far, project-based learning has been utilized as a major mechanism to teach different modules in Computer Science Education (CSEd) in a collaborative manner. Many experts have applied this method in teaching programming languages intending to give students hands-on experience to overcome the challenges in real-world problems.

As per recent research conducted by Sobral, many students like the strategy of learning programming through a real-world project. However, it was found that the average of responses was always decreasing as the semester progressed and several issues related to working groups were also identified based on the peer evaluation conducted at the end of the study. Even though, the author has been concluded that project-based learning is a mechanism to improve and develop the skills of motivated students while others have a lot of difficulty in following up [13].

Another study was conducted to compare the effectiveness of Team-Based Learning (TBL) over the Mixed Active Traditional (MATL) learning for an introductory programming course [14]. The authors have used two groups of novice undergraduates learning C programming following the same course content and textbook to identify the effectiveness of the two approaches mentioned above. Student feedback shows that students were satisfied with both the learning approaches nevertheless, they noticed that the performance of students following TBL was better than the ones who followed the MATL approach.

Likewise, other studies have investigated the success of using project-based learning as a comprehensive approach to teaching and learning in computing-related modules [15–17]. With the existing literature, it was evident that project-based learning was used extensively in CSEd as a mechanism to improve teaching and learning among many modules.

2.2 Project-Based Learning and Project Modules

According to Bell Project-based learning is an innovative technique for learning important strategies for success and acquiring skills needed to survive in the twenty-first century [18]. It encourages students to work cooperatively to produce projects that represent their knowledge and guides them towards self-learning through inquiry. Furthermore, project-based learning aids in the development of the technical skills of students while improving their other soft skills. Several researchers have discussed how project-based learning has been applied in the project modules among undergraduates, specifically in the capstone project. A capstone project is the concluding project of a course that requires prior coursework knowledge and various skills to be utilized in an aggregated manner.

A paper by McManus and Costello discussed the application of project-based learning in computer science education from the perspective of a student and the research advisor [19]. Overall, they have concluded that project-based learning is a suitable mechanism to improve students' critical thinking and several professional skills like time management, communication, etc. The authors of a similar study described the gradual transition from solving closed-ended problems to addressing real-world open-ended problems, as well as the project experience, which looked at the technical challenges and challenges of working collaboratively with students from different countries in an international capstone project [20].

Another study has evaluated the structural relationships among distinct factors that affect learner satisfaction and achievement in project-based learning [21]. They have analyzed data from 363 university students who were enrolled in a capstone project and concluded that problem-solving efficacy, teamwork competency, task authenticity, and satisfaction factors have a significant effect on the learning outcomes. In a recent study, the authors have designed a project-based learning strategy to improve the critical thinking skills of students which were run online with the new developments of the COVID-19 pandemic [22]. They have carried out an intervention with 834 students with five activities focusing on developing critical thinking skills. They have used a project-based learning strategy with both the controlled and experimental groups however, the experimental group was provided with scaffolding for a socially shared regulation process. Both the groups have performed well posttest however following the socially shared regulation process have shown further improvement. This recent study has proved that online project-based learning has a positive impact on critical thinking.

The work-in-progress study has described the experience in managing a project in the German Teaching Quality Pact, from 2011 to 2021 to improve the first-year teaching in STEM modules [23]. They have discussed the change management process used to sustain the impact of the project during the period. The evaluation has shown how project-based learning has made an impact on their overall education. Another study integrated PBL in an online environment and investigated critical issues, dynamics, and challenges related to PBL using 49students in an online course. As per the students' perspective, the online PBL course supported their professional development providing practical knowledge, enhanced project development skills, self-confidence, and research capability [24].

Similarly, this study explains the experience of conducting a project module for a set of undergraduate students following a degree in Information Technology. The ultimate motivation behind this study is to identify the student perspective on module-based learning and project-based learning. The next sections of the paper reveal the results section with an analysis of survey data carried out with undergraduate students and lecturers. Finally, an overall discussion is provided with detailed insights from the study followed by concluding remarks and future research directions.

3   Methodology

"Information Technology Project" (ITP) is a second-year, second-semester undergraduate project module offered for IT undergraduates which runs during a single semester. Figure 1 shows the flow diagram of the ITP module delivery.

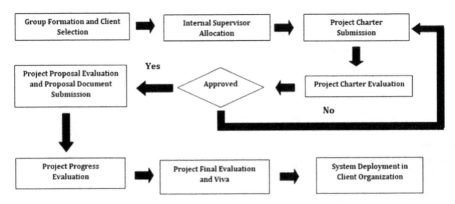

Fig. 1. Information Technology Project Process Flow.

At the initial phase of the "Information Technology Project" (ITP) module, students form a group of eight members and find an external client with specific business requirements that require a software solution. They have the freedom to select their peer members for the group, appoint a leader, and select the client project. The project requires to accommodate sufficient depth of work (i.e., project scope) to cover up the workload of eight members. Also, the students were instructed to follow the full software development life cycle (SDLC) [25] during the project. Each member has the responsibility to complete an entire function of the system (e.g., authentication sub system, report generation sub system, etc..) and should complete his/her function from user interfaces and visual aspects or front-end up to back-end where data processing happens. Here, the intention is to give each student the experience of full SDLC. Moreover, each function should contain all the create, read, update, and delete (CRUD) operations with meaningful reports via a reporting tool.

In the first stage, students need to form their project group and appoint a leader. Then the lecturer-in-charge assigns an internal supervisor for each group. After that, the group needs to find an external client who would provide them a problem that needs an IT-related solution. They have the freedom to select any government or private organization as the client such as a government school, bank, office, or a private organization such as a hotel/restaurant or a supermarket. To increase the authenticity of the scenario and to provide a real-world experience, the groups are given the entire responsibility of handling clients, meet them frequently and conduct meetings to gather requirements. Also, they must report the progress of the project to the client and need to get feedback frequently. After the requirement gathering phase, each project group needs to submit a project charter for their project which is a short description of the entire project including the objectives. Then the project charter will be evaluated through an internal panel. If the project charter gets approved, they can go to the next step, Proposal evaluation. If not, they have to re-submit an updated or new charter document.

As the next step, proposal presentations are conducted by a lecturer panel to evaluate the project objectives and the scope. Each member is evaluated individually as well as a group to check whether the group has identified the client requirements properly. Each student should have a proper understanding of the overall project and a broad idea about the function allocated for him/her. After the proposal presentation, the approved projects could directly start the project while other groups must do modifications to their project as per the panel recommendations before starting the project. The student groups have the freedom to use any suitable technological stack for the project as per the client's advice. At the end of the proposal presentation, the group should submit a project proposal report by including all the project-related information.

Next the project groups design, develop and test the project by applying the theoretical knowledge they gained during the first one and half years as an undergraduate. they are able to gain a hands-on experience on the previously studied modules. The project leader is advised to organize and have internal group meetings every week to monitor the contribution of each member and to meet the internal supervisor every fortnight to discuss the progress of the project. All the members are advised to participate in these project meetings and attendance is recorded to find out any absentees.

In the middle of the semester, a progress presentation is conducted to evaluate the progress. At the progress presentation, students are asked to complete at least 75% of the project and at this level, the panel evaluates the aspects such as business logic of the functions with appropriate validation, look and feel of the proposed system, usability, and proper navigation. This evaluation helps to direct the project group towards the desired goal and guide them back to the correct track if they have gone wrong. Hence, a progress presentation is extremely useful for the successful completion of a project.

At the end of the semester, the project group must complete the project and each project group participates in the final presentation followed by an individual viva session. Also, they must submit a final project report including every detail about their work.. At the final presentation, the completed project is evaluated by the panel and each member must present their individual function allocated. At the viva session, each member is questioned on programming-related aspects to verify the desired members' actual involvement/contribution in the project. During all these evaluations, the students are evaluated from both group and individual perspectives. In the assessment criteria, a larger weight is assigned for individual contribution to avoid free-riding or social loafing. The project final report is evaluated by the internal supervisor of the project and some marks out of the final marks are allocated for that as well. Finally, the project group can deploy the system in the client organization. Each evaluation has a prior allocated number of marks of the project and at the end of the semester, each student is awarded a total mark for their work.

3.1 Activity-Based Learning Through In-Class Activities

As a traditional module, ITP too had face-to-face lecture sessions on topics like SCRUM practices, interface designing, etc… Lecturers noticed a lack of student interest and attendance for such face-to-face sessions. As a solution, the researchers have introduced a set of project-based learning activities during lecture time. Those activities were handled as in-class activities and all the group members of the group needed to join the activity. All the activities were related to their respective project and all the group members were involved in the activities. One week before the lecture session, the lecturers were published all the instructions and materials they need to bring to do the in-class activities via the course web. Then, at the beginning of the lecture, the students were asked to get into their groups and the lecturer give a brief introduction to the activity. Then students were asked to start the activity and the lecturer was monitored their progress throughout the session. Also, the student queries were answered related to the activities. All the activities were designed in a manner that students have the freedom to come up with creative outputs. Meanwhile, attendance of each group was taken, and the students were asked to complete all the activities during the time of the lecture. Finally, they had to submit the output as a softcopy by taking photographs of the created output at the end of each session. Some marks were also allocated for the members involved in the activity to motivate students. The following table shows the activities introduced during the lecture time.

Table 1. Lecture related activities.

Activity	The intention of the activity
SCRUM activity	Introducing SCRUM framework
Interface designing activity	Help students to develop logical interfaces
Database activity	Help students to finalize the database of the project
User testing activity	Help students to test the system

SCRUM Activity. In the SCRUM-related activity, each member of the project group was asked to identify and write user stories of the main function that was assigned for him/her. Then they had to generate a presentation by adding those user stories and other related information of their project (Fig. 2). Finally, each group was asked to present their project presentation to the rest of the groups. This activity was conducted during the second week of the semester, and this was helped students to get to know about what kind of projects doing with their colleagues. Also, they got the chance to share their knowledge and experiences with others. There was a Q and A session at the end of each presentation. Project groups were given the freedom to select their templates for their presentation and to make the presentation unique by adding creative ideas. This activity also helped them to improve their communication skills and time management. At the end of the session, each group was submitted their final presentation to the course web through the given submission link.

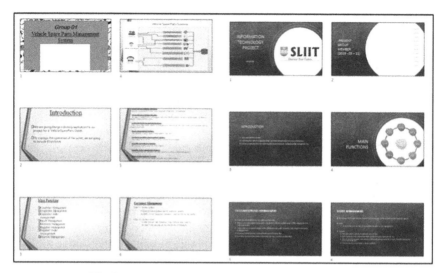

Fig. 2. Sample outputs produced for "SCRUM" activity.

Interface Designing Activity. This activity was designed to guide the students to come up with logical interfaces. Like the other activities, students were informed about the one week in advance via the course web. The students were given the chance to design their interfaces as a paper prototype or sketch the interfaces using paper and pencil. So, the students were instructed to bring the necessary materials if they plan to make paper prototypes. Then as in the previous activity, the lecturer was given instructions to create logical interfaces. Then the student groups were given the freedom to come up with interfaces for their proposed project (Fig. 3).

Fig. 3. Project groups involving in "interface designing" in-class activity [5].

Each member was asked to sketch interfaces for his/her function. Finally, they had to compile a pdf document by combining the photographs of interfaces developed by each member and submit via the submission link (Fig. 4).

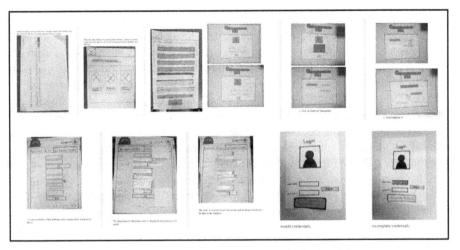

Fig. 4. Sample outputs of "interface designing" activity.

Database Activity. This activity was aimed at designing and developing the ER diagram and designing the structure of the database of the group project. Students were expected to design and draw the ER diagram, write schemas for the ER diagram, normalize the schemas, and finally draw a database diagram for their project database. As in the previous activities, student groups were asked to submit the final document via the course web link at the end of the session (Fig. 5).

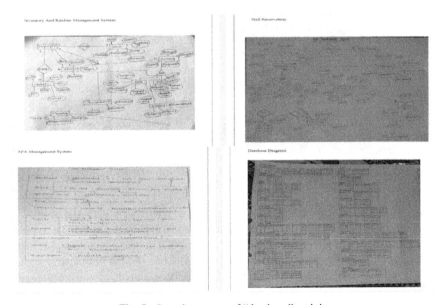

Fig. 5. Sample outputs of "database" activity.

User Testing Activity. This activity was conducted as the last activity at the latter part of the project when they had almost completed the system. To do this activity, each group was asked to bring the working system of their project running on a laptop/s. Then a printed testing template was provided to each group, and they had to complete the fields in the given testing template such as project ID, project title, and names of the functions to be tested. Then each group was assigned a random peer group to evaluate the system. Then the testing template and the system were handed over to the peer testing group for evaluation. The peer testing group was asked to test each function mentioned in the sheet thoroughly under aspects such as interface logic, validation with proper error messages, DB connectivity, user-friendliness of the system, proper navigation, and overall completeness of each function and the system. Then the peer testing group was asked to complete the test template with their comments for improvements. The sheet was handed over to the system owner group and finally, all the groups were received the user testing feedback for their respective system. At last, each group was asked to take images of the filled Test Template and upload them to respective links in the course web at the end of the session (Fig. 6).

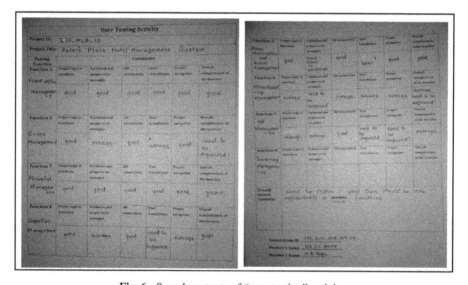

Fig. 6. Sample outputs of "user testing" activity.

Data related to student perspectives on in-class activities were gathered in two consecutive years 2019 and 2020. In 2019 students were involved in the in-class activities in a physical face-to-face learning environment and in 2020 they conducted in-class activities in an online learning setting. At the end of the ITP module in both years, students were asked to comment on their perspectives, experiences, and ideas on lecture sessions with in-class activities compared to traditional lecture sessions. In 2019 students were asked to give their comments written on a piece of paper and in 2020 a google form was shared among students. The observations and experiences of the lecturers while conducting in-class activities were also taken into consideration to analyze the successfulness of activity-based learning in project-based modules.

3.2 Mechanisms to Reduce Freeriding/Social Loafing

Considering the masses in the classes (average 120 to 200 in a batch, per year), social loafing was identified as a critical factor over the years, especially in the group projects. The main intention of this study is to explore the effectiveness of a set of mechanisms proposed to reduce social loafing among group members. Moreover, to identify which aspects are the most effective in controlling free riding as per the student's perspective.

A student group with 140 students taking Information Technology Project as a module during their second year second semester was selected for this study. The students were asked to group into 5–8 member groups of their own and find a client for their project as of their interest. Twelve distinct approaches to minimize social loafing were introduced based on the past experiences of social loafing incidents that were identified among group members during previous academic years. Then the approaches were developed to implement those mechanisms and the students were given all the instructions about new evaluation approaches, scaffolding mechanisms around the module that they must follow, and task distribution among members at the beginning of the module. Figure 1 shows how students were engaged in completing project activities collaboratively within groups. The selected approaches and scaffolding mechanisms were experimented with during one semester period.

At the end of the project, a questionnaire was distributed as an optional submission. The questionnaire included five-point Likert scale questions where each participant expressed how much they agree or disagree with a particular mechanism. 50 responses representing 18 project groups were received and used in the analysis. Moreover, lecturer experience was considered also to evaluate the outcomes of these twelve approaches. Table 1 shows the twelve different mechanisms that were experimented with for reducing social loafing among members of IT group projects.

Implementation of Mechanisms. These mechanisms were explained in detail in the previous research studies published by the same authors [5, 6]. Following table (Table 2) shows a summary of mechanisms used and the effect of each mechanism to reduce social loafing.

All required instructions related to project work and evaluation criteria were given to all the students at the beginning of the project. At the end of the project, the student's experience and their perspective on the mechanisms used for reducing free-riding were collected via a questionnaire. The students' scores were taken for each mechanism used and scoring was based on a five-point Likert scale ranging from 'strongly disagree' to 'strongly agree'. 50 students answered the questionnaire, and the answers were analyzed to get a better understanding of the most effective mechanisms according to the student perspective. Also, the number of complaints against the free-riding effect was logged and compared with the previous years. Finally, lecturers' perspectives and experiences were gathered using interviews and discussion to come up with a final decision.

Table 2. Twelve mechanisms used to reduce social loafing in IT group projects [5, 6].

No	Mechanism	Influence in reducing social loafing
M1	Allowing students to select members for their group by themselves reduce the ability of free riding	By selecting members with similar interests and values as their peers may lead to reduction of free riding
M2	Allowing students to select client/project by the group which interests them more rather than assigned by the lecturer reduce the free riding	The expectation was that when students engage in something of their interest as a group, the members may not try to free-ride and give their fullest support to succeed
M3	Maintaining a moderate group size (not too large groups) reduce the ability of free riding	Too small groups may increase the frustration of members, since they must be responsible for huge workload of the project while too large groups would support free riding too, since everyone has very small responsibility. Therefore, moderate group size (5–8 members) will reduce social loafing attempt by students
M4	Assign individual functionalities for each member and give whole responsibility of that component reduce the ability of free riding	This method makes it difficult for a member to free ride because absence of that component of the project will be clearly visible and the whole responsibility lies with that member
M5	Assign similar responsibilities (responsibility of entire unit from design to testing) to all members to reduce the ability of free riding	When each member of the group was given similar responsibilities where each member must complete from front-end and to back-end of the respective function, student may not tend to compare their workload with peers and get frustrated. Hence, less likely to free-ride in the projects
M6	Assess the individual contribution of each member in evaluations reduces the ability of free riding	Group members face a series of evaluations throughout the semester (e.g., initial product proposal, prototype stage, final product presentation). Such evaluations are assessed both individually and as a group with more weight to individual contributions in the assessment rubric. Hence, they find it difficult to free-ride within the project
M7	Checking the overall understanding of each member about the project reduces the ability of free riding	The intention was to check their overall understanding about the project apart from their individual component, which might be a good indicator to identify free riders

(continued)

Table 2. (*continued*)

No	Mechanism	Influence in reducing social loafing
M8	Conducting individual viva session reduce the ability of free riding	At the end of the presentation, each member was asked product implementation related questions from their software program to assess whether they have the required technical knowledge that they claim to have in their respective portion allocated for them. The members, who failed to explain most portions with related to their own code, were noted as presenting the work done by someone else after providing several opportunities to explain themselves. This might influence to reduce social loafing
M9	Checking individual contributions in document preparation reduces the ability of free riding	Students were asked to mention a sub section explaining the individual contributions of group members when preparing various project documents like project proposal, progress reports, final report etc. and the group leaders were advised to equally distribute document. Allocating similar workload in document preparation may also leads to reduce social loafing
M10	Regular group meetings with the supervisor and marking attendance reduce the ability of free riding	Members with less attendance were asked to explain the reasons and were evaluated thoroughly in evaluations. This may avoid students from free-riding in projects
M11	Peer review (all students grade the contribution of other members in the group confidentially) reduces the ability of free riding	Students were asked to grade their colleagues in the group confidentially via a Google form. The responses and comments given by peers were used to identify free riders, if any
M12	Lecturer involvement and supervision in task distribution and group communication when there are conflicts within the group reduce the ability of free riding	If any group conflicts were identified during the period of project, lecturers were closely monitoring the group and all the formal communication among group members were done under the guidance and supervision of the lecturer. This method may force stop free-riding by the group members

In 2020, students have conducted project activities in an online teaching-learning environment via the Microsoft Teams platform. Hence, an online questionnaire was distributed to get their perspective on free-riding online learning environments. 72 students were responded to the questionnaire.

3.3 Identifying Most Suitable Teaching-Learning Environment for Project-Based Modules

Until 2019, the project-based activities of the ITP module were conducted in a physical face-to-face teaching-learning environment. Due to Covid 19 pandemic in 2020, the ITP module was shifted to an online teaching-learning environment where all the lecture activities and project activities were conducted using online platforms. Microsoft Teams environment was used as the main online platform to conduct project evaluations and viva. Other than that, in both years the learning designing and the ITP curriculum including all the guidelines, instructions, and evaluations were the same for the entire project.

The in-class activities were published on the course web with pre-recorded instruction lectures and students were asked to complete the activities offline and submit the output via the course web submission links. They were given more time to submit the activities since the student group members need to communicate via online methods. Also, only the first three activities SCRUM, Interface designing, and database diagram activities were conducted under an online environment. User testing activity was disregarded due to the inability of conducting the activity using online methods…

In 2020, at the end of the second semester, an online questionnaire was shared with students to get their perspective on performing project activities such as evaluations and group discussions in an online teaching-learning environment rather than in a physical environment. Students were asked to rate their perspective on a five-point Likert scale and to comment on their experience as an open-ended question. Sixty students out of 120 students were responded to the questionnaire.

3.4 Student Perspective on Importance of Project Modules to Improve Technical Skills of IT Undergraduates

In 2019, students conducted the project activities in physical face-to-face classroom environments, and in 2020 they conducted the project in an emergency remote teaching-learning environment. At the end of the ITP module in 2019 and 2020, a questionnaire was distributed among students who completed the project to identify their experience and perspective on project-based learning over the module-based learning. In 2019 a printed questionnaire (physical form) was distributed among students and in 2020, a google form was used to gather student perspectives. Thus, researchers were able to explore whether there is any change in the student perspective on project-based modules under online education than physical in-class education. In 2019, 55 students out of 15 groups responded to the printed questionnaire and in 2020, 70 students were responded to the online questionnaire. Then, the obtained responses were analyzed from both groups to identify the student perspective on the successfulness of project-based learning over the module-based learning to enhance technical skills among IT, undergraduates.

4   Results and Discussion

4.1   Impact on Activity-Based Learning in Enhancing Students' Passion and Attendance in Project Related Lecture Sessions

As per observations, introducing project-related in-class activities was highly influenced to improve the students' passion for project module lectures as well as student attendance. The student attendance was increased by more than 85% and it was a huge increase compared to traditional lectures. Also, the lecturers observed that students were engaged in in-class activities very enthusiastically. As per lecturers, students were able to share their knowledge among colleagues and it was a good opportunity for low-performing students to improve their weaknesses. Also, the student got the chance to improve their soft skills such as communication skills, time management, teamwork, leadership, and creative thinking via in-class activities.

As per students' comments in 2019, they preferred activity-based lectures to traditional theoretical lecture sessions. As the reason, some students have mentioned that "since we designed our interfaces and database as an in-class activity, it reduced our homework and saved out time. Also, we were able to use those documents throughout the project design and implementation…". Another comment was that " in-class activities helped us to share our knowledge and identify our mistakes…". Some students were asked to add more in-class activities to write project reports as well.

Even in 2020, under the online teaching-learning environment students recommend having in-class activities rather than having traditional lectures. Anyway, most of the students (66.7%) preferred to involve with in-class activities in a physical environment rather than performing activities online (Fig. 7). The reason may be lack of infrastructure, poor internet connections, the inability of knowledge sharing, and frustration due to not meeting their peers.

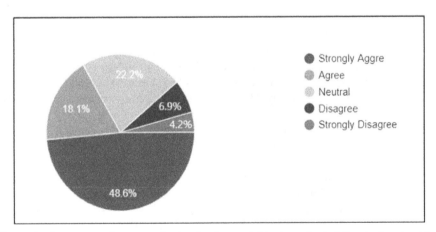

Fig. 7. Student perspective on conducting in-class activities in a physical face to face teaching-learning environment

While considering all the facts it is evident that having activity-based lectures in project modules is much better than having traditional lectures. Further, in-class activities help students for knowledge sharing and minimize mistakes by learning from peers.

4.2 Reducing Social Loafing in Group Projects

Students' ratings for each mechanism to reduce free-riding were analyzed to get the overall idea of students' perspectives on social loafing and reducing mechanisms.

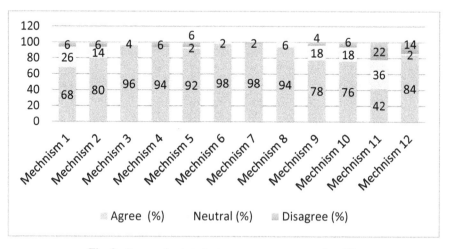

Fig. 8. Summarized student responses on approaches [5].

Figure 8 shows the summarized results of students' perspectives on approaches used. Here both agreed and strongly agreed responses were concatenated as the students confirmed that the proposed approach is successful and the same goes for disagreeing and strongly disagreeing. Neutral responses were considered separately.

According to the results, both M6 and M7 have the highest (98%) approval by students for reducing social loafing. Then students had selected M3 (96%), M4 and M8 (94%), and M5 (92%) respectively. The next highest agreed approaches are M12 (84%) and M2 (80%). Mechanisms 9 and 10 had moderate approval rates as 78% and 76%. Mechanism 1 has a considerably low percentage (68%) compared to others. The lowest approval is for mechanism 11 which is 42%.

When considering disagreed percentages, mechanism 11 (22%), and mechanism 12 (14%) have higher disagreed percentages than other approaches. All the other percentages are low than 10%. Out of those M1, M2, M4, M5, and M10 have a similar disagreed percentage of 6%. M7 and M9 have low percentages of 2% and 4%. None of the students disagreed for M3, M6, and M8 where the percentage indicated as 0% and is the lowest.

According to the lecturers' experience, the number of free-riding complaints was reduced after applying the aforementioned mechanisms. They found only two free-riding complaints from two groups and those were too under control after close monitoring. Also, they were able to identify free riders as early as possible which gave them the chance to take corrective actions like issuing warnings and close monitoring, etc. As another approach, they suggested not to mix regular students who take the module for the first time with those who are repeating the module. They have seen that, in most of the situations, repeat students in the group maintain minimal communication and involvement in the work with other members and try to free ride. From past experiences, it was identified that most of the free-riding complaints were noticed in these mixed groups. But one student gave an opposite idea on this matter in the survey where he/she suggested not to group only repeating students together. Sometimes most of the members in repeating groups may not work and the hardworking students in those groups may find it difficult to carry out the work. So that approach may have both pros and cons.

4.3 Most Suitable Teaching-Learning Environment for Project-Based Modules

From per student perspective, most of the students (73.6%) agreed with the fact that conducting project evaluations and group discussions in a physical face-to-face teaching-learning environment is better than conducting project activities in an online teaching-learning environment (Fig. 9).

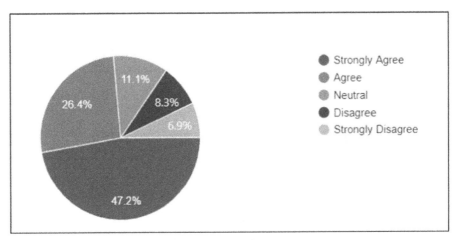

Fig. 9. Student perspective on conducting project evaluations and group discussions in a physical face to face teaching-learning environment.

Since Sri Lanka is a low-resource country, some students may find it difficult to involve in project activities properly due to the lack of technological infrastructure and poor internet connections they have at home. Additionally, the mental stress of not being able to meet their peers and share their experiences during the Covid19 pandemic may also severely affect the students to gain the actual advantage from the project. Hence, they may perceive physical settings as better for project-based learning modules. On the other hand, there are around 28% of students who neutral or disagreed with the above fact. Sometimes those students may not have any problems related to internet resources or infrastructure at home or may be able to complete their project successfully without any burden.

As per the challenges in conducting project activities in an online environment, the students mentioned that ineffective group communication, difficulties in time management, lack of technological infrastructure, poor internet connections, power failures, lack of knowledge sharing, lack of encouragement and motivation, lack of bond between peer group members, unavailability of group members at the same time and lack of feedback. Consequently, as per the experiences of students, it was evident that the physical learning environment is most suitable for project-based modules.

4.4 Importance of Project Modules to Improve Technical Skills of IT Undergraduates

Following Fig. 8, shows the student perspective summary for both years 2019 and 2020. This clearly shows that most of the students agreed with the fact that their technical skills and programming knowledge improved after following the IT project module. Subsequently, many of them also agreed with the fact that project-based learning is better than module-based learning to improve their programming knowledge (Fig. 10).

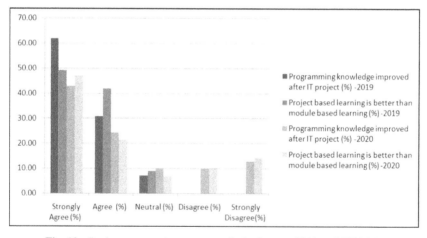

Fig. 10. Student perspective summary for both years 2019 and 2020 [6].

In addition to the above facts, the students were also asked to comment on the extra skills that they have gained through the IT project in the two open-ended questions in the questionnaire. Most of the students were mentioned teamwork, time management, leadership as additional skills they gained through the project which is an indication of students' perspective towards positive outcomes of project-based learning.

Furthermore, some students also mentioned this project module as their most favorite module which shows the success of project-based learning in undergraduate education. Moreover, several students suggested including more project-based modules in the curriculum in the future. In Addition, some students commented on their experience in the IT project module. One student mentioned that "it was an amazing experience for me since we built a fully functional system as professionals". Another student's idea was that "it was a fantastic project experience and I agree that the experience gained from the IT Project will help us with our final year projects as well". Another comment was that "doing the ITP Project helped me a lot for improve my programming skills, thinking ability and working with the team in a client project." A Few students have given some suggestions too, to improve the module such as adding more knowledge on Agile development and to reduce the group size.

According to the responses for the online questionnaire in 2020, 45.8% of the students agreed that free-riding attempts were identified in their project groups under an online environment. As the reasons, they have mentioned that, lack of knowledge sharing, inability to contact peer group members on time, lack of technology infrastructure, difficulties in monitoring team members by the leaders, poor project inspections, easy access to online resources, and loopholes in online evaluations (Fig. 11).

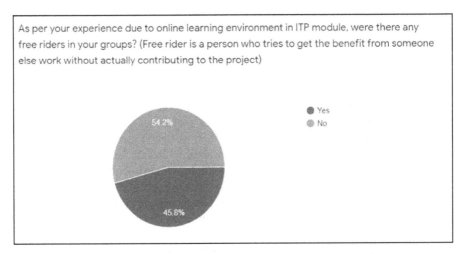

Fig. 11. Student perspective on free-riding attempts in an online environment.

On the other hand, 54.2% of students did not agree with the above fact. According to them, free-riding attempts were not identified in their groups in an online environment and they have given reasons for their ideas as well. As per their ideas and experiences, free riding is a common problem in both physical and online learning environments. So, they perceive online learning as an advantage giving them more free time for studies. Also, they found it easy to communicate via online platforms such as Whatsapp, Zoom, and Microsoft Teams platform. On the other hand, twelve mechanisms for preventing free-riding were also applied in the year 2020 in an online environment. So that may be another reason for the reduction of free-riding attempts as per students. Anyway, it seems that the free-riding attempts in 2020 were a bit higher than in 2019. That may be due to limitations in student monitoring in an online environment.

While considering all the facts it is evident that project-based learning allows students to gain more practical exposure and improve technical abilities than a module that is purely based on heavy theoretical delivery of content. So the research team would like to suggest including more project-based modules specifically in the IT undergraduate curriculum and also to include at least some project-based assignments in theory-based programming modules.

An overall, the study shows that students preferred project-based learning over module-based learning. Hence, we can improve the quality of project-based learning by introducing activity-based learning with project-related in-class activities. This will lead to high student satisfaction and attendance for lectures. Social leafing is a major challenge in group base projects. Hence, we need proper mechanisms to reduce social loafing. Further, as per student perspective, physical learning environment is more suitable for project-based modules rather than online learning environments.

5 Conclusion

This study shows the utility of applying activity-based learning to the redesign of a project-based module over two cohorts of undergraduate students enrolled in an Information Technology degree program. We were interested in finding different approaches to overcome the inherent challenges in group projects-based modules and collaborative work to redesign the delivery of the project-based module to increase the enthusiasm of the students towards the module. The main objective of the study was to study the previously introduced diverse mechanisms to enhance the quality of the project-based module by improving student enthusiasm during project-related lecture sessions and to reduce social loafing in group projects. Further, we investigated the most appropriate teaching-learning environment for project modules and the student perspective on project-based modules over the module-based learning to improve technical skills among IT undergraduates.

The paper presents a detailed discussion about various activities implemented during the delivery of the module along with the most effective mechanisms for reducing free riding. We discovered that introducing project-related in-class activities was highly influenced to improve the students' enthusiasm towards the project module where the student attendance in conducted lecture sessions got increased by 85% compared to the previous cohorts. Considering the students response, it was evident that having activity-based lectures in project modules is much better than having traditional lectures. Majority of the students were agreed with the introduced mechanisms to reduce free riding beside having peer reviews, which was quite an interesting fact to consider in the future. Furthermore, the lecturer experience was also similar to the students where they agreed with the provided mechanisms and having a smaller number of complaints compared to the previous cohorts was the evidence to prove that it was a success in applying the mechanisms in assessing the projects. Further, we identified that majority of the students preferred having a face-to-face learning environment in doing the project-based module. However, this was an actual challenge in delivering the module, as face-to-face learning was moved to online learning in 2020 due to the wide spread of COVID-19 pandemic. Also, the introduced user testing activity was not conducted during 2020 with the move to online learning which can be considered as a limitation we came across while adapting the methodology of the study with the situation.

The evidence presented showed that project-based learning allows students to gain more practical exposure and improve technical abilities than a module that is purely based on heavy theoretical delivery of content.In conclusion, we recommend this approach to be utilized in project related modules to overcome the common challenges in delivering project-based module. Further, we highlight the need of applying the approach fully in an online environment to understand how to adapt them accordingly.

Acknowledgement. This study was conducted at Sri Lanka Institute of Information Technology, Sri Lanka and we would like to express our gratitude to the management, all the lecturers, non-academic staff, and especially the students who have been involved in this study.

References

1. Woodward, B.S., Sendall, P., Ceccucci, W.: Integrating soft skill competencies through project-based learning across the information systems curriculum. Inf. Syst. Educ. J. **8**(8), n8 (2010)
2. Jun, H.: Improving undergraduates' teamwork skills by adapting project-based learning methodology. In: 2010 5th International Conference on Computer Science Education, pp. 652–655. IEEE, August 2010
3. Vogler, J.S., Thompson, P., Davis, D.W., Mayfield, B.E., Finley, P.M., Yasseri, D.: The hard work of soft skills: augmenting the project-based learning experience with interdisciplinary teamwork. Instr. Sci. **46**(3), 457–488 (2018)
4. Aggarwal, P., O'Brien, C.L.: Social loafing on group projects: structural antecedents and effect on student satisfaction. J. Mark. Educ. **30**(3), 255–264 (2008). https://doi.org/10.1177/0273475308322283
5. Samarakoon, U., Imbulpitiya, A., Manathunga, K.: Say no to free riding: student perspective on mechanisms to reduce social loafing in group projects. In: CSEDU (1), pp. 198–206 (2021)
6. Samarakoon, U., Manathunga, K., Imbulpitiya, A.: Theory or practice: student perspective on project based learning versus module based learning to improve technical skills among it undergraduates. In: Auer, M.E., Hortsch, H., Michler, O., Köhler, T. (eds.) ICL 2021. LNNS, vol. 390, pp. 968–979. Springer, Cham (2022). https://doi.org/10.1007/978-3-030-93907-6_103
7. Pucher, R., Lehner, M.: Project based learning in computer science–a review of more than 500 projects. Proc. Soc. Behav. Sci. **29**, 1561–1566 (2011)
8. Kanigolla, D., Corns, S.M., Cudney, E.A.: Enhancing engineering education using project-based learning for lean and six sigma. Int. J. Lean Six Sigma **5**(1), 46–51 (2014)
9. Schwering, R.E.: Optimizing learning in project-based capstone courses. Acad. Educ. Leadersh. J. **19**(1), 90–104 (2015)
10. Blumenfeld, P.C., Soloway, E., Marx, R.W., Krajcik, J.S., Guzdial, M., Palincsar, A.: Motivating project-based learning: sustaining the doing, supporting the learning. Educ. Psychol. **26**(3–4), 369–398 (1991)
11. Rohm, A.J., Stefl, M., Ward, N.: Future proof and real-world ready: the role of live project-based learning in students' skill development. J. Mark. Educ. **00**, 1–12 (2021)
12. Bilgin, I., Karakuyu, Y., Ay, Y.: The effects of project-based learning on undergraduate students' achievement and self-efficacy beliefs towards science teaching. Eurasia J. Math. Sci. Technol. Educ. **11**(3), 469–477 (2015)
13. Sobral, S.R.: Project based learning with peer assessment in a introductory programming course. Int. J. Inf. Educ. Technol. **11**(7), 337–341 (2021)

14. Kesterson, C., Joyce, C.L., Ahamed, M.M., Selim, M.Y.: Mixed active-traditional learning versus team-based learning: a comparative study for a Freshman programming course. In: EDULEARN21: 13th Annual International Conference on Education and New Learning Technologies Proceedings, pp. 1–2. IATED, Spain (2021)
15. Dietrich, S.W., Urban, S.D.: Database theory in practice: learning from cooperative group projects. SIGCSE Bull. **28**(1), 112–116 (1996)
16. Podeschi, R.J.: Building I.S. professionals through a real-world client project in a database application development course. Inf. Syst. Educ. J. **14**(6), 34–40 (2016)
17. Scorce, R.: Perspectives concerning the utilization of service learning projects for a computer science course. J. Comput. Sci. Coll. **25**(3), 75–81 (2010)
18. Bell, S.: Project-based learning for the 21st century: skills for the future. Clear. House: J. Educ. Strateg. Issues Ideas **83**(2), 39–43 (2010)
19. McManus, J.W., Costello, P.J.: Project based learning in computer science: a student and research advisor's perspective. J. Comput. Sci. Coll. **34**(3), 38–46 (2018)
20. Sanger, P.A., Ziyatdinova, J.: Project based learning: real world experiential projects creating the 21st century engineer. In: 2014 International Conference on Interactive Collaborative Learning (ICL), pp. 541–544. IEEE, United Arab Emirates (2014)
21. Cortázar, C., et al.: Promoting critical thinking in an online, project-based course. Comput. Hum. Behav. **119**, 1–18 (2021)
22. Young, J.J., So, Y.L.: Project-based learning in capstone design courses for engineering students: factors affecting outcomes. Issues Educ. Res. **29**(1), 123–140 (2019)
23. Winzker, M., Bounif, N., Luppertz, C.: Improving first-year teaching with project-based learning and support of STEM modules. In: Proceedings Frontiers in Education Conference, pp. 1–3. IEEE, Sweden (2020)
24. Yukselturk, E., Baturay, M.H.: Online project-based learning: students' views, concerns and suggestions. In: Student Satisfaction and Learning Outcomes in E-Learning: An Introduction to Empirical Research, pp. 357–374. IGI Global (2011)
25. Scroggins, R.: SDLC and development methodologies. Glob. J. Comput. Sci. Technol. (2014)

Problem-Based Multiple Response Exams for Students with and Without Learning Difficulties

Panos Photopoulos[1]([✉]) [ID], Christos Tsonos[2] [ID], Ilias Stavrakas[1] [ID],
and Dimos Triantis[1] [ID]

[1] Department of Electrical and Electronic Engineering, University of West Attica, Athens,
Greece
pphotopoulos@uniwa.gr
[2] Department of Physics, University of Thessaly, Lamia, Greece

Abstract. Objective computer-assisted examinations (CAA) are considered a
preferable option compared to constructed response (CR) ones because marking is done automatically without the intervention of the examiner. This publication compares the attitudes and perceptions of a sample of engineering students
towards a specific objective examination format designed to assess the students'
proficiency to solve electronics problems. Data were collected using a 15-item
questionnaire which included a free text question. Overall the students expressed
a preference for the objective-type examination format. The students who self-
reported to face learning difficulties (LD) were equally divided between the two
examination formats. Their examination format preference was determined by the
details of their learning difficulties, indicating that none of the two assessment
formats effectively solves the assessment question for these students. For the rest
of the respondents, examination format preference was accompanied by opposing views regarding answering by guessing, having the opportunity to express
their views, selecting instead of constructing an answer, having the opportunity to
demonstrate their knowledge, and having control of the exam answers.

Keywords: Multiple choice · Constructed response · Learning disabilities ·
Power relations · Managerialism

1 Introduction

The administration of objective examinations using a computer system increases the
scoring process's efficiency and objectivity [1–4]. However, answering by guessing is
a weak point regarding the reliability of objective examinations [2, 3]. Uncued and
extended matching tests [5, 6], two-tied multiple-choice (MC) tests [7], adaptive MC
tests [2], and weighted scoring of objective tests [3] are promising attempts to enhance
objective assessment reliability. Problem-solving in physics and engineering does not
result merely from applying a specific theory and requires multiple skills [8, 9]. It is
a complicated and time-consuming process [10]. Problem-based questions need much

© Springer Nature Switzerland AG 2022
B. Csapó and J. Uhomoibhi (Eds.): CSEDU 2021, CCIS 1624, pp. 359–377, 2022.
https://doi.org/10.1007/978-3-031-14756-2_18

more time to be answered compared to single-answer MC questions. Therefore, given fatigue and time constraints, an objective examination designed to assess the students' problem-solving proficiency can include a small number of problems to be answered. For this reason, turning problems into objective-type questions is a challenge for engineering teachers.

Introducing a new examination format requires information concerning its administrative efficiency, suitability to the course's objectives, reliability, and students' acceptance [11]. The questions of fairness, easiness, anxiety, guessing, the expectancy of success, and preparation for the exams [11, 12] are usually included in surveys recording the students' perceptions of the various assessment types. However, past qualitative research findings have focused attention on issues of power and control [13]. Do objective exams disempower the students? Would they perform better had they been allowed to express their view in an essay examination? Do they feel disempowered not having control of their answers? [13]. The increasing number of university students [14, 15] and the underfinance of Higher education [16, 17] are two of the reasons academic teachers feel obligated to adopt efficient solutions in teaching [18], such as CAA [1]. The efficiency imperative overlooks the specific learning needs of the students who face learning difficulties (LD) [19]. Accommodation models view LDs as a problem to be fixed within the competitive, efficiency-oriented university [20–22]. Adopting inclusive practices can enhance all the students' feelings of belonging, including LD ones [23].

2   Theoretical Context

2.1   The Student, the Teacher and the Efficient University

Public investment in Higher Education has been significantly reduced after 2008, while the number of students in full-time education has dramatically increased [24–26]. The demand for high-quality education with fewer resources is a challenge for academic teachers to achieve more with less [17, 27, 28]. Misconceptions about the political role of technology give rise to a constant search for technological fixes to emerging quality and efficiency problems [29].

The wide adoption of CAA, usually in conjunction with the MC format, has been attributed, among other reasons, to its cost-effectiveness [1, 30–33]. Online and blended courses are probable scenarios of future education concerning the various teaching modalities because of their flexibility, ease of access, and cost-effectiveness [34–36]. Efficiency and quality are the two promises of the "marketized" university, although efficiency improvements, sometimes, degrade the point and purpose of higher education. Temporary and part-time academic staff, online learning, and CAA reduce costs, but they also create new problems [28].

Back in 1979, Henry Mintzberg described universities as examples of what he called professional bureaucracies. The academic professors derived their power from the scarcity of their skills and complexity of their work which could not be supervised by managers or standardized by analysts. The professors enjoyed mobility, which enabled them to insist on considerable autonomy in their work [37]. Nowadays, not so many professors can see themselves in Mintzberg's description of academic professionals. The number of Ph.D. holders has increased. Temporary and part-time academic staff play a central role in the modern university [28], which means that the services of the professors are not rare anymore. The audit culture, which is supposed to measure good teaching and research quality, has shifted the authority from academics to the managers [28]. Moreover, the academic teachers have accepted the imperative of efficiency as a legitimate reason to accommodate the limited resources with oversized classes, and professionalism, reminiscence, and a remnant of Mintzberg's analysis. The managerial paradigm removes authority from the teachers and reduces their status to that of teaching assistants [27].

Assessment embodies power relations between institutions, teachers, and students [38]. MC exams disempower the students because they are not allowed to express their answers in their own words. They demonstrate their proficiency to the subject matter assessed, not by building an argument or making original interpretations but by selecting the correct answer, among those predetermined by the tutor. By limiting the choices, MC exams disturb the power relation between the student and the examiner in favor of the latter [13]. MC tests affect the originality of student's answers and obscure student's actual knowledge. Even when the students select the correct answers, the teacher remains uncertain about the student's knowledge and understanding.

Although a solution to the significant amount of time required to grade CR exam papers [39], MC exams have raised worries for guesswork [1, 12] and rote learning. When the examination is of objective format, teachers do not spend time marking exam papers but making the assessment valid and reliable [40]. The pressures for efficient operation despite limited inputs disempower both teachers and students. Although MC exams assess a wide range of learning objectives, they are also accommodations for efficiency pressures despite reduced resources [18]. Therefore, a view that considers power relations in the universities to be exhausted in the teacher-student power balance is inadequate. A three actors power relationship that considers the power transfer from the students and the teachers to management is more suitable [41].

2.2 Students' Acceptance of Objective Examinations

The objective type of examinations has been studied in terms of fairness, easiness, anxiety invoking, and performance [39, 42, 43]. In one of the first publications, Zeidner [11] studied the attitudes of two groups of students toward MC and CR exams. The data collected recorded students' perceptions on difficulty, complexity, clarity, interest, trickiness, fairness, value, success expectancy, anxiety, and ease for the two examination formats. The research concluded that MC exams are perceived more favorably compared to CR exams.

Tozoglu et al. [44] collected information on the preference of the two examination formats, success expectancy, knowledge, perceived facility, feelings of comfort with the format, perceived complexity, clarity, trickiness, fairness, and perceived anxiety. A recent survey found that 65% of the respondents preferred MC exams based on their ease of taking and the possibility of guessing [12].

Forty years of research has shown that the students prefer objective-type examinations compared to CR ones [12, 44–48]. The most frequently stated reasons are fairness, easiness, and lower exam invoked anxiety. However, Traub and McRury [47] have attributed students' preference to the belief that MC exams are easier to take and prepare for. Thus, they tend to consider that they will get easier and presumably higher scores.

Perceived Easiness. The question of the easiness of the examination format is a multi-faced one related to a) The expected performance, b) the time and effort needed to finish the examination, c) the opportunity to identify the correct answer by guessing, and d) the effort put into getting prepared for the exams. Students expect to perform better on MC than CR exams [39]. A recent publication showed that students consider MC exams more accessible because they require less effort to identify the correct answer and are shorter than essay exams [12]. The possibility of identifying the correct answer by guessing has also been included in the factors that make MC exams easier [49]. Comparison of the study methods and the preparation of the students showed that the students employ surface learning approaches when preparing for MC exams and deep learning approaches when preparing for essay examinations [50]. Chan & Kennedy [51] compared the performance of 196 students on MC questions and "equivalent" CR questions and found that students scored better on the MC examination.

Test Anxiety. A situation is perceived as stressful when it requires abilities outside the resources available to a student. Students with LDs report higher stress levels than their peers, leading to feelings of frustration and insecurity [19, 20, 52]. Extended time to complete the assessment is the most common accommodation for students with LDs [53]. The students report less anxiety, improved grades, and the opportunity to demonstrate their actual knowledge [54]. However, extended time makes learning difficulties visible to the rest of the students and stigmatizes the receivers of such accommodations [21, 55].

On the positive side, objective exams invoke less anxiety than CR ones because they provide the students with immediate feedback [56]. Students with high test anxiety tend to prefer objective tests, while those with low anxiety prefer essay exams [48]. On the negative side, negative marking is a source of anxiety during MC exams [43]. A recent publication reported increased math anxiety to affect students' performance during MC tests [42].

Fairness. Perceived fairness of exams has three aspects: a) fairness of outcomes, i.e., getting the same grade with another student who produces a paper of the same quality, b) fairness of procedures, i.e., establishing clear and unambiguous criteria for allocating marks, and c) openness in discussing and explaining to the students mark allocation [57]. Marking of MC exams is automatic, and it is free of tutor intervention. For this reason, students consider that they are fairer in terms of outcomes and procedures [39]. However, judgments regarding the fairness of exams are somewhat complicated. For example, Emeka & Zilles [57] found that, in the case of version exams, students expressed concerns about fairness which did not have to do with the difficulty of the version they received but with their overall course performance. For students with LDs, the question of fairness is more substantial. Objective exams require the exercise of faculties like short-term memory, reading comprehension, and discriminatory visual ability, in which students with LDs are more vulnerable than the rest of the students [19, 58].

Surface Learning. The character of studying can be reproductive or transformative. Reproductive studying aims to increase knowledge, memorization, or efficient application of facts and rules. Transformative studying increases understanding, focuses on viewing things from different perspectives and transforms learners' thinking. However, some students adopt a strategic approach where reproduction and transformation are used interchangeably to successfully respond to the assessment criteria [59]. The format and content of exams influence students' preparation and studying. Preparation for MC exams fosters memorization of detailed but usually fragmented knowledge, while CR exams demand concept and in-depth learning [1, 50, 60, 61]. The students were asked whether they consider that objective exams assess surface learning.

Making Knowledge Demonstrable. Bloom's taxonomy of educational objectives is widely used to categorize questions of objective format examinations based on the competencies they require [62]. Some commentators consider problem-solving to demand the competence of application in Bloom's taxonomy, i.e., it is a case of applying methods, theories, and concepts [1]. Except for a small number of cases, the solution of problems in electronics does not result from the application of circuit theory alone neither from plugging the correct numbers in an equation. It requires understanding and interpreting a circuit that is not merrily a two-dimensional representation of interconnected devices [63]. Understanding a circuit involves good knowledge about the functioning of the various components and the ability to translate between valid approximations [64]. Problem-solving proceeds in stages; therefore, the number-answer is usually a poor indicator of a student's abilities and learning. Engineering students may feel disempowered if they cannot demonstrate their original thinking. What matters is the process to derive the solution, not the number answer itself. [65]. One of the questions asked the students if they felt disempowered for not demonstrating their knowledge.

Qualitative research findings indicate that students consider objective tests disempowering because they cannot express their understanding using their own words. The students are not encouraged to think independently, but they have to confine their thinking to the options given. There is no room for alternative interpretations or assumptions that lead to a relevant argument [13]. In engineering education, the students may feel disempowered for not having the opportunity to present their original solution to a problem. Past research has shown that one may find striking differences between MC and CR equivalent questions [66, 67]. However, in physics, engineering, and mathematics, where an objective test consists of eight to ten items, individual grade differences affect the overall grade. Objective type tests do not reward imaginative thinking if it does not arrive at the correct number answer. The questionnaire asked the participants whether they felt disempowered for not being allowed to:
demonstrate what they have learned,
construct their solution,
having control over the answer they submit,
express their view,
being forced to select the answer between given options.

Answering by Guessing. One of the drawbacks of MC examinations is answering by guessing because the examinees increase the probability of selecting the correct alternative by eliminating some distractors [40, 68]. In order to alleviate this problem, various strategies have been proposed [40, 69, 70].

2.3 Solving Problems in Engineering Education

Problem-solving plays a central role in engineering education and physics [71]. It requires understanding and interpreting the information contained in circuits and drawing conclusions based on approximations and rigorous reasoning [8]. Mobilizing and synchronizing these abilities is a complex and time-consuming process [10, 72] that goes beyond mere recall and application of a specific theory or equation [73]. A problem in electronics can take the form of an objective question. The stem, usually accompanied by a figure, states the problem to be solved, followed by the key and several distracters. Students' answers are marked automatically, while most of the student's work remains uncommunicated to the examiner and not rewarded.

One of the merits of objective tests is that they provide rapid feedback to the student. Feedback is meaningful when it clarifies the strong and weak points of the solution [75]. Feedback on the strategy followed to solve the problem improves student's knowledge. Informing the student that the correct answer was 2 V instead of any other value is meaningless if some explanation does not follow the right answer.

On the other hand, getting no marks for selecting the wrong number answer is unfair to those students who have done substantial though incomplete work. Getting no marks because of miscalculations and depending heavily on the number-answer may generate feelings of unfairness, increased stress, or increased difficulty. Additionally, it remains an open question how students evaluate a situation where they cannot show their original thinking towards the objectives stated by the problem.

However, a particular type of objective examination introduced during 2019–2020 allowed the assessment of the problem-solving proficiency of the students while tracking their effort. The test consisted of 5 to 8 items that evaluated students' competence in problem-solving [1]. Figure 1 shows the example of a problem in electronics having the form of a problem-based multiple-response (PBMR) item. The stem describes a problem in electronics followed by a circuit. The stem/problem addresses three inter-related questions, i.e., the students cannot answer the question (i+1) unless they have successfully responded to the question (i). There are three options per question; therefore, the probability of answering all three questions by guessing is 11%. This type of objective examination reduced the marking load, discouraged guessing, and rewarded the students who answered all the questions correctly.

There was no negative marking [74], but a specific scoring rule was applied, not giving marks for answers after guessing. The students had to solve the problem and then select the correct answers. An average of 5 min was allowed for each item. Question 1 (the correct value of VE) was worth 0,2 marks, question 2 (VB) counted for 0,3 marks and question 3 (RB) for 0,5 marks. According to the scoring rule applied, a student who selected the correct answer to questions 1 and 2 would get 0,5 marks. If a student answered questions 1 and 3, s/he would get marks only for question 1 because one cannot

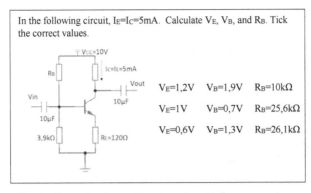

In the following circuit, $I_E=I_C=5mA$. Calculate V_E, V_B, and R_B. Tick the correct values.

$V_E=1,2V$ $V_B=1,9V$ $R_B=10k\Omega$

$V_E=1V$ $V_B=0,7V$ $R_B=25,6k\Omega$

$V_E=0,6V$ $V_B=1,3V$ $R_B=26,1k\Omega$

Fig. 1. Example of a problem-based multiple response item.

calculate the correct answer for question 3 without knowing the answer to question 2. Selecting the correct answers for questions 2 and 3 while making a wrong selection for question 1, is a clear result of guessing and gets zero marks. The students had taken similar tests during the semester and knew the tricks and details of this examination format.

The computer-assisted examinations decreased the marking load, but the time spent preparing the examination increased [1]. CAA results in time savings when marks enter automatically in the student record system. The lack of an appropriate interface between the assessment and the student record system hinders increasing administrative efficiency.

3 Research and Data Collection

The data were collected in September 2020 using an anonymous questionnaire administered to the students via the Open e-Class platform. E-Class is an Integrated Course Management System to support asynchronous e-learning services. The participants were full-time students of the Department of Electrical and Electronic Engineering, University of West Attica. The researchers invited the students who followed the course during 2019–2020 (163 persons) to answer the survey. Most of these students had already taken six objective tests of the type described above (PBMR) during the course delivery from February 2020 to June 2020. The students had taken the tests distantly before the COVID crisis and after the university's closure in March 2020. The July and September 2020 examinations in "Electronics" were taken remotely and in the same format. Before the publication of the questionnaire, a pilot study tested its appropriateness. A total number of 112 students responded to the questionnaire.

The research collected information on the following questions: 1. Do the students with LDs consider one of the two examination formats more suitable to their learning characteristics? 2. What is the students' preference regarding the two examination formats? 3. How do they perceive the two examination formats in terms of anxiety, fairness, and easiness? 4. How do the students respond to particular characteristics of objective exams such as not having to construct a solution, selecting an answer instead of constructing one, not having the opportunity to express their view, and having the opportunity to answer by guessing? The demographic questions collected information on gender, age, and the number of PBMR exams taken. One question asked the students whether they faced any learning difficulties offering the "no answer" option. The questionnaire included one open-ended question: "Describe the positive or negative aspects of the objective as compared to CR exams. Feel free to make any comments you like".

One hundred and twelve students filled the questionnaire. The responses retrieved were those from students who had already taken at least four PBMR examinations, i.e., 105 students. Seventeen of them were female and 88 males. Eighteen (18) out of the 112 respondents reported facing some learning difficulty (15%). Statistics on LDs say that for Greek students in secondary education, dyslexia, one of the most frequent LDs, is estimated at 5.5%, consistent with the data from other countries [76].

The students replied on a 3-point Likert scale, "I disagree," "Neither Agree nor Disagree," and "Disagree." The main drawback of three-point Likert scales is that they suffer from what is called the rounding error. Nonetheless, in cases where the researcher is interested in mean perceptions, the importance of the rounding error is reduced, and 3-point scales are considered good enough [77]. Matell and Jacoby [78] suggested that in 3-point Likert scales, validity and reliability do not decrease compared to 5 or 7-points. Following Lehmann and Hulbert [77], this research adopted a 3-point Likert scale.

4 Results and Discussion

4.1 Quantitative Findings

From the N = 105 valid answers collected, 53% expressed a preference for PBMR exams, 32% for CR exams, another 10% chose "any of the two," and four students did not answer this question. The students with LDs expressed an equal preference for the two assessment formats (47%). Among the students with no LDs, a percentage equal to 57% preferred objective PBMR tests, another 31% CR tests, and 12% of these students replied "any of the two formats."

An Exploratory Factor Analysis was conducted in order to find out the factor structure of the questionnaire. In order to verify whether the assumptions underlying the factor analysis hold, the Kaiser-Meyer-Olkin test was performed, resulting in sampling adequacy of .85 and a Barlett test of sphericity of $\chi 2 = 343$, df $= 45$, p $< 0 .000$. Two Principal Components had eigenvalues greater than one (F1 $= 4,257$ and F2 $= 1,297$). Two factors were identified under a Direct Oblimin Rotation, accounting for 55% of the total variance. The first factor included four items: Fairness, easiness, anxiety, and answering by guessing. The second factor subsumed six items that had to do with the eagerness of the students to show what they had learned (Table 2). The two factors were named "Comfort" and "Expressivity." The calculated Cronbach's coefficient alpha for the first factor was 0,67 and for the second 0,82.

The first factor, "Comfort," comprises questions usually included in preference surveys: easiness, fairness, anxiety, and guessing [11, 12, 44]. This dimension combines questions related to the perceived comforts of each examination format. It shows how the students evaluate the environment of the PBMR and the CR examinations. Quite significantly, the students who prefer PBMR tests adopted a middle position regarding easiness, fairness, and anxiety. Although these students consider PBMR exams not to improve easiness, fairness, and anxiety, they still prefer this assessment format. However, their views regarding the benefits from guessing are opposite to those of the students who prefer CR exams. A percentage equal to 67% of the students who preferred PBMR examinations expected benefits from guessing. In comparison, 65% of the students who preferred CR examinations expressed the opposite view. Therefore, for the sample studied, easiness, fairness, and anxiety are not substantial reasons explaining their preference towards objective format examinations.

The second factor, "Expressivity," shows how the students perceive themselves as participants in the assessment process. It shows if the examinees want to control their answers, construct a solution instead of selecting an answer, demonstrate what they have learned, and express their views. These questions were included in the questionnaire following the findings of Paxton's qualitative research [13]. The data recorded show that the students of the two subgroups, PBMR test or CR preference, express opposing views on these questions. The Cronbach alpha value indicates the meaningfulness of these questions that capture essential differences in the students' attitudes regarding the two examination formats.

Table 1 shows the mean values and the standard deviation for each examination format preference and the average scores for each one of the items included in the "Comfort" dimension, namely: whether objective exams are perceived as fairer, objective exams invoke less anxiety, objective exams are easier, whether the respondents consider that they will manage to pass the objective exams based on some knowledge and guessing (1: agree, 3: disagree).

Table 1. "Comfort" dimension.

| | All | | Preference for | | | |
| | (N = 105) | | PMBR (N = 56) | | CR (N-43) | |
	M	SD	M	SD	M	SD
Fairness	2.13	.80	1.91	.61	2.50	.96
Anxiety	1.92	.70	1.66	.58	2.41	.74
Easiness	1.79	.78	1.64	.62	2.18	.87
Guessing	1.99	.93	1.61	.87	2.47	.79

The statistical analysis showed a significant difference in the scoring of the composite variable "Comfort" depending on the preference of the examination format. The students with a preference for objective exams were associated with the numerically lowest score in "Comfort" (M = 1.70, SD = 0.37). In contrast, the students with a preference for the CR exam were associated with a high score in "Comfort" (M = 2.39, SD = 0.65).

The percentages of the answers show that only 25% of the respondents consider objective exams to be easier, 16% disagreed, and the majority (59%) expressed the view that objective tests are as easy as the CR ones. Regarding the fairness of the objective exams, only 9% of the students agreed they were fairer than CR, 35% disagreed, and

56% adopted a middle position. Only 30% of the respondents considered that objective exams invoke less anxiety than CR, 22% expressed the opposite view, and another 49% believed that the two examination formats invoke equal anxiety. Finally, 43% of the total number of the respondents reported favouring the possibility to answer by guessing. However, this is a question where the two subsets of the respondents, students without LDs who prefer MC vs. CR exams, expressed opposing views. 41% of the students who face LDs consider MC tests easier, fairer (47%), less anxiety invoking (44%), while only 39% replied that guessing makes them feel less worried during MC exams.

The findings of this research are in contrast to those reported in previous studies [11], where 80% of the respondents considered MC exams to be easier, 56% to be fairer, and 83% to be less anxiety invoking. Our findings show that perceptions of fairness, easiness, or less anxiety do not influence the preference for objective tests. At the same time, the possibility of guessing appears to play some critical role in their preference. The specific objective examination left little room for guessing, which may explain the respondents' relatively little reliance on guessing compared to other publications (77% in [12]). If the possibility to answer by guessing influences students' preference for MC exams, then the low expectation of guessing may affect perceptions of easiness and anxiety [47].

Table 2 shows the mean values and the standard deviation for each one of the items included in the "Expressivity" dimension. This dimension included the degree to which the respondents like to select between the given options (1: agree, 3: disagree), constructing a solution, having control of the answer they submit, having the opportunity to demonstrate what they have learned, expressing their view and whether they considered that MC exams assess only surface knowledge.

Table 2. "Expressivity" dimension.

| | All (N = 105) | | Preference for | | | |
| | | | PBMR (N = 56) | | CR (N = 34) | |
	M	SD	M	SD	M	SD
Surface knowledge	1.56	.87	1.32	.72	2.00	.95
Express own view	1.87	.92	1.45	.76	2.53	.83
Demonstrate learning	2.06	.95	1.52	.83	2.85	.44
Construct solution	1.75	.89	1.21	.49	2.59	.70
Select an answer	1.67	.92	1.34	.75	2.15	.96
Control of the answer	1.79	.95	1.25	.61	2.53	.83

The statistical analysis of the data showed that the respondents with a preference for the objective exams gave the lowest score in "Expressivity" (M = 1.35, SD = 0.39). In contrast, the students who preferred the CR exam reported the highest score in "Expressivity" (M = 2.44, SD = 0.50). The statistical analysis showed a significant difference in the scoring of the composite variable "Expressivity" based on the preference of the examination format.

The quantitative data indicate that the examination format preference is associated with two student profiles. The first profile is that of the students with no LDs who prefer objective examinations. The following attitudes characterize these students: not constructing a solution (81%) and not having control over the answer they submit (85%) is not a problem. They do not feel powerless for not having the opportunity to demonstrate their actual knowledge (69%), and they prefer to select an answer (85%) instead of constructing a solution. The second profile is that of the students who face no-LDs and prefer CR exams. These students perceive a problem as not being allowed to build a solution (69%). Not having control over the answer, they submit (69%). They feel powerless for not having the opportunity to demonstrate their knowledge (92%), and they do not like the fact that with the objective exams, they have to select an answer from a set of predetermined options (58%).

4.2 Qualitative Findings

An overall percentage equal to 52% answered the open-ended question. For the students with LD, this percentage was 78%, for the no-LD students and a preference for CR exams, this percentage was 81%, and it dropped to 42% for the students with no LD and a preference for MC tests.

Students with LDs. The text answers of these students showed that the specificity of their learning difficulties largely determines their preference towards the two examination formats. As Table 3 [4] shows, the students who reported problems in writing coherent texts preferred objective exams. The students who reported having difficulty understanding the "language" of the MC questions (tell the difference between the options following the stem or the meaning of stem itself) preferred CR exams. Therefore, their preference is dictated by the details of their learning difficulties. The equal preference for objective and CR exams reflects that half of them encounter problems with text writing, and the other half comprehending the text format of the objective test questions.

Table 3. Students with LDs.

	Preference	
	MC (8)	CR (8)
Write comprehensive texts	4	1
Time management	3	2
Difficulty to understand options	1	3
Difficulties with PBMR	3	–
Anxiety	3	0
Selecting answers randomly	2	0

The long text answers submitted exhibited problems related to incoherent writing, inadequate organization of ideas, fragmented views, and lack of clarity. The following text answers received explain their difficulties with text writing.

"With CR exams, I lose marks because I cannot put in writing what I have in my mind. I need an examination format that will allow me to explain what I want to say".

"With MC exams, I can give an exact answer; text exams make me anxious because I cannot put in text things I do know."

"With text exams, I am treated unfairly compared to my colleagues because I cannot write text answers that make sense."

Some students who preferred MC tests explained that they are struggling even with this examination. Some students noted the difficulties they face to tell the difference between the options given or to understanding the stem itself: "I cannot tell the difference between the options given." In such cases, they reported that they selected the answers randomly. The students who preferred objective tests reported test anxiety and poor time management [54]. Three of the students who preferred CR exams characterized objective exams as "good," "easier," and "helpful."

Students with LDs are not efficient in faculties necessary in academic life and examinations like holding in short-term memory pieces of information, making active comparisons, clarifying meaning, organizing ideas [19], processing speed, motor skills, and anxiety [54]. Therefore, they are more vulnerable than the rest of the students and more likely to achieve lower scores, have low self-efficacy, and are more likely to give up their efforts [20].

Accommodations reflect the assumption that the students with LDs are a problem to be fixed within the competitive ableist educational system [21]. However, such accommodations may be seen as an unfair advantage by the rest of the students. Students with LDs do not want to be singled out or labeled by their peers because of these accommodations [54]. Moreover, research has challenged the effectiveness of such blanket provisions that do not consider each student's individual needs and specific difficulties [53]. The development of inclusive practices can give rise to a learning environment where accommodations for LD students will be unnecessary.

Students Without LDs. The issues raised by these students were different from those raised by the LD students, and they had to do with: The duration of the exams (15 comments), the opportunity to compare the number-answer they found with the options given (12 comments), forced guessing whenever the duration of the exam was short (5 comments), negative marking (4 comments), and shorter time of the objective exams compared to CR ones. Table 4 [4] shows the issues raised for the two preference groups.

The time allotted for single answer items is approximately 1,5 min/item [1]. In the case of problem-based multiple response tests, this time framework is not enough. Comparison of average latency showed that the average time to submit was the same for high and medium performers. However, low performers were more likely to submit their answers earlier (~90% of high performers' average latency). Therefore, if the test duration is short, it will have the most potent effect on high and medium performers, affecting the measured facility factor of the test items and the discriminatory power of the examination.

Indeed, in some objective tests taken during the semester, students with an excellent overall performance selected a wrong option in the last one or two items they answered, indicating that they probably selected their answers randomly. The examiners adjusted the marks to ensure that no student was penalized because of the short test duration. It

Table 4. Students with no-LDs.

	Preference for	
	MC (48)	CR (26)
Not enough time given	9	6
MCQ exams are shorter	1	2
Combined format prop	2	4
Confirmation of results	7	5
Negative marking	3	1
Solution is more important	0	5
Forced to guess	2	3

appears that for problem-based objective tests, the average time per item needed varies between 5 and 15 min, depending on the complexity of the question.

Five students raised this issue independently by explaining that they made random choices whenever the time available was limited. If the students feel that the time given to answer the exam questions is not enough may generate feelings of frustration, reduced fairness, and increased anxiety. Increased stress, short duration, and difficulty of exam questions contribute to the students' negative experiences of distance examinations during the COVID-19 period [83, 84].

During movement restrictions and social distancing, all the exams were remote, and the likelihood of unfair practices increased. Only a few universities have software that detects cheating and plagiarism during exams or assignment submission, while examination redesign has been proposed as an alternative to minimize cheating [85]. OECD suggested that setting stricter time limits would prevent dishonest behaviors [86]. Although setting stricter time limits precludes students from finding the answer from other sources or sharing their answers with colleagues [87, 88], it may also affect the examinees' performance.

Some respondents commented on comparing the answer they calculated to the options given to see whether their solution is correct. Some of the students who preferred CR exams considered that objective examinations are not suitable in engineering courses because, in the case of a problem, it is the solution that matters, not the answer itself.

Although there was no negative marking, some students commented on its fairness, raising the following points: Increase in anxiety and negative marking reduces the marks obtained from correct answers assessing different learning objectives. As the students commented: "There is no logic in negative marking because both of them (to find the correct answer and eliminate a distractor) take the same time." "Negative marking makes me anxious because a wrong answer will reduce the marks from the correct answers I gave.", "I understand why negative marking has been introduced but is it necessary?", "I do not like negative marking, especially when there are many options per item; it takes much time to evaluate each one of them." The students considered that a mix of the two formats would assess more effectively students' learning. As one student noticed,

"I would like to have the opportunity to explain my answers but not necessarily in all the questions."

5 Conclusions

Problem-based multiple response exams are an objective type of examination suitable for assessing the problem-solving proficiency of engineering students. Answering by guessing is not an effective strategy for such a type of examination. This factor affects students' preference for the specific type of examination, as shown in the percentages recorded. Overall, 53% of the respondents preferred MC exams compared to CR ones. This percentage is much lower than those reported in previous publications, e.g., 83% by Zeidner [11] and 65,5% by Kaipa [12]. Although past publications claim that easiness, anxiety, and fairness explain students' preference for objective tests, our research found that the students may consider that objective tests do not improve these factors and still prefer them. The questions of easiness, fairness, and guessing insinuate a utilitarian perspective which assumes that the students prefer a particular type of examination because it is more useful or convenient to them. The students surveyed prefer the objective examinations because they avoid writing text answers and select an answer instead. Overall only 43% of the students considered guessing an effective strategy. Of course, this percentage was higher among the students without LDs who prefer objective tests (67%). Whenever guessing becomes an ineffective strategy, the students are more likely to adopt an in-depth studying style to prepare for the exams. The majority of the respondents replied that the specific type of objective examination does not assess surface knowledge.

The students expressed opposing views in questions related to themselves as participants of the examination. The issues identified by Paxton's qualitative research [13] are relevant to our students as well. The data collected show that the respondents hold opposing views on constructing the answer, avoid writing texts, control the answers they submit, express their point of view, or select an answer. The replies received were differentiated according to the preference towards the two examination formats. Students with a preference for PBMR format also liked to select the answer. They were not interested in showing what they had learned or construct their solutions. On the contrary, students with a preference for CR exams want to construct a solution and control the answer they submit. They feel powerless because they cannot show their actual knowledge, and they select an answer.

Students with learning difficulties showed an equal preference for objective and CR examination formats. However, difficulties persist for both examination formats. Our findings showed that neither of the two examinations formats is suitable for these students. Depending on the LDs, each student needs a specific examination format, which is not necessarily one of the two studied here. As Leach et al. suggest [80], students with LDs must have a role in an assessment partnership. An assessment suited to their LDs would make studying a meaningful effort, and assessment would play a motivating role instead of a source of anxiety.

Students with learning difficulties have to invest extra effort to perform tasks easily carried out by their non-LD colleagues. They have higher drop-out rates and longer time to complete their studies [20]. Their personal and professional life is affected as well. They have fewer friends, are lower-income, and have difficulty integrating into the labor market [81]. These difficulties result in less life satisfaction and happiness [79]. Designing inclusive learning environments for all is considered a more effective strategy to fit students with LD in the academic environment without appealing to disability accommodations. Although tensions and fragmentation characterize the conceptualization of inclusive pedagogies, introducing inclusive practices [82] may offer a more immediate alternative.

References

1. Bull, J., McKenna, C.: Blueprint for Computer-Assisted Assessment, 1st edn. Routledge Falmer, London (2004)
2. Stavroulakis, P., Photopoulos, P., Ventouras, E., Triantis, D.: Comparison of electronic examinations using adaptive multiple-choice questions and constructed-response questions. In: Proceedings of the 12th International Conference on Computer Supported Education, Volume 1: CSEDU, pp. 358–365 (2020)
3. Photopoulos, P., Tsakiridis, O., Stavrakas, I., Triantis, D.: Weighted scoring of multiple-choice questions based exams: expert and empirical weighting factors. In: Proceedings of the 12th International Conference on Computer Supported Education, Volume 1: CSEDU, pp. 382–387 (2020)
4. Photopoulos, P., Tsonos, C., Stavrakas, I., Triantis, D.: Preference for multiple choice and constructed response exams for engineering students with and without learning difficulties. In: Proceedings of the 13th International Conference on Computer Supported Education, Volume 1: CSEDU, pp. 220–231 (2021)
5. Case, S.M., Swanson, D.B.: Extended-matching items: a practical alternative to free-response questions. Teach. Learn. Med. **5**(2), 107–115 (1993). https://doi.org/10.1080/10401339309539601
6. Fenderson, B.A., Damjanov, I., Robeson, M.R., Veloski, J.J., Rubin, E.: The virtues of extended matching and uncued tests as alternatives to multiple choice questions. Hum. Pathol. **28**(5), 526–532 (1997)
7. Gero, A., Stav, Y., Wertheim, I., Epstein, A.: Two-tier multiple-choice questions as a means of increasing discrimination: case study of a basic electric circuits course. Glob. J. Eng. Educ. **21**(2), 139–144 (2019)
8. Duffy, G., O'Dwyer, A.: Measurement of first year engineering students' cognitive activities using a spatial skills test and an electrical concepts test: implications for curriculum design. In: Proceedings of the Research in Engineering Education Symposium, Dublin, Ireland (2015)
9. Duffy, G., Sorby, S., Bowe, B.: An investigation of the role of spatial ability in representing and solving word problems among engineering students. J. Eng. Educ. **109**, 424–442 (2020). https://doi.org/10.1002/jee.20349
10. Wasis, Kumaidi, Bastari, Mundilarto, Wintarti, A.: Analytical weighting scoring for physics multiple correct items to improve the accuracy of students' ability assessment. Eurasian J. Educ. Res. **18**(76), 187–202 (2018)
11. Zeidner, M.: Essay versus multiple-choice type classroom exams: the student's perspective. J. Educ. Res. **80**(6), 352–358 (1987). https://doi.org/10.1080/00220671.1987.10885782
12. Kaipa, R.M.: Multiple choice questions and essay questions in curriculum. J. Appl. Res. High. Educ. **13**(1), 16–32 (2021). https://doi.org/10.1108/JARHE-01-2020-0011
13. Paxton, M.: A linguistic perspective of multiple-choice questioning. Assess. Eval. High. Educ. **25**(2), 109–119 (2000). https://doi.org/10.1080/713611429

14. Finn, J.D., Pannozzo, G.M., Achilles, C.M.: The "Why's" of class size: student behavior in small classes. Rev. Educ. Res. **73**(3), 321–368 (2003). https://doi.org/10.3102/00346543073003321

15. Bettinger, E., Doss, C., Loeba, S., Rogers, A., Taylor, E.: The effects of class size in online college courses: experimental evidence. Econ. Educ. Rev. **58**, 68–85 (2017). https://doi.org/10.1016/j.econedurev.2017.03.006

16. Kauppi, N.: Waiting for Godot? On some of the obstacles for developing counter-forces in higher education. Globalizations **16**(5), 745–750 (2019). https://doi.org/10.1080/14747731.2019.1578100

17. Grummell, B., Lynch, K.: New managerialism: a political project in Irish education. In: Murphy, M.P., Dukelow, F. (eds.) The Irish Welfare State in the Twenty-First Century, pp. 215–235. Palgrave Macmillan UK, London (2016). https://doi.org/10.1057/978-1-137-57138-0_10

18. Lynch, K.: Control by numbers: new managerialism and ranking in higher education. Crit. Stud. Educ. **56**(2), 190–207 (2015). https://doi.org/10.1080/17508487.2014.949811

19. Trammell, J.: Accommodations for multiple choice tests. J. Postsecond. Educ. Disabil. **24**(3), 251–254 (2011)

20. Niazov, Z., Hen, M., Ferrari, J.R.: Online and academic procrastination in students with learning disabilities: the impact of academic stress and self-efficacy. Psychol. Rep. (2021). https://doi.org/10.1177/0033294120988113

21. Nieminen, J.H., Pesonen, H.V.: Politicising inclusive learning environments: how to foster belonging and challenge ableism? High. Educ. Res. Dev. (2021). https://doi.org/10.1080/07294360.2021.1945547

22. Liasidou, A.: Critical disability studies and socially just change in higher education. Br. J. Spec. Educ. **41**(2), 120–135 (2014). https://doi.org/10.1111/1467-8578.12063

23. Gravett, K., Ajjawi, P.: Belonging as situated practice. Stud. High. Educ. (2021). https://doi.org/10.1080/03075079.2021.1894118

24. Benson, W., Probst, T., Jiang, L., Olson, K., Graso, M.: Insecurity in the Ivory Tower: direct and indirect effects of pay stagnation and job insecurity on faculty performance. Econ. Ind. Democr. **41**(3), 693–708 (2020). https://doi.org/10.1177/0143831X17734297

25. Li, A.Y.: Dramatic declines in higher education appropriations: state conditions for budget punctuations. Res. High. Educ. **58**(4), 395–429 (2016). https://doi.org/10.1007/s11162-016-9432-0

26. Krug, K.S., Dickson, K.W., Lessiter, J.A., Vassar, J.S.: Student preference rates for predominately online, compressed, or traditionally taught university courses. Innov. High. Educ. **41**(3), 255–267 (2015). https://doi.org/10.1007/s10755-015-9349-0

27. Holley, D., Oliver, M.: Pedagogy and new power relationships. Int. J. Manag. Educ. (2000). https://www.researchgate.net/publication/238721033_Pedagogy_and_New_Power_Relationships/citations

28. Watts, R.: Public Universities, Managerialism and the Value of Higher Education, 1st edn., p. 20, 22–23, 230–233. Palgrave Macmillan, London (2017)

29. Teräs, M., Suoranta, J., Teräs, H., Curcher, M.: Post-Covid-19 education and education technology 'Solutionism': a seller's market. Postdigit. Sci. Educ. **2**(3), 863–878 (2020). https://doi.org/10.1007/s42438-020-00164-x

30. Mandel, A., Hörnlein, A., Ifland, M., Lüneburg, E., Deckert, J., Puppe, F.: Cost analysis for computer supported multiple-choice paper examinations. GMS Z. Med. Ausbild. **28**(4), Doc.55 (2011). https://doi.org/10.3205/zma000767. https://www.researchgate.net/publication/51970103_Cost_analysis_for_computer_supported_multiple-choice_paper_examinations. Accessed 30 Nov 2020

31. Loewenberger, P., Bull, J.: Cost-effectiveness analysis of computer-based assessment. ALT-J. – Assoc. Learn. Technol. J. **11**(2), 23–45 (2003)

32. Bull, J.: Computer-assisted assessment: impact on higher education institutions. J. Educ. Technol. Soc. **2**(3), 123–126 (1999). https://www.jstor.org/stable/jeductechsoci.2.3.123

33. Topol, B., Olson, J., Roeber, E.: The Cost of New Higher Quality Assessments: A Comprehensive Analysis of the Potential Costs for Future State Assessments. Stanford University, Stanford, CA (2010). Stanford Center for Opportunity Policy in Education

34. Collins, R.: Social distancing as a critical test of the micro-sociology of solidarity. Am. J. Cult. Sociol. **8**, 477–497 (2020). https://doi.org/10.1057/s41290-020-00120-z

35. Rahman, A., Arifin, N., Manaf, M., Ahmad, M., Mohd Zin, N.A., Jamaludin, M.: Students' perception in blended learning among science and technology cluster students. J. Phys.: Conf. Ser. **1496**, 012012, 1–11 (2020). https://doi.org/10.1088/1742-6596/1496/1/012012

36. Vivitsou, M.: Digitalisation in education, allusions and references. Center Educ. Stud. J. **9**(3) 117–136, (2019). https://doi.org/10.26529/cepsj.706. Robotisation, Automatisation, the End of Work and the Future of Education

37. Mintzberg, H.: The Structuring of Organizations, pp. 352–354. Prentice Hall, Englewood Cliffs (1979)

38. Tan, K.H.K.: How teachers understand and use power in alternative assessment. Educ. Res. Int. 11 (2012). https://doi.org/10.1155/2012/382465. Article ID 382465

39. Simkin, M.G., Kuechler, W.L.: Multiple-choice tests and student understanding: what is the connection? Decis. Sci. J. Innov. Educ. **3**, 73–98 (2005). https://doi.org/10.1111/j.1540-4609. 2005.00053.x

40. Scharf, E.M., Baldwin, L.P.: Assessing multiple choice question (MCQ) tests - a mathematical perspective. Act. Learn. High. Educ. **8**(1), 31–47 (2007). https://doi.org/10.1177/146978740 7074009

41. Wong, M.-Y.: Teacher–student power relations as a reflection of multileveled intertwined interactions. Br. J. Sociol. Educ. **37**(2), 248–267 (2016). https://doi.org/10.1080/01425692. 2014.916600

42. Núñez-Peña, M.I., Bono, R.: Math anxiety and perfectionistic concerns in multiple-choice assessment. Assess. Eval. High. Educ. **46**(6), 865–878 (2021). https://doi.org/10.1080/026 02938.2020.1836120

43. Pamphlett, R., Farnill, D.: Effect of anxiety on performance in multiple choice examination. Med. Educ. **29**, 297–302 (1995). https://doi.org/10.1111/j.1365-2923.1995.tb02852.x

44. Tozoglu, D., Tozoglu, M. D., Gurses, A., Dogar, C.: The students' perceptions: essay versus multiple-choice type exams. J. Baltic Sci. Educ. **2**(6), 52–59 (2004). http://oaji.net/articles/ 2016/987-1482420585.pdf

45. Gupta, C., Jain, A., D'Souza, A.S.: Essay versus multiple-choice: a perspective from the undergraduate student point of view with its implications for examination. Gazi Med. J. **27**, 8–10 (2016). https://doi.org/10.12996/GMJ.2016.03

46. van de Watering, G., Gijbels, D., Dochy, F., van der Rijt, J.: Students' assessment preferences, perceptions of assessment and their relationships to study results. High. Educ. **56**, 645–658 (2008). https://doi.org/10.1007/s10734-008-9116-6

47. Traub, R.E., MacRury, K.: Multiple choice vs. free response in the testing of scholastic achievement. In: Ingenkamp, K., Jager, R.S. (eds.) Tests und Trends 8: Jahrbuch der Pädagogischen Diagnostik, pp. 128–159. Weinheim und Basel, Beltz (1990)

48. Birenbaum, M., Feldman, R.A.: Relationships between learning patterns and attitudes towards two assessment formats. Educ. Res. **40**(1), 90–97 (1998)

49. Parmenter, D.A.: Essay versus multiple-choice: student preferences and the underlying rationale with implications for test construction. Acad. Educ. Leadersh. **13**(2), 57–71 (2009)

50. Scouller, K.: The influence of assessment method on students' learning approaches: multiple choice question examination versus assignment essay. High. Educ. **35**, 453–472 (1998). https://doi.org/10.1023/A:1003196224280

51. Chan, N., Kennedy, P.E.: Are multiple-choice exams easier for economics students? A comparison of multiple-choice and "equivalent" constructed-response exam questions. South. Econ. J. **68**(4), 957–971 (2002)
52. Heiman, T., Precel, K.: Students with learning disabilities in higher education: academic strategies profile. J. Learn. Disabil. **36**(3), 248–258 (2003)
53. Gelbar, N., Madaus, J.: Factors related to extended time use by college students with disabilities. Remedial Spec. Educ. (2020). https://doi.org/10.1177/0741932520972787
54. Slaughter, M.H., Lindstrom, J.H., Anderson, R.: Perceptions of extended time accommodations among postsecondary students with disabilities. Exceptionality (2020). https://doi.org/10.1080/09362835.2020.1727339
55. Nieminen, J.H.: Disrupting the power relations of grading in higher education through summative self-assessment. Teach. High. Educ. (2020). https://doi.org/10.1080/13562517.2020.1753687
56. DiBattista, D., Gosse, L.: Test anxiety and the immediate feedback assessment technique. J. Exp. Educ. **74**(4), 311–327 (2006)
57. Emeka, Ch., Zilles, C.: Student perceptions of fairness and security in a versioned programming exams. In: ICER 2020: Proceedings of the 2020 ACM Conference on International Computing Education Research, pp. 25–35 (2020). https://doi.org/10.1145/3372782.3406275
58. Duncan, H., Purcell, C.: Consensus or contradiction? A review of the current research into the impact of granting extra time in exams to students with specific learning difficulties (SpLD). J. Furth. High. Educ. **44**(4), 439–453 (2020). https://doi.org/10.1080/0309877X.2019.1578341
59. Entwistle, A., Entwistle, N.: Experiences of understanding in revising for degree examinations. Learn. Instr. **2**, 1–22 (1992). https://doi.org/10.1016/0959-4752(92)90002-4
60. Martinez, M.E.: Cognition and the question of test item format. Educ. Psychol. **34**(4), 207–218 (1999)
61. Biggs, J.B., Kember, D., Leung, D.Y.P.: The revised two factor study process questionnaire: R-SPQ-2F. Br. J. Educ. Psychol. **71**, 133–149 (2001)
62. Sobral, S.R.: Bloom's taxonomy to improve teaching-learning in introduction to programming. Int. J. Inf. Educ. Technol. **11**(3), 148–153 (2021)
63. Beichner, R.J.: Testing student interpretation of kinematics graphs. Am. J. Phys. **62**, 750–784 (1994)
64. Trotskovsky, E., Sabag, N.: The problem of non-linearity: an engineering students' misconception. Int. J. Inf. Educ. Technol. **9**(6), 449–452 (2019)
65. Gipps, C.V.: What is the role for ICT-based assessment in universities? Stud. High. Educ. **30**(2), 171–180 (2005). https://doi.org/10.1080/03075070500043176
66. Lukhele, R., Thissen, D., Wainer, H.: On the relative value of multiple-choice, constructed response, and examinee selected items on two achievement tests. J. Educ. Meas. **31**(3), 234–250 (1994)
67. Bridgeman, B.: A comparison of quantitative questions in open-ended and multiple-choice formats. J. Educ. Meas. **29**, 253–271 (1992)
68. Bush, M.: A multiple choice test that rewards partial knowledge. J. Furth. High. Educ. **25**(2), 157–163 (2001)
69. McKenna, P.: Multiple choice questions: answering correctly and knowing the answer. Interact. Technol. Smart Educ. l **16**(1), 59–73 (2018)
70. Ventouras, E, Triantis, D., Tsiakas, P., Stergiopoulos, C.: Comparison of oral examination and electronic examination using paired multiple-choice questions. Comput. Educ. **56**(3), 616–624 (2011). https://doi.org/10.1016/j.compedu.2010.10.003
71. Redish, E.F., Scherr, R.E., Tuminaro, J.: Reverse engineering the solution of a "simple" physics problem: why learning physics is harder than it looks. Phys. Teach. **44**, 293–300 (2006). https://doi.org/10.1119/1.2195401

72. Adeyemo, S.A.: Students' ability level and their competence in problem-solving task in physics. Int. J. Educ. Res. Technol. **1**(2), 35–47 (2010)
73. McBeath, R.J. (ed.): Instructing and Evaluating in Higher Education: A Guidebook for Planning Learning Outcomes. Educational Technology Publications, Englewood Cliffs (1992)
74. Holt, A.: An analysis of negative marking in multiple-choice assessment. In: Mann, S., Bridgeman, N. (eds.) 19th Annual Conference of the National Advisory Committee on Computing Qualifications (NACCQ 2006), Wellington, New Zealand, pp. 115–118 (2006). https://citeseerx.ist.psu.edu/viewdoc/download?doi=10.1.1.679.2244&rep=rep1&type=pdf
75. Brown, E., Glover, C.: Evaluating written feedback. In: Bryan, C., Clegg, K. (eds.) Innovative Assessment in Higher Education, pp. 81–91. Routledge, London (2006)
76. Vlachos, F., Avramidis, E., Dedousis, G., Chalmpe, M., Ntalla, I., Giannakopoulou, M.: Prevalence and gender ratio of dyslexia in Greek adolescents and its association with parental history and brain injury. Am. J. Educ. Res. **1**(1), 22–25 (2013). https://doi.org/10.12691/education-1-1-5
77. Lehmann, D.R., Hulbert, J.: Are three-point scales always good enough? J. Mark. Res. **9**(4), 444–446 (1972)
78. Matell, M.S., Jacoby, J.: Is there an optimal number of alternatives for Likert scale items? Study 1: reliability and validity. Educ. Psychol. Meas. **31**, 657–674 (1971)
79. Kalka, D., Lockiewicz, M.: Happiness, life satisfaction, resiliency and social support in students with dyslexia. Int. J. Disabil. Dev. Educ. **65**(5), 493–508 (2018). https://doi.org/10.1080/1034912X.2017.1411582
80. Leach, L., Neutze, G., Zepke, N.: Assessment and empowerment: some critical questions. Assess. Eval. High. Educ. **26**(4), 293–305 (2001). https://doi.org/10.1080/02602930120063457
81. McLaughlin, M.J., Speirs, K.E., Shenassa, E.D.: Reading disability and adult attained education and income. J. Learn. Disabil. **47**(4), 374–386 (2014). https://doi.org/10.1177/0022219412458323
82. Thomas, L.: Developing inclusive learning to improve the engagement, belonging, retention, and success of students from diverse groups. In: Shah, M., Bennett, A., Southgate, E. (eds.) Widening Higher Education Participation, pp. 135–159. Elsevier (2016)
83. Elsalem, L., Al-Azzam, N., Jum'ah, A.A., Obeidat, N., Sindiani, A.M., Kheirallah, K.A.: Stress and behavioral changes with remote E-exams during the Covid-19 pandemic: a cross-sectional study among undergraduates of medical sciences. Ann. Med. Surg. **60**, 271–279 (2020). https://doi.org/10.1016/j.amsu.2020.10.058
84. Clark, T.M., Callam, C.S., Paul, N.M., Stoltzfus, M.W., Turner, D.: Testing in the time of COVID-19: a sudden transition to unproctored online exams. J. Chem. Educ. **97**(9), 3413–3417 (2020). https://doi.org/10.1021/acs.jchemed.0c00546
85. Munoz, A., Mackay, J.: An online testing design choice typology towards cheating threat minimisation. J. Univ. Teach. Learn. Pract. **16**(3) (2019). Article 5. https://ro.uow.edu.au/jutlp/vol16/iss3/5. Accessed 15 June 2020
86. OECD: Remote online exams in higher education during the COVID-19 crisis (2020). oecd.org/education/remote-online-exams-in-higher-education-during-the-covid-19-crisis-f53e2177-en.htm
87. Ladyshewsky, R.K.: Post-graduate student performance in supervised in-class vs. unsupervised online multiple-choice tests: implications for cheating and test security. Assess. Eval. High. Educ. **40**(7), 883–897 (2015)
88. Schultz, M., Schultz, J., Round, G.: Online non-proctored testing and its affect on final course grades. Bus. Rev. Cambr. **9**, 11–16 (2008)

An Assessment of Statistical Classification for Gamification and Socially Oriented Methodologies in Undergraduate Education

M. E. Sousa-Vieira[✉], J. C. López-Ardao, and M. Fernández-Veiga

Atlantic Research Centre, Universidade de Vigo, Vigo, Spain
estela@det.uvigo.es

Abstract. We present in this paper an analysis of the effect of gamification and social learning/teaching on the performance of students. We identify course variables that quantitatively explain the improvements reported when using these methodologies integrated in the course design, and apply techniques from the machine learning domain to conduct success/failure classification methods. We found that, generally, very good results are obtained when an ensemble approach is used, that is, when we blend the predictions made by different classifiers.

Keywords: Gamification · Social learning · Learning analytics · Success prediction

1 Introduction

The use of gamification strategies and socially oriented methodologies within a formal educational context has been extensively studied in the literature recently [10,53,62], Generally, the studies found consistent, sound evidence that introducing gamified tasks (i.e., the dynamics of game playing and reward) into the learning tasks has a positive effect in keeping the interest of students up and in improving their academic achievements in the course. However, measuring exactly the positive correlations between gamification and performance has remained a topic for further discussion and investigation, for several reasons. One is that the design of the gamified activities and the reward system can influence the perception of students, so it is not clear which incentives work better in a given academic context. A second reason is that, even if the environment of the experiment is tightly controlled, there are typically many variables or features that can be analyzed as predictors or early signals of performance. Hence, it is essential to discern which of the variables acting in a gamified strategy (type and amount of rewards, actions from the students, quality of the responses, etc.) are really significant for extracting useful quantitative information from the class. A third reason is that assessing the effectiveness of a learning activity is in itself a difficult task which cannot be taken in isolation from the whole academic design and the background of the students.

In this chapter, an extension of [48], we attempt to contribute to the design of good gamification strategies by synthesizing a predictive methodology constructed from

© Springer Nature Switzerland AG 2022
B. Csapó and J. Uhomoibhi (Eds.): CSEDU 2021, CCIS 1624, pp. 378–404, 2022.
https://doi.org/10.1007/978-3-031-14756-2_19

the observations and measurements taken in a typical gamified learning context, in a university-level course. Our approach is to view the prediction problem as a semi-supervised classification problem. This shifts the main question to the identification of the relevant features involved in accurate prediction of the final outcomes in the course, where the data samples and full features are the interactions among the students in the learning platform. For our predictive methodology, we test the performance of standard classifiers (logistic regression, support vector machines, and linear discriminant analysis) and quantify the best technique in this educational domain. Our results can be seen primarily as a feature extraction problem, in addition to being a classical machine learning task, since the relationships between learning outcomes and the input features can be quite complex. In this work, we have tested all the subsets of features related to forums participation, so exploring fully the search space, in order to discover the statistically significant features for our domain. Related to classification, we have found that the best results are obtained when an ensemble approach is used, that is, when we blend the predictions made by the different classifiers. In the chapter we describe the experimental setup, the data collection, the methodology for analysis, and the experimental results obtained in our tests.

The rest of the paper is organized as follows. Section 2 summarizes some recent related work. The methodology employed in the courses under study are reported in Sect. 3. Section 4 contains the main results of the analysis applied to the datasets. The proposed learning success/failure prediction methodology is explained in Sect. 5. Finally, some concluding remarks are included in Sect. 6.

2 Review of the Literature

2.1 Social Learning

The structure of several natural and artificial complex systems has been extensively analyzed in the research literature, and as a result many of the properties of these objects have been discovered. As pointed out in the previous section, in the field of education, OSNs arise quite naturally when information technology is used in the classroom as an inherent part of the learning activities. The network is just a depiction of the existence and strength of interaction among the students, or among the students with the instructors. It has long been recognized that the structure of such interactions (of its underlying graph) is key to a deep comprehension of the information flow within the students' group, and that it can ultimately be used to assess the quality of the learning process and to infer students' performance directly from their pattern of interactions. In this section, we review the literature related to the application of statistical techniques [37] to disentangle the relationships taking place among social actors in a SLE, and for understanding the distinctive patterns arising from these interactions. A more extensive compilation can be found in [10].

An empirical investigation of the relationships between communication styles, social relationships and performance in a learning community was addressed in [8]. According to it, not only individual but also structural factors determine the way the students build up collaborative learning social networks whose structure strongly shapes their academic performance. Related to the latter, the focus of the study reported in [30] is to highlight the advances that social network analysis can bring, on combination with other methods, when studying the nature of the interaction patterns within a networked learning community, and the way its members share and construct knowledge. A structural analysis of student networks has been done in [12] too, to see if a student's position in the social learning network could be a good predictor of the sense of community. Empirical evidence was found that such position is indicative of the strength of their subjective sense of community, and also of the type of support required by students for advancing as expected through the course. The study addressed in [17] focuses on communities of students from a social network point of view, and discusses the relations taking place in those communities, the way media can influence the formation of online relationships, and the benefits that can emerge after actively maintaining a personal learning network. Understanding the information flow that really occurs in SLEs is obviously of key importance to improve the learning tools and methodologies. For instance, [33] argues that detection of high-valued posts is fundamental for good social network analysis. In [42], authors examine the impact of academic motivation on the type of discourse contributed and on the position of the learner in the social learning network, and concludes that intrinsically motivated learners become central and prominent contributors to the cognitive discourse. In contrast, extrinsically motivated learners only contribute on average and are positioned randomly throughout the social graph. The work [18] investigates the patterns and the quality of online interactions during project-based learning showing its correlation with project scores. The identification of social indices that actually are related to the experience of the learning process is addressed in [57], showing that some popular measures such as density or degree centrality are meaningful or not depending on the characteristics of the course under study. The structure of two distributed learning networks is given in [6] in order to understand how students' success can be enhanced. Degree centrality is proposed as the basic predictor for effectiveness of learning in the course under study in [20]. In addition to structural properties, the influence of cognitive styles and linguistic patterns of self-organizing groups within an online course have also been the focus of some works, such as [47].

More recently, the work [45] examines relationships between online learner self- and co-regulation. The observations reveal that students with high levels of learner presence occupy more advantageous positions in the network, suggesting that they are more active and attract more reputation in networks of interaction. The authors of [51] study the patterns of network dynamics within a multicultural online collaborative learning environment. The experiment tests a set of hypothesis concerning tendencies towards homophily/heterophily and preferential attachment, participant roles and group work. The article [54] measures the impact of OSN software versus traditional LMSs, showing that with the first one students experience higher levels of social interaction, course community, engagement and satisfaction. In [60], social network analysis techniques are used to examine the influence of moderator's role on online courses. The main conclusion is that when students are assigned to the moderator position their participation quantity, diversity and interaction attractiveness increases significantly, and their lack of participation influences the group interaction. In [9], authors investigate the association between social network properties, content richness in academic learning discourse and performance, concluding that these factors cannot be discounted in the learning process and must be accounted for in the learning design. The study addressed in [15] demonstrates that social network indices can offer valuable insight into the creative collaboration process. The article [32] compares the impact of social-context and knowledge-context awareness on quantitative and qualitative peer interaction and learning performance, showing that with the first one the community had significantly better learning performance, likely related to the more extensive and frequent interactions among peers. And [46] investigates the discourses involving student participation in fixed groups and opportunistic collaboration. They find that actively participating and contributing high-level ideas were positively correlated with students' domain of knowledge. The existence of a positive relationship between centralization and cohesion and the social construction of knowledge in discussion forums is the main conclusion in [55]. In [40] the authors present a new model for students' evaluation based on their behavior during a course and its validation through an analysis of the correlation between social network measures and the grades obtained by the students. The work [50] focuses on the quantitative characterization of social learning methodologies. To this end, authors use one custom software platform, SocialWire, for discovering what factors or variables have significant correlation with the students' academic achievements. The dataset was collected along several consecutive editions of an undergraduate course. The paper [24] investigates the influence of learning design and tutor interventions on the formation and evolution of communities of learning, employing social network analysis to study three differently designed discussion forums. Finally, the study addressed in [59] focuses on an advanced course on statistics. Authors investigate how a learning community based on social media evolved with the implementation of the activities and how different relational ties and SNA measures are associated with students outcomes, finding that students' outdegree and closeness centralities have the most consistent predictive validity of their learning outcomes.

2.2 Gamification

Gamification is defined as the use of game design elements in non-game contexts [27]. It can be applied in several situations to influence the behaviors of individuals, mainly to increase engagement, to motivate action or to promote learning. Due to the fact that all these are major issues faced by teachers of all educational levels, in recent years multiple implementations of gamification in educational contexts have emerged. In this section, we present a literature review of those focused on higher education. Again, a more extensive compilation can be found in [53,62].

The work [31] studies the effect of a gamified version of an online wiki-based project in an industrial/organizational psychology course, showing an increase of the interactions. Moreover, results indicate that time-on-tasks predicts learning outcomes. The results explained in [22] show positive effects on the engagement of C-programming students toward gamified learning activities and a moderate improvement in learning outcomes. The purpose of the study [16] is to investigate if achievements based on badges can be used to affect the behavior of students of a data structures and algorithms course, even when the badges have not impact on the grading. Statistically significant differences in students' behavior are observed with some badges types, while other types do not seem to have such an effect. The study [41] examines college undergraduates enrolled in first-year writing courses, where badges represent essential course outcomes. Participants are categorized as having either high or low expectancy-values, and intrinsic motivation to earn badges is measured repeatedly during the 16-week semester. Findings suggest that incorporating digital badges as an assessment model could benefit learners who have high expectations for learning and place value on learning tasks, but it could also disenfranchise students with low expectations. In [52], authors conduct an experimental study to investigate the effectiveness of gamification of an online informatics course on computer graphics. Results show that students enrolled in the gamified version of the course achieved greater learning success. The article [19] reports the effects of game mechanics on students cognitive and behavioral engagement through an experience conducted in a design questionnaires course. The results of the study show positive effects of gamification on motivating students to engage with more difficult tasks, to contribute in the discussion forums and to increase the quality of the artifacts. The article [28], reports the application of gamification to an online context for academic promotion and dissemination. Both quantitative and qualitative data were collected and analyzed, revealing that gamification has the potential to attract, motivate, engage and retain users. The research addressed in [3] examines the impact of different learning styles and personality traits on students' perceptions of engagement and overall performance in a gamified business course. Findings suggest that students who are oriented towards active or global learning as well as extroverted students have a positive impression of gamification. The effect of gamified instructional process to ICT students engagement and academic performance is studied in [7]. Conclusions show that using the proposed combination of elements provided a positive motivational impact on engagement and indirectly affected the academic results. Similar results are observed in [34], where authors found evidence that gamification can be used to improve the overall academic performance in practical assignments and to promote social interaction in a qualification for ICT users course. However, findings also raise that the creation of gamified experiences for higher education requires a deep knowledge of the motivational preferences of students and a careful design of the rewards that are perceived by students and that eventually stimulate participation. The work [13] describes the application of gamification in an operations research/management science course, where it was possible to observe an increase of participation in class, better results and a good assessment of the course made by the students. In the study [44], authors vary different configurations of game design elements and analyze them in regard to their effect on the fulfillment of basic psychological needs. The article [36] presents a gamification experience within prospective primary teachers in a general science course. In an effort for promoting collaborative dynamics rather than competitive ones, a new variable called game index, that takes into account the scoring of the whole class, was introduced. A positive correlation among scoring and academic marks was confirmed. The experiment described in [61] was conducted to determine the effect of gamification-based teaching practices on achievements and attitudes towards learning, in a course about teaching principles and methods. The results show positive attitudes towards the lessons and a moderate effect on achievements. Although there was not difference between the final grades of the gamified and the control groups, students regard wiki and gamified activities positively.

Recently, the article [1] presents the gamification process, iterations made into the game elements and their features and students' perceptions in a gamified teacher education course. The study [11] investigates college students' experiences of a gamified informatics course, showing positive trends with respect to students' perceptions of gamification's impact on their learning, achievement, and engagement in the course material. The authors of [14] explore the effects of gamification on students' engagement in online discussions. Conclusions and interviews with students and teachers suggest a positive effect of the game-related features of the platform. The focus of the study reported in [21] is to explore whether gamification could be a good strategy to motivate students to participate in more out-of-class activities without decreasing the quality of work. Results from two experiments conducted in two master level courses on statistics reveal that the gamified classes completed significantly more activities and produced higher quality work. In [25], authors test the effectiveness of adding educational gaming elements into the online lecture system of a flipped classroom, as a method to increase interest in online preparation before class, obtaining good results and better academic achievements of mid-upper level students. The work [26] analyses the effect of using gamification elements in a course related to software development. The study confirms that students' grades and motivation can increase as a result of applying gamification to their learning process. In the experiment described in [29] students were randomly assigned to three different conditions: no badges, badges visible to peers and badges only visible to students themselves. Contrarily to expectations, the last one was evaluated more positively than the second one. The article [35] describes the use of a gamified platform in programming courses that allows students to face different types of challenges and obtain badges and points which are added to the ranking of the course. Authors conclude that gamification is useful for teaching programming but they need further validation to corroborate the findings obtained and to investigate the influence on students' performance. The effects of using gamification elements in courses that make use of a wiki environment on the participation rates as well as on student academic success is addressed in [39]. Authors conclude that wiki activities positively contribute to student academic success, while gamification increases student participation. In [43], authors measure the possible evolution of motivational levels in response to the interaction with the game elements used in a university course. The findings illustrate the significance of the individual nature of motivational processes, the importance of sensitive longitudinal motivation measurements and the relevance of the implemented game elements' design characteristics. The work addressed in [58] indicates, from a cohort of undergraduate business students, that course performance was significantly higher among those students who participated in the proposed gamified system than in those who engaged with the traditional delivery. The article [4] describes an advanced learning environment that detects and responds to computer programming students' emotions by using ML techniques, and incorporates motivation strategies by using gamification methods. In [2], authors develop a scale to measure the factors that may affect the gamification process via kahoot in a pre-service teachers undergraduate course. Conclusions of the study suggest that the achievement intrinsic in the gamification process need to be regularly improved and changed. In [5], authors try to contribute to determine if a gamified social e-learning platform can improve the learning performance an engagement of students enrolled in a master level subject about design of mobile applications. Results show that social gamification is a suitable technique to improve the learning outcome of students, at least for those skills related to programming, providing more communication skills and producing more engagement and motivation while using the platform. The study presented in [38] describes the positive effect of gamification, based on leaderboards, on learning performance in an introductory computer programming course. In [56] authors propose a solution to help instructors to plan and deploy gamification concepts with social network features in learning environments. A case study over a programming course reveals that the implemented gamified strategies achieved positive acceptance among teachers and students. And finally, in [49] authors report the analysis of the impact of enhanced gamification methodologies, included in the custom made software platform SocialWire, on master level students' engagement and performance. Specifically, the use of the virtual classroom was rewarded by the automatic scoring of different actions carried out in the platform related to the normal activity along the term. These scores allowed to obtain badges and rewards helpful to pass the subject.

3   Educational Context and Dataset

We have taken as our educational environment the 2020 and the 2021 editions of a course on Computer Networks directed to undergraduates of the second year of the Telecommunications Technologies Engineering bachelor degree. This course has a weekly schedule that spans 14 weeks, between January and May. The classroom activities are organized as follows:

- Lectures, that blend the presentation of concepts, techniques and algorithms with the practice of problem-solving skills and discussion of theoretical questions.
- Laboratory sessions, where the students design and analyze different network scenarios with different protocols, using real or simulated networking equipment. Moreover, in some of these sessions students make a programming assignment.

In both editions the activities are supported by a tailored Moodle site to which the students and teachers have access, and wherein general communication about the topics covered takes place. To encourage networked learning and collaborative work, different activities are planned and carried out in the platform. The students may gain different kinds of recognition by completing or participating in these activities. In the editions analyzed in this work, the following online activities were proposed:

1. Homework tasks, to be worked out before the in-class or the laboratory sessions. With this activity teachers encourage the students to prepare some of the material in advance.
2. Quizzes, proposed before the midterm exams for self-training.
3. Collaborative participation in forums. Several forums were created in Moodle to allow the students to post questions, doubts or puzzles related to the organization of the course, the content of the in-class lectures or the laboratory sessions and the programming assignments.
4. Optional activities, such as games, peer assessment of tasks, etc.

The score of tasks (and their peer assessment) and quizzes is measured in so-called merit points, and represents the total score gained for accomplishment of these activities in the modality B of the continuous assessment (a 10% of the final grade). It is possible to obtain extra merit points by doing the optional activities in order to compensate for low scores or late submissions of some of the tasks or quizzes. Well-done peer assessments and the best scores in tasks and quizzes are rewarded with virtual coins and badges.

Participation in forums, solving doubts or sharing resources, is also valued with points or votes granted by the teachers or the classmates. As new points or votes are obtained, the so-called karma level of each student increases, depending on different factors that take into account the quality of the student's actions and the comparison with that of his/her classmates. As the karma level increases, students get coins.

The use of the virtual classroom is also rewarded by the automatic scoring of different actions carried out in the platform related to the normal activity unfolded along the term, like posting or viewing resources, posting new threads, replying to posts, completing tasks, etc. The so-called experience points are organized into levels and are awarded in a controlled environment with maximum values and their frequency set by the teachers. When students level up, they get coins too.

At any time, a student can check his/her accumulated merit points, karma level and accumulated experience points and level. Moreover, students can check their positions in the global rankings and the averages values of the course. And occasionally, the best students of a ranking can be made public to the group.

The coins accumulated at the end of the course can be converted into benefits helpful to pass the subject.

In the 2020 edition of the course the final exam was online for the pandemic and consisted of 6 exercises; each exercise scores 2 points and the maximum score is 10 points:

- 32 coins can be changed by the *extra exercise wildcard*: the 6 exercises of the exam are corrected and their mark is added up to a maximum of 10 points.
- 22 coins can be changed by the *remove worse exercise wildcard*: the 6 exercises of the exam are corrected and the worse is not taken into account.
- 16 coins can be changed by the *remove exercise wildcard*: students choose the exercise whose score is not taken into account.
- 4 or 6 coins can be converted into one or two pages of notes for the final exam.
- 3 coins can be changed by 5 bonus merit points up to a maximum of 25.

For students who do not have a wildcard, the 6 exercises are corrected and the score of each one is scaled by $10/6$.

In the 2021 edition of the course:

- 8 or 16 coins can be converted into 10 or 20 min of extra time in the final exam.
- 12 or 24 coins can be converted into one or two pages of notes for the final exam.
- 8 coins can be changed by 5 bonus merit points up to a maximum of 25.

In both editions, it is clear that the students that follow the modality B of the continuous assessment can get more benefit from the gamification strategy.

Students may pass the course after a single final examination covering all the material (provided the programming assignment meets the minimum requirements), but they are encouraged to follow the continuous assessment. In continuous assessment we allow two modalities, A and B.

Depending on the edition, 2020 or 2021, we weighed 40% or 50% the final exam, but the rest was split as follows: 36% or 30% in modality A and 24% or 20% in modality B from the midterm exams, 24% or 10% from the programming assignment and 12% or 10% (only in modality B) coming out from the merit points obtained by accomplishing the online activities (task, quizzes and optional tasks) described previously, devised as a tool to increase the level of participation. Students have two opportunities to pass the exam (non-exclusive), May and July.

To finish our description, among all students enrolled in the 2020 edition of the course, 121 students did not drop out (i.e. they attended the final exam). Among these 121 students, of the 114 students who followed the continuous assessment (18 in modality A and 96 in modality B), 84 finally passed the course (13 (72%) in modality A and 71 (74%) in modality B). And none of the 7 students that followed the single final exam modality was able to pass. In the 2021 edition of the course, 102 students attended the final exam. Among them, of the 101 students who followed the continuous assessment (14 in modality A and 87 in modality B), 87 finally passed the course (5 (35%) in modality A and 82 (94%) in modality B). The student that followed the single final exam modality did not pass the course.

At this point, it is important to highlight that the adaptation of this subject to the lockdown caused by the pandemic in the 2020 edition, that affected the second part of the term, was fast and without incidents for the vast majority of students, because they were already used to the platform and the blended learning methodology since January.

4   Analysis of the Datasets

4.1   Merit Points

Table 1 shows the estimated mean value and standard deviation of the individual merit points (merit points form tasks and quizzes(tqmp), extra merit points (emp) and total merit points (tmp)) and the slope of total merit points (tmps), that is, the coefficient of a linear regression model of the total merit points per time graph, of all the students that followed the modality B of the continuous assessment in both editions of the course.

In order to check the relationship among the patterns of engagement along the term and knowledge acquisition, we have measured the statistical correlations between the individual merit points and the performance in the final exam (in the 2020 edition, taking into account the sum of the scores of the 6 exercises before applying wildcards) of the students that followed the modality B of the continuous assessment and did not drop out the course. For this purpose, the sample correlations $\hat{\rho}$ were computed and the linear regression statistical test was used to quantify such correlations. The estimated linear coefficient is denoted by $\hat{\beta}$. Under the null hypothesis (meaning that there is not such linear dependence) the test statistic follows a t-distribution and high values are very unlikely to be observed empirically [23]. In Table 2 we can see statistically significant positive dependencies.

Table 1. Individual merit points.

	2020 edition		2021 edition	
	$\hat{\mu}$	$\hat{\sigma}$	$\hat{\mu}$	$\hat{\sigma}$
tqmp	58.5253	18.1286	59.5679	28.2998
emp	13.8074	6.1131	11.3762	5.8953
tmp	75.8786	21.5459	75.5389	31.1586
tmps	0.7023	0.2013	0.5619	0.2016

Table 2. Correlation between individual merit points and student's performance in the final exam.

	2020 edition		2021 edition	
	$\hat{\rho}$	$(\hat{\beta}, t, P(>|t|))$	$\hat{\rho}$	$(\hat{\beta}, t, P(>|t|))$
tqmp	0.4783	$(0.0689, 5.3112, 7.21 \cdot 10^{-7})$	0.3683	$(0.0262, 3.9823, 1.29 \cdot 10^{-4})$
emp	0.1677	$(0.0559, 1.6591, 1.01 \cdot 10^{-1})$	0.2499	$(0.0771, 2.5953, 1.09 \cdot 10^{-2})$
tmp	0.3654	$(0.0483, 3.8261, 2.33 \cdot 10^{-4})$	0.3469	$(0.0197, 3.7182, 3.31 \cdot 10^{-4})$
tmps	0.3944	$(5.2706, 4.1831, 6.41 \cdot 10^{-5})$	0.3318	$(2.6681, 3.5362, 6.15 \cdot 10^{-4})$

4.2   Forums Activity

We have applied standard SNA techniques [37] to analyze the data collected in forums. For this task, we have recorded the events that took place in each forum, users who posted new threads, users who replied and the average valuations they received. This information is represented as a graph where two nodes, i.e., the users, are connected by an edge if one has given a reply to an entry posted by the other. Moreover, self-edges represent new threads. The weight of each edge is related to the points or votes obtained by the reply. Orange edges identify useful or the best replies based on the opinion of the owner of the question and/or the teachers. An illustration of the graphs related to each forum is given in Figs. 1 and 2. The node with label 0 corresponds to the instructors.

Measures. Next, we report some of the typical measures of a graph that can be obtained globally or individually for each node, and their values in our datasets. Notice that for some measures we consider simplified versions of the graphs, where the weight of each edge is the sum of the weights of all the edges between the underlying pair of nodes. Moreover, including self-edges means including the opening of new forum threads in the analysis.

For the case of degree centrality, we considered separately the in-degree and out-degree centralities. In this application, considering the simplified version of the graphs, the in-degree centrality is the number of neighbors whose replies a student receives, and the out-degree centrality is the number of neighbors that receive the replies given by a student. The results in Tables 3, 4 and 5 reveal that the in-degree centrality values are moderate, but the out-degree centrality is noticeable, indicating a non-homogeneous distribution of the number of neighbors that receive the replies submitted by the par-

Fig. 1. Forums activity graphs in the 2020 edition of the course. Lessons graph (left), programming graph (middle) and organization graph (right).

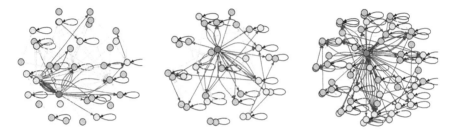

Fig. 2. Forums activity graphs in the 2021 edition of the course. Lessons graph (left), programming graph (middle) and organization graph (right).

ticipants, higher in the 2020 edition of the course. A subset of few nodes act as very active participants in forums (among them the teachers). Nevertheless, more nodes act as generators of new threads and recipients of information.

For the closeness centrality, that measures how easily a node can reach other nodes computing the inverse of the average length of the shortest paths to all the other nodes in the graph, the high values shown in Tables 3, 4 and 5 for the 2020 edition of the course are again indicative of the existence of few very active contributors. In the 2021 edition, values are more homogeneous among the participants.

In the case of the betweenness centrality, that tries to capture the importance of a node in terms of its role in connecting other nodes, computing the ratio between the number of shortest paths that a node lies on and the total number of possible shortest paths between two nodes, the high values observed in Tables 3, 4 and 5 suggest that few nodes act as bridges between different parts of the graph.

Eigenvector centrality is a measure based on the premise that a node's importance is determined by how important or influential its neighbors are. The scores arise from a reciprocal process in which the centrality of each node is proportional to the sum of the centralities of the nodes it is connected to. Considering the version of the graph with self-edges, Tables 3, 4 and 5 show that the measured eigenvector centrality values are noticeable. Again, this clearly means that there are substantial differences among the nodes in their patterns of participation in this activity.

A clique is a completely connected subgraph of a given graph. So, cliques represent strongly tied subcommunities where each member interacts with any other member.

Table 3. Summary of parameters of the lessons graph.

		2020 edition	2021 edition
degree	in	0.2136	0.2901
	out	0.6244	0.2687
closeness		0.6026	0.4605
betweenness		0.6716	0.4777
eigenvector		0.8454	0.8191
	Size		
# cliques	2	110	60
	3	68	29
	4	15	9
	4	0	1
number new threads ($\mu - \sigma$)		0.4608–0.9485	0.3333–1.5564
number new threads (mod. B) ($\mu - \sigma$)		0.5–0.9765	0.2183–0.7986
number replies ($\mu - \sigma$)		0.9043–2.0475	0.5784–1.3455
number replies (mod. B) ($\mu - \sigma$)		1.0408–2.1821	0.5632–1.1475
points replies ($\mu - \sigma$)		20.5652–46.1802	1.0392–2.2731
points replies (mod. B) ($\mu - \sigma$)		24–49.2351	1.0574–2.1205

Table 4. Summary of parameters of the programming graph.

		2020 edition	2021 edition
degree	in	0.3298	0.2241
	out	0.7161	0.4058
closeness		0.6827	0.1471
betweenness		0.7489	0.4449
eigenvector		0.8425	0.7853
	Size		
# cliques	2	84	50
	3	43	16
	4	4	1
number new threads ($\mu - \sigma$)		0.3217–0.7199	0.2058–0.6501
number new threads (mod. B) ($\mu - \sigma$)		0.3673–0.7651	0.2183–0.6892
number replies ($\mu - \sigma$)		0.7304–1.8746	0.2352–0.5482
number replies (mod. B) ($\mu - \sigma$)		0.8163–2.0017	0.2413–0.5491
points replies ($\mu - \sigma$)		14.2434–33.9929	0.4117–0.9476
points replies (mod. B) ($\mu - \sigma$)		15.6531–35.9713	0.4137–0.9219

Table 5. Summary of parameters of the organization graph.

		2020 edition	2021 edition
degree	in	0.2185	0.0487
	out	0.5832	0.5886
closeness		0.6071	0.1139
betweenness		0.6417	0.8091
eigenvector		0.8721	0.8832
	Size		
# cliques	2	141	72
	3	99	13
	4	42	1
	5	12	0
	6	1	0
number new threads ($\mu - \sigma$)		0.6086–1.2333	0.4509–1.0012
number new threads (mod. B) ($\mu - \sigma$)		0.6428–1.2701	0.4137–0.6743
number replies ($\mu - \sigma$)		1.2696–2.7859	0.2941–0.6977
number replies (mod. B) ($\mu - \sigma$)		1.3061–2.6879	0.2988–0.6308
points replies ($\mu - \sigma$)		27.9304–58.5248	0.3627–0.8764
points replies (mod. B) ($\mu - \sigma$)		28.1428–54.4638	0.3675–0.7941

And the crossclique number accounts for the number of cliques a node belongs to. Tables 3, 4 and 5 list the number of cliques in the graphs by their size.

Finally, if we consider the non-simplified version of the graphs, the in-degree centrality is the number of replies a student receives, and the out-degree centrality is the number of replies given by a student. Moreover, the number of self-edges accounts for the number of new threads opened by each student. In addition to the intensity of interactions, another important factor is their quality that can be measured taking into account the weights of the edges. The results in Tables 3, 4 and 5 show the mean value and the standard deviation of these measures for all the students that did not drop out the course and for the students that followed the modality B of the continuous assessment.

Correlations with Final Results. In order to check the relationship among the patterns of participation in the forums and the achievements of the course, we have measured the statistical correlations between the features under study in this section and the final grades in the final exam (in the 2020 edition of the course taking into account the sum of the scores of the 6 exercises before applying wildcards) and in the course of the students that followed the continuous assessment and did not drop out the course.

The results in Tables 6, 7, 8, 9, 10 and 11 show a statistically significant positive dependence between many of the considered factors and the students' performance in the lessons graph in both editions of the course and in the programming graph in the last edition.

Table 6. Correlation between individual features and student's performance in the final exam (lessons graph).

	2020 edition		2021 edition	
	$\hat{\rho}$	$(\hat{\beta}, t, \mathbb{P}(> |t|))$	$\hat{\rho}$	$(\hat{\beta}, t, \mathbb{P}(> |t|))$
in degree	0.2218	$(0.2441, 2.4192, 1.72 \cdot 10^{-2})$	0.6171	$(0.1311, 1.3521, 1.81 \cdot 10^{-1})$
out degree	0.1181	$(0.1298, 1.2643, 2.09 \cdot 10^{-1})$	0.1494	$(0.2934, 2.0071, 4.75 \cdot 10^{-2})$
betweenness	0.1151	$(0.0082, 1.2332, 2.22 \cdot 10^{-1})$	0.0547	$(0.0032, 0.5491, 5.84 \cdot 10^{-1})$
closeness	0.1593	$(1.4924, 1.7153, 8.91 \cdot 10^{-2})$	0.3322	$(3.0517, 3.5221, 6.41 \cdot 10^{-4})$
eigenvector	0.1918	$(2.9671, 2.0719, 3.99 \cdot 10^{-2})$	0.2188	$(2.1558, 2.2431, 2.71 \cdot 10^{-2})$
crossclique #	0.1473	$(0.0401, 1.5831, 1.16 \cdot 10^{-1})$	0.1808	$(0.0491, 1.8391, 6.89 \cdot 10^{-2})$
# new threads	0.1728	$(0.3498, 1.8652, 6.47 \cdot 10^{-2})$	0.0095	$(0.0107, 0.0952, 9.24 \cdot 10^{-1})$
# replies	0.0974	$(0.0913, 1.0404, 3.01 \cdot 10^{-1})$	0.2157	$(0.2811, 2.2112, 2.94 \cdot 10^{-2})$
points replies	0.1156	$(0.0048, 1.2376, 1.77 \cdot 10^{-1})$	0.2713	$(0.2092, 2.8192, 5.81 \cdot 10^{-3})$

Table 7. Correlation between individual features and student's performance in the course (lessons graph).

	2020 edition		2021 edition	
	$\hat{\rho}$	$(\hat{\beta}, t, \mathbb{P}(> |t|))$	$\hat{\rho}$	$(\hat{\beta}, t, \mathbb{P}(> |t|))$
in degree	0.1654	$(0.1951, 1.7832, 7.72 \cdot 10^{-2})$	0.6124	$(0.0981, 0.9091, 3.66 \cdot 10^{-1})$
out degree	0.1244	$(0.1466, 1.3331, 1.85 \cdot 10^{-1})$	0.0995	$(0.3093, 1.9121, 5.89 \cdot 10^{-2})$
betweenness	0.1526	$(0.0116, 1.6424, 1.03 \cdot 10^{-1})$	0.0516	$(0.0034, 0.5171, 6.06 \cdot 10^{-1})$
closeness	0.1072	$(1.0054, 1.3943, 1.66 \cdot 10^{-1})$	0.2932	$(2.9779, 3.0671, 2.78 \cdot 10^{-3})$
eigenvector	0.1201	$(1.9893, 1.2851, 2.01 \cdot 10^{-1})$	0.1545	$(1.6837, 1.5651, 1.21 \cdot 10^{-1})$
crossclique #	0.1249	$(0.0363, 1.3381, 1.83 \cdot 10^{-1})$	0.1411	$(0.0424, 1.4261, 1.57 \cdot 10^{-1})$
# new threads	0.1286	$(0.2791, 1.3791, 6.47 \cdot 10^{-2})$	−0.0307	$(-0.0382, -0.3083, 7.59 \cdot 10^{-1})$
# replies	0.1064	$(0.1071, 1.1385, 2.57 \cdot 10^{-1})$	0.1994	$(0.2871, 2.0351, 4.45 \cdot 10^{-2})$
points replies	0.1268	$(0.0056, 1.3591, 1.77 \cdot 10^{-1})$	0.2581	$(0.2201, 2.6724, 8.82 \cdot 10^{-3})$

Table 8. Correlation between individual features and student's performance in the final exam (programming graph).

	2020 edition		2021 edition	
	$\hat{\rho}$	$(\hat{\beta}, t, \mathbb{P}(> |t|))$	$\hat{\rho}$	$(\hat{\beta}, t, \mathbb{P}(> |t|))$
in degree	0.0461	$(0.0575, 0.4908, 6.25 \cdot 10^{-1})$	0.1799	$(0.2685, 1.7481, 8.35 \cdot 10^{-2})$
out degree	0.0233	$(0.1101, 0.2481, 8.05 \cdot 10^{-1})$	0.0474	$(0.6028, 1.9213, 5.77 \cdot 10^{-2})$
betweenness	−0.0005	$(-0.0004, -0.0001, 9.99 \cdot 10^{-1})$	0.2122	$(0.0331, 2.1721, 3.22 \cdot 10^{-2})$
closeness	0.1162	$(1.0895, 1.2442, 2.16 \cdot 10^{-1})$	0.0641	$(1.2051, 0.6421, 5.22 \cdot 10^{-1})$
eigenvector	0.0567	$(0.9791, 0.6041, 0.54 \cdot 10^{-1})$	0.2461	$(2.8183, 2.5461, 1.24 \cdot 10^{-2})$
crossclique #	0.0598	$(0.0242, 0.6381, 5.25 \cdot 10^{-1})$	0.1678	$(0.1201, 1.7032, 9.17 \cdot 10^{-2})$
# new threads	0.1042	$(0.2781, 1.1142, 2.68 \cdot 10^{-1})$	0.1801	$(0.4855, 1.6881, 7.01 \cdot 10^{-2})$
# replies	0.0408	$(0.0417, 0.4342, 6.65 \cdot 10^{-1})$	0.1885	$(0.6028, 1.9234, 3.58 \cdot 10^{-2})$
points replies	0.0055	$(0.0003, 0.0591, 9.53 \cdot 10^{-1})$	0.1519	$(0.2811, 1.5381, 1.27 \cdot 10^{-1})$

Table 9. Correlation between individual features and student's performance in the course (programming graph).

	2020 edition		2021 edition	
	$\hat{\rho}$	$(\hat{\beta}, t, \mathbb{P}(>|t|))$	$\hat{\rho}$	$(\hat{\beta}, t, \mathbb{P}(>|t|))$
in degree	0.0921	$(0.1232, 0.9832, 3.27 \cdot 10^{-1})$	0.3655	$(0.3056, 1.8012, 7.47 \cdot 10^{-2})$
out degree	0.0785	$(0.1101, 0.8381, 4.04 \cdot 10^{-1})$	0.0589	$(0.7355, 2.1281, 3.58 \cdot 10^{-2})$
betweenness	0.0361	$(0.0037, 0.3857, 7.01 \cdot 10^{-1})$	0.2046	$(0.0352, 2.0912, 3.91 \cdot 10^{-2})$
closeness	0.1012	$(1.5546, 1.6656, 9.88 \cdot 10^{-2})$	0.0715	$(1.4884, 0.7171, 4.75 \cdot 10^{-1})$
eigenvector	0.1098	$(2.0321, 1.1742, 0.24 \cdot 10^{-1})$	0.1882	$(2.3471, 1.9171, 5.81 \cdot 10^{-2})$
crossclique #	0.1835	$(0.0435, 1.0751, 2.85 \cdot 10^{-1})$	0.1861	$(0.1472, 1.8942, 6.11 \cdot 10^{-2})$
# new threads	0.1555	$(0.4446, 1.6743, 9.69 \cdot 10^{-2})$	0.1664	$(0.4961, 1.6881, 9.45 \cdot 10^{-2})$
# replies	0.0948	$(0.1041, 1.0132, 3.13 \cdot 10^{-1})$	0.2881	$(0.7355, 2.1281, 3.58 \cdot 10^{-2})$
points replies	0.0794	$(0.0048, 0.8472, 3.99 \cdot 10^{-1})$	0.1754	$(0.3588, 1.7831, 7.77 \cdot 10^{-2})$

5   Predicting the Outcomes

To assess the power of the above selected measures in predicting students success/failure, we have considered several statistical learning classifiers, namely logistic regression (LR), linear discriminant analysis (LDA) and support vector machines (SVM). These classifiers function in two phases: during the training phase they are fed with a set of labeled input-output pairs. Each classifier then adjusts its internal parameters so as to minimize a given loss function, and subsequently during the testing phase they are presented with new input data to predict the outputs. If actual output values are available, the comparison with the predicted ones is used to measure the performance of the classifier. Details of implementation of each classifier can be found in [23].

5.1   Background

This Section gives a very brief review of the selected classifiers. For a full reference on the topic, there exist excellent textbooks on the matter such as [23]. Let $\mathcal{T} = \{(\mathbf{x}_1, y_1), (\mathbf{x}_2, y_2), \dots, (\mathbf{x}_N, y_N)\}$ where $\mathbf{x}_i \in \mathbf{R}^n$, for $i = 1, \dots, N$, are n-dimensional the data points—each point has n distinct features—and $y_i \in \{-1, 1\}$ are the labels or indicators of a binary classification system.

   Logistic regression is the generalized non-linear regression over the data $\mathcal{T}$ with the logit function $\text{logit}(p) = \log p/(1-p)$, the inverse of the well-known sigmoid function. The regression (predicted) values are in $[0, 1]$, thus they naturally bear the meaning of probabilities for a binary classification problem.

   Linear discriminant analysis assumes that the points in each subset $y_i = +1$ and $y_i = -1$ have a Gaussian prior distribution with mean μ_i and covariance matrix Σ_i. A Bayesian hypothesis test optimally separates both classes according to a threshold criterion for the log-likelihood function

Table 10. Correlation between individual features and student's performance in the final exam (organization graph).

	2020 edition		2021 edition	
	$\hat{\rho}$	$(\hat{\beta}, t, \mathbb{P}(> |t|))$	$\hat{\rho}$	$(\hat{\beta}, t, \mathbb{P}(> |t|))$
in degree	−0.0203	$(-0.0156, -0.2176, 8.29 \cdot 10^{-1})$	0.0501	$(0.1002, 0.5021, 6.17 \cdot 10^{-1})$
out degree	−0.0279	$(-0.0252, -0.2971, 7.67 \cdot 10^{-1})$	0.1437	$(0.3872, 1.4531, 1.49 \cdot 10^{-1})$
betweenness	−0.0044	$(-0.0002, -0.0487, 9.62 \cdot 10^{-1})$	−0.0144	$(-0.0014, -0.1451, 8.85 \cdot 10^{-1})$
closeness	0.0362	$(0.3258, 0.3861, 7.01 \cdot 10^{-1})$	0.1028	$(2.0253, 1.0341, 3.04 \cdot 10^{-1})$
eigenvector	−0.0139	$(-0.2144, -1.4801, 8.83 \cdot 10^{-1})$	0.2019	$(2.1651, 2.0621, 4.18 \cdot 10^{-2})$
crossclique #	−0.0308	$(-0.0039, -0.3281, 7.43 \cdot 10^{-1})$	0.0998	$(0.0703, 1.0041, 3.18 \cdot 10^{-1})$
# new threads	−0.0347	$(-0.0541, -0.3691, 7.13 \cdot 10^{-1})$	−0.0052	$(-0.0091, -0.0523, 9.58 \cdot 10^{-1})$
# replies	−0.0795	$(-0.0547, -0.8487, 3.98 \cdot 10^{-1})$	0.1261	$(0.3165, 1.2706, 2.07 \cdot 10^{-1})$
points replies	−0.0686	$(-0.0022, -0.7324, 7.95 \cdot 10^{-1})$	0.1468	$(0.2935, 1.4841, 1.41 \cdot 10^{-1})$

Table 11. Correlation between individual features and student's performance in the course (organization graph).

	2020 edition		2021 edition	
	$\hat{\rho}$	$(\hat{\beta}, t, \mathbb{P}(> |t|))$	$\hat{\rho}$	$(\hat{\beta}, t, \mathbb{P}(> |t|))$
in degree	−0.0065	$(-0.0053, -0.0692, 9.45 \cdot 10^{-1})$	0.0507	$(0.1123, 0.5081, 6.12 \cdot 10^{-1})$
out degree	0.0095	$(0.0092, 0.1021, 9.19 \cdot 10^{-1})$	0.1639	$(0.4883, 1.6621, 9.96 \cdot 10^{-2})$
betweenness	0.0301	$(0.0014, 0.3221, 7.56 \cdot 10^{-1})$	−0.0592	$(-0.0066, -0.5941, 5.54 \cdot 10^{-1})$
closeness	0.0432	$(0.8162, 0.9044, 3.68 \cdot 10^{-1})$	0.0994	$(2.1653, 1.0012, 3.21 \cdot 10^{-1})$
eigenvector	−0.0068	$(-0.1128, -0.0732, 9.42 \cdot 10^{-1})$	0.1295	$(1.5156, 1.3071, 1.94 \cdot 10^{-1})$
crossclique #	−0.0225	$(-0.0031, -0.3281, 8.11 \cdot 10^{-1})$	0.0808	$(0.0629, 0.8112, 4.18 \cdot 10^{-1})$
# new threads	−0.0129	$(-0.0216, -0.1381, 7.13 \cdot 10^{-1})$	−0.0411	$(-0.0795, -0.4111, 6.82 \cdot 10^{-2})$
# replies	−0.0519	$(-0.0383, -0.5522, 5.82 \cdot 10^{-1})$	0.1321	$(0.3666, 1.3321, 1.86 \cdot 10^{-1})$
Points replies	−0.0245	$(-0.0008, -0.2611, 7.95 \cdot 10^{-1})$	0.1299	$(0.2873, 1.3112, 1.93 \cdot 10^{-1})$

$$(\mathbf{x} - \boldsymbol{\mu}_1)^T \Sigma_1^{-1}(\mathbf{x} - \boldsymbol{\mu}_1) + \log |\Sigma_1| - (\mathbf{x} - \boldsymbol{\mu}_{-1})^T \Sigma_{-1}^{-1}(\mathbf{x} - \boldsymbol{\mu}_{-1}) + \log |\Sigma_{-1}| > T$$

for some T.

A *support vector machine* uses the training points T for seeking the hyperplane that best separates the two classes. This is mathematically formulated as

$$\min_{\mathbf{w},b} ||A\mathbf{x} + \mathbf{1}b||^2$$

where $A \in \mathbf{R}^{N \times n}$ is the matrix having the observations $\mathbf{x}_i$ as rows and $\mathbf{1}$ is a column vector with all its elements equal to one. The resulting hyperplane defined by $\mathbf{w}$ and b linearly separates both sets of points, those from class $+1$ from those in class -1. A nonlinear version of a SVM is readily computed if one replaces the linear term $A\mathbf{x}$ with a general non-linear function (kernel) $K(\mathbf{x}, A)$ and solves the resulting optimization problem.

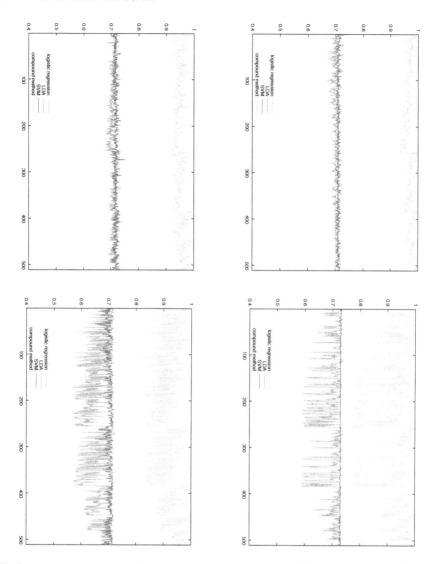

Fig. 3. Accuracy of each classifier for each subset of predictors of the lessons graphs. 1920 (top-left), 2021 (top-right), 1920 → 2021 (bottom-left) and 2021 → 1920 (bottom-right).

5.2 Methodology

In our application, the training sets consist of the selected student data of the two offerings of the course considered in the study (we have selected these datasets due to the high similarities in the methodology along the whole term in both offerings). The output is the binary variable that represents the success or failure of the students in the course, and the input is a combination of the features described in the previous section. k-fold cross validation is used to consider multiple training/testing set partitions. If the set of

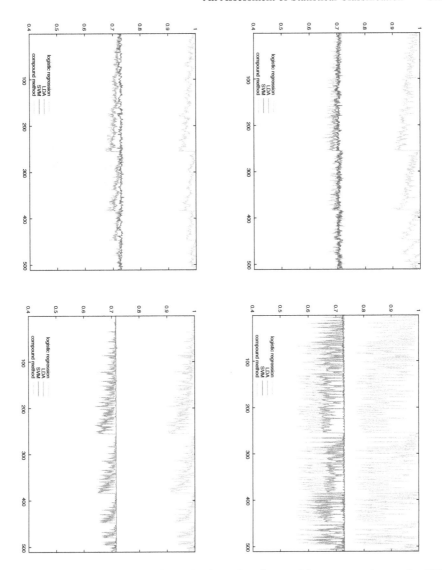

Fig. 4. Accuracy of each classifier for each subset of predictors of the programming graphs. 1920 (top-left), 2021 (top-right), 1920 → 2021 (bottom-left) and 2021 → 1920 (bottom-right).

observations is the same for training and testing, this approach involves randomly divide it into k groups of approximately equal size. The procedure is repeated k times and each time $k - 1$ different groups of observations are treated as the training set and the other one as the testing set. If one set of observations is used for training and another different for testing, the first one is divided into k groups of approximately equal size and in each repetition of the procedure $k - 1$ different groups are treated as the training set. In any case, as this procedure results in k values, the performance results are computed by

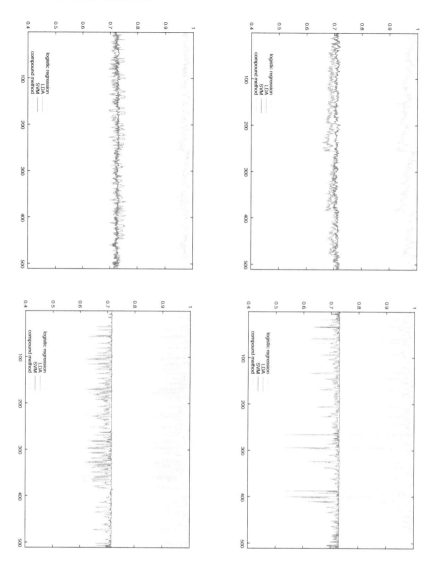

Fig. 5. Accuracy of each classifier for each subset of predictors of the organization graphs. 1920 (top-left), 2021 (top-right), 1920 → 2021 (bottom-left) and 2021 → 1920 (bottom-right).

averaging these values. We have selected $k = 5$ in our tests and, in order to increase the accuracy, we have repeated the procedure 10 times, being the final performance values obtained by averaging again the 10 resulting values.

To evaluate the performance of decision we have used three different criteria, which estimate the accuracy, the sensitivity and the precision. Let PF be the predicted failures, PS the predicted successes, TPF the correct predicted failures, TPS the correct predicted successes, FPF the incorrect predicted failures and FPS the incorrect predicted successes.

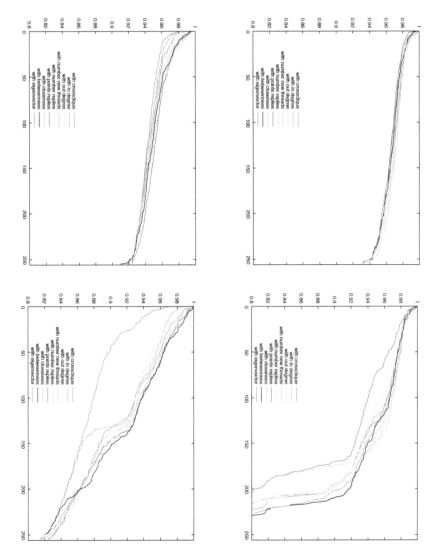

Fig. 6. Accuracy of the compound classifier for each subset of predictors of the lessons graphs. 1920 (top-left), 2021 (top-right), 1920 → 2021 (bottom-left) and 2021 → 1920 (bottom-right).

– The accuracy of a classifier is measured as the total proportion of the students whose final status, failing or passing the course, was correctly predicted

$$\text{Accuracy} = \frac{\text{TPF} + \text{TPS}}{\text{PF} + \text{PS}}.$$

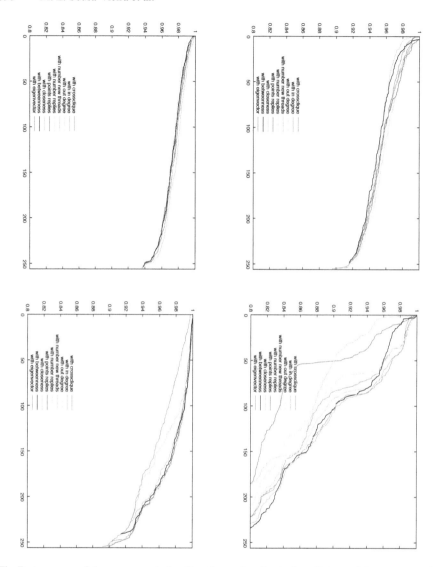

Fig. 7. Accuracy of the compound classifier for each subset of predictors of the programming graphs. 1920 (top-left), 2021 (top-right), 1920 → 2021 (bottom-left) and 2021 → 1920 (bottom-right).

– The sensitivity is measured as the proportion of the students whose final status, failing (or passing) the course, was correctly predicted

$$\text{Sensitivity} = \frac{\text{TPF}}{\text{TPF} + \text{FPS}} = \frac{\text{TPS}}{\text{TPS} + \text{FPF}}.$$

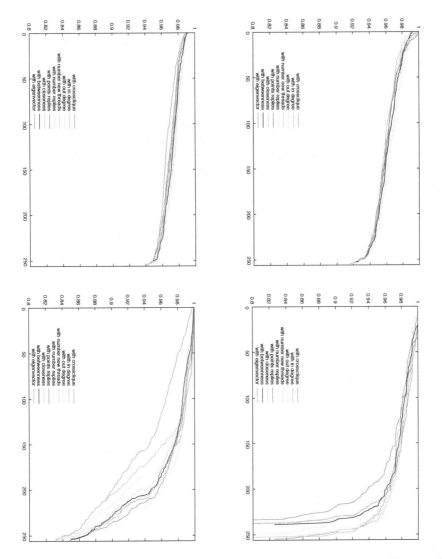

Fig. 8. Accuracy of the logistic regression classifier for each subset of predictors of the organization graphs. 1920 (top-left), 2021 (top-right), 1920 → 2021 (bottom-left) and 2021 → 1920 (bottom-right).

– The precision determines the proportion of the students that actually failed (or passed) the course, among all those that the method predicted as such.

$$\text{Precision} = \frac{\text{TPF}}{\text{TPF} + \text{FPF}} = \frac{\text{TPS}}{\text{TPS} + \text{FPS}}.$$

In Figs. 3, 4 and 5 we show the results obtained for the accuracy with each classifier (considering the prediction of successes), taking into account as predictors the 2^9 –

1 combinations of the 9 measures under study in this paper (crossclique number, in degree, out degree, number new threads, number replies, points replies, betweenness, closeness and eigenvector).

In each Figure, the two top graphs consider the same dataset for training and testing (19/20 and 20/21, respectively) and the two bottom graphs consider one of the datasets for training and the other one for testing (19/20 → 20/21 and 20/21 → 19/20, respectively). We can see that all the studied classifiers show similar results (in terms of accuracy), above 70%, for most of the combinations of predictor variables if we consider the same edition for training and texting. Nevertheless, accurate rates are a little lower and more variable, between 60% and 70%, if we consider and edition for training and the other one for testing.

In order to increase the level of accuracy, we propose a compound method where a student is predicted to fail the subject if at least one technique (logistic regression, LDA or SVM) has classified it as such. As we can see in Figs. 3, 4 and 5 with this strategy we obtain values of accuracy above 90% in most cases.

A closer look into the results unveils the better predictor variables, either individually or in combination with others. This can be seen in Figs. 6, 7 and 8, where we depict the results obtained for the accuracy of the compound classifier, taking into account as predictors all the $2^8 - 1$ combinations including one of the forum measures under measure, in order to identify the best predictors.

Therefore, for further study in a future work, we leave the task of analyzing the quality of classification when the best variables related to forums are taken jointly with other variables sampled along the course, such as the merit points, the karma level, etc.

6   Conclusions

Discovering the interactions and features which most influence the success of a gamification experience in teaching is difficult due to the variety of dynamics and rewards that can be designed and valued within a learning platform. In this study, we have posed this question as an automated classification problem, and then applied techniques from the machine learning domain to conduct the feature extraction and rank the classification methods. We found that, generally, the best results are obtained when an ensemble approach is used, that is, when blending the predictions made by the different classifiers. This is an interesting observation for our future work, which is focused on constructing a highly accurate ensemble classifier for the learning outcomes.

References

1. Aldemir, T., Celik, B., Kaplan, G.: A qualitative investigation of student perceptions of game elements in a gamified course. Comput. Hum. Behav. **78**, 235–254 (2018). https://doi.org/10.1016/j.chb.2017.10.001
2. Baydas, O., Cicek, M.: The examination of the gamification process in undergraduate education: a scale development study. Technol. Pedagogy Educ. **28**(3), 269–285 (2019). https://doi.org/10.1080/1475939x.2019.1580609

3. Buckley, P., Doyle, E.: Individualising gamification: an investigation of the impact of learning styles and personality traits on the efficacy of gamification using a prediction market. Comput. Educ. **106**, 43–55 (2017). https://doi.org/10.1016/j.compedu.2016.11.009
4. Cabada, R.Z., Estrada, M.L.B., Félix, J.M.R., Hernández, G.A.: A virtual environment for learning computer coding using gamification and emotion recognition. Interact. Learn. Environ. **28**(8), 1048–1063 (2018). https://doi.org/10.1080/10494820.2018.1558256
5. Cabot, A.G., López, E.G., Álvaro, S.C., Martínez, J.M.G., de Marcos, L.: Measuring the effects on learning performance and engagement with a gamified social platform in an MSc program. Comput. Appl. Eng. Educ. **28**(1), 207–223 (2019). https://doi.org/10.1002/cae.22186
6. Cadima, R., Ojeda, J., Monguet, J.M.: Social networks and performance in distributed learning communities. J. Educ. Technol. Soc. **15**, 296–304 (2012)
7. Cakirogli, U., Basibuyuk, B., Guler, M., Atabay, M., Memis, B.Y.: Gamifying an ICT course: influences on engagement and academic performance. Comput. Hum. Behav. **69**, 98–107 (2017). https://doi.org/10.1016/j.chb.2016.12.018
8. Cho, H., Gay, G., Davidson, B., Ingraffea, A.: Social networks, communication styles, and learning performance in a CSCL community. Comput. Educ. **49**(2), 309–329 (2007). https://doi.org/10.1016/j.compedu.2005.07.003
9. Chung, K.S.K., Paredes, W.C.: Towards a social networks model for online learning and performance. J. Educ. Technol. Soc. **18**, 240–253 (2015)
10. Dado, M., Bodemer, D.: A review of methodological applications of social network analysis in computer-supported collaborative learning. Educ. Res. Rev. **22**, 159–180 (2017). https://doi.org/10.1016/j.edurev.2017.08.005
11. Davis, K., Sridharan, H., Koepke, L., Singh, S., Boiko, R.: Learning and engagement in a gamified course: investigating the effects of student characteristics. J. Comput. Assist. Learn. **34**(5), 492–503 (2018). https://doi.org/10.1111/jcal.12254
12. Dawson, S.: A study of the relationship between student social networks and sense of community. J. Educ. Technol. Soc. **11**, 224–238 (2008)
13. Dias, J.: Teaching operations research to undergraduate management students: the role of gamification. Int. J. Manag. Educ. **15**(1), 98–111 (2017). https://doi.org/10.1016/j.ijme.2017.01.002
14. Ding, L., Er, E., Orey, M.: An exploratory study of student engagement in gamified online discussions. Comput. Educ. **120**, 213–226 (2018). https://doi.org/10.1016/j.compedu.2018.02.007
15. Gaggioli, A., Mazzoni, E., Milani, L., Riva, G.: The creative link: investigating the relationship between social network indices, creative performance and flow in blended teams. Comput. Hum. Behav. **42**, 157–166 (2015). https://doi.org/10.1016/j.chb.2013.12.003
16. Hakulinen, L., Auvinen, T., Korhonen, A.: The effect of achievement badges on students' behavior: an empirical study in a university-level computer science course. Int. J. Emerg. Technol. Learn. **10**(1), 18 (2015). https://doi.org/10.3991/ijet.v10i1.4221
17. Haythornthwaite, C.: Learning relations and networks in web-based communities. Int. J. Web Based Communities **4**(2), 140–158 (2008). https://doi.org/10.1504/IJWBC.2008.017669
18. Heo, H., Lim, K.Y., Kim, Y.: Exploratory study on the patterns of online interaction and knowledge co-construction in project-based learning. Comput. Educ. **55**(3), 1383–1392 (2010). https://doi.org/10.1016/j.compedu.2010.06.012
19. Hew, K.F., Huang, B., Chu, K.W.S., Chiu, D.K.: Engaging Asian students through game mechanics: findings from two experiment studies. Comput. Educ. **92–93**, 221–236 (2016). https://doi.org/10.1016/j.compedu.2015.10.010
20. Hommes, J., Rienties, B., de Grave, W., Bos, G., Schuwirth, L., Scherpbier, A.: Visualising the invisible: a network approach to reveal the informal social side of student learning. Adv. Health Sci. Educ. **17**(5), 743–757 (2012). https://doi.org/10.1007/s10459-012-9349-0

21. Huang, B., Hew, K.F.: Implementing a theory-driven gamification model in higher education flipped courses: effects on out-of-class activity completion and quality of artifacts. Comput. Educ. **125**, 254–272 (2018). https://doi.org/10.1016/j.compedu.2018.06.018

22. Ibáñez, M.B., Di-Serio, A., Kloos, C.D.: Gamification for engaging computer science students in learning activities: a case study. IEEE Trans. Learn. Technol. **7**(3), 291–301 (2014). https://doi.org/10.1109/tlt.2014.2329293

23. James, G., Witten, D., Hastie, T., Tibshirani, R.: An Introduction to Statistical Learning. Springer, New York (2013). https://doi.org/10.1007/978-1-4614-7138-7

24. Jan, S.K., Vlachopoulos, P.: Influence of learning design of the formation of online communities of learning, **19**(4) (2018). https://doi.org/10.19173/irrodl.v19i4.3620

25. Jo, J., Jun, H., Lim, H.: A comparative study on gamification of the flipped classroom in engineering education to enhance the effects of learning. Comput. Appl. Eng. Educ. **26**(5), 1626–1640 (2018). https://doi.org/10.1002/cae.21992

26. Jurgelaitis, M., Ceponiene, L., Ceponis, J., Drungilas, V.: Implementing gamification in a university-level UML modeling course: a case study. Comput. Appl. Eng. Educ. **27**(2), 332–343 (2018). https://doi.org/10.1002/cae.22077

27. Kapp, K.M.: The Gamification of Learning and Instruction: Game-Based Methods and Strategies for Training and Education. PFEIFFER & CO., May 2012

28. Kuo, M.S., Chuang, T.Y.: How gamification motivates visits and engagement for online academic dissemination - an empirical study. Comput. Hum. Behav. **55**, 16–27 (2016). https://doi.org/10.1016/j.chb.2015.08.025

29. Kyewski, E., Kramer, N.C.: To gamify or not to gamify? an experimental field study of the influence of badges on motivation, activity, and performance in an online learning course. Comput. Educ. **118**, 25–37 (2018). https://doi.org/10.1016/j.compedu.2017.11.006

30. de Laat, M., Lally, V., Lipponen, L., Simons, R.J.: Investigating patterns of interaction in networked learning and computer-supported collaborative learning: a role for social network analysis. Comput. Support. Learn. **2**(1), 87–103 (2007). https://doi.org/10.1007/s11412-007-9006-4

31. Landers, R.N., Landers, A.K.: An empirical test of the theory of gamified learning, **45**(6), 769–785 (2014). https://doi.org/10.1177/1046878114563662

32. Lin, J.-W., Mai, L.-J., Lai, Y.-C.: Peer interaction and social network analysis of online communities with the support of awareness of different contexts. Int. J. Comput.-Support. Collab. Learn. **10**(2), 139–159 (2015). https://doi.org/10.1007/s11412-015-9212-4

33. Manca, S., Delfino, M., Mazzoni, E.: Coding procedures to analyse interaction patterns in educational web forums. J. Comput. Assist. Learn. **25**(2), 189–200 (2009). https://doi.org/10.1111/j.1365-2729.2008.00296.x

34. de Marcos Ortega, L., Cabo, A.G., López, E.G.: Towards the social gamification of e-learning: a practical experiment. Int. J. Eng. Educ. **33**, 66–73 (2017)

35. Marín, B., Frez, J., Lemus, J.C., Genero, M.: An empirical investigation on the benefits of gamification in programming courses. ACM Trans. Comput. Educ. (TOCE) **19**(1), 1–22 (2019). https://doi.org/10.1145/3231709

36. Martín, J.S., Cañada, F.C., Acedo, M.A.D.: Just a game? Gamifying a general science class at university, **26**, 51–59 (2017). https://doi.org/10.1016/j.tsc.2017.05.003

37. Newman, M.: Networks. Oxford University Press, Oxford (2018)

38. Ortiz-Rojas, M., Chiluiza, K., Valcke, M.: Gamification through leaderboards: an empirical study in engineering education. Comput. Appl. Eng. Educ. **27**(4), 777–788 (2019). https://doi.org/10.1002/cae.12116

39. Ozdener, N.: Gamification for enhancing web 2.0 based educational activities: the case of pre-service grade school teachers using educational wiki pages. Telematics Inf. **35**(3), 564–578 (2018). https://doi.org/10.1016/j.tele.2017.04.003

40. Putnik, G., Costa, E., Alves, C., Castro, H., Varela, L., Shah, V.: Analysing the correlation between social network analysis measures and performance of students in social network-based engineering education. Int. J. Technol. Des. Educ. **26**(3), 413–437 (2015). https://doi.org/10.1007/s10798-015-9318-z

41. Reid, A.J., Paster, D., Abramovich, S.: Digital badges in undergraduate composition courses: effects on intrinsic motivation. J. Comput. Educ. **2**(4), 377–398 (2015). https://doi.org/10.1007/s40692-015-0042-1

42. Rienties, B., Tempelaar, D., den Bossche, P.V., Gijselaers, W., Segers, M.: The role of academic motivation in computer-supported collaborative learning. Comput. Hum. Behav. **25**(6), 1195–1206 (2009). https://doi.org/10.1016/j.chb.2009.05.012

43. van Roy, R., Zaman, B.: Need-supporting gamification in education: an assessment of motivational effects over time. Comput. Educ. **127**, 283–297 (2018). https://doi.org/10.1016/j.compedu.2018.08.018

44. Sailer, M., Hense, J.U., Mayr, S.K., Mandl, H.: How gamification motivates: an experimental study of the effects of specific game design elements on psychological need satisfaction. Comput. Hum. Behav. **69**, 371–380 (2017). https://doi.org/10.1016/j.chb.2016.12.033

45. Shea, P., et al.: Online learner self-regulation: learning presence viewed through quantitative content- and social network analysis. Int. Rev. Res. Open Distrib. Learn. **14**(3), 427 (2013). https://doi.org/10.19173/irrodl.v14i3.1466

46. Siqin, T., van Aalst, J., Chu, S.K.W.: Fixed group and opportunistic collaboration in a CSCL environment. Int. J. Comput.-Support. Collab. Learn. **10**(2), 161–181 (2015). https://doi.org/10.1007/s11412-014-9206-7

47. Smith, P.V., Jablokow, K.W., Friedel, C.R.: Characterizing communication networks in a web-based classroom: cognitive styles and linguistic behavior of self-organizing groups in online discussions. Comput. Educ. **59**, 222–235 (2012). https://doi.org/10.1016/j.compedu.2012.01.006

48. Sousa, M.E., Ferreira, O., López, J.C., Fernández, M.: Effectiveness of gamification in undergraduated education. In: Proceedings of the 13th International Conference on Computer Supported Education, vol. 1, pp. 376–385 (2021). https://doi.org/10.5220/0010495603760385

49. Sousa, M.E., López, J.C., Fernández, M., Ferreira, O.: The interplay between gamification and network structure in social learning environments: a case study. Comput. Appl. Eng. Educ. **28**(4), 814–836 (2020)

50. Sousa, M.E., López, J.C., Fernández, M., Rodríguez, M., López, C.: Mining relationships in learning-oriented social networks. Comput. Appl. Eng. Educ. **25**(5), 769–784 (2017)

51. Stepanyan, K., Mather, R., Dalrymple, R.: Culture, role and group work: a social network analysis perspective on an online collaborative course, **45**(4), 676–693 (2013). https://doi.org/10.1111/bjet.12076

52. Strmecki, D., Bernik, A., Radoevic, D.: Gamification in e-learning: introducing gamified design elements into e-learning systems. J. Comput. Sci. **11**, 1108–1117 (2015). https://doi.org/10.3844/jcssp.2015.1108.1117

53. Subhash, S., Cudney, E.A.: Gamified learning in higher education: a systematic review of the literature. Comput. Hum. Behav. **87**, 192–206 (2018). https://doi.org/10.1016/j.chb.2018.05.028

54. Thoms, B., Eryilmaz, E.: How media choice affects learner interactions in distance learning classes. Comput. Educ. **75**, 112–126 (2014). https://doi.org/10.1016/j.compedu.2014.02.002

55. Tirado, R., Hernando, Á., Aguaded, J.I.: The effect of centralization and cohesion on the social construction of knowledge in discussion forums. Interact. Learn. Environ. **23**(3), 293–316 (2012). https://doi.org/10.1080/10494820.2012.745437

56. Toda, A.M., do Carmo, R.M., da Silva, A.P., Bittencourt, I.I., Isotani, S.: An approach for planning and deploying gamification concepts with social networks within educational con-

texts. Int. J. Inf. Manage. **46**, 294–303 (2019). https://doi.org/10.1016/j.ijinfomgt.2018.10. 001

57. Toikkanen, T., Lipponen, L.: The applicability of social network analysis to the study of networked learning. Interact. Learn. Environ. **19**(4), 365–379 (2011). https://doi.org/10.1080/ 10494820903281999

58. Tsay, C.H.H., Kofinas, A., Luo, J.: Enhancing student learning experience with technology-mediated gamification: an empirical study. Comput. Educ. **121**, 1–17 (2018). https://doi.org/ 10.1016/j.compedu.2018.01.009

59. Wu, J.Y., Nian, M.W.: The dynamics of an online learning community in a hybrid statistics classroom over time: implications for the question-oriented problem-solving course design with the social network analysis approach. Comput. Educ. **166**, 104120 (2021). https://doi. org/10.1016/j.compedu.2020.104120

60. Xie, K., Yu, C., Bradshaw, A.C.: Impacts of role assignment and participation in asynchronous discussions in college-level online classes. Internet High. Educ. **20**, 10–19 (2014). https://doi.org/10.1016/j.iheduc.2013.09.003

61. Yildirim, I.: The effects of gamification-based teaching practices on student achievement and students' attitudes toward lessons. Internet High. Educ. **33**, 86–92 (2017). https://doi.org/10. 1016/j.iheduc.2017.02.002

62. Zainuddin, Z., Chu, S.K.W., Shujahat, M., Perera, C.J.: The impact of gamification on learning and instruction: a systematic review of empirical evidence. Educ. Res. Rev. **30**, 100326 (2020). https://doi.org/10.1016/j.edurev.2020.100326

Social Context and Learning Environments

Guided Inquiry Learning with Technology: Community Feedback and Software for Social Constructivism

Clif Kussmaul[1](✉) ⓘ and Tammy Pirmann[2] ⓘ

[1] Green Mango Associates, LLC, Bethlehem, PA 18018, USA
clif@kussmaul.org, trp74@drexel.edu
[2] College of Computing and Informatics, Drexel University, Philadelphia, PA 19104, USA

Abstract. To meet current and future demands, education needs to become more effective and more scalable. One evidence-based, social constructivist approach is Process Oriented Guided Inquiry Learning (POGIL). In POGIL, teams of learners work on specifically designed activities that guide them to practice key skills and develop their own understanding of key concepts. This paper expands a prior conference paper, and describes a series of investigations of how technology might enhance POGIL to be more effective and more scalable. The investigations include a survey and structured discussions among POGIL community leaders, a UI mockup and a working prototype, and experiences piloting the prototype in a large introductory course at a university. These investigations reveal that instructors are interested in using such tools. Tools can support richer learning experiences for students and provide better reporting to help instructors monitor progress and facilitate learning. The course pilot demonstrates that a prototype can support a large hybrid class. These investigations also identified promising areas for future work.

Keywords: Computer Supported Collaborative Learning · Process Oriented · Guided Inquiry Learning · POGIL · Social constructivism

1 Introduction

Worldwide, the demand for education continues to increase. However, education is traditionally labor-intensive and therefore costly. Thus, we need to make education more *scalable*. For example, *Massive Open Online Courses (MOOCs)* seek to provide open access to unlimited numbers of learners.

At the same time, the diversity of learner backgrounds and aptitudes also continues to increase. Educational approaches that might have worked for highly motivated, prepared, and talented learners are often less effective for broader audiences. Thus, we need to make education more *effective*. The ICAP Model [1] describes how learning outcomes improve as student behaviors progress from *passive* (P) to *active* (A) to *constructive* (C) to *interactive* (I). Thus, students should *interact* with each other and *construct* their own understanding of key concepts. A variety of such *social constructivist* approaches

© Springer Nature Switzerland AG 2022
B. Csapó and J. Uhomoibhi (Eds.): CSEDU 2021, CCIS 1624, pp. 407–428, 2022.
https://doi.org/10.1007/978-3-031-14756-2_20

have been developed and validated, including *Peer Instruction* [2–4], *Peer-Led Team Learning* [5, 6], *Problem-Based Learning* [7, 8], and *Process Oriented Guided Inquiry Learning (POGIL)* (see below).

Ideally, approaches should be both effective and scalable. This paper explores ways that technology could make social constructivism more effective and scalable.

This paper is organized as follows. Section 2 briefly summarizes relevant background on computing to support education and POGIL. Section 3 describes feedback from POGIL community leaders via a survey and structured discussions. Section 4 describes UI mockups, and a working prototype. Section 5 describes experiences with the prototype in an introductory computing course. Section 6 discusses insights, limitations, and future directions. Section 7 concludes.

This paper is an updated and expanded version of a short paper presented at a virtual conference on Computer Supported Education (CSEDU) [9]. It includes more background, including a description of a sample POGIL activity. It also includes more detail on the prototype (including more screenshots), detail on data visualization and accessibility, more data from community feedback, and an expanded discussion of insights, limitations, and future directions. A later, related conference paper focuses on monitoring team progress and responses [10].

2 Background

2.1 Computing to Support Education

For over 50 years, developments in computing have been applied to education [e.g., 11–13]. Typically, a *Computer Assisted Instruction (CAI)* or *Computer Based Training (CBT)* system presents a question or problem, evaluates responses, provides feedback, and chooses subsequent questions. Several subareas are particularly relevant for the current investigations.

Intelligent Tutoring Systems (ITS) seek to give more immediate and customized guidance, often using artificial intelligence methods [e.g., 14–19]. However, Baker [20] notes that widely used ITS are often much simpler than early visions or current research systems, and proposes a focus on systems "that are designed intelligently and that leverage human intelligence" (p. 608).

Computer Supported Collaborative Learning (CSCL) seeks to use technology to help students learn collaboratively [e.g., 21–24]. Jeong and Hmelo-Silver [25] describe desirable affordances for CSCL that align with social constructivism: joint tasks; ways to communicate; shared resources; productive processes; co-construction; monitoring and regulation; and effective groups.

More recently, *Learning Analytics (LA)* seeks to understand and optimize learning and learning systems using data from a variety of sources, including students' background and prior experiences, and interactions with a learning system [26]. *LA Dashboards* seek to aggregate and visualize such data, usually to support instructors [27]. However, LA tools "are generally not developed from theoretically established instructional strategies" and should seek to identify insights into learning processes and gaps in student understanding, not just measures of student performance [28, p. 65]. Perhaps for such reasons, a recent literature review and analysis concluded that LA has not yet demonstrated benefits in learning [29].

2.2 Process Oriented Guided Inquiry Learning (POGIL)

Process Oriented Guided Inquiry Learning (POGIL) is an evidence-based approach to teaching and learning that involves a set of synergistic practices [30, 31].

In POGIL, students work in teams of three to five to interact and construct their own understanding of key concepts. At the same time, students practice *process skills* (also called *professional* or *soft skills*) such as teamwork, communication, problem solving, and critical thinking.

To help focus attention on specific skills and perspectives, each team member has an assigned role; e.g., the *manager* tracks time and monitors team behavior, the *recorder* takes notes, the *presenter* (or *speaker*) interacts with other teams and the instructor, and the reflector (or *process analyst*) considers how the team might work more effectively. The roles rotate each class period so that all students experience each role and practice all skills. Roles also help to engage all students and prevent one or two students from dominating the team.

A POGIL activity consists of a set of *models* (e.g., tables, graphs, pictures, diagrams, code) each followed by a sequence of questions. Each team works through the activity, ensuring that every member understands and agrees with every answer; when students explain answers to each other, all of them understand better. POGIL activities are specifically designed to use *explore-invent-apply (EIA) learning cycles* in which different questions prompt students to *explore* the model, *invent* their own understanding of a concept, and then *apply* this learning in other contexts. (The next subsection describes a sample activity.)

In POGIL, the instructor is not a lecturer or passive observer, but an active facilitator who observes teams, provides high-level direction and timing, responds to student questions, guides teams that struggle with content or with process skills, and leads occasional short discussions of key questions and concepts. For example, if a team is stuck and unable to answer a question, the instructor might provide a hint. If several teams seem confused by a concept, the instructor might clarify it with a mini-lecture. An instructor who notices one student dominating a team might ask "Who is the reflector? Is everyone on this team contributing equally?". When most teams have answered a key question, the instructor might ask a few presenters to share their team's answer, and if answers vary the instructor might moderate a short discussion.

POGIL was initially developed for college level general chemistry, and has expanded across a wide range of disciplines [e.g., 32–36]. POGIL is used in small (<30) to large (>200) classes, and was adapted for virtual classes during the COVID pandemic. In a literature review [37], 79% (34 of 43) studies found positive effects and one only found negative effects.

The POGIL Project is a non-profit organization that works to improve teaching and learning by fostering an inclusive community of educators. The Project reviews, endorses, and publishes learning activities, and runs workshops, conferences, and other events. The Project has been identified as a model "community of transformation" for STEM education [38].

2.3 Sample POGIL Activity

This section briefly describes a POGIL-style activity on models of disease and vaccines. The first few sections of the activity do not assume programming experience, and students who complete them should be able to: describe general characteristics and examples of *compartmental models*; and explain how rates of infection and immunity can affect the course of a disease. Later sections of the activity do assume programming experience and provide pseudocode and starter code (in Python), and students who complete them should be able to design, code, test, and debug simulations that use compartmental models.

The activity starts with a few sentences of background, including the impacts of historic pandemics such as the Black Death, the Spanish flu, and HIV/AIDS. Next, a table prompts the team to assign a role to each member.

Section A of the activity shows two simple compartmental models. The first few questions (shown in Fig. 4 below) prompt students to *explore* the models; students count the number of stages and transitions, recognize terms such as "infected", "suspectable", and "recovered", and notice differences between the models. Next, students *invent* their own understanding and explain why the models are called SIS and SIR. Finally, students *apply* their new understanding and draw or describe a new model, and (if time permits) describe other situations where such models could be used.

Section B briefly explores a compartmental model with over twice as many stages and transitions. Teams who have worked well together on section A generally have little difficulty applying their new understanding to this model.

Section C presents three factors that can affect the how quickly a disease might spread (the length of the sickness, the number of daily contacts, and the probability of infection). Questions prompt students to *explore* the impact of each factor, and then use them to *invent* (write) an equation for the *basic reproductive number R_0*. Next, questions prompt students to *explore* the impacts of vaccine-based immunity, and then *invent* (write) their own definition for *herd immunity*.

Later sections prepare students to write a computer simulation based on a compartmental model. Section D focuses on pseudocode, section E on a set of data structures used in the simulation, and sections F and G on starter code. Thus, an instructor could select sections based on students' programming background, and the amount of expected effort.

In another activity for introductory computer science (CS), the first model describes a simple game, and questions guide teams to identify and analyze strategies to play the game, leading teams to discover a tradeoff between algorithm complexity and performance [39]. Websites have sample activities for a variety of disciplines (http://pogil.org), and numerous CS activities (http://cspogil.org).

2.4 Enhancing POGIL with Technology

POGIL practitioners are exploring how technology could enhance the POGIL experience in traditional classrooms as well as blended or online settings.

Prior to COVID, POGIL was primarily used in face-to-face settings, and students wrote or sketched answers on paper. Activities are usually distributed as PDFs or printed workbooks. Some instructors use tools (e.g., clickers, phone apps, learning management systems) to collect and summarize student responses, particularly in larger classes. Some instructors, particularly in computer science, give each team a collaborative document (e.g., a Google Doc) which make it easier to copy code or data to and from other software tools.

The COVID pandemic forced instructors and students to adapt to hybrid and online learning [e.g., 40–42], often using video conferencing tools (e.g., Zoom, Google Hangouts, Skype) and collaborative documents. It is harder for an instructor to monitor team progress when students are not physically present. For example, an instructor can visually scan a classroom in a few seconds to notice which teams are working well and which are struggling; this takes much longer when the instructor must cycle through breakout rooms. Thus, the pandemic has highlighted the importance of social presence, personal connections, and interactive learning for students, and has thus raised awareness and interest in social constructivism and supporting tools. Software tools also have the potential to leverage interactive models (e.g., simulations, data-driven documents, live code, collaborative documents). Networked tools have the potential to provide near real time data and feedback to students and instructors, which is common in CAI, but less common in social constructivism.

3 Community Feedback

This section describes feedback from POGIL community leaders, including a 2019 survey and a 2020 structured discussion.

3.1 Survey of POGIL Community Leaders

In June 2019, community leaders at the *POGIL National Meeting* were surveyed about available technology in their classes, and their interest in features for online activities. The response rate was 70% (47 of 65). Respondents included college (n = 36) and K-12 (n = 9) instructors. Disciplines included chemistry (n = 38), biology (n = 8), and others (n = 7); some taught multiple subjects. Typical class sizes were < 25 (n = 23), 25–50 (n = 10), and > 50 (n = 6). These values seem typical of the POGIL community, except that the latter involves a larger fraction of K-12 instructors.

Respondents rated the availability of three categories of technology. Figure 1 (top) summarizes responses, from least to most common. Half of respondents (n = 25) were at institutions that never or rarely provide computers, and less than a third (n = 14) often or always provide computers. In contrast, about half (n = 24) had students who often or always bring their own devices, and less than a third (n = 15) had students who never or rarely bring devices. Nearly all had reliable internet access.

These results show that (at US institutions) internet access was broadly available, but that most instructors expected students to bring a variety of devices (smartphones, tablets, and laptops, presumably) and could not assume that all students had access to institutional devices. Thus, to be widely adopted, technological tools should be web-based and not require specific hardware.

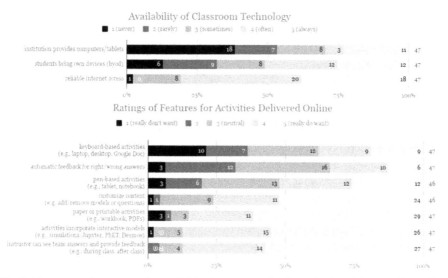

Fig. 1. Summary of responses for available classroom technology (top) and features for online activities (bottom). Items are listed on the left, from least to most common. The stacked bars show the number of instructors with each response, from 1 (most negative) on the left to 5 (most positive) on the right. Total responses are shown on the right. This figure is from the prior short paper [9].

Respondents also rated their interest in seven potential features. Figure 1 (bottom) summarizes responses, from least to most popular. Respondents were split on keyboard-based activities; 17 (36%) didn't want them, and 18 (38%) did. This might reflect POGIL's traditional use of paper activities, including activities where students draw or label diagrams and other content. Respondents were also split on automated feedback; 15 (32%) didn't want it, and 16 (34%) did. This might reflect unfamiliarity with such tools, or a strong belief in the instructor as an active facilitator. Pen-based activities were more popular; 9 (20%) didn't want them, and 24 (52%) did; this might reflect the similarities between a pen-based interface and traditional paper activities.

Respondents were strongly in favor of the other four features. Most (n = 35) wanted the ability to customize content, although this could result in activities that don't meet all POGIL criteria. Most (n = 40) wanted the option of paper or printable activities; again, this is common in POGIL. Nearly all (n = 41) wanted interactive models and ways to monitor responses and provide feedback. Thus, these became high priorities for the prototype (see below).

3.2 Discussions Among POGIL Community Leaders

In June 2020, a POGIL National Meeting session invited participants to "explore the opportunities and constraints that technology can provide", "explore a prototype ... for POGIL-style activities", and "have structured discussions about the opportunities and challenges". Twelve participants were selected to provide diverse perspectives, and worked in three groups. In each of three segments, they considered potential benefits and risks for three audiences: (a) students, (b) instructors and authors, and (c) The POGIL Project. Groups then discussed what they had written and identified themes and insights, which were then shared with the larger group. Segments (a) and (b) were preceded by demos of the web-based prototype (described below).

All feedback was copied into a *Freeplane* mind map (http://freeplane.org). Statements with multiple items were split, and similar items were clustered. Tables 1, 2 and 3 summarize the most common benefits and risks for each audience; the numbers in parentheses indicate the number of references to each idea. These tables add detail to a single table from the prior short paper [9].

Table 1. Potential benefits and risks of an online POGIL platform for students.

	Students
Benefits	(12) Flexibility for face-to-face, synchronous, asynchronous, and hybrid settings (during pandemic)
	(9) Variety of models and representations,
	including simulations (e.g., PhET), easier use of color
	(3) Individual students could be prompted for answers, receive feedback,
	and review content, for more engagement and less free-riding
	(3) Easy to use and familiar to students
	(6) Other: multiple response types, accessibility, lower cost flexibility,
	and reduced impact of "loud people"
Risks	(14) Less student-student interaction, less discussion and collaboration
	(6) More cognitive load and less student focus
	(6) Technology issues (rural access, devices), accessibility
	(4) Less emphasis on process skills
	(5) Concerns about system mechanics (e.g., saving, printing, drawing answers)
	(4) Other: cost, multiple formats, learning curve

Table 1 summarizes the top benefits and risks for students. The top benefits included flexible learning activities in a variety of settings, and a wider variety of models and representations. Less common benefits included student engagement and ease of use. The top risks included less interaction among students, more cognitive load, less emphasis on process skills, and technology, accessibility, and usability problems. Other risks included a reduced emphasis on process skills and concerns about systems mechanics.

Table 2 summarizes the top benefits and risks for instructors and activity authors. The top benefit was access to data, and specifically real time monitoring of student progress, and data to improve activities, compare classes, and support research. Other benefits included a single integrated platform and sharing activities with other instructors. The top risks included less interaction with students, and added effort to learn new tools and practices and to develop activities and facilitate them in classes. Other risks included system cost and reliability, as well as student privacy.

Table 2. Potential benefits and risks of an online POGIL platform for instructors and activity authors.

	Instructors & activity authors
Benefits	(24) Data generally (including 6 less specific responses)
	(9) Monitor student progress and answers in real time,
	and provide feedback, particularly in large classes
	(9) Access to student responses to compare classes, improve activities,
	study outcomes more broadly
	(12) Integrated platform with activity, responses, feedback, reporting out, etc
	(vs. using multiple tools)
	(9) Less work overall, compared to instructors creating and adapting
	their own online materials and tools
	(5) Avoids classroom limitations, especially in big classes
	(4) Predefined feedback to help students self-correct
	(3) Mix and match activities from different sources
Risks	(12) Less student-instructor interaction,
	due to watching dashboard instead of students
	(10) Learning curve for tools and facilitation practices
	(7) Increased effort to author or adapt materials using new tools
	(especially answer-specific feedback)
	(5) Cost and reliability of platform
	(4) Privacy issues with collecting, sharing, storing student data
	(3) Increased emphasis on final (correct) answers vs. process to reach answers
	(3) Not enough (or too much) flexibility in system

Table 3 summarizes the top benefits and risks for The POGIL Project and community. In general, these were similar to those for instructors and authors. Added benefits included broader access to materials, possible revenue to support The Project, and a general shift to digital delivery, particularly given COVID. (The POGIL Project is a non-profit, and revenue enables The Project to develop and distribute materials, and provide support for instructors to attend workshops and other events.) Added risks included concerns about equity and accessibility, and the risk that tools might go against the philosophy of collaborative learning.

Table 3. Potential benefits and risks of on online POGIL platform for The POGIL Project and community.

	The POGIL project & community
Benefits	(13) Data generally (including 4 less specific responses)
	(5) Data to review and improve activities
	(4) Data useful for research
	(12) Broader access to POGIL materials,
	including more adopters, faster dissemination
	(8) Increased revenue (without publishers in middle)
	(7) Push to digital given COVID and other trends
	(4) Other: accessibility / ADA, community support
Risks	(16) Time and cost to create system and convert materials
	(8) Ongoing costs, updating system and resources
	(10) Security and privacy for student data (and activities)
	(8) Equity and accessibility
	(5) Goes against philosophy of collaborative learning
	(3) Doesn't meet needs of diverse POGIL community
	(3) Misuse by untrained adopters
	(3) Other: copyright, subscriptions, third party platform

4   Software Development

This section describes a user interface mockup and a web-based prototype. These are similar to many CAI systems and ITS, and seek to follow the advice (summarized above) from Baker [20] and Jeong and Hmelo-Silver [25]. However, POGIL provides key differences, including structured student teams, the learning cycle structure, and active instructor facilitation. Teams typically respond to a question every minute or so, so the speed and correctness of their responses should provide valuable insights into how they work and learn, and how POGIL could better support student learning.

4.1   User Interface Mockup

In 2018, the first author developed a user interface (UI) mockup (in HTML and JavaScript) to stimulate discussion and reflection, and gather informal feedback from the POGIL community.

Figure 2 shows a view for a student team. (The model is a placeholder and the question text is lorem ipsum to focus on visual form rather than content.) The header (blue) has the title of the activity and a list of sections, for easy navigation. A timer shows the time left in the activity, to help teams manage time more effectively. The header also has a status bar (green) showing how much of the activity the team has completed, and how often they responded correctly. Below the header is the section title, which could also include a countdown timer. The section starts with a model; the figure shows a Data-Driven Document (D3) [43], but models could also use text, static figures, or other interactive tools.

Below the model is a sequence of questions. The questions can take several forms: multiple choice, checkboxes, numeric value, short or long text, etc. When the student team gives a correct response, they see the next question. They can also receive feedback, perhaps with a hint to help them find a better response (green). In some cases, incorrect responses might branch to other questions to guide the team towards the correct answer.

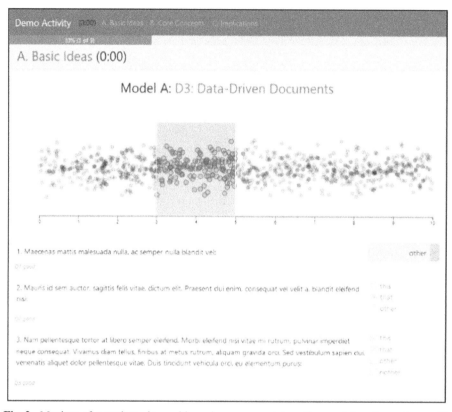

Fig. 2. Mockup of a student view, with pacing cues, an interactive model, and questions with auto-feedback.

Figure 3 shows a view for an instructor. This view shows a header, titles, models, and questions, similar to the student team view. As described above, in POGIL the instructor is an active facilitator, who continually monitors progress, assists teams that have problems with content or process, and leads short discussions. Thus, this view adds bar graphs (red, yellow, and green) and/or histograms (grey) for each question to show the distribution of responses, and timing. An instructor could then drill down to see which teams have the most difficulty, and might need help. Similarly, an activity author could find out which questions are most difficult or confusing (based on response time and incorrect responses) in order to revise them, or add questions or hints to help students succeed. If demographic data is available, an author could identify questions that are difficult for certain audiences (by gender, ethnicity, etc.).

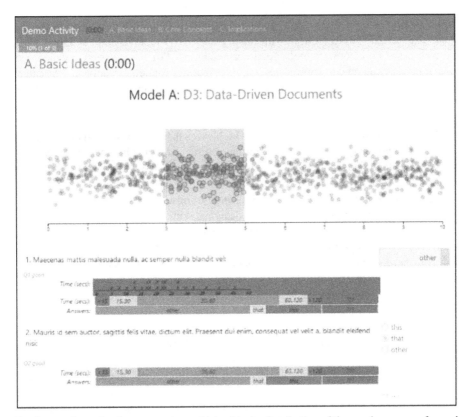

Fig. 3. Mockup of an instructor view, which adds the distribution of time and responses for each question. This is an expanded version of a figure from the prior short paper [9].

This mockup was not evaluated formally, but was shared informally with POGIL practitioners and other interested people, to stimulate conversation and reflection about the potential impacts of such tools.

A working system based on this mockup would not require new research or technology; most learning management systems (LMSs) have similar features. However, POGIL provides several key differences, including the learning cycle structure, students who work and learn in teams, and active facilitation by the instructor. Thus, the system should closely monitor team progress as the activity guides them through learning cycles to develop concepts and practice skills. Teams typically respond to a question every minute or so, and data on the speed and correctness of their responses should provide valuable insights into how students think and teams work, and how POGIL activities can be used and refined to better support student learning.

4.2 Web-Based Prototype

Based on the mockup and instructor feedback, a web-based platform is being developed to support POGIL and similar forms of social constructivism, and to help instructors create, facilitate, assess, and refine learning activities. *Guided Inquiry Learning with Technology (GILT)* builds on work with social constructivism, POGIL, CAI, ITS, CSCL, and web-based collaboration tools. As described below, GILT focuses on key elements used in POGIL and related approaches, includes features for collaboration and research, and leverages existing tools (when possible) to avoid overdesign or duplication of effort. (GILT is a single page application using the *Mongo, Express, Angular, Node (MEAN)* software stack.)

With GILT, teams work in a browser, which saves their responses and questions for later review and analysis. In hybrid or online settings, students interact virtually (e.g., in Zoom or Hangouts). An instructor can manage teams and activities, monitor team progress, and review and comment on team responses. An activity author can create and edit activities, and review student responses and timings.

In POGIL, each activity is deliberately designed with a sequence of questions about a model. In GILT, models can be also dynamic and interactive, including videos, simulations such as *NetLogo* [44] or *PhET* [45], coding environments such as *repl.it* (https:// repl.it) or *Scratch* [46], and other components, such as *Data-Driven Documents (D3)* [43]. Questions can take varied forms including plain text, multiple choice, numeric sliders, etc. GILT could also include questions to help assess process skills, mood, and effectiveness.

In POGIL, teams discuss each question and agree on a response; if it is incorrect, the instructor might ask a leading question or - if several teams have the same difficulty - lead a short class discussion. The instructor's notes for an activity might include sample answers and discussion prompts. GILT can provide predefined feedback for common team responses to partially automate this process (particularly when teams are remote or asynchronous), and an author can review a report on the most common responses. Defining feedback can be time consuming but one author's work could benefit many instructors.

In GILT, an instructor can quickly see the status of each team and the class as a whole, lead classroom discussions on difficult questions, and check in with teams having difficulty. Similarly, an activity author can explore the most difficult questions, the most common wrong responses, and useful correlations (e.g., by institution, gender/ethnicity).

Figure 4 shows a student team view of a model and the first few questions in an activity on models of disease (described above). The model has some text with background information and two compartmental models. The questions include text, sliders for numeric responses, and checkboxes for multiple choice.

Figure 5 shows an instructor or author view of the same activity. In this view, each question includes possible responses and advice specified by the activity author, and a history of all student responses. Question 1a has one correct response, and two incorrect responses with advice to students. One team tried several times before giving the correct response. An instructor could also add team-specific feedback, particularly in settings where the instructor can't easily speak to individual teams.

Thus, GILT seeks to help balance key tensions in the learning experience. Guiding students to discover concepts leads to deeper, broader understanding but can take more time. Student teams enhance learning but can allow some students to be passive and let others do the work. Automated feedback can help students who are stuck, but can

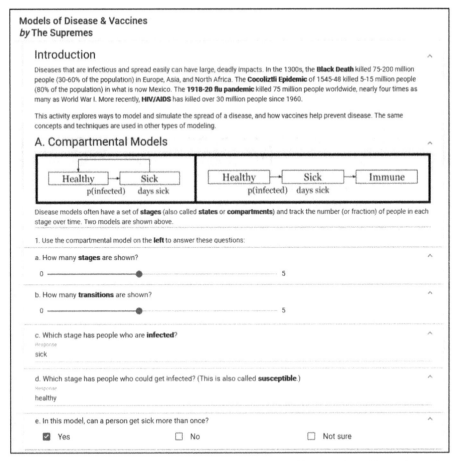

Models of Disease & Vaccines
by The Supremes

Introduction

Diseases that are infectious and spread easily can have large, deadly impacts. In the 1300s, the **Black Death** killed 75-200 million people (30-60% of the population) in Europe, Asia, and North Africa. The **Cocoliztli Epidemic** of 1545-48 killed 5-15 million people (80% of the population) in what is now Mexico. The **1918-20 flu pandemic** killed 75 million people worldwide, nearly four times as many as World War I. More recently, **HIV/AIDS** has killed over 30 million people since 1960.

This activity explores ways to model and simulate the spread of a disease, and how vaccines help prevent disease. The same concepts and techniques are used in other types of modeling.

A. Compartmental Models

| Healthy | → | Sick | | Healthy | → | Sick | → | Immune |
| p(infected) | days sick | | p(infected) | days sick | |

Disease models often have a set of **stages** (also called **states** or **compartments**) and track the number (or fraction) of people in each stage over time. Two models are shown above.

1. Use the compartmental model on the **left** to answer these questions:

a. How many **stages** are shown?

0 ——————●———————————— 5

b. How many **transitions** are shown?

0 ——————●———————————— 5

c. Which stage has people who are **infected**?
Response
sick

d. Which stage has people who could get infected? (This is also called **susceptible**.)
Response
healthy

e. In this model, can a person get sick more than once?

☑ Yes   ☐ No   ☐ Not sure

Fig. 4. Student view of an activity in GILT, including a model with text and graphics, and questions with numeric sliders, multiple choice, and text.

also encourage random guessing. An instructor can provide valuable guidance and support, but might not always be available (e.g., if teams work outside of scheduled class meetings).

GILT tracks every student response, when it occurred, and when each team starts and stops work on each part of an activity. This can provide near real time feedback to teams and instructors, summary reports for instructors and activity authors, and rich evidence for researchers. For example, authors could examine the distribution of responses and timings to see the impact of adding, editing, or removing elements of an activity. An author could use AB testing to test different forms of a question, different sequences of questions, or different models, and decide which best support student learning. Similarly, researchers could track team and instructor behavior using minute-by-minute data on which views, elements, or subcomponents are used, similar to established classroom observation protocols [e.g., 47–49].

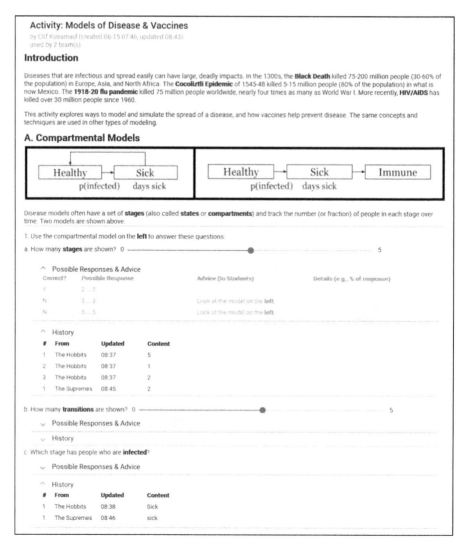

Fig. 5. Instructor and author view of a GILT activity, including possible correct and incorrect responses (shaded); and history of student responses. This is an expanded version of a figure from the prior short paper [9].

4.3 Data Visualization

The full history of team responses (as shown in Fig. 5) can be useful, but can also be difficult to interpret, particularly when there are more than a few teams, and different teams are working on different questions. Thus, GILT includes a variety of data visualizations to help instructors monitor team responses and provide appropriate support.

Figure 6 shows a summary table of team responses to a specific set of questions. Each row corresponds to a student team, and the column to the right of the team names shows the number of responses from that team. The remaining columns correspond to individual questions, and show each team's response to that question. The column heading includes an abbreviated version of the question (e.g., "stages" rather than "how many stages are shown?"), the correct response, and the number of teams that have responded. Thus, an instructor can quickly see that one team ("The Ducks") has multiple incorrect responses, while the other two teams are responding correctly. In the future, this table could be enhanced to highlight correct and incorrect responses (e.g., using different colors), and to allow an instructor to sort the table by correctness.

Team	Question: responses	a. stages? [2] 3	b. transitions? [2] 3	c. infected? [sick] 3	d. could be? [healthy] 3
1 The Ducks	5	5	4	Infected	Susceptible
2 The Hobbits	5	2	2	Sick	healthy
3 The Supremes	5	2	2	sick	healthy

Fig. 6. Summary table showing each team's responses to each question.

Figure 7 shows bar and pie charts to summarize the distribution of student responses (to the first two questions in the summary table). These help an instructor to see that two-thirds of teams gave one response and one-third gave another response, even if there were many teams. In both charts, tooltips show the actual responses and a list or count of teams with that response. The numeric sliders on the left allow an instructor to adjust the range of response counts shown in the charts; increasing the minimum count would hide the least frequent responses, decreasing the maximum would hide the most frequent responses so an instructor could more easily explore the remainder. Again, these charts could be enhanced to highlight correct and incorrect responses.

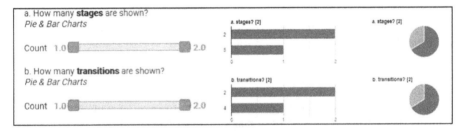

Fig. 7. Bar chart and pie chart showing distribution of team responses to each question.

A separate conference paper [10] describes other more complex visualizations, include timelines to show when each team responded to each question, Sankey diagrams to show how the changing distribution of team responses to a question (as teams revise their responses), and word trees to show common words and sequences in responses that consist of phrases or sentences.

4.4 Accessibility

Technology has the potential to enhance (or restrict) access and independence for people with disabilities, and increasing the diversity of computing professionals also benefits technical fields and society. Thus, accessibility of software and learning materials (for persons with disabilities) is a critical issue, particularly for legacy systems [50–53].

GILT is built on the Angular framework which includes accessibility support, and uses tools such as Pa11y and WAVE [53–55] to help identify and correct problems in the platform and individual learning activities. Currently, GILT meets most criteria for Web Content Accessibility Guidelines (WCAG) 2 Level AAA [56]. Most of the changes were minor, such as tagging navigation elements, adding `alt` attributes for images, and using `id` and `headers` attributes in tables.

5   Course Pilot

In Fall 2020, GILT was piloted in weekly sessions of an introductory computing course with five sections of ~80 students (i.e., ~20 teams per section). The instructor and the GILT developer worked together to migrate learning activities, identify and resolve problems, and clarify priorities for future development. Migrating activities was usually straightforward, and could probably be supported by undergraduate assistants. Most teams used videoconferencing (e.g., Zoom) to see a shared screen and each other. The user interface was revised to improve clarity for students, and to add information for instructors. It was helpful for the instructor to be able to quickly see the range of student responses, especially for questions that prompted students to develop insights or conclusions.

The instructor reviewed and analyzed end-of-term student course evaluations. Feedback on GILT focused on the POGIL-style class and activities, more than the software platform, and was similar to that in traditional POGIL classes – some students articulate how social constructivism helps them understand concepts and develop skills, and a few dislike teams or believe they would learn more through lectures. Some students wanted smaller teams, and the use of roles was polarizing – some students found them very helpful, others disliked them. As in traditional POGIL classes, it might help for the instructor to more frequently articulate and demonstrate the advantages of POGIL over lecture [57–59].

Feedback on specific learning activities included ambiguous wording and some repetitive questions. Feedback on the GILT platform focused on some problems saving responses, linking to external sites, and navigating within GILT. Some issues were addressed during the term, and others are in progress.

Here are quotes (from four different students) that are generally positive, lightly edited for spelling and grammar. All of these students clearly recognize the value of working in teams and constructing their own understanding.

> *"… one aspect that should absolutely be retained … is the GILT worksheet activities. I feel like they were robust and … where I learned the most … so I ended up learning all that information in a fairly reasonable amount of time."*

> *"I think that the GILT activities … should be retained … These activities were easy to do and introduced us to various new topics while also giving us experience working in a team. I liked how our roles during these activities rotated each week and allowed us to contribute to our group in a different way each time. …*

the GILT activities were effective because they allowed us to learn new concepts while discussing questions among team members."

"I believe that the GILT team activities not only teach the students about concepts ..., but they also teach us teamwork and how to play a role ..., which will be very important in the working field."

"... completing questions [in] GILT is amazing. Not only because it would put us through a new realm of knowledge, but because it forced us to involve in a healthy and fruitful discussion with our peers and helps us to create a new bonding. In fact, I also made a few close friends from this course."

Here are quotes (from two different students) that are more critical, lightly edited for spelling and grammar. The first focuses on occasional technical issues, which are being addressed in more recent versions of the software. The second is more about the POGIL approach than the GILT platform, and is similar to feedback from more traditional POGIL courses. Research generally shows that students learn more from social constructivism than from lecture, although this might not be true for every student.

"My main issue with this class was the GILT assignments. There were a few issues my group would run into sometimes: whether it be answers not being saved or not having access to the assignment at all."

"I feel as though the lecture sessions should be more lecture based and not work-sheet filling based. ... I might have learned more if I was being told the information, and had to get the answers down on a sheet to show I was paying attention. Because doing the GILT's helped me learn the concepts, but I feel as though if there was a secondary way of learning the information that would have helped me learn better."

6   Discussion

The work described in this paper demonstrates how POGIL and related approaches to teaching and learning could benefit from technology. GILT leverages human intelligence, as suggested by Baker [20]; is based on an established instructional strategy and seeks to identify useful insights, as suggested by Gašević, Dawson, and Siemens [28]; and provides at least some of the affordances identified by Jeong and Hmelo-Silver [25].

GILT also demonstrates the power of a modern software framework (MEAN), which supports navigation and routing, database interaction, accessibility, and provides tools for unit testing, coverage, style, accessibility.

As with any research, particularly involving education, the investigations described in this paper have some limitations. The survey (Sect. 3.1) and structured discussions (Sect. 3.2) involved experienced or expert POGIL instructors, many with 10–20 years of experience creating materials and using social constructivist practices. Instructors with little or no POGIL experience might have more (or less) trouble considering how tools might affect their classes, but might also be more open to new approaches. In contrast, relatively few instructors had used (or developed) CAI tools like GILT; it would be useful to more have feedback from instructors who are familiar with other tools. The survey pre-dated the COVID pandemic, and the discussions during the first summer (2020) when many instructors and institutions were scrambling to figure out how to provide effective learning environments for 2020–21. After a year or more of online and hybrid learning, instructors might have quite different opinions and insights. The discussions were coded and analyzed by one person (the first author), not a calibrated team; other researchers might identify different themes.

The UI mockup (Sect. 4.1) and web-based prototype (Sect. 4.2) are very much works in progress, and will benefit from further refinement. The instructor feedback was mostly informal and based on a brief introduction. It would be useful to have a more thorough, rigorous, and theory-based assessment, to see how well GILT aligns with principles from CAI, CSCL, and POGIL, and to identify potential problems and enhancements.

The course pilot (Sect. 5) involved one course with one instructor, although other instructors have used GILT for that course more recently. Also, the student feedback was extracted from end-of-term course evaluations. It will be useful to pilot GILT with other instructors, courses, disciplines, and institutions, and to collect more targeted data on student and instructor perceptions of specific features. For example, learning activities and courses where students draw diagrams or write complex equations might motivate additional features.

7 Conclusions

Technology, thoughtfully applied, has the potential to enhance education practices, including social constructivist approaches such as POGIL. This paper has described a set of investigations on how technology could support such approaches, including UI mockups, a survey and structured discussion among leading practitioners in the POGIL community, a working prototype, and a pilot in a large course.

These investigations have produced a variety of useful insights. POGIL instructors are interested in tools that can support or enhance POGIL practices, particularly for classes that are large or physically distributed (e.g., due to the COVID pandemic). Tools can provide near real time data and reports to help instructors facilitate learning, and enhance communication among student teams. Compared to traditional paper-based activities, software tools could also support more diverse contexts and a variety of interactive models. However, these benefits are tempered by concerns about reduced interactions among students and with teachers. Additional concerns include the time required to learn new tools, migrate learning activities, and adapt teaching and learning practices.

Piloting a web-based prototype in a large introductory course demonstrated that the prototype could support teams and provide useful data for instructors and activity authors. Feedback from instructors and students is encouraging. The pilot also identified a variety of areas for improvement and numerous potential enhancements.

We continue to implement, test, and refine technology-based tools to support POGIL. As in many software projects, a key challenge is to avoid feature creep and focus on features that are most useful for the students, instructors, activity authors, and researchers who will use the system. Current priorities include:

- Better error handling and logging to help diagnose and correct problems.

- Improving system performance, by reducing the number and size of messages between the server and clients.
- Enhance reporting and dashboards with charts and natural language processing.
- Support responses using tables, matching, sorting, and perhaps sketches.

Acknowledgements. This material is based in part upon work supported by the US National Science Foundation (NSF) grant #1626765. Any opinions, findings and conclusions or recommendations expressed are those of the author(s) and do not necessarily reflect the views of the NSF.

The POGIL Project (http://pogil.org) and the broader POGIL community have provided invaluable advice, encouragement, and support.

References

1. Chi, M.T.H., Wylie, R.: The ICAP framework: linking cognitive engagement to active learning outcomes. Educ. Psychol. **49**(4), 219–243 (2014). https://doi.org/10.1080/00461520.2014.965823

2. Mazur, E.: Peer instruction: getting students to think in class. In: AIP Conference Proceedings, vol. 399, no. 1, pp. 981–988 (1997). https://doi.org/10.1063/1.53199

3. Porter, L., Bouvier, D., Cutts, Q., et al.: A multi-institutional study of peer instruction in introductory computing. In: Proceedings of the ACM Technical Symposium on CS Education, pp. 358–363 ACM, New York (2016). https://doi.org/10.1145/2839509.2844642

4. Porter, L., Bailey Lee, C., Simon, B., Zingaro, D.: Peer instruction: do students really learn from peer discussion in computing? In: Proceedings of the International Workshop on Computing Education Research (ICER), pp. 45–52. ACM, Providence (2011). https://doi.org/10.1145/2016911.2016923

5. Gafney, L., Varma-Nelson, P.: Peer-Led Team Learning: Evaluation, Dissemination, and Institutionalization of a College Level Initiative. Springer, Cham (2008). https://doi.org/10.1007/978-1-4020-6186-8

6. Horwitz, S., Rodger, S.H., Biggers, M., et al.: Using peer-led team learning to increase participation and success of under-represented groups in introductory computer science. In: Proceedings of the ACM Technical Symposium on CS Education (SIGCSE), pp 163–167. ACM, Chattanooga (2009)

7. Hmelo-Silver, C.E. Problem-based learning: what and how do students learn? Educ. Psychol. Rev. **16**, 3, 235–266 (2004). https://doi.org/10.1023/B:EDPR.0000034022.16470.f3

8. Amador, J.A., Miles, L., Peters, C.A.: The Practice of Problem-Based Learning: A Guide to Implementing PBL in the College Classroom. Anker Publishing Company Inc., Bolton (2007)

9. Kussmaul, C., Pirmann, T. Guided inquiry learning with technology: investigations to support social constructivism. In: Proceedings of the International Conference on Computer Supported Education (CSEDU) (2021a). https://doi.org/10.5220/0010458104830490

10. Kussmaul, C., Pirmann, T.: Monitoring student team progress and responses in guided inquiry learning with technology. In: Proceedings of the International Conference on Advanced Learning Technologies (ICALT) (2021b). https://doi.org/10.1109/ICALT52272.2021.00046

11. Rath, G.J.: The development of computer-assisted instruction. IEEE Trans. Hum. Factors Electron. **HFE-8**(2), 60–63 (1967). https://doi.org/10.1109/THFE.1967.233312

12. Buck, G., Hunka, S.: Development of the IBM 1500 computer-assisted instructional system. IEEE Ann. Hist. Comput. **17**(1), 19–31 (1995). https://doi.org/10.1109/85.366508
13. Johnson, D.A., Rubin, S.: Effectiveness of interactive computer-based instruction: a review of studies published between 1995 and 2007. J. Organ. Behav. Manag. **31**(1), 55–94 (2011). https://doi.org/10.1080/01608061.2010.541821
14. Sleeman, D., Brown, J.S. (eds.): Intelligent Tutoring Systems. Academic Press, Cambridge (1982)
15. Graesser, A.C., Conley, M.W., Olney, A.: Intelligent tutoring systems. In: APA Educational Psychology Handbook, vol. 3: Application to Learning and Teaching, pp. 451–473. American Psychological Association (2012). https://doi.org/10.1037/13275-018
16. Paviotti, G., Rossi, P.G., Zarka, D. (eds.): Intelligent Tutoring Systems: An Overview. Pensa Multimedia (2012)
17. Ma, W., Adesope, O.O., Nesbit, J.C., Liu, Q.: Intelligent tutoring systems and learning outcomes: a meta-analysis. J. Educ. Psychol. **106**(4), 901–918 (2014). https://doi.org/10.1037/a0037123
18. Steenbergen-Hu, S., Cooper, H.: A meta-analysis of the effectiveness of intelligent tutoring systems on college students' academic learning. J. Educ. Psychol. 106(2), 331–347 (2014). https://doi.org/10.1037/a0034752
19. Kulik, J.A., Fletcher, J.D.: Effectiveness of intelligent tutoring systems: a meta-analytic review. Rev. Educ. Res. **86**(1), 42–78 (2016). https://doi.org/10.3102/0034654315581420
20. Baker, R.S.: Stupid tutoring systems, intelligent humans. Int. J. Artif. Intell. Educ. **26**(2), 600–614 (2016). https://doi.org/10.1007/s40593-016-0105-0
21. Goodyear, P., Jones, C., Thompson, K.: Computer-supported collaborative learning: instructional approaches, group processes and educational designs. In: Spector, J.M., Merrill, M.D., Elen, J., Bishop, M.J. (eds.) Handbook of Research on Educational Communications and Technology, pp. 439–451. Springer, Cham (2014). https://doi.org/10.1007/978-1-4614-3185-5_35
22. Chen, J., Wang, M., Kirschner, P.A., Tsai, C.-C.: The role of collaboration, computer use, learning environments, and supporting strategies in CSCL: a meta-analysis. Rev. Educ. Res. **88**(6), 799–843 (2018). https://doi.org/10.3102/0034654318791584
23. Jeong, H., Hmelo-Silver, C.E., Jo, K.: Ten years of computer-supported collaborative learning: a meta-analysis of CSCL in STEM education during 2005–2014. Educ. Res. Rev. **28**, 100284 (2019). https://doi.org/10.1016/j.edurev.2019.100284
24. Stahl, G., Koschmann, T., Suthers, D.: Computer-supported collaborative learning. In: Sawyer, R.K. (ed.) Cambridge Handbook of the Learning Sciences, 3rd edn. Cambridge University Press (2021)
25. Jeong, H., Hmelo-Silver, C.E.: Seven affordances of computer-supported collaborative learning: how to support collaborative learning? How can technologies help? Educ. Psychol. **51**(2), 247–265 (2016). https://doi.org/10.1080/00461520.2016.1158654
26. Siemens, G., Gašević, D.: Guest editorial: learning and knowledge analytics. Educ. Technol. Soc. **15**(3), 1–163 (2012)
27. Schwendimann, B.A., Rodriguez-Triana, M.J., Vozniuk, A, et al.: Perceiving learning at a glance: a systematic literature review of learning dashboard research. IEEE Trans. Learn. Technol. **10**(1), 30–41 (2017). https://doi.org/10.1109/TLT.2016.2599522
28. Gašević, D., Dawson, S., Siemens, G.: Let's not forget: learning analytics are about learning. TechTrends **59**(1), 64–71 (2014). https://doi.org/10.1007/s11528-014-0822-x
29. Viberg, O., Hatakka, M., Balter, O., Mavroudi, A.: The current landscape of learning analytics in higher education. Comput. Hum. Behav. **89**, 98–110 (2018). https://doi.org/10.1016/j.chb.2018.07.027
30. Moog, R.S., Spencer, J.N. (eds.): Process-Oriented Guided Inquiry Learning (POGIL). In: ACS Symposium Series, vol. 994. American Chemical Society (2008)

31. Simonson, S.R. (ed.): POGIL: An Introduction to Process Oriented Guided Inquiry Learning for Those Who Wish to Empower Learners. Stylus Publishing, Sterling (2019)

32. Farrell, J.J., Moog, R.S., Spencer, J.N.: A guided-inquiry general chemistry course. J. Chem. Educ. **76**(4), 570 (1999). https://doi.org/10.1021/ed076p570

33. Straumanis, A., Simons, E.A.: A multi-institutional assessment of the use of POGIL in Organic Chemistry. In: Moog, R.S., Spencer, J.N. (eds.) Process Oriented Guided Inquiry Learning (ACS Symposium Series), vol. 994, pp. 226–239. American Chemical Society (2008). https://doi.org/10.1021/bk-2008-0994.ch019

34. Douglas, E.P., Chiu, C.-C.: Use of guided inquiry as an active learning technique in engineering. In: Proceedings of the Research in Engineering Education Symposium, Queensland, Australia (2009)

35. Lenz, L.: Active learning in a math for liberal arts classroom. Primus **25**(3), 279–296 (2015). https://doi.org/10.1080/10511970.2014.971474

36. Hu, H.H., Kussmaul, C., Knaeble, B., Mayfield, C., Yadav, A.: Results from a survey of faculty adoption of POGIL in computer science. In: Proceedings of the ACM Conference on Innovation and Technology in CS Education, Arequipa, Peru, pp. 186–191 (2016). https://doi.org/10.1145/2899415.2899471

37. Lo, S.M., Mendez, J.I.L.: Learning—the evidence. In: Simonson, S.R. (ed.) POGIL: An Introduction to Process Oriented Guided Inquiry Learning for Those Who Wish to Empower Learners, pp. 85–110. Stylus Publishing (2019)

38. Kezar, A., Gehrke, S., Bernstein-Sierra, S.: Communities of transformation: creating changes to deeply entrenched issues. J. High. Educ. **89**(6), 832–864 (2018). https://doi.org/10.1080/00221546.2018.1441108

39. Kussmaul, C.: Patterns in classroom activities for process oriented guided inquiry learning (POGIL). In: Proceedings of the Conference on Pattern Languages of Programs, Monticello, IL, pp. 1–16 (2016). https://dl.acm.org/doi/abs/10.5555/3158161.3158181

40. Flener-Lovitt, C., Bailey, K., Han, R.: Using structured teams to develop social presence in asynchronous chemistry courses. J. Chem. Educ. **97**(9), 2519–2525 (2020). https://doi.org/10.1021/acs.jchemed.0c00765

41. Reynders, G., Ruder, S.M.: Moving a large-lecture organic POGIL classroom to an online setting. J. Chem. Educ. **97**(9), 3182–3187 (2020). https://doi.org/10.1021/acs.jchemed.0c00615

42. Hu, H.H., Kussmaul, C.: Improving online collaborative learning with POGIL practices. In: Proceedings of the ACM Technical Symposium on CS Education (SIGCSE) (2021). https://doi.org/10.1145/3408877.3439600

43. Bostock, M., Ogievetsky, V., Heer, J.: D3 data-driven documents. IEEE Trans. Vis. Comput. Graph. **17**(12), 2301–2309 (2011). https://doi.org/10.1109/TVCG.2011.185

44. Wilensky, U., Stroup, W.: Learning through participatory simulations: network-based design for systems learning in classrooms. In: Proceedings of the Conference on Computer Support for Collaborative Learning (CSCL), Stanford, CA, 80-es (1999). https://dl.acm.org/doi/10.5555/1150240.1150320

45. Perkins, K., et al.: PhET: interactive simulations for teaching and learning physics. Phys. Teach. **44**(1), 18–23 (2005). https://doi.org/10.1119/1.2150754

46. Resnick, M., et al.: Scratch: programming for all. Commun. ACM **52**(11), 60–67 (2009). https://doi.org/10.1145/1592761.1592779

47. Sawada, D., Piburn, M.D., Judson, E.: Measuring reform practices in science and mathematics classrooms: the reformed teaching observation protocol. Sch. Sci. Math. **102**(6), 245–253 (2002). https://doi.org/10.1111/j.1949-8594.2002.tb17883.x

48. Smith, M.K., Jones, F.H.M., Gilbert, S.L., Wieman, C.E.: The classroom observation protocol for undergraduate STEM (COPUS): a new instrument to characterize university STEM

classroom practices. CBE—Life Sci. Educ. **12**(4), 618–627 (2013). https://doi.org/10.1187/cbe.13-08-0154

49. Frey, R.F., Halliday, U., Radford, S., Wachowski, S.: Development of an observation protocol for teaching in interactive classrooms (OPTIC). Abstracts of Papers of the American Chemical Society, vol. 257 (2019)
50. Billingham, L.: Improving academic library website accessibility for people with disabilities. Libr. Manag. **35**(8/9), 565–581 (2014). https://doi.org/10.1108/LM-11-2013-0107
51. Shawar, B.A.: Evaluating web accessibility of educational websites. Int. J. Emerg. Technol. Learn. (iJET) **10**(4), 4–10 (2015)
52. Babin, L.A., Kopp, J.: ADA website accessibility: what businesses need to know. J. Manag. Policy Pract. **21**(3), 99–107 (2020). https://doi.org/10.33423/jmpp.v21i3.3144
53. Rysavy, M.D.T., Michalak, R.: Assessing the accessibility of library tools and services when you aren't an accessibility expert: part 1. J. Libr. Adm. **60**(1), 71–79 (2020). https://doi.org/10.1080/01930826.2019.1685273
54. Manning, R., et al.: Pa11y (2021). https://pa11y.org
55. Smith, J., Whiting, J.: WAVE Web Accessibility Evaluation Tool (2021). https://wave.webaim.org
56. Caldwell, B. et al. (eds.) Web Content Accessibility Guidelines (WCAG) 2.0. (2008). https://www.w3.org/TR/WCAG20/
57. Carpenter, S.K., Wilford, M.M., Kornell, N., Mullaney, K.M.: Appearances can be deceiving: instructor fluency increases perceptions of learning without increasing actual learning. Psychon. Bull. Rev. **20**(6), 1350–1356 (2013). https://doi.org/10.3758/s13423-013-0442-z
58. Menekse, M., Stump, G.S., Krause, S., Chi, M.T.H.: Differentiated overt learning activities for effective instruction in engineering classrooms. J. Eng. Educ. **102**(3), 346–374 (2013). https://doi.org/10.1002/jee.20021
59. Deslauriers, L., McCarty, L.S., Miller, K., Callaghan, K., Kestin, G.: Measuring actual learning versus feeling of learning in response to being actively engaged in the classroom. Proc. Natl. Acad. Sci. **116**(39), 19251–19257 (2019). https://doi.org/10.1073/pnas.1821936116

A Model for the Analysis of the Interactions in a Digital Learning Environment During Mathematical Activities

Alice Barana$^{(\boxtimes)}$ (ID) and Marina Marchisio (ID)

Department of Molecular Biotechnology and Health Sciences, University of Turin,
Via Nizza 52, 10126 Turin, Italy
{alice.barana,marina.marchisio}@unito.it

Abstract. Learning environments are the core of education and the recent digitalization of the learning processes brought the diffusion of Digital Learning Environments (DLE). In this paper we propose a definition and a conceptualization of DLEs and provide a model to analyze the interactions occurring among the members of a DLE during Mathematics activities. The conceptualization is rooted on results from the literature about digital education. The model is applied to analyze the activity of 299 8th grade students when tackling two tasks of algebraic modelling with automatic formative assessment in two different modalities: online and classroom-based. Two episodes have been selected and discussed, with the aim of identifying the interactions through which formative assessment strategies emerge. The results show that all the Black and Wiliam's strategies of formative assessment are developed during the activities through the interactions among the members of the learning community or with the technologies. This study can be a basis for extending the research in the learning analytics direction, to analyze the interactions during formative activities in large online courses.

Keywords: Automatic formative assessment · Digital learning environment · Interactions · Mathematics education · Online learning

1 Introduction

The learning environment a key element of education. According to Wilson [1], a learning environment is *"a place where learning is fostered and supported"*. It includes at least two elements: the learner, and a *"setting or space wherein the learner acts, using tools and devices, collecting and interpreting information, interacting perhaps with others, etc."* [1]. Accepting this definition, we agree that a learning environment is not limited to the physical place but includes also at least a learner. The traditional learning environment that everyone knows is the classroom, where the teacher teaches, students learn, individually or with their peers, using tools such as paper, pen, and a blackboard. The diffusion of technology transformed this traditional learning environment by adding digital tools, as tablets or computers, and the IWB (Interactive White Board). Besides equipping physical places with technologies, the technological revolution brought to the

© Springer Nature Switzerland AG 2022
B. Csapó and J. Uhomoibhi (Eds.): CSEDU 2021, CCIS 1624, pp. 429–448, 2022.
https://doi.org/10.1007/978-3-031-14756-2_21

creation of a new learning environment, situated in a non-physical dimension: that of the Internet, accessible from everywhere via computers, tablets, or even smartphones. This is the essence of the "Digital Learning Environment" (DLE); besides the learner and a setting, which can be virtual, a device is needed to access the activities.

Today, due to the COVID-19 pandemic, online platforms have known increasing popularity, supporting smart-schooling, and class attendance from home [2, 3]. They have been invaluable to permit students from all social and cultural backgrounds to carry on their education. The interest in DLEs in the research has increased accordingly, making different theories and models come to life [4, 5].

This paper is an extension of the paper "Analyzing interactions in Automatic Formative Assessment activities for Mathematics in a Digital Learning Environment" [6] presented by the authors at the 13th International Conference on Computer Supported Education (CSEDU 2021), which intends to contribute to the discussion about the essence of DLEs providing a definition and a model for analyzing learning interactions in a DLE. In this revised and extended paper, the theoretical framework, which is the core of the paper, has been developed further, including a review of various studies on DLEs and discussing a proposal of definition. Particular characteristics of DLEs for Mathematics are considered, based on theories on formative assessment and Automatic Formative Assessment (AFA). Then, a model for the interactions among the members of a DLE is proposed, to highlight the interactions during AFA activities. Moreover, the results have been expanded, adding the analysis of a second AFA activity for grade 8 Mathematics in an online context, so that the model is applied in two different contexts (online and classroom-based teaching). The discussion about the kinds of interactions that can support formative assessment strategies is consequently enriched. The conclusions suggest how these findings could be used in learning analytics research.

2   Theoretical Framework

2.1   Definition of Digital Learning Environment

The concept of "Digital Learning Environment" has a long history, and it has known several developments and many different names over the years. Suhonen [7] defined Digital Learning Environments as *"technical solutions for supporting learning, teaching and studying activities."* Some years before, Abdelraheem [8] spoke about "Computerized Learning Environments" (CLEs), which are *"systems that provide rich databases, tools, and resources to support learning and information seeking and retrieval, as well as individual decision making."* Abdelraheem's definition is more detailed than Suhonen's one, but the essence is similar: disseminating learning materials through the Internet. CLEs, or DLEs, emphasize empowerment through metaknowledge, which individuals invoke and refine while attempting to use their learning tasks. Other authors use the term "Online Learning Environments" as Khan [9], who defined them as *"hypermedia based instructional [systems], which utilizes the attributes and resources of the World Wide Web to create a meaningful learning environment where learning is fostered and supported"*. Other scholars speak about "Virtual Learning Environments" referring to a particular type of Learning Environment where *"students interact primarily with other networked participants, and with widely disseminated information tools"*[10]. Here the interaction among learners is emphasized, and it is considered the most powerful key to learning.

The common factor among all these definitions is the use of the Internet and its tools to provide an environment where learning is supported, generally represented by a Learning Management System (LMS). An LMS, according to Watson and Watson [11], is *"the infrastructure that delivers and manages instructional content, identifies and assesses individual and organizational learning or training goals, tracks the progress towards meeting those goals, and collects and presents data for supervising the learning process of an organization as a whole."* While similar environments are mainly used to support online educational processes, we are convinced and have proof of the fact that web-based platforms can also be successfully adopted in classroom-based settings: in our conception, DLEs should not only be confined to distance education [12–15].

More recently, many authors have developed an interest in conceptualizing digital learning environments as ecosystems, borrowing the term from ecology [2, 16–19]. According to Encyclopaedia Britannica (www.britannica.com), an ecosystem is *"a complex of living organisms, their physical environment, and all their interrelationships in a particular unit of space."* The natural ecosystem, constituted by a biological community in a physical environment, is the fundamental example; however, this definition can be applied to any domain, even artificial environments, by specifying the living community, environment, and space unit.

There are several models of learning or e-learning ecosystems in the literature, which vary for the components included based on the theoretical assumptions considered. In general, they contemplate individuals, computer-based agents, communities, and organizations in a network of relations and exchanges of data that supports the co-evolutions and adaptations of the components themselves [17].

Following this trend, in this paper, we chose to use the term "Digital Learning Environment" to indicate a learning ecosystem in which teaching, learning, and the development of competence are fostered in classroom-based, online or blended settings. It is composed of a human component, a technological component, and the interrelations between the two.

The human component consists of one or more learning communities whose members can be: teachers or tutors, students or learners, and their peers, the administrators of the online environment.

The technological component includes:

- a Learning Management System, together with software, other tools, and integrations which accomplish specific purposes of learning (such as web-conference tools, assessment tools, sector-specific software, and many others);
- activities and resources, static or interactive, which can be used in synchronous or asynchronous modality;
- technological devices through which the learning community has access to the online environment (such as smartphones, computers, tablets, IWB);
- systems and tools for collecting and recording data and tracking the community's activities related to learning (such as sensors, eye-trackers, video cameras).

The interrelations between the two components can be:

- the interactions and learning processes activated within the community and through the use of the technologies, including dialogues between the members of the learning community, human-technology interactions, and so on;
- pedagogies and methodologies through which the learning environment is designed.

Figure 1 shows a graphical representation of the components of a DLE. The community is in the middle in a human-centered approach to learning. In the ecology metaphor, it is the complex of living organisms, while the technological component surrounds the community, as the physical environment. The arrows linking the community and the technologies represent the learning processes as well as the pedagogies and methodologies used to design the learning materials and interpret data from the digital environment. They are double-ended to indicate the reciprocal relationships between the two components, which bring to the development of all the parts.

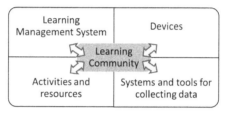

Fig. 1. Schema of the components of a DLE.

Independently of the fact that the DLSs are based on a web-based platform, teaching and learning can occur in one of the following modalities:

- face to face, in the classroom or a computer lab, with students working autonomously or in groups through digital devices, or solving tasks displayed on the IWB with paper and pen or other tools;
- entirely online, using the DLE as the only learning environment in online courses or MOOCs;
- in a blended approach, using online activities to integrate classroom work, such as asking students to complete them as homework.

These three modalities can be adapted to different situations, grades, aims, and needs. For example, the face-to-face modality can be suitable with students of the lowest grades and in scholastic situations where the classroom work is predominant [15]. The blended approach can offer useful support to the face-to-face lessons at secondary school or university [12, 20]. Online courses are generally used for training and professional courses, university courses, or learning in sparse communities, where face-to-face meetings are difficult to organize [8, 21, 22].

In this conceptualization, the DLE is not limited to technological artifacts, even if they play a crucial role. The learning community takes a prominent place: it can include, according to the kind of DLE, students and peers, teachers and tutors (who are facilitators of learning activities), designers of educational materials, and administrators of the digital environment. There can also be more communities involved or a community of communities: it happens, as an example, in the Italian National Problem Posing and Solving Project [23] where, hosted in an integrated LMS, there are many communities of students, one for each class participating to the project, and the community of all the classes' teachers [24]. In this case, the students' communities are based on learning and teaching intentions, while the teachers' community pursues the development of competence related to teaching with innovative technologies and didactic methodologies. The teachers' work in their students' communities allows them to practice the competences which are fostered through training courses in the teachers' community. The students' communities work in a blended modality, and their online activities are mainly accomplished asynchronously, while the teachers participate in online synchronous and asynchronous training [25, 26].

This definition of DLE does not disagree with the other definitions collected from the literature. However, it is more comprehensive: it is not limited to the web-based platform, which conveys the activities and is a relevant and essential part of a DLE; it also includes a "human" part.

The use of these technologies, such as web-based platforms, assessment tools, and other systems such as sensors or eye-trackers, allows for collecting, recording, and using learning data. These data can be elaborated within the DLE to provide information useful to make decisions and take action. In the following paragraphs, we will explain how these data can be used to improve learning, teaching, and the development of competences.

There is an in-depth discussion on the real effectiveness of DLEs (and their synonyms) that involves many researchers. For instance, in the paper "Media will never influence learning", Clark [27] claims that the use of technologies, per se, is not more effective than traditional learning unless a learning theory supports it. The chosen learning theory should be coherent with the aims of the materials or the course and should guide the materials' design. Clark and Mayer [28] analyzed the effect sizes gained in several studies that compare digital and traditional education. The average effect size is not much different from zero, meaning that digital tools are not better than paper and pen. However, they noticed that there are many cases where the effect size is considerably large: this means that digital technologies have great potential. When they are used following suitable principles, they can make a difference in education.

2.2 Formative Assessment

Formative assessment is one of the key principles which, according to the majority of scholars, should be included in the design of a learning environment, being it physical or virtual [29, 30]. In this study, we refer to Black and Wiliam's definition and framework of formative assessment [31]. According to them, "*a practice in a classroom is formative to the extent that evidence about student achievement is elicited, interpreted, and used by teachers, learners, or their peers, to make decisions about the next steps in instruction that are likely to be better, or better founded, than the decisions they would have taken in the absence of the evidence that was elicited*". They identified three agents that are principally activated during formative practices: the teacher, the student, and peers.

Moreover, they theorized five key strategies enacted by the three agents during the three different processes of instruction:

- KS1: clarifying and sharing learning intentions and criteria for success;
- KS2: engineering effective classroom discussions and other learning tasks that elicit evidence of student understanding;
- KS3: providing feedback that moves learners forward;
- KS4: activating students as instructional resources; and
- KS5: activating students as the owners of their own learning.

2.3 DLEs for Mathematics

In this paper, we consider particular DLEs for working with Mathematics through suitable technologies and methodologies. The LMS that we use is based on a Moodle platform and it is integrated with an Advanced Computing Environment (ACE), which is a system for doing Mathematics through symbolic computations, geometric visualization, and embedding of interactive components [32], and with an Automatic Assessment System based on the ACE engine [33]. In particular, we chose Maple ACE and Moebius AAS. Through this system, we create interactive activities for Mathematics based on problem solving and Automatic Formative Assessment (AFA), which are the main methodologies used in the DLE, and that we have better defined and characterized in previous works [30, 34]. In detail, the characteristics of the Mathematics activities that we propose are the following:

- availability of the activities for a self-paced use, allowing multiple attempts;
- algorithm-based questions and answers, so that at each attempt different numbers, formulas, graphs, and texts are displayed, computed on the base of random parameters;
- open mathematical answers, accepted for its Mathematical equivalence to the correct one;
- immediate feedback, returned when the student is still focused on the task;
- interactive feedback, which provides a sample of a correct solving process for the task, which students can follow step-by-step;
- contextualization in real-life or other relevant contexts.

2.4 Modelling Interactions in a Digital Learning Environment

The technological apparatus of a DLE, particularly when the LMS is integrated with tools for automatic assessment, has a mediating role in the learning processes. In particular, we can identify the following functions through which it can support the learning activities [34]:

- **Creating and Managing:** supporting the design, creation, editing, and managing of resources (e.g., interactive files, theoretical lessons, glossaries, videos), activities (e.g., tests, chats for synchronous discussions, forums for asynchronous discussions, questionnaires, submission of tasks) and more generally of the learning environment by teachers, but also by students or peers;

- **Delivering and Displaying:** making the materials and activities available to the users;
- **Collecting:** collecting all the quantitative and qualitative data concerning the actions of the students (such as movements and dialogues), the use of the materials (for example, if a resource has been viewed or not, how many times and how long), and the participation in the activities (such as given answers, forum interventions, number of tasks delivered, number of times a test has been performed, evaluations achieved);
- **Analyzing and Elaborating:** analyzing and elaborating all the data collected through the technologies related to teaching, learning, and the development of competences;
- **Providing Feedback:** giving the students feedback on the activity carried out and providing teachers, as well as students, with the elaboration of learning data.

To schematize these functions, we propose the diagram shown in Fig. 2. The external cycle represents the five functions; the black dashed arrows represent how data are exchanged within the DLE through automatic processes. The technologies of a DLE, to accomplish one function, uses the data or the outputs resulting from the previous one: the learning materials, created through the LMS or other sector-specific software through the "creating and managing function", are displayed via devices through the "delivering and displaying function". Information about the students' activities is collected by the LMS, other software, or tools through the "collecting function" and it is analyzed by these systems, which may use mathematical engines, learning analytics techniques, algorithms of machine learning, or artificial intelligence, through the "analyzing and elaborating" function. The results of the analysis are feedback in the sense of Hattie's definition (i.e., "information provided by an agent regarding aspects of one's performance or understanding") [35]. They can be returned to students and teachers through the "providing feedback" function, and they can be used to create new activities or edit the existing ones. This circle represents a perfect adaptive system from the technological perspective [36–38].

In a human-centered approach, at the center of the DLE, there is the learning community, composed of students, teachers, and peers (who are the agents in the Black and Wiliam's theory of formative assessment [31]): they can interact with the DLE through its functions receiving and sending information. The blue dotted arrows represent the interactions between the community and the digital systems that occur through human actions, such as reading, receiving, inserting, providing, digiting. For example, the teacher, or designer, or tutor can create the digital activities through the "creating and managing" functions of the DLE; tasks are displayed ("delivering and displaying" function) and received, seen, or read by the students through some device. The students, individually or with their peers, can insert their answers or work. The technology collects them through the "collecting" function. The system analyzes the students' answers and provides feedback ("providing feedback" function) returned to the student. Simultaneously, the information about the students' activity is returned to the teacher through the "providing feedback" function; the teacher can use it to edit the existing task or create new ones. The continuous double-ended orange arrows represent the interactions among students, teachers, and peers, which in classroom-based settings can be verbal while in online settings can be mediated by the technology.

The diagram in Fig. 2 is a proposal of schematization of the interactions among the components of a DLE. It helps us understand how data are shared among the components

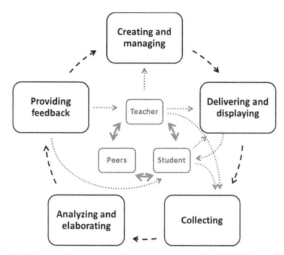

Fig. 2. Diagram of the interactions among the components of a DLE through the functions of the technology [6].

of a DLE, elaborated, and used; for this reason, it can be useful in the perspective of learning analytics.

In the diagram, one can follow the arrows along close paths and identify particular situations. For example, we can focus on the upper-left side of the circle as in Fig. 3 and describe the process of content creation by a teacher or an instructor: she receives information about the activities completed by the students and uses them for the creation of new ones. If we focus on the right part of the circle as in Fig. 4, we can examine the process of the fruition of static digital resources: the teacher makes them available to the students through the delivering function, and the students read/observe/study them. If we consider the lower-left part of the diagram as in Fig. 5, we have interactive activities: the student can insert answers in the system, they are automatically analyzed and feedback is returned to the student. Moreover, including or excluding interactions among peers, individual, and collaborative activities are identified. Disregarding the arrows linking the teacher, the student, and the peers, we have a scenario of individual and self-paced online learning.

In the end, the model that we have analyzed in this section allows us to identify some outcomes that the adoption of a similar DLE with AFA, through the functions previously shown, makes it possible to achieve:

- **To Create an Interactive Learning Environment:** all the materials for learning and assessment can be collected in a single environment and be accessible at any time. They can activate the students who can be engaged in the navigation of the learning path, solve the tasks and receive feedback;
- **To Support Collaborative Learning**, through specific activities, delivered to groups of students, which enhance the communication and sharing of materials, ideas, understanding;

- **To Promote Formative Assessment**, by offering immediate feedback to students about their results, their knowledge and skills acquired, and their learning level. Feedback can also be returned to the teachers on the students' results and their activities, supporting decision-making.

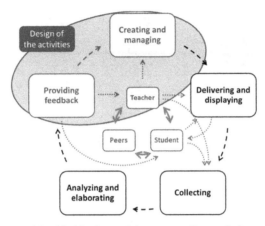

Fig. 3. Paths identified in the model corresponding to design activities.

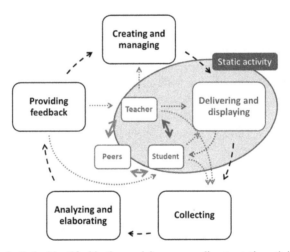

Fig. 4. Paths identified in the model corresponding to static activities.

The identification and classification of a DLE's functions can allow us to identify the interactions in a DLE, to analyze their nature and the contribution of technology that mediates them. The information gained is useful from a learning analytics perspective since it allows us to identify the role of data during the learning processes. Moreover, this model helps us identify the functions and outcomes of technology in learning processes. Individuating and separating functions and outcomes is necessary to have a clear frame and find causal connections, especially when analyzing large data quantities.

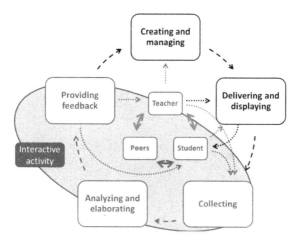

Fig. 5. Paths identified in the model corresponding to interactive activities.

3 Methodology

In this study, we aim at showing how the diagram of the interactions among the components of a DLE can be used to model learning processes, and in particular to understand how formative assessment can be enacted in a DLE for Mathematics.

To this purpose, we analyzed two AFA activities, both concerning symbolic computations for students of grade 8, experimented in two different contexts: an online context and a classroom-based one. Both tasks ask students to formulate, represent, and compare different functions derived from several geometrical figures. The first one is shown in Fig. 6. Firstly, students are asked to write a formula for the area of a geometrical figure whose lengths are given through a variable and write the formula in the blank space. The geometrical figure is not standard, but students can decompose it in simpler parts, such as rectangles or squares; they can use several decompositions to reach different forms of the same formula. Thanks to the Maple engine, the system can recognize the formula's correctness independently of its form, so every formula obtained through different reasoning is considered correct. Students have three attempts to provide the formula: they can self-correct mistakes and deepen their reasoning if a red cross appears. After three attempts, either correct or not, a second section appears, showing a table that students have to fill in with the values of the figure's area when the variable assumes specific values. In this part, students have to substitute in the formula different values of the variable; the purpose is to increase the awareness that variables are symbols that stand for numbers and that a formula represents a number, which has a particular meaning in a precise context. The table is a bridge to the last part of the question, where students are asked to sketch the graph of the function, using an interactive response area of Moebius Assessment that accepts answers within a fixed tolerance, without manual intervention. A second task repeats similar questions with the perimeter of the same figure. The activity is algorithmic: at every attempt the numbers, and consequently the figure and the function, change. This activity was tested in a bigger project involving 299 students of 13 classes of grade 8; they were asked to complete it from home individually.

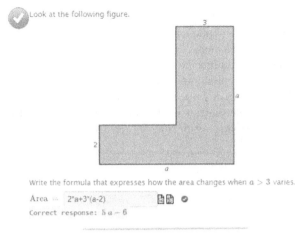

Look at the following figure.

Write the formula that expresses how the area changes when $a > 3$ varies.

Area $=$ 2*a+3*(a-2)

Correct response: $5a - 6$

Fill in the following table computing the area of the figure when a varies.

$a(\text{cm})$	$\text{Area}\left(\text{cm}^2\right)$	
4	14	✓
5	19	✓
6	24	✓
7	29	✓

Sketch the graph of the function that expresses how the area changes when a varies.

Pay attention to the domain of the function! Eliminate the parts and the points that do not belong to the domain.

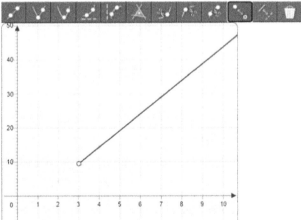

Fig. 6. First activity on symbolic computations.

The second activity (Fig. 7) is similar; it involves a different figure and students are asked to find as many formulas as they can to express its area. In the first section, they have 3 attempts to write a formula for the area. In the second one, they are asked to fill other 4 response areas with different formulas expressing the same area. The intent is to make them explore the symbolic manipulation of an algebraic formulas through the geometric context, to confer a more concrete meaning to the technical operations. In the last part, students have to substitute the variable with a given value and compute the area. This activity was tested in a classroom-based setting in another experiment involving 97 students of 4 different classes of grade 8. In the classrooms there were the teacher and 2 researchers of the group; the students worked in pairs using a computer or a tablet. The work and discussions of some pairs of students were recorded through a video camera.

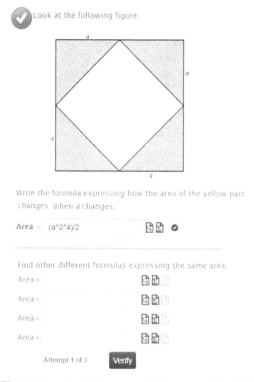

Fig. 7. Second activity on symbolic computations [6].

In both cases, data from the platform were analyzed through the diagram of the interactions in a DLE presented in the previous section. For the second activity, the video recordings were analyzed as well.

4   Results

To analyze the interactions and the development of formative assessment strategies during individual automatic assessment activities in online modality (first activity), we used data from the AAS gradebook, where all the students' results are collected.

We analyzed the attempts made by all the students to the questions. We noticed that some students made more than one attempt at this assignment; we analyzed Erica's answers, who made 3 attempts, obtaining 64%, 83%, and 100%, respectively. In her first attempt, which lasted 19 min, she answered correctly to the formula's request, although she wrote the formula in a different form, without the simplifications; she obtained the simplified form as feedback. The attempt is shown in Fig. 8. Then, she correctly used the formula to fill the table with the values of the area for the given values of the variable, but she failed the choice of the graph, choosing the graph of the function $f(a) = 5a$ instead of $f(a) = 5a-4$. Then she moved on to the second question, related to the perimeter. The figure was similar to the previous one, with just a numerical value changing. At this point, she gave an incorrect answer to the first part, asking the formula for the perimeter. Erica received the correct formula as feedback and correctly filled the table with the perimeter's values for the given values of a, then she correctly sketched the graph of the function. The system allows one to sketch a line clicking on two passing points on a cartesian plane: she chose as points the first two points of the table. After that, she submitted the assignment. She obtained the percentage of correct answers (64%) as final feedback, together with all her answers paired with the correct ones, which she had already seen after each step during the test.

After 5 min, Erica started a new attempt, which lasted 11 min: the previous reasonings helped her accelerate the procedure. In the second attempt, she found a new figure, similar to the previous ones but having different numbers. She correctly answered the first two items; then, she failed the choice of the graph. In the second question, related to the perimeter, she correctly inserted the formula. We can notice that she did not write 4a as the answer, still not noticing the invariance for all the figures of the same kind. Instead, she added all the measures of the sizes expressed through the parameter a. Then she correctly answered the following parts, except for a number in the table, which probably was a distraction mistake.

She ran the third attempt just 2 min after finishing the second one, and it lasted 7 min, further reducing the duration. She answered correctly to all the items; in particular, she inserted 4a as perimeter, meaning that she understood the invariance of the perimeter for the class of figures.

Here, the human component of DLE is Erica, who is the only member of the learning community; the technological components are the LMS integrated with the AAS, the digital activity, and the computer used to access it.

The interactions among these components can be schematized with repeated cycles of AFA, as shown in yellow in Fig. 9. The assignment she opened was displayed through the "delivering and displaying" function of the technology. The AAS accepted the answers she inserted through the "collecting" function. The mathematical engine of the AAS processed the answers through its "analyzing and elaborating" function. Feedback was provided ("providing feedback" function) to the student, who could enter a new cycle, moving on to the following item, or running a new attempt. Erica was activated as the owner of her learning (KS5) every time she ran a new attempt through the "delivering and displaying" function and inserted her answers through the "collecting" function. The KS3 ("providing feedback which moves the learner forward") was activated every time she received feedback.

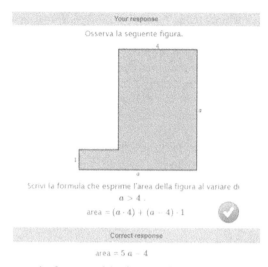

Fig. 8. Erica's answers to the first part of the first question in her first attempt. The question asks to write a formula for the area of the given figure. Erica's answer on the left matches the correct one, on the right.

We analyzed the videos realized during the second activity in the classrooms, to understand how the interactions among the components of the DLE changed and how the formative assessment strategies took place during a group activity. We choose some episodes which we considered most significant. Here, the learning community includes a class of students and a teacher; the digital activities are created in a LMS integrated with an AAS, and the devices used to access them are an Interactive White Board (IWB) and computers.

The first episode involves the teacher (T) who illustrates the task to the students (S). The teacher was at the IWB and was pointing at the figure shown.

T: Look at this figure. Write the formula which expresses how the area of this figure varies when *a* varies. That is, [pointing at one of the sizes of the yellow triangles] how long is this side?

S: *a*.

T: Well, you have to calculate the area of this figure using *a*. Those sides measure *a*. What does it mean? What is *a*?

S: A variable.

In this excerpt, the teacher introduced the activity and explained to the students what their task was. The explanation took the form of a dialogue, as he engaged the students with questions to make sure that they were following the discourse. The teacher exploited the "delivering and displaying" function of the technology to display the task and, in particular, the figure; then, she interacted with the students. If we consider the diagram, we are in the right part; the parts of the model involved in this excerpt are shown in yellow in Fig. 10. While explaining the tasks, she developed the KS1 "clarifying and sharing learning intentions and criteria for success". The KS2 "engineering effective classroom discussions and other learning tasks that elicit evidence of student understanding" was

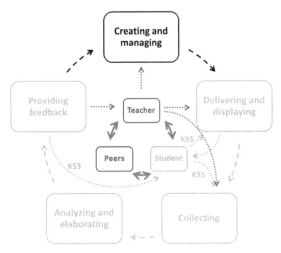

Fig. 9. Diagram of the strategies of formative assessment enacted during individual online activities.

accomplished during the phase of the creation of this activity by the researchers (that we can include in the "Teacher" subject of our analysis) through the "creating and managing" function of the technologies; it is also activated when the teacher asks questions to the class aimed at making students reason in the correct direction.

The second episode involves Marco (M) and Giulia (G), two students of medium level who were trying to solve the first part of the activity, working together. In the beginning, they observed the figure displayed on the screen of their computer and tried to understand the task.

M: We have to compute the area, but we don't have any data!

G: But we have *a*.

M: But *a* is not a number!

G: Ok, but we can compute the area using *a*.

M: Teacher, how can we compute the area without numbers? Can we use *a*?

T: Yes, it is like a generic number.

G: We have to write a formula using *a*, isn't it?

T: That's right.

The two students started reasoning together on the figure trying a way to compute the area. After about 15 min, they came up with a quite complex formula, built subtracting the area of the inner white square to that of the external square. They used the Pythagorean theorem to compute the length of the white square's side. They inserted the formula in the response area and the system returned a green tick with positive feedback. They passed to the following part, which asked them to find other 4 formulas for the same area. For the first two formulas, they reasoned algebraically, manipulating the original formula. For the other two, they reasoned geometrically, developing new ways to compute the area. The peer discussion allowed them to correct mistakes before entering the formulas in the response areas, so their answers were marked as correct at their first attempt.

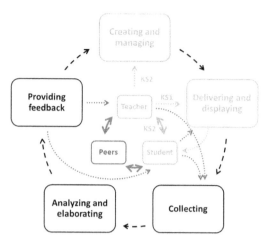

Fig. 10. Diagram of the formative assessment strategies enacted in the first episode of activity 2 through the interactions in the DLE [6].

In this episode, the students look at the task displayed on the screen through the "delivering and displaying" function, then interact among them discussing the task. They also interact with the teacher asking questions about their doubts. Then they insert their answers in the system, which collects them through the "collecting" function, analyzes them, and returns feedback. They repeat the same cycle several times. The students activate KS4 "activating students as instructional resources" when discussing in pair. KS5 "activating students as the owners of their own learning" is enacted when they insert their answers in the AAS, and KS3 is developed when they receive feedback from

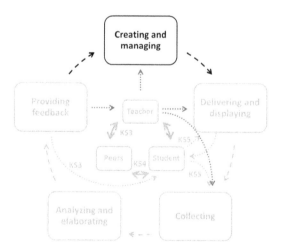

Fig. 11. Diagram of the formative assessment strategies enacted in the second episode of activity 2 through the interactions in the DLE [6].

the AAS, but also by the teacher. The yellow parts in Fig. 11 schematize the interactions that occurred in this episode and the formative assessment strategies developed.

5 Discussion and Conclusions

The episodes presented in the previous section help clarify how the interactions among the members of a DLE occur during AFA Mathematics activities in online and classroom-based settings. In the first case, the computerized interactive feedback has a key role in providing feedback and in engaging the student, who is working individually. The design of the activity enables KS3 and KS5, which keep Erica engaged with the task until its full comprehension, demonstrated by the repeated attempts and the correctness of the last one. In the second case, the main feedback is provided by the social interactions within the learning community and especially among peers; in fact, Marco and Giulia reasoned more time on the tasks and they tended to answer correctly at the first attempt. We would like to underline that also these interactions, which are mainly dialogues among the learning community, are part of the Digital Learning Environment as we have conceptualized it. Similar activities can lead to a deep understanding of fundamental Mathematics concepts; the technologies and methodologies used – in particular, an AAS based on a mathematical engine and AFA – supported the design and implementation of interesting activities for the development of mathematical competences.

The diagram used for the analyses helped clarify what functions of the technologies and through which kinds of interactions the formative assessment strategies are elicited in different situations. In particular, we can see that all the Black and Wiliam's strategies of formative assessment can be enacted through AFA activities, and all of them are identified and located along the arrows of our diagram, that is during the interactions among the human components of the DLE or between human and technological components. Thus, we can include a fourth agent in Black and Wiliam's framework: in the DLEs that we consider, the technology is also an agent of the formative assessment strategies, especially for providing feedback and engaging students (KS3 and KS5).

Through the analysis of the interactions among the members of these DLEs, we can also point out that the three outcomes mentioned in our framework are achieved. In particular:

- the analyzed learning environments are interactive, since students are actively engaged in the activities, they are stimulated to reflect and have the opportunity to achieve important understanding;
- the formative assessment is promoted by the activities, as all the 5 key strategies are enacted;
- collaboration among students is supported, especially in the classroom-based setting where students are asked to work together.

The diagram used in the analyses allows us to conceptualize the DLE as an ecosystem: we can see that the human and technological components are strictly related, and the interrelations among them cause the development of the learning community, in terms of learning processes, knowledge, and competences gained; but also an improvement of the learning activities on the base of the results obtained.

The analyses conducted in this study have a qualitative nature: they are aimed at showing how the schema of the interactions among the components of a DLE can be used to model formative assessment practices, especially when the AFA is adopted. However, they can be a starting point for extending the research about learning analytics for formative assessment. This model can be used to create a taxonomy of the interactions occurring in a DLE, identifying which support formative assessment or other learning processes. Since interactions in a DLE can be described using log data, this model can also be used with extensive learning data to identify the formative assessment strategies or other learning processes occurring in large online courses. This would allow us to identify the learning activities which are better related to the development of knowledge, abilities, and competences or the elicitation of interactions and engagement. The results of similar analyses could help adjust and improve the digital materials in online courses. Using other technologies and different learning methodologies to build suitable activities, this model of analysis could also be adapted to other disciplines.

References

1. Wilson, B.G.: Metaphors for instruction: why we talk about learning environments. Educ. Technol. **35**, 25–30 (1995)
2. Giovannella, C., Passarelli, M., Persico, D.: The effects of the covid-19 pandemic on Italian learning ecosystems: the school teachers' perspective at the steady state. Interact. Des. Architect. J. **45**, 264–286 (2020)
3. Darling-Hammond, L., Schachner, A., Edgerton, A.K.: Restarting and Reinventing School: Learning in the Time of COVID and Beyond. Learning Policy Institute, Palo Alto, CA (2020)
4. Fissore, C., Marchisio, M., Rabellino, S.: Secondary school teacher support and training for online teaching during the covid-19 pandemic. In: European Distance and E-Learning Network (EDEN) Proceedings, pp. 311–320. European Distance and E-Learning Network, Timisoara (2020)
5. OECD: Lessons for Education from COVID-19: A Policymaker's Handbook for More Resilient Systems. OECD, Paris, (2020). https://doi.org/10.1787/0a530888-en
6. Barana, A., Marchisio, M.: Analyzing interactions in automatic formative assessment activities for mathematics in digital learning environments. In: Proceedings of the 13th International Conference on Computer Supported Education, CSEDU 2021, pp. 497–504. SCITEPRESS (2021). https://doi.org/10.5220/0010474004970504
7. Suhonen, J.: A formative development method for digital learning environments in sparse learning communities (2005). http://epublications.uef.fi/pub/urn_isbn_952-458-663-0/index_en.html
8. Abdelraheem, A.Y.: Computerized learning environments: problems, design challenges and future promises. J. Interact. Online Learn. **2**, 1–9 (2003)
9. Khan, B.H.: Web-based instruction: what is it and why is it? In: Web-Based Instruction, pp. 5–18. Educational Technology Publications, Englewood Cliffs (1997)
10. Wilson, B.G.: Constructivist Learning Environments: Case Studies in Instructional Design. Educational Technology Publications, Englewood Cliffs (1996)
11. Watson, W.R., Watson, S.L.: An argument for clarity: what are learning management systems, what are they not, and what should they become? Techtrends Tech. Trends **51**, 28–34 (2007). https://doi.org/10.1007/s11528-007-0023-y
12. Marchisio, M., Remogna, S., Roman, F., Sacchet, M.: Teaching mathematics in scientific bachelor degrees using a blended approach. In: Proceedings - 2020 IEEE 44th Annual Computers, Software, and Applications Conference, COMPSAC 2020, pp. 190–195 (2020)

13. Borba, M.C., Chiari, A., de Almeida, H.R.F.L.: Interactions in virtual learning environments: new roles for digital technology. Educ. Stud. Math. **98**(3), 269–286 (2018). https://doi.org/10.1007/s10649-018-9812-9

14. Barana, A., Marchisio, M., Miori, R.: MATE-BOOSTER: design of tasks for automatic formative assessment to boost mathematical competence. In: Lane, H.C., Zvacek, S., Uhomoibhi, J. (eds.) CSEDU 2019, vol. 1220, pp. 418–441. Springer, Cham (2020). https://doi.org/10.1007/978-3-030-58459-7_20

15. Barana, A., Boffo, S., Gagliardi, F., Garuti, R., Marchisio, M.: Empowering engagement in a technology-enhanced learning environment. In: Rehm, M., Saldien, J., Manca, S. (eds.) Project and Design Literacy as Cornerstones of Smart Education, vol. 158, pp. 75–77. Springer, Singapore (2020). https://doi.org/10.1007/978-981-13-9652-6_7

16. García-Holgado, A., García-Peñalvo, F.J.: Human interaction in learning ecosystems based on open source solutions. In: Zaphiris, P., Ioannou, A. (eds.) Learning and Collaboration Technologies. Design, Development and Technological Innovation. LNCS, vol. 10924, pp. 218–232. Springer, Cham (2018). https://doi.org/10.1007/978-3-319-91743-6_17

17. Guetl, C., Chang, V.: Ecosystem-based theoretical models for learning in environments of the 21st century. Int. J. Emerg. Technol. Learn. **3**, 50–60 (2008). https://doi.org/10.3991/ijet.v3i1.742

18. Uden, L., Wangsa, I.T., Damiani, E.: The future of e-learning: e-learning ecosystem. In: Inaugural IEEE International Conference on Digital Ecosystems and Technologies, pp. 113–117 (2007)

19. Väljataga, T., Poom-Valickis, K., Rumma, K., Aus, K.: Transforming higher education learning ecosystem: teachers' perspective. Interact. Des. Architect. J. **46**, 47–69 (2020)

20. Cavagnero, S.M., Gallina, M.A., Marchisio, M.: School of tasks. Digital didactics for the recovery of scholastic failure. Mondo Digitale **14**, 834–843 (2015)

21. Brancaccio, A., Esposito, M., Marchisio, M., Sacchet, M., Pardini, C.: Open professional development of math teachers through an online course. In: Proceedings of the International Conference on e-Learning 2019, pp. 131–138. IADIS Press (2019). https://doi.org/10.33965/el2019_201909F017

22. Bruschi, B., et al.: Start@unito: a supporting model for high school students enrolling to university. In: Proceedings 15th International Conference CELDA 2018: Cognition and Exploratory Learning in Digital Age, pp. 307–312 (2018)

23. Brancaccio, A., Marchisio, M., Demartini, C., Pardini, C., Patrucco, A.: Dynamic interaction between informatics and mathematics in problem posing and solving. Mondo Digitale **13**, 787–796 (2014)

24. Demartini, C.G., et al.: Problem posing (& solving) in the second grade higher secondary school. Mondo Digitale **14**, 418–422 (2015)

25. Barana, A., Fissore, C., Marchisio, M., Pulvirenti, M.: Teacher training for the development of computational thinking and problem posing and solving skills with technologies. In: Proceedings of eLearning Sustainment for Never-ending Learning, Proceedings of the 16th International Scientific Conference ELearning and Software for Education, vol. 2, pp. 136–144 (2020)

26. Barana, A., et al.: Online asynchronous collaboration for enhancing teacher professional knowledges and competences. In: The 14th International Scientific Conference eLearning and Software for Education, pp. 167–175. ADLRO, Bucharest (2018). https://doi.org/10.12753/2066-026x-18-023

27. Clark, R.E.: Media will never influence learning. ETR&D. **42**, 21–29 (1994). https://doi.org/10.1007/BF02299088

28. Clark, R.C., Mayer, R.: E-Learning and The Science of Instruction. Pfeiffer (2008)

29. Gagatsis, A., et al.: Formative assessment in the teaching and learning of mathematics: teachers' and students' beliefs about mathematical error. Sci. Paedagog. Exp. **56**, 145–180 (2019)
30. Barana, A., Marchisio, M., Sacchet, M.: Interactive feedback for learning mathematics in a digital learning environment. Educ. Sci. **11**, 279 (2021). https://doi.org/10.3390/educsci11 060279
31. Black, P., Wiliam, D.: Developing the theory of formative assessment. Educ. Assess. Eval. Account. **21**, 5–31 (2009). https://doi.org/10.1007/s11092-008-9068-5
32. Barana, A., et al.: The role of an advanced computing environment in teaching and learning mathematics through problem posing and solving. In: Proceedings of the 15th International Scientific Conference eLearning and Software for Education, Bucharest, pp. 11–18 (2019). https://doi.org/10.12753/2066-026X-19-070
33. Barana, A., Fissore, C., Marchisio, M.: From standardized assessment to automatic formative assessment for adaptive teaching: In: Proceedings of the 12th International Conference on Computer Supported Education, pp. 285–296. SCITEPRESS - Science and Technology Publications, Prague (2020). https://doi.org/10.5220/0009577302850296
34. Barana, A., Conte, A., Fissore, C., Marchisio, M., Rabellino, S.: Learning analytics to improve formative assessment strategies. J. e-Learn. Knowl. Soc. **15**, 75–88 (2019). https://doi.org/10. 20368/1971-8829/1135057
35. Hattie, J., Timperley, H.: The power of feedback. Rev. Educ. Res. **77**, 81–112 (2007). https:// doi.org/10.3102/003465430298487
36. Di Caro, L., Rabellino, S., Fioravera, M., Marchisio, M.: A model for enriching automatic assessment resources with free-text annotations.In: Proceedings of the 15th International Conference on Cognition and Exploratory Learning in the Digital Age, CELDA 2018, pp. 186–194 (2018)
37. Di Caro, L., Fioravera, M., Marchisio, M., Rabellino, S.: Towards adaptive systems for automatic formative assessment in virtual learning communities. In: Proceedings of 2018 IEEE 42nd Annual Computer Software and Applications Conference (COMPSAC), pp. 1000–1005. IEEE, Tokyo (2018). https://doi.org/10.1109/COMPSAC.2018.00176
38. Barana, A., Di Caro, L., Fioravera, M., Marchisio, M., Rabellino, S.: Ontology development for competence assessment in virtual communities of practice. In: Penstein Rosé, C., et al. (eds.) Artificial Intelligence in Education. LNCS (LNAI), vol. 10948, pp. 94–98. Springer, Cham (2018). https://doi.org/10.1007/978-3-319-93846-2_18

Ubiquitous Learning

Pedagogical Scenario Design and Operationalization: A Moodle-Oriented Authoring Tool for Connectivist Activities

Aïcha Bakki[(⊠)] and Lahcen Oubahssi

LIUM, EA 4023, Le Mans University, Avenue Messiaen, 72085 CEDEX 9 Le Mans, France
aicha.bakki@univ-lemans.fr

Abstract. Recent studies on Massive Open Online Courses highlighted the importance of learning design in order to avoid different issues of these environments. Research also emphasizes the challenges experienced by teachers and instructional designers regarding the implementation of pedagogical scenarios in Learning Management System (LMS), due to the lack of tools and models to support them on these aspects. These difficulties are mainly related to the operationalization of pedagogical scenarios and the use of the corresponding tools, especially in a connectivist context. This research work addressees a large part of theses issues and focuses on the study of learning design model for Massive Open Online Course (MOOC) environments, and more specifically on assisting teachers in the design and implementation of pedagogical scenarios for connectivist MOOCs (cMOOCs). It also focuses on Moodle Platform and proposes for this specific platform. The major contribution of this work is a visual authoring tool, based on business workflows for the design and deployment of cMOOC-oriented scenarios on the Moodle platform. The tool was also evaluated, primarily from the point of view of utility and usability. The findings confirm that our tool can provide all the elements needed to formalize and operationalize such courses on the Moodle platform.

Keywords: MOOC · Connectivism · Moodle · Authoring tool · Pedagogical workflow · Operationalization · Instructional design · LMS

1 Introduction

MOOCs may be considered as contributing to social inclusion by democratization of higher education [34, 39]. Nevertheless, several studies [12, 27] have shown that MOOC economic model limits their development and their generalization due to their high production costs, the fact that they must be sufficiently broadly based to be profitable, and therefore exclude an important segment of knowledge. Despite this, educational institutions increasingly rely on MOOCs as a new form of pedagogical support and to modernize their educational practice. Beyond the supporting role of MOOCs, a new pedagogy is emerging that is enhancing or redynamising face-to-face (F2F), distance and

© Springer Nature Switzerland AG 2022
B. Csapó and J. Uhomoibhi (Eds.): CSEDU 2021, CCIS 1624, pp. 451–471, 2022.
https://doi.org/10.1007/978-3-031-14756-2_22

hybrid teaching. However, because of their structure, MOOCs rarely allow for personalized/adapted teaching, learning, supervision and pedagogical methods, thereby hindering their appropriation by teaching teams or educational managers. The adaptation question and the learning individualization (either face-to-face or distance learning) have been addressed in many research studies. Indeed, the boundaries between supervised and self-directed learning are increasingly blurring, leading to a redefinition of the teacher's and learner's roles. Indeed, studying in an open and networked environment such as a MOOC is challenging, since control of educational activities is handed over from educational institutions to individuals, who are generally isolated learners [24]. On the one hand, the tasks that were previously carried out by a teacher, such as pedagogical objectives setting and student's progress evaluating, can now be assigned to learners. These tasks can be overwhelming for learners who are unaccustomed to learning environments that require them to be self-directed and self-regulated [29]. On the other hand, the modalities of access to educational resources must be defined, because they are externalized and accessible anytime and anywhere, which brings into question of the transmissive model. Indeed, discussion about the transmissive pedagogical approach, its advantages, its disadvantages and mainly its adequacy with this new mode of digitalized and connected teaching, have been a subject of wide discussion in the educational field. In this sense, Siemens introduced a new learning approach in 2005 called connectivism, which he described as *"a learning theory for the digital age"* [49]. In his opinion, this new learning approach addressed the limitations of previous learning theories within a world driven by Web 2.0 technologies.

With the advent of connectivism, the specific needs of the digital world, and particularly of online platforms and their influence on learning, are increasingly taken into account. cMOOCs are learner-centered courses with a dynamic course structure that emphasize creativity, cooperation, autonomy, and social connections within the learning community [21, 22, 50]. According to the co-founders of the first connectivist course (cMOOC), the structure of such courses is based on four practices [20, 29]: aggregation, remixing, repurposing, and feed forwarding. A fifth category that involves evaluating activities was added to this categorization [10].

Many research studies have shown that the current model of MOOCs, and the mechanisms developed for distance learning platforms in universities, are not adapted to the appropriation needs expressed by teachers and learners. For the particular case of the cMOOC, the design of these courses, though a learning management system (LMS) presents several challenges that are related to the assimilation of the design features of such courses and to the operationalization of scenarios by teachers. In addition, pedagogical practices related to connectivism are not explicitly embedded in the pedagogical model of an LMS. Some platforms, such as Moodle[1] or OpenEdx[2], do not reduce this complexity; nevertheless, the provision of adequate support and facilities is difficult, despite the large and active communities involved. In addition, since each platform is based on a specific instructional design paradigm and a specific pedagogy, practitioners are often unfamiliar with this implicit type of instructional design method [1, 31, 32]. The engineering process to be privileged must better involve the teachers in charge of

---

[1] https://moodle.com/.

[2] https://open.edx.org/.

the courses and allow them to appropriate the contents and the pedagogical modalities, in order to develop hybrid teaching and thus the intertwining of the use of MOOCs and face-to-face teaching. This will allow promoting the implementation of new pedagogical practices and modes of interaction through the instrumentation of the different actors of a pedagogical situation. All these points raise new challenges concerning: (1) The pedagogical approach to be implemented, (2) The appropriation of the Learning Management System (LMS), its technical and pedagogical features. (3) The implementation of the designed pedagogical scenario on the LMS.

The research work presented in this paper is part of the Pastel research project. A key objective of this project is the design of a process for incorporating new technologies into pedagogy and teaching. This process can be divided into three main phases of capitalization, scenario design and operationalization, which is required to set up an editorial process. This paper focuses on the first two phases of the editorial process and provides solutions for designing and operationalizing pedagogical scenarios (Sect. 2.2). In order to promote and facilitate the implementation of cMOOCs by teachers and instructors using the Moodle platform, we propose a visual editor that includes concepts closely related to the Moodle platform. We provide graphical notations that are more user-friendly and which better address the needs of the user. This paper focuses on learning design models and address ways of constructing an LMS-centric language that combines a pedagogical model for a particular platform with a specific pedagogical approach.

The paper is structured as follows: in Sect. 2, we present some our research and local context, and analyze the relationships between LMSs, instructional design and MOOCs. We will also draw up a state of the art on the current state of MOOCs and research related to our problematic. In Sect. 3 we will describe our approach. This section will describe our principal contributions, namely the visual authoring tool and the operationalization service. Section 4 presents the result of the experiment that we had conducted of the proposed tool. Finally, Sect. 5 draws some conclusions and outlines some directions for future work.

2 Literature Review and Research Aims

2.1 MOOC: A Decade Afterward

Massive open online courses (MOOCs) are tightly related to online education, higher education, sustainable education, open and distance education. Their main characteristics as their name suggests are availability, openness and massiveness. These environments are distinguished by several dimensions, such as massive numbers of learners, openness to all, accessibility on a large scale, the nature of the qualifications and content, the evaluation modalities, etc.

Nowadays, educators and institutions pay increasingly more attention to the application of MOOCs in higher education [17, 52]. Since their emergence, MOOCs have been adopted by a significant number of educational institutions. [47] states that in 2018, more than 900 universities worldwide announced or launched more than 11,400 MOOCs, and 101 million students signed up to study a wide range of topics such as technology, economics, social sciences and literature [13]. The purpose of these courses is to contribute to the generalization of learning, both for students and for individuals who want to undertake lifelong learning. It is also to extend education to persons who, for social or geographical reasons, presently lack access to training [7]. They are based on existing LMSs such as OpenEdx[3] and Moodle[4].

Considerable interest has been devoted to MOOCs from an institutional, educational and research perspective. Since their emergence, many studies [5, 15, 44] have claim that they have become a consolidated reality in education [26]. This has been particularly marked in the last year, when the global pandemic has increased the prominence of online learning and training as well as self-directed learning. In fact, providers gained over 60M new learners and around 2.8 million courses were added in 2020 alone (Figs. 1 and 2) [38, 48]. In addition, during the peak month of the pandemic in April 2020, the MOOC network traffic of the two major MOOC service providers Coursera and edX rose to 74.6 million and 34.9 million, respectively [48]. Overall, these environments have undergone considerable change regarding the number of users, the pedagogical methods, the economic model, etc.

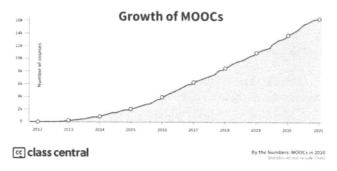

Fig. 1. Growth of MOOCs [48].

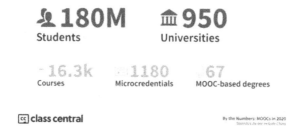

Fig. 2. MOOCs according to numbers [48].

From a research perspective, studies on MOOC is increasingly diverse and evolving [53]. Several studies have addressed specific issues related to MOOC characteristics from different perspectives. Furthermore, a large number of research studies on MOOCs have

---

[3] www.edx.org, https://www.fun-mooc.fr./en/.

[4] http://mooc-culturels.fondationorange.com/.

been essentially learner-centered, and have addressed various issues related to drop-out rates, engagement or motivation using various approaches such as trace analysis, for different purposes, such as adaptation, personalization, etc. To summarize, these researches include concept, case studies, educational theory, motivations and challenges of using MOOCs, student engagement and learning success, methodological approaches in MOOC research, etc.

The above non-exhaustive list provides a global overview of research related to MOOCs. We are particularly interested in two of them. The first one concern assistance to teachers for the implementation of such courses by exploiting the potential and func-tionalities of online learning platforms (which in our case study is Moodle). The second one concerns the integration of new technologies in learning and training process. On this last point, as we mentioned in the introduction, this research work is part of PAS-TEL project (Sect. 2.2). It constitutes in fact one of the technological contributions, being part of the main objective of an editorial chain for the development of a MOOC in academic context. All these points raised are part of a research problematic concerning instructional design and its several challenges.

The role and importance of MOOCs is no longer in question. However, the peda-gogical design of these massive environments raises questions. According to Rizvi *et al.* [43] a number of studies on formal online environments [18, 41, 42]; and MOOC envi-ronments [43] show the importance of learning design. These studies highlight that a better understanding of the nature of engagement with various learning activity types may lead to ways, which we can use to improve persistence in these environments [43]. Indeed, different research raised this problematic. A study conducted by Fianu *et al.* [23] emphasized the importance of course design in MOOC. In the same direction, Kim *et al.* [28] has conducted a study on the importance of instructional design and its impact on different parameters of a MOOC including, course completion. The authors mentioned the paper work of Fianu *et al.* [23] which found that instructional quality, including course design, is a significant positive predictor of students' satisfaction and usage of MOOC. A recent systematic review conducted by Goopio and Cheung [25] confirmed that a poor course design contributes to the high dropout rate in MOOCs. In other words, improving course design in a MOOC may prevent students from dropping [28].

Getting concerned with learning design implies two main stakeholders: the learner and the teacher. Nevertheless, few works have addressed issues related to teachers. In this regard, several research questions can be highlighted, for example: what role does the teacher play in these massive and open environments? What are the teacher's needs when implementing these environments? What tools and methods have been proposed to support the teacher's activities (design, deployment, monitoring, analysis, etc.)? In our work, we are particularly interested in the second of these research areas. We focus mainly on the design of pedagogical scenarios, and especially in the process of scenario design for cMOOCs, and their deployment on specific LMSs.

An analysis of the environments currently used to implement MOOCs led to the identification of Moodle and OpenEdx as the two LMSs that are most widely used to support these types of learning environments. Based on this analysis, we examined and compared the functionalities of these two LMSs. The objective was to identify the connectivist-oriented functionalities embedded in these two LMSs. As shown in

Table 1, Moodle provides a more diverse range of solutions for teachers who intend to adopt connectivism as a pedagogical approach. In addition, Moodle has been used for about 13 years in the educational system. Moodle is also a free LMS system that is predominantly used in universities. In fact, according to statistics from Moodle, the number of Moodle users is currently approximately 297,000,000. Conceived as an open source platform, it has a community of developers and technological contributors who have created plugins for a variety of needs [19]. A plugin is a software component that adds or expands a specific feature to complete a software application. It may also enable the customization of an interface or other features depending on the user's needs, such as accessibility. For these reasons, we chose the Moodle platform as our area of study and analysis.

2.2 Local and Research Context

As mentioned in the previous section, this research work was carried out as part of PAS-TEL (Performing Automated Speech Transcription for Enhancing Learning) research project. The main objective of this project is to provide and validate instrumented learning and teaching environments for both distance and face-to-face synchronous pedagogical situations. In order to achieve this goal, this project intends to elaborate an editorial chain starting from the design of the pedagogical scenario up to its implementation on the learning management system. The editorial chain is designed to allow the integration and organization of online pedagogical content; and develop prototype plugins that allow the editorial chain to be operationalized. The process is structured into four major phases as illustrated in Fig. 3.

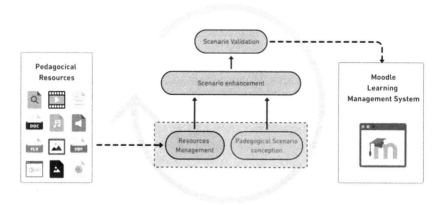

Fig. 3. PASTEL editorial chain process.

The first phase of the process, "*Scenario Design*" phase, allows for the design and modeling of a scenario based on an existing solution for an adapted educational content scenarization for the editorial chain. The contribution presented in this article concerns this phase in particular. Subsequently, we present the way our authoring tool has been extended to assist teachers in designing and implementing Moodle-oriented Connectivist pedagogical activities.

The second phase of the process, *"Resource management"*, allows the management of resources that have been instrumented in the project. The resources instrumentation can concern addition, modification and adaptation. Some of the functionalities will be based on the exploitation of automatic speech transcription. The general approach of the project consists in studying and instrumenting two examples of pedagogical resources: oral presentation such as a synchronous lecture, either face-to-face or distance one, and collaborative work such as synchronous practical work. Other resources will also be used to enrich the teacher's course in order to enable him/her to create his/her cMOOC on the Moodle platform. It should be noticed that the *"Scenario design"* and *"Resource management"* phases could be performed in parallel.

The third phase of the process, *"Scenario enhancement"*, allows the teacher to extend the scenario with the materials that were processed in phase 2. The scenario enhancement can be performed either in the authoring tool or Moodle platform. The fourth and last phase *"Scenario Validation"* allows the teacher to test and validate his scenario and to check whether his course has been successfully operationalized on the targeted platform.

As we had noticed below, our main contribution concerns the first and last phase. The main purpose is then to define a pedagogical scenario model that we reify on an authoring tool. However, the approach adopted in this research work is a platform oriented approach, and more specifically, a Moodle-oriented approach. Thus, abstracting the pedagogical model of Moodle platform is primordial. In other words, we need to examine the pedagogical aspects of Moodle platform allowing, in our case, to build connectivist courses. Before discussing these aspects in depth in Sect. 3, it is essential to examine the relationship between learning management systems and pedagogical scenario building from a scientific point of view. This is the purpose of Sect. 2.3.

2.3 Learning Design and Learning Management System

The emergence of technology is constantly expanding the possibilities for online learning, and continues to contribute greatly to the evolution of e-learning. The current educational context is undeniably changing due to, among other reasons, the emergence of new technological paradigms [16, 45]. Nowadays, learning and teaching processes cannot be understood without technology [26], notably through the use of LMSs or learning support systems (LSSs), which are one of these technologies and are defined as online learning technologies for the creation, management and delivery of online content. Thus, in this context, the use of Learning Design (LD) is a key factor for teachers who wish to improve the results of their educational practices [30]. The term *"Learning Design"* began to appear in the late 90's [40]. In today's ubiquitous digital environment, LMSs also play an important role in improving and facilitating these educational practices in distance teaching and learning. An LMS improve and increase the quality of learning in a collaborative environment, by providing solutions to the delivery of digital instruction and resources. They also allow teachers to focus on designing their teaching activities. We note that the design of pedagogical situations on learning devices such as educational platforms or LMS systems is not a straightforward task. A number of constraints faces a large number of teacher-designers when using these platforms to design pedagogical scenarios [51]. They are not accustomed to the implicit pedagogical design language used [31], and are not able to implement the scripts required by the platforms [35]. The main challenges relate to the specification of functionalities, based on their knowledge about the LMS and their skills in terms of pedagogical conception. This is especially important since pedagogical designs on LMS platforms are not sufficiently flexible, and impose a specific paradigm. Despite the existence of standards [31, 35], approaches [55], languages [6], standards [31, 35], tools [56] and architectures [4] that aim to promote and improve the use of platforms through the specification of graphical instructional languages and platform-centric authoring tools, these are generally incompatible with the platforms. In addition, they do not facilitate the operationalization of the designs that are produced. This means that several modifications to the initial scenario are required. Resulting in a loss of information and semantic during the operationalization of the scenarios described outside of the platforms [1].

As part of their research work, Abedmouleh *et al.* [1] and El Mawas *et al.* [32] have described a process for identifying and formalizing the pedagogical practices embedded in distance learning platforms, based on a metamodelling approach. The aim is to provide a method in terms of necessary analysis and steps for the identification and the formalization of such LMSs' instructional design languages. The method takes into account three different viewpoints: a viewpoint centered on the LMS macro-HMIs (Human-Machine Interfaces), a functional viewpoint and a micro viewpoint [32].

The advantage of this proposal is that the identified language can be used as a basis for the development of new pedagogical conceptions and authoring tools. Solutions that rely on the definition of the platform's pedagogical model have a second purpose. They also provide a communication bridge between authoring tools and the platforms concerned. In addition, adopting a platform-centric language can preserve the semantics of the pedagogical scenarios, meaning that these scenarios can be implemented with limited information loss. As part of our work, and in order to develop a connectivist-oriented scenario model centered on the Moodle platform (as described in Sect. 3.1), we relied on this process when identifying and formalizing the embedded pedagogical aspects in the LMS. Our main objective was to identify cMOOC-oriented pedagogical concepts embedded in Moodle in order to provide solutions for assisting and supporting teachers interested in adopting Moodle as a platform for delivering cMOOC-based courses. Further work is detailed in Sect. 3 of this paper.

3   A Moodle-Oriented Authoring Tool for Connectivist Context

3.1   A Moodle-Oriented Pedagogical Model for Connectivist MOOC

In this paper, we are specifically interested in the identification and formalization of the implicit cMOOC-centric instructional design language used in LMSs. This language will form the basis for the development of binding solutions that will simplify instructional design on the Moodle platform. These solutions must ensure that pedagogical scenarios formalized in conformance with a proposed language can be operationalized in the LMS, with a reduced semantic loss. In order to identify the pedagogical core of the Moodle platform, and more specifically to identify connectivist-oriented pedagogical concepts, we adopt a platform-centric approach.

We do not intend to enhance the semantics of the pedagogical model embedded into the Moodle platform. According to Abedmouleh [3], a LMS is not pedagogically neutral but embeds an implicit language that is used to describe the process of designing learning activities. Thus, our proposition is based aims to identify the connectivist language embedded in the Moodle platform and then to explicitly formalize this language in a computer-readable format. This format can be used as a binding format for various external tools with different design aspects. Our approach involves carrying out a functional analysis of the Moodle platform, in which we rely on the work conducted by El Mawas *et al.* [32].

We have previously conducted a study of the current state of the art in terms of the pedagogical scenario design aspects of cMOOCs, and have put together a compendium of teachers' needs related to the design and deployment of such courses, resulting in a set of criteria and elements that regulate scenario design in a cMOOC course [10]. This exploratory work also allowed for the abstraction of a pedagogical model from existing cMOOCs, and some of these elements are presented in Table 1.

Table 1. Connectivism vs. OpenEDX - Moodle elements mapping [8].

	Associated bloom taxonomy	Activities	Moodle	OpenEdx
Aggregation	Read, search, categorize, quote, read, etc	Consultation Cognition	Page, URL, resource	Video, File, HTML
Remixing	Select, identify, argue, criticize, justify, adapt, discuss, illustrate, summarize, interpret, etc	Metacognition Sharing Communication	Chat, Forum, LTI, Wiki, Glossary, Journal	Forum, HTML page
Feed forwarding	Share	Sharing	Page	HTML
Repurposing	Compose, construct, create, elaborate, plan, reorganize, represent, schematize, write, etc	Production Collaboration	Workshop, LTI	xBlock LTI
Evaluation	Examine, test, evaluate	Evaluation	Quiz, Workshop	Quiz

The next step is therefore to combine this abstract pedagogical model with the requirements identified for the cMOOC scenario design language, and the main characteristics of a connectivist course (Fig. 4). This led us to the proposal of a cMOOC oriented scenario model. Nevertheless, this model is independent from any learning platform.

Concurrently, on the basis of the work performed by El Mawas *et al.* [32] we have conducted a functional analysis of the Moodle platform. The purpose of this analysis is to identify the pedagogical aspect embedded in this platform. This permitted us to identify the pedagogical model of Moodle platform, independently of any pedagogical approach.

These two models, namely the connectivist pedagogical scenario model and the Moodle-centered pedagogical model, lead to specifications for these models from two different perspectives.

A comparison (confrontation) between these two models is then carried out (Figs. 4), which allows us to formalize the final conceptual language. The objective is to combine the platform's pedagogical architecture with the elements of the connectivist pedagogical scenario, in order to build a representative model. This phase essentially involves comparing, factorizing and structuring the elements of both models. Some parts of the models may be identical, complementary or at different levels of abstraction. The methodology consists of verifying the elements of the respective models based on several points, including the definition of similar elements, the non-existence of certain elements, or the generality or specificity of the relationships between particular elements. The general concept of this comparison is as follows: (i) we first verify the non-existence of one or several elements if applicable; (ii) we identify a specification or generalization relationship between the model's elements; (iii) otherwise, we verify the difference of the definition of the element in each model. More specifically, we verify whether elements are at different levels of abstraction. When all elements have been verified, we obtain the Moodle-centric pedagogical scenario model illustrated in Fig. 5.

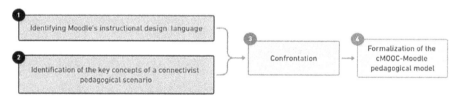

Fig. 4. Construction process of our Moodle-centric scenario model [8].

3.2 Proposed Authoring Tool

In this section, we will present our visual authoring tool for pedagogical design, which allows specifying learning situations and then implement the pedagogical scenarios to the Moodle Platform using the deployment service. In our context, we have opted for providing teachers with a visual modeling tool. This choice is motivated by several research works that are especially concerned with the use of graphical notation in different domains and contexts. Indeed, the use of graphical notation to provide a visual syntax for modeling languages has been developed and put into practice in many different domains, and graphical notations have also been developed to reduce the cognitive load when working with complex semantic models. In this vein, we opted for an extended BPMN graphical notation [37] for the design of our pedagogical scenarios.

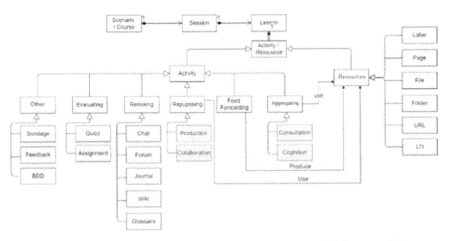

Fig. 5. Moodle-centric pedagogical scenario model for cMOOC context [8].

The use of the BPMN was motivated through an exploratory study of existing modeling languages. We compared the BPMN language to other pedagogical modeling languages according to two prevalent classifications, presented by Botturi *et al.* [14] and Nodenot [36]. Regarding the page limit of this paper, more details on the conducted study can be found in [10]. BPMN meets our requirements from the technical and descriptive aspects contained in Botturi *et al.*'s [14] classification and according to the pedagogical scenario aspects (such as role representation, sequencing, collaboration, etc.) presented in Nodenot's classification [36].

However, in order to meet the requirements to provide the teacher with support in terms of designing cMOOC-oriented pedagogical scenarios, we cannot use BPMN as it stands. The specifications for BPMN not only involve the use of graphical notations for process descriptions, but also definitions of abstract metamodels for these domains. We therefore propose an extension to the concepts underlying BPMN, which takes into account the specificities of a cMOOC scenario by defining an abstract model and a particular graphical notation for this model. We then embed this extended notation and model into our tool. The notations include elements describing the roles of participants, the learning sessions, the different categories of activities, and the resources and sequencing of activities (Fig. 4).

The objective is not to build a new platform, but to start with an existing tool and extend it. We therefore selected the BPMN. io tool[5], an open-source web application that uses BPMN 2.0 notation [37]. Developing an extension to BPMN notation is not the purpose of the current paper [9]. Instead, we mainly focus on the presentation of the tool's functionalities, the extension of the visual notation and the development of a Moodle operationalization service. In the following section, we will take an example in order to illustrate our proposal. This example involves a week's activities as part of a MOOC, on the topic of digital identity, as shown in Table 2.

---

[5] http://bpmn.io/.

Table 2. Example of a textual pedagogical scenario [8].

Lesson	Activities	Type
Discovering the subject	Consulting a collection of resources to discover the subject of the week	Aggregation
	Conducting a web search on digital identity	Aggregation
	Examining the key elements that build a user's digital identity	Remixing
Exploring and discovering the interests and ideas of learners	Writing a blog post on the topic	Remixing
	Discovering the publications of others	Aggregation
	Exchanges on the forum	Remixing
	Explaining and discussing acquired ideas with peers; interacting proactively in the chat room	Remixing

Pedagogical Scenario Conception Using Pedagogical Workflow. We propose an authoring tool for teacher-designers using the Moodle learning platform. This tool provides a graphical interface for the design of pedagogical scenarios by combining the elements of a connectivist pedagogical scenario with a pedagogical design language that is specific to the Moodle platform. It is a web application that reifies the pedagogical model presented in Sect. 3.1.

Once the teachers are connected, they can either create a new scenario or modify an existing one. In the following, we consider that teachers have chosen to create a new scenario. After specifying the name of the scenario and choosing the blank model, the teacher is directed to the conception page. When starting the scenario conception process, the teachers first create a learning session. They access this via the toolbox on the left of Fig. 6, in the learning session block.

In order to support teachers, we ensured that the modeling space was not empty when creating a new scenario (Fig. 6), and an initial learning session is therefore created by default. In a MOOC, one session typically represents one week. The teachers are then provided with an interface containing a pool, which can be renamed or deleted. They can use the properties section (Fig. 6) to specify the duration of this session (the start and end dates).

After creating their first session and specifying the roles, the teachers can start creating different lessons. We assume that a lesson encompasses a number of activities. The teachers can continue modeling by dragging the activities they want to include from the toolbox into the model. In order to facilitate the identification of activities, according to the four principles of a connectivist course, we classify them into four blocks with different color codes.

Each activity has its own properties; for example, for a consultation activity, the teachers specify whether to use a resource, a page or a URL that describes the activities to be carried out or presents a description of the progress of this activity. If teachers want to set a resource, they specify its type and the link to access it. The example presented in Table 2 illustrates this process. Once the teacher has designed all the activities, s/he will produce a workflow.

Once modeling is complete, the teachers can save the scenario in different formats or deploy it on an online platform using the *"Export to"* button (Fig. 6). This action transforms the BPMN file into one that can be imported by the Moodle platform.

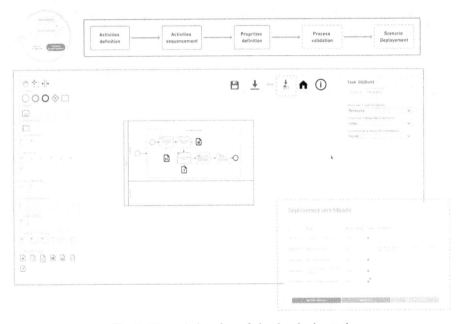

Fig. 6. The main interface of visual authoring tool.

Operationalization of Pedagogical Scenarios into Moodle Learning Environment. In order to support the teacher, a service allowing for the deployment of pedagogical scenarios was developed. Operationalization represents an intermediate phase between learning and scenario design, and the aim of this step is to ensure that the scenario described by the teacher can be used and manipulated in a LMS while preserving the pedagogical semantics [2]. Our contribution uses hybrid approaches based on processes and tools inspired by and/or applied in model-driven engineering [13].

We implement an deployment service that allows teachers to automatically deploy their pedagogical scenarios on the Moodle platform, using the transformation described in Sect. 3.2. To do so, we provide a solution that allows the pedagogical workflow to be transformed into a deployable scenario. As illustrated in Fig. 7, we propose a two-phase approach. Each one of the proposed approaches consists on 3 steps.

(1) Transformation/Pretreatment. The aim at this step is to propose a confrontation (comparison) between the pedagogical scenario and the elements of Moodle, in

order to resolve any ambiguities and to match each concept in the scenario with a concept in the chosen platform. The general idea of the transformation algorithm follows that described in Sect. 3.2. After the BPMN pedagogical workflow has been generated, the scenario is transformed into the format required by Moodle, which in this case is the JSON format (Fig. 7). The teacher can also export the course in MBZ format (Fig. 7), via a process in which the BPMN scenario is transformed into a set of XML files. In practice, this transformation involves making a backup of the Moodle course using the platform's import functionality; our transformation engine then modifies the backup archive with the information contained in the BPMN file for the new scenario.

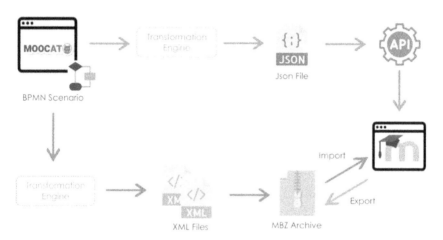

Fig. 7. Moodle operationalization process.

Finally, the teacher can import the modified course manually into the Moodle platform. In our example, after designing the lesson, the teacher can deploy this scenario on the Moodle platform. Pressing the Deploy button applies a pre-treatment phase that runs in the backend and generates a JSON file containing all the activities, by transforming the BPMN file using the mechanism described below.

(2) Deployment. The operationalization module allows a communication between our authoring tool and the Moodle LMS. For this purpose, we developed a web service as a Moodle plugin. The JSON file resulting from the preceding phase is sent via an HTTP request; this is received and interpreted by the developed plugin, which is automatically connected to the platform, and is then automatically executed on the LMS.

The Rest API that we have developed connects to the Moodle platform and deploys the different activities after generating the JSON file. Note that the transformation phase (i.e. the generation of the JSON file) and the Rest API are executed in the backend.

4 Experiments and Evaluation

4.1 Experiment Objective, Protocol and Data Collection

In this research, the contributions were evaluated and tested as they were specified through simulations and user tests. Although experiments in a real-world situation in which the tool was integrated into a cMOOC project would have been valuable, this was not possible, as it is risky for MOOC designers to rely on a research prototype.

Two evaluations were conducted with two distinct groups. A total of 24 participants have participated in the experiment, 12 for the first one and 12 for the second. Our main object is to evaluate the utility and usability of our authoring tool. Indeed, the aim is to verify the ability of the proposed tool to model cMOOC scenarios. The second objective is to evaluate the Moodle deployment service and its effectiveness.

The participants had prior knowledge and experience with pedagogical scenario design, had previously designed pedagogical situations and had used different instructional design tools. For our experiment, we follow a similar experimental protocol as presented in [9] with a separate group of participants.

This experiment was carried out during a pedagogical scenario design workshop. Our evaluation protocol consisted of three steps, as follows:

(1) Preparation: We provided participants with a user's guide that explained the philosophy and described the functionalities of the tool, and an experimental guide that described the different steps to be performed during this evaluation and the scenario to be deployed.
(2) Conception and deployment: The aim of this step was for the participants to design a pedagogical scenario for a cMOOC according to the instructions provided during the preparation phase, and then to deploy this scenario on the Moodle platform provided.
(3) Results: In this step, we asked participants to complete a questionnaire at the end of the evaluation, in order to validate the utility and usability of our tool and to obtain more information on their experiences.

The methodology used to collect the data in this experiment involved opinion data collected from the participants via a questionnaire. At the end of the experiment, participants were asked to complete an online questionnaire containing 25 closed-ended questions, which was evaluated using a six-point Likert scale. The first part of the questionnaire focused mainly on the expressivity of the notation and the deployment service. The second part measured the usability of the tool, and for this part we used the System Usability Scale (SUS) questionnaire [11].

4.2 Experimental Results

As mentioned below, the aim of the experiment is to evaluate usability and utility of our contributions. For usability measurement, we rely on SUS questionnaire. It is a popular and effective tool for assessing the usability of various systems (Bangor *et al.* 2009). SUS is based on a 10 closed-ended questions, where each item is rated on five-point

Likert scale ranging from 1 (strongly disagree) to 5 (strongly agree). Although only 24 participants were involved in this experiment, this was sufficient to detect any major problems with usability [54].

The calculation of the SUS score can be influenced by misunderstandings of the negative statements in the questionnaire. In order to avoid these errors, we pre-processed the participants' responses before calculating the score. For this purpose, we used the grid presented by McLellan *et al.* [33], which considers all responses provided with a score greater than three for all negative statements as incorrect. Of the 24 responses received, seven were withdrawn.

Overall, the average SUS score for all participants was 70, with a SD of 12.74 (Fig. 8). This corresponds to the 52th percentile, according to the standardization presented by Sauro and Lewis [46].

In accordance with the empirical rule for interpreting SUS scores [11], scores of less than 50 were considered unacceptable, scores of between 50 and 70 were marginally acceptable, and scores of above 70 were acceptable. Using this acceptability scale, an average SUS score of 70 indicates that our tool is "acceptable", and a result between "OK" and "good" was obtained for the notation.

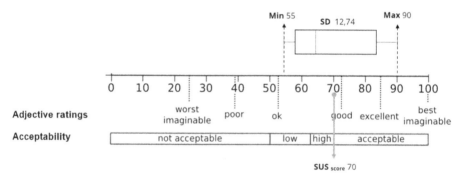

Fig. 8. Authoring tool usability: SUS score.

The other part of the questionnaire aims to determine whether the tool allowed the participant to produce a simple design for a connectivist course; and to assess the usability and utility of the deployment service. This part also assessed whether participants were satisfied with the tool and whether the notation used in the tool was easy to understand. The final aim was to evaluate the potential of the tool in terms of designing a cMOOC course.

As result, 16 of 24 participants affirm to have succeeded in formalizing all the concepts of their scenarios. We asked the participants whether the tool offered all the concepts required when designing a cMOOC course, 21 of the participants confirmed this statement. 17 participants agreed that the layout of the toolbox allowed them to identify the conceptual elements. The participants were also asked about the visual design and the use of visual notation, which facilitate the task of scenario conception. 16 participants over the 24 agreed with this statement.

For deployment service evaluation, all of the participants completed this phase. In total, 18 participants reported that the operationalization service was very useful, and 10 found the automated operationalization service easy to use. A review of the feedback from participants revealed that their scenarios had been successfully deployed: between 75% and 100% of the designed concepts have been successfully deployed on the platform. These results are generally consistent with those of the scenario analysis. We noticed that some users did not fully complete the metadata, which explains the rates obtained.

5 Conclusion

In this paper, a specific problem has been addressed, concerning the pedagogical scenario design in these massive and open environments. We have accordingly centered on the pedagogical practices related to the connectivist approach. The need for models and tools to design and deploy such courses is evident.

This research work is part of the Pastel research project. As we mentioned before the main objective of this project is the design of a process for incorporating new technologies into pedagogy and teaching. This work is a continuity of the work realized in [9] whose is based instructional design approach and [8] whose based on a centered approach. However, in the Pastel project, we are interested in a platform-oriented approach and more particularly in the Moodle platform. Designing and implementing such pedagogical scenarios is a complex task for a teacher, moreover, the literature review revealed a lack of models and tools to support the teacher in designing such courses. Addressing this problem involves the use of methodologies integrated within the process of instructional engineering. The purpose is to explore both the theoretical aspects of the connectivist approach from the existing literature and the functional and pedagogical analysis of the Moodle platform. Thus, we have presented in this article our main contribution, which allows the design and deployment of cMOOC-oriented activities. Our approach is based on a two-step process, from the design to the operationalization of pedagogical activities. The first step allows teachers to model their pedagogical activities (via our visual authoring tool) without specific technical knowledge. The second step involves the automatic deployment of a scenario designed using two methods: a Moodle web service API and the importation of MBZ archive. Our educational scenarios oriented cMOOC model as well as our visual authoring tool are developed, evaluated and tested according to an agile approach, in order to validate them against the real needs teachers. These evaluations were performed through simulations, technical testing and end-user evaluation. The final evaluation was carried out in the context of a pedagogical design workshop involving 24 participants in order to evaluate the utility and usability of our authoring tool. Our findings confirm that the developed tool allows to easily design connectivist pedagogical scenarios and their deployment on the Moodle platform. In the cMOOC context, a course is initially designed by the teacher, and learners are then encouraged to adapt it based on their learning objectives. The contributions presented in this paper have a twofold purpose: firstly, our aim is to assist teachers in conceiving MOOCs, and secondly, we seek to bridge the gap between the design and deployment phases by providing technical solutions to teachers or educational institutions working with the Moodle platform.

Using Moodle platform is motivated by several points. The first one concerns the wide use of this platform. Indeed, Moodle platform module is recognized as one of the most widely used platforms at the national and international levels, with more than 296 million users and 39 million courses delivered in over 240 countries. According to Moodle's statistics[6] their LMS is currently used by over 60% of higher education institutions all over the world. The pedagogical active community (teachers and pedagogical engineers), whose exchanges are centered on the pedagogical model embedded into the Moodle platform, constitute one of the major strength of Moodle platform. We also highlight tech-oriented community (Moodle API developers), who focus on providing technical solutions to Moodle users. Both communities could contribute significantly to the adoption of our solutions to educational institutions, and especially universities, which have started to deploy MOOCs over the last few years.

Acknowledgements. The current work is supported by the ANR project PASTEL <ANR-16-CE38-0007>. The authors want to thank all the persons who have contributed to this project.

References

1. Abedmouleh, A., Laforcade, P., Oubahssi, L., Choquet, C.: Operationalization of learning scenarios on existent learning management systems the moodle case-study. In: Proceedings of the 6th International Conference on Software and Database Technologies, ICSOFT 2011, pp. 143–148 (2011). https://doi.org/10.5220/0003486001430148
2. Abedmouleh, A.: Approche Domain-Specific Modeling pour l'opérationnalisation des scénarios pédagogiques sur les plateformes de formation à distance (Doctoral dissertation) (2013)
3. Alario-Hoyos, C., Bote-Lorenzo, M.L., GóMez-SáNchez, E., Asensio-PéRez, J.I., Vega-Gorgojo, G., Ruiz-Calleja, A.: GLUE! : an architecture for the integration of external tools in virtual learning environments. Comput. Educ. **60**(1), 122–137 (2013)
4. Alario-Hoyos, C., Pérez-Sanagustín, M., Cormier, D., Delgado-Kloos, C.: Proposal for a conceptual framework for educators to describe and design MOOCs. J. Univ. Comput. Sci. **20**(1), 6–23 (2014). https://doi.org/10.3217/jucs-020-01-0006
5. Baggetun, R., Rusman, E., Poggi, C.: Design patterns for collaborative learning: from practice to theory and back. In: EdMedia + Innovate Learning, pp. 2493–2498 (2004)
6. Bakki, A., Oubahssi, L.: A moodle-centric model and authoring tool for cmooc-type courses. In: CSEDU (1), pp. 545–556 (2021)
7. Bakki, A., Oubahssi, L., Cherkaoui, C., George, S.: Motivation and engagement in MOOCs: how to increase learning motivation by adapting pedagogical scenarios? In: Conole, G., Klobučar, T., Rensing, C., Konert, J., Lavoué, E. (eds.) Design for teaching and learning in a networked world, vol. 9307, pp. 556–559. Springer, Cham (2015). https://doi.org/10.1007/978-3-319-24258-3_58
8. Bakki, A., Oubahssi, L., George, S., Cherkaoui, C.: MOOCAT: a visual authoring tool in the cMOOC context. Educ. Inf. Technol. **24**(2), 1185–1209 (2018). https://doi.org/10.1007/s10639-018-9807-2

---

[6] https://stats.moodle.org/

9. Bakki, A., Oubahssi, L., George, S.: Design and operationalization of connectivist activities: an approach through business process management. In: Scheffel, M., Broisin, J., Pammer-Schindler, V., Ioannou, A., Schneider, J. (eds.) Transforming Learning with Meaningful Technologies, vol. 11722, pp. 251–265. Springer, Cham (2019). https://doi.org/10.1007/978-3-030-29736-7_19

10. Bangor, A., Kortum, P., Miller, J.: Determining what individual SUS scores mean : adding an adjective rating scale. J. Usability Stud. **4**(3), 114–123 (2009)

11. Blagojević, M., Milošević, D.: Massive open online courses : EdX vs Moodle MOOC. In: Proceedings of 5th International Conference on Information Society and Technology, Kopaonik, Serbia, pp. 346–351 (2015)

12. Bonk, C.J., Zhu, M.: MOOC Instructor Motivations, Innovations, and Designs : Surveys, Interviews, and Course Reviews Curtis J. Bonk, Meina Zhu, Annisa Sari, Indiana University (2018)

13. Botturi, L., Derntl, M., Boot, E., Figl, K.: A classification framework for educational modeling languages in instructional design. In: 6th IEEE International Conference on Advanced Learning Technologies (ICALT 2006) (2006)

14. Buhl, M., Andreasen, L.B.: Learning potentials and educational challenges of massive open online courses (MOOCs) in lifelong learning. Int. Rev. Educ. **64**(2), 151–160 (2018). https://doi.org/10.1007/s11159-018-9716-z

15. Cabero, J.: Visiones educativas sobre los MOOC. RIED **18**(2), 39–60 (2015). https://doi.org/10.5944/ried.18.2.13718

16. Castells, M., Cardoso, G.: The network society: from knowledge to policy, pp. 3–21. Center for Transatlantic Relations, Paul H. Nitze School of Advanced International Studies, Johns Hopkins University, Washington, DC (2005)

17. Černý, M.: Connectivism in the phenomenological-pragmatist tradition. e-Pedagogium **20**(2), 7–24 (2020). https://doi.org/10.5507/epd.2020.017

18. Chatwattana, P.: A MOOC system with self-directed learning in a digital university. Global J. Eng. Educ. **23**(2), 134–142 (2021)

19. Conde, M.Á., Hernández-García, Á.: Learning analytics for educational decision making. Comput. Hum. Behav. **47**, 1–3 (2015)

20. Da Costa, J.: BPMN 2.0 pour la modélisation et l'implémentation de dispositifs pédagogiques orientés processus [PhD Thesis]. University of Geneva (2014)

21. De Vries, F., Tattersall, C., Koper, R.: Future developments of IMS learning design tooling. J. Educ. Technol. Soc. **9**(1), 9–12 (2006)

22. Dougiamas, M., Taylor, P.: Moodle: using learning communities to create an open-source course management system. In: EdMedia + Innovate Learning, pp. 171–178 (2003)

23. Downes, S.: Places to go: connectivism and connective knowledge. Innov.: J. Online Educ. **5**(1), 6 (2008)

24. El Mawas, N., Oubahssi, L., Laforcade, P.:.A method for making explicit LMS instructional design languages. Technol. Instr. Cogn. Learn. **10**(3) (2016)

25. Ersoy, N.S., Kumtepe, E.G.: Transcultural elements in connectivist massive open online courses. TOJET **20**(4), 159–166 (2021)

26. Fianu, E., Blewett, C., Ampong, G.O.A., Ofori, K.S.: Factors affecting MOOC usage by students in selected Ghanaian universities. Educ. Sci. **8**(2), 70 (2018)

27. Fournier, H., Kop, R., Durand, G.: Challenges to research in MOOCs. MERLOT J. Online Learn. Teach. **10**(1) (2014)

28. Goopio, J., Cheung, C.: The MOOC dropout phenomenon and retention strategies. J. Teach. Travel Tour. **21**(2), 177–197 (2021)

29. Gonzalez, C.: The role of blended learning in the world of technology, 10 December 2004

30. Hidalgo, F.J.P., Abril, C.A.H.: MOOCs: origins, concept and didactic applications: a systematic review of the literature (2012–2019). Technol. Knowl. Learn. **25**(4), 853–879 (2020). https://doi.org/10.1007/s10758-019-09433-6

31. Huin, L., Bergheaud, Y., Codina, A., Disson, E.: When a university MOOC become a professional training product. In: Proceedings of the European MOOC Stakeholder Summit, EMOOCS 2016, p. 351 (2016)

32. Kim, D., et al.: Exploring the structural relationships between course design factors, learner commitment, self-directed learning, and intentions for further learning in a self-paced MOOC. Comput. Educ. **166**, 104171 (2021)

33. Kop, R.: The challenges to connectivist learning on open online networks: learning experiences during a massive open online course. Int. Rev. Res. Open Distance Learn. **12**(3), 19–37 (2011). https://doi.org/10.19173/IRRODL.V12I3.882

34. Lee, G., Keum, S., Kim, M., Choi, Y., Rha, I.: A study on the development of a MOOC design model. Educ. Technol. Int. **17**(1), 1–37 (2016)

35. Lockyer, L., Bennett, S., Agostinho, S., Harper, B.: Handbook of Research on Learning Design and Learning Objects: Issues, Applications, and Technologies (2 Volumes). IGI Global, Hershey (2009)

36. Martinez-Ortiz, I., Sierra, J.-L., Fernandez-Manjon, B.: Authoring and reengineering of IMS learning design units of learning. IEEE Trans. Learn. Technol. **2**(3), 189–202 (2009)

37. Mclellan, S., Muddimer, A., Peres, S.C.: The effect of experience on system usability scale ratings. J. Usability Stud. **7**(2), 56–67 (2012)

38. Mekpiroona, O., Tammarattananonta, P., Buasrounga, N., Apitiwongmanita, N., Pravalpruka, B., Supnithia, T.: SCORM in open source LMS: a case study of LEARNSQUARE. In: ICCE2008, Taipei, Taiwan, pp. 166–170 (2008)

39. Nodenot, T.: Scénarisation pédagogique et modèles conceptuels d'un. EIAH: Que peuvent apporter les langages visuels? Revue Internationale Des Technologies En Pédagogie Universitaire (RITPU)/Int. J. Technol. High. Educ. (IJTHE), **4**(2), 85–102 (2007)

40. Omg, O., Parida, R., Mahapatra, S.: Business process model and notation (BPMN) version 2.0. Object Management Group **1**(4) (2011)

41. Ossiannilsson, E.: MOOCS for lifelong learning, equity, and liberation. In: MOOC (Massive Open Online Courses). IntechOpen (2021)

42. Belleflamme, P., Jacqmin, J.: An economic appraisal of MOOC platforms: business models and impacts on higher education. CESifo Econ. Stud. **62**(1), 148–169 (2016). https://doi.org/10.1093/cesifo/ifv016

43. Persico, D., et al.: Learning design Rashomon I - supporting the design of one lesson through different approaches. Res. Learn. Technol. **21** (2013)

44. Pilli, O., Admiraal, W.: A taxonomy of massive open online courses. Contemp. Educ. Technol. **7**(3), 223–240 (2016)

45. Ramírez-Donoso, L., Rojas-Riethmuller, J.S., Pérez-Sanagustín, M., Neyem, A., Alario-Hoyos, C.: MyMOOCSpace: a cloud-based mobile system to support effective collaboration in higher education online courses. Comput. Appl. Eng. Educ. **25**(6), 910–926 (2017). https://doi.org/10.1002/cae.21843

46. Rienties, B., Nguyen, Q., Holmes, W., Reedy, K.: A review of ten years of implementation and research in aligning learning design with learning analytics at the Open University UK. Interact. Des. Architect. **33**, 134–154 (2017)

47. Rienties, B., Toetenel, L.: The impact of learning design on student behaviour, satisfaction and performance: a cross-institutional comparison across 151 modules. Comput. Hum. Behav. **60**, 333–341 (2016)

48. Rizvi, S., Rienties, B., Rogaten, J., Kizilcec, R.F.: Beyond one-size-fits-all in MOOCs: variation in learning design and persistence of learners in different cultural and socioeconomic contexts. Comput. Hum. Behav. **126**, 106973 (2021)

49. Sagar, C.: TICs y aprendizaje de idiomas: ¿existe algún sistema existe de aprendizaje digital y conectado? In: Roig-Vila, R. (ed.) Tecnología, innovación e investigación en los procesos de enseñanzaaprendizaje, pp. 1840–1847. Octaedro, Barcelona (2016)
50. Sanz, G., et al.: Guía para la observación nivometeorológica (2015)
51. Sauro, J., Lewis, J.R.: When designing usability questionnaires, does it hurt to be positive? In: Proceedings of the SIGCHI Conference on Human Factors in Computing Systems, pp. 2215–2224 (2011)
52. Shah, D.: Year of MOOC-based degrees: a review of MOOC stats and trends in 2018. Class Central (2019)
53. Shah, D.: By The Numbers: MOOCs in 2020. Class Central Report (2020). https://www.classcentral.com/report/mooc-stats-2020
54. Siemens, G.: Connectivism: a learning theory for the digital age. Int. J. Instr. Technol. Distance Learn. (2004). http://www.elearnspace.org/Articles/connectivism.html
55. Steel, C.: Reconciling university teacher beliefs to create learning designs for LMS environments. Australas. J. Educ. Technol. **25**(3) (2009)
56. Stylianakis, G., Arapi, P.: CoLearn: real time collaborative learning environment. e-Learning and e-\ldots, c (2013). http://ieeexplore.ieee.org/xpls/abs_all.jsp?arnumber=6644340
57. Tan, M., Yu, P., Gong, F.: The development path of MOOCs for China's higher education and its applications in engineering and technology education. World Trans. Eng. Technol. Educ. **14**(4), 525–530 (2016)
58. Toven-Lindsey, B., Rhoads, R.A., Lozano, J.B.: Virtually unlimited classrooms : pedagogical practices in massive open online courses. Internet High. Educ. **24**, 1–12 (2015)
59. UNESCO: making sense of MOOCs: a guide for policymakers in developing countries (2016). http://unesdoc.unesco.org/images/0024/002451/245122E.pdf
60. Virzi, R.A.: Refining the test phase of usability evaluation : how many subjects is enough? Hum. Factors **34**(4), 457–468 (1992)
61. Wang, Z., Anderson, T., Chen, L.: How learners participate in connectivist learning : an analysis of the interaction traces from a cMOOC. Int. Rev. Res. Open Distrib. Learn. **19**(1) (2018)
62. Weegar, M.A., Pacis, D.: A Comparison of two theories of learning-behaviorism and constructivism as applied to face-to-face and online learning. In: Proceedings e-leader Conference, Manila (2012)
63. Zedan, H., Al-Ajlan, A.: E-learning (Moodle) based on service oriented architecture. In: Proceedings of the EADTU's 20th Anniversary Conference (2007)
64. Zheng, S., Wisniewski, P., Rosson, M.B., Carroll, J.M.: Ask the instructors: motivations and challenges of teaching massive open online courses. In: Proceedings of the 19th ACM Conference on Computer-Supported Cooperative Work and Social Computing - CSCW 2016, pp. 205–220 (2016). https://doi.org/10.1145/2818048.2820082

Students Perceptions of Mobile Apps: Learning Features Impacts on EFL Vocabulary Learning

Zeng Hongjin[1,2](✉)

[1] Princeton University, Princeton, NJ 08544, USA
1152087038@qq.com
[2] Department of Foreign Language, Tianjin Normal University, Binshuixi Street, Tianjin, China

Abstract. Although mobile apps have been used for many educational purposes, little is known about how effective these apps are in EFL (English as a Foreign Language) vocabulary learning. Students' expectations on their usage of mobile apps are also lacking. To fill this gap, further improve the app development and teaching, this study centered on the effectiveness of mobile apps on EFL vocabulary learning from students' perspectives. A total of 25 articles were collected from 3 selected databases—Web of Science, Eric, and Academic Search Complete. The findings were analyzed through content analysis. Also, an empirical study was conducted between 2 classes of the same grade and same average proficiency level of English in Tianjin No.21 high school. The participants were asked to take part in pretest and mid-term diagnostic test to specify their vocabulary level proficiency and words retentions. Questionnaire and selected interview were adopted to analyze students' perceptions of mobile app--*Bubei*. The results provide a profile of using contexts of mobile applications for EFL vocabulary learning and the impacts of using mobile apps on EFL vocabulary learning outcomes. Mobile applications are mostly used in informal learning contexts and adopted in higher education for EFL vocabulary learning. The studies also identified 8 categories of impacts, including vocabulary acquisition and retention, administration for learning, pronunciation feature, usage frequency, learners' perceptions and attitudes, motivation and interest, feedback and evaluation, and learning environments. Implications are discussed, and suggestions for future research are provided.

Keywords: Mobile apps · EFL · Vocabulary learning · Students' perceptions · Impact · Mobile learning

1 Introduction

In recent years, the rapid development in communications and wireless technologies has resulted in mobile devices (e.g., PDAs, cell phones) becoming widely available, more convenient, and less expensive. More importantly, each successive generation of devices has added new features and applications, such as Wi-Fi, e-mail, productivity software, music player, and audio/video recording. Mobile devices could open new doors with their unique qualities such as "accessibility, individualization, and portability" [55] (p. 253). One of the main current trends of educational applications for new technologies is mobile

© Springer Nature Switzerland AG 2022
B. Csapó and J. Uhomoibhi (Eds.): CSEDU 2021, CCIS 1624, pp. 472–496, 2022.
https://doi.org/10.1007/978-3-031-14756-2_23

learning. Editors [50] have defined mobile learning as taking place when the learner is not at a fixed, predetermined location or when the learner takes advantage of learning opportunities offered by mobile technologies. Author [38] defined mobile learning as being concerned with learner mobility in the sense that learners should be able to engage in educational activities without being tied to a tightly-delimited physical location. Thus, mobile learning features engage learners in educational activities, using technology as a mediating tool for learning via mobile devices accessing data and communicating with others through wireless technology.

The new generation, as digital natives [53] or the Net generation [60], enjoy using the latest technology such as online resources, cell phones, and applications. Author [53] conceptualizes digital natives as a young generation of learners who have grown up engrossed in recent digital technological gadgets. The young generation is "surrounded by and using computers, videogames, digital music players, video cams, cell phones, and all the other toys and tools of the digital age" [3]. The advocates of digital natives believe that educational communities must quickly respond to the surge of the technology of the new generation of students [30]. Along with the surge of the device, ownership is a growing obsession with smartphone applications (apps). As a result, most young adults have an assess to smartphone applications. Meanwhile, 90% of users' mobile time has been spent on using apps [13] that encompass all aspects of our lives, such as books, business, education, entertainment, and finance. The ownership and use of mobile devices generate and facilitate more non-formal language learning opportunities for learners [40]. Mobile technologies allow students to access learning content of all types anywhere and at any time [39, 41, 51]. Likewise, most university students are equipped with touchscreen smartphones [60].

However, offering students mobile devices does not guarantee their effective use to acquire language knowledge [15, 59]. As authors [18] argue, students' learning outcomes are not merely determined by the technology itself. Learners use the same technology differently to achieve their learning aims [42], but many may fail to effectively use the resources available due to a lack of digital literacy skills [18]. To address this issue, a study on the effectiveness of mobile apps is of great importance.

With the increasing popularity of mobile learning, language learning assisted with mobile technologies is becoming a new focus of educational research. This phenomenon has prompted educators and researchers to take a pedagogical view toward developing educational applications for mobile devices to promote teaching and learning. As a result, research on mobile learning has expanded significantly [41]. However, this growing body of literature has focused on several broad areas of inquiry, such as the development of mobile learning systems and how mobile technologies assist learning a language (e.g., exploring the widely-used commercial L2 learning apps like Duolingo) [45], paying less attention to the effectiveness of mobile-assisted EFL vocabulary learning. It is identified that among those students who use mobile apps to learn a language, most of them use mobile apps for vocabulary learning. Although statistics suggest that the number of students learning vocabulary through mobile apps is increasing [48], little is known about how effective these apps are in EFL vocabulary learning. To fill this gap, this study focused on the effectiveness of mobile apps on EFL vocabulary learning. It is unreasonable to expect any single study to tell us to what extent mobile applications assisted EFL vocabulary learning is effective in improving language learning. However, a comprehensive review of the existing studies can get us closer to an answer [9]. Moreover, an empirical study is effective to verify the results of the review.

To this end, this study centers around EFL vocabulary learning assisted with mobile applications. The specific research questions that this study aims to address are as follows:

- RQ1: In what contexts have mobile apps been used for EFL vocabulary learning?
- RQ2: What are the impacts (if any) of using mobile apps on EFL vocabulary learning outcomes?

This study is significant in several aspects. It was determined that research on vocabulary learning strategies is related to the indirect vocabulary learning strategy [4, 57]. When, why, and how mobile apps are used by EFL learners to learn vocabulary has been researched. However, there is a lack of empirical review of the effectiveness of mobile apps assisted EFL vocabulary learning. Furthermore, the acquisition of mobile apps is of great importance for students with limited vocabulary and language skills in academic and professional lives. In the teaching and learning processes, mobile devices could create new models with their unique qualities, and the physical characteristics (e.g., size and weight), input capabilities (e.g., keypad or touchpad), output capabilities (e.g., screen size and audio functions), file storage and retrieval, processor speed, and the low error rates" [2] (p. 179). EFL learners, one of the leading mobile user groups, are facing a "transitional period" from formal teacher-led English learning to non-formal self-directed English learning [49]. Against such a background, this study aims to figure out the practicability of mobile apps to assist vocabulary learning, which helps to enlarge EFL learners' English vocabulary and diverse cultural knowledge in helping them to acquire a high level of English and culture understanding. Moreover, vocabulary teaching is at the heart of developing proficiency and achieving competence in the target language. This study evaluated the impact of using mobile apps on EFL vocabulary learning outcomes, which affords teachers an overall dialectical view to improve their teaching methods. Also, there is a need to determine exactly what strategies are employed by mobile developers on apps and their effects on vocabulary learning. In this sense, this study can offer apps developers a fundamental review of learners' needs of vocabulary learning and also credible data to test the results.

2   Methods

2.1   Data Sources and Search Process

Data were collected from three databases, including the Web of Science, Academic Search Complete, and Eric. The reason for selecting these three databases was that they were the most commonly cited databases for educational research. Particularly, the Web of Science is generally deemed to be one of the most reliable databases for scholars in social science research [7]. Common search key words "apps (applications) vocabulary learning" was applied in the databases for search any publication which contains "apps (applications) vocabulary learning" in its content.

As of October 24th, 2021, after the initial literature search, a total of 151 results were produced in the 3 databases, including 30 duplicates that were deleted. The author read through the abstracts of the remaining 121 articles and determined whether they were appropriate to be included in the review by inspecting carefully to find whether it met the inclusion criteria. A total of 39 articles were determined consequently. Then the author further examined these articles by full-text scrutinizing and excluded 13 articles. Finally, a total of 25 papers were reviewed and analyzed for this study. The following Fig. 1 demonstrates the search process of the literature.

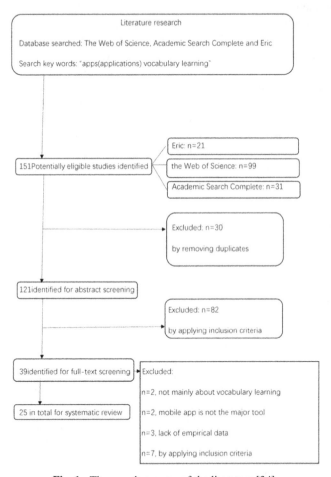

Fig. 1. The search process of the literature [34].

2.2   Inclusion and Exclusion Criteria

Guided by the research questions, the following inclusion criteria were applied:

(1) The empirical research must be conducted with mobile apps (applications). Studies that deal with other kinds of apps, such as computer apps, were excluded.
(2) The empirical research must be conducted with vocabulary learning. Using the word like the level of vocabulary or vocabulary acquisitions is also acceptable. Articles that deal with other educational purposes such as grammar, writing, or listening were excluded.
(3) The empirical research must deal with learning English as the second language. Articles that stress learning English as mother language or learning other languages such as Chinese, Korean, Japanese were excluded.
(4) The empirical research must include empirical findings with actual data. Articles that present personal opinions and theoretical argumentations were excluded.
(5) The empirical research must be published in a peer-reviewed journal. Books, book chapters, and conference proceedings were excluded. However, review articles on mobile apps assisted vocabulary learning were read. The information from these reviewed articles was used as background information.
(6) The empirical research must be written in English. All other languages were excluded.

2.3   Data Coding and Analysis

To address the first research question concerning in what contexts mobile apps have been used for EFL vocabulary learning, data were coded in an inductive way using content analysis [17]. Contexts in this study were defined from different dimensions, including geographical information, grades of students, learning contexts [21]. Moreover, research methods used in the articles reviewed were also analyzed.

To explore the impacts of using mobile applications on EFL vocabulary learning outcomes, content analysis [17] was employed again. First, units of analysis such as "peer pressure could encourage Chinese EFL learners' interests and motivation in language learning" were identified by scrutinizing the results of the section of each study for open coding. To complete open coding, preliminary codes appearing from the articles (such as "motivation" or "interests") were decided, and then all the results were coded with these codes. When data did not adapt to an existing code, new codes were added. Next, similar codes were grouped and placed into categories that were revised, refined, and checked until they were mutually exclusive to form the final categories (such as "motivation and interests"). Impacts of using mobile applications to learn vocabulary were recorded and numbered in notes first after each article had been read and then were compared cross articles to find common patterns for theme generation [17]. Themes from the categories were developed through a qualitative design through a grounded theory [32].

All papers were scrutinized gingerly and completely by the author. In order to intensify the validity of the results, two stages were adopted. First, the literature on the impacts of mobile applications assisted EFL learning was reviewed thoroughly for theoretical validity. Moreover, an expert was invited to examine the categories of impacts of mobile applications assisted EFL vocabulary learning that emerged from the data analysis by the author and to confirm the results by reviewing the main findings of the 25 studies identified. The inclusion criteria for the expert reviewer were based on his academic impact, including publications, citations, H-index, and i10-Index. An agreement rate of 62% was yielded in that the author and the expert agreed on eight categories of the impacts out of thirteen. Distinctions were resolved through discussion until consent was reached. At last, eight categories of the impacts of using mobile apps in EFL vocabulary learning were explicated. No prior assumptions were generated before the analysis. The results emerged inductively from inspecting and interacting with real data.

3   Research

3.1   Research Methodology and Data Collection Instruments Following

This research adopted a mixed-methods. The mixed-method type [22] used in this study was "questionnaire survey with follow-up interview or retrospection" (p. 170). For the quantitative part the data was collected through questionnaires while semi-structured interviews were utilized to collect qualitative data.

The pre-test and mid-term diagnostic test were designed discreetly. Pre-test was scrutinized to precisely evaluate different proficiency levels. The pre-test and mid-term diagnostic test mainly include words on text book students have learned.

Bubei as the targeted vocabulary learning application was another instrument to be installed on students' smartphones and used for a month. In this app English words on text books are divided into three groups based on their difficulty level (easy, medium, and hard). Each word includes pronunciation, meaning, contextualization in a sentence, semantic relation (synonym, antonym), morpheme for better memorization. Quizzes are provided for review. Users can choose in what level have they acquired the word. Meanwhile, hints are provided to help strengthen memory. Feedback which specifies all answers as right or wrong, and in the latter case the correct option is provided instantly. Learners' usage over app includes selecting the words to be ordered alphabetically or randomly, and opting to be shown either all words from the selected level or alternatives such as seen words, new words, and learnt words. Words can be bookmarked for easy access by tapping on a star symbol (see Fig. 2.).

The main instruments for collecting data were a questionnaire and an interview, the purpose of which was evaluating contexts and impacts of using *Bubei* app for vocabulary learning. Questionnaire piloting was conducted to get rid of any ambiguities and pitfalls and evaluating its appearance, clarity, and answering time [21]. Interview questions (6 items) were closely related to the questionnaire but in an indirect way and they are not identical.

3.2   Participants

The participants are 64 senior two students from 2 classes of TJ No.21 high school. The average English levels of 2 classes are almost identical. Students were arranged into 3 levels (A, B, C) according to numbers of assessments they had at school, and they

Fig. 2. Learning and review functions from *Bubei* app.

are taught by the same teacher. They were instructed to use the app for one month after which the researcher sent them questionnaire. 6 students were selected to be interviewed randomly including two representatives (one male and one female) from each group (A, B, C). The following Table 1 gives a summary of different categories:

3.3 Data Collection

The participants engaged in a textbook-based vocabulary list learning for one month but reflecting a transition from the traditional way to mobile assisted vocabulary learning, they divided into the two groups based on the accessibility conditions: *Bubei* app available for both iOS and Android system. The learning contents, learning processes, and the course evaluation systems were almost identical between these two groups.

Table 1. Information of the questionnaire participants.

Gender	N	Vocabulary proficiency level	N	Age range	
Male	36	A	12	Max	14
Female	28	B	34	Min	17
		C	18	Mean	15

3.4 Findings

Quantitative: To figure out the impact of *Bubei* app. An evaluation criterion was taken into consideration to tap into participants' perceptions of vocabulary learning using the app. Data were collected from the questionnaire mainly about: (1) Whether they are content with functions of *Bubei* app? (2) Do they frequently use the app? (3) If the app stimulates their motivations in vocabulary learning? (4) After quizzes and mid-term test, do they find an increase in their vocabulary retention? The degree of students' attitudes is leveled from 1 to 5 (Fig. 3).

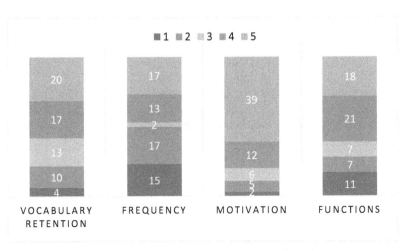

Fig. 3. Students' attitudes based on the evaluation criterion.

Qualitative: Students' perceptions and attitudes towards mobile apps assisted vocabulary learning were further studied by interview as shown in Table 2.

Table 2. Descriptive summary of selected interview.

Questions:	Items	Chinese Translations
What features do you find most effective?	1) I like review, I find it better than traditional one. It's more regular and ubiquitous. 2) I prefer pronunciation feature which helps to improve my accent.	1）我喜欢复习功能，因为相比传统复习方式，这个功能更有规律性，并且可以随时随地进行。 2）我喜欢语音功能，它让我的口音更地道。
In what context do you usually use the app? Do you use the app frequently?	1) I incline to use it after class, for I am easily distracted when I use the phone. I have little free time that I use it not very often. 2) Teachers still teach vocabulary in traditional way. We can only use it to supplement learning. Teachers assign many vocabulary learning tasks for us which occupy a lot of time. Thus, I cannot use the app frequently.	1）我更倾向于在课下使用，因为手机上很多其他软件很容易使自己分心。但是因为空闲时间少，所以并不经常使用。 2）老师还是使用的传统的单词教学方式，手机软件只能自己用作课下的补充。老师其他的单词学习任务占据了大量时间，所以没有富余时间使用软件。
3. What need to be proved by this app? Any suggestion?	1) Quiz is sometimes very boring, adding some pictures will be helpful. 2) Sometimes, it's easy to forget review. Noticing is quite important. I think it should be periodic, suggestive. It cannot be forceful. 3) If I can see other students learning process, I will be more encouraged to learn. Also, it's a good idea to design a feature that 2 students can study vocabulary together in a shared room online.	1）Quiz相对比较枯燥，最好配一些图片来帮助记忆 2）有些时候会忘记复习，提醒可以是定期的，并且不能太过强制性 3）如果可以看到班上同学的学习进度，并且一起定时完成相应任务，会更见激发我的学习动力。

4   Results

4.1   RQ1: In What Contexts Have Mobile Apps Been Used for EFL Vocabulary Learning?

The geographical distribution of relevant studies was examined (Fig. 4). The results indicated that the majority of the studies were conducted in Asia (11), which includes Turkey (1), China (6), and Japan (4) respectively; five studies conducted in Middle East countries with three in Iran and two in Arab countries; three studies conducted in Europe included Spain (1), Serbia (1) and Czech (1); one conducted in South Africa and five studies did not indicate country and region.

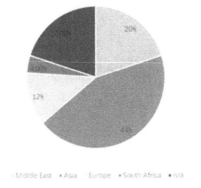

Fig. 4. Numbers of studies distributed based on geographical information. *n/a = not applicable.

The studies were conducted in different grades (see Fig. 5.). The studies covered ranges from primary school to college and university. The results concluded that most studies were conducted in colleges and universities (15), three conducted in primary school and one in senior high school. There are six studies that did not indicate the grades of students.

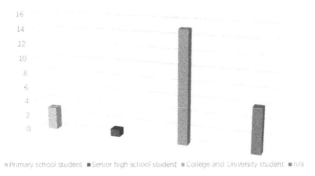

Fig. 5. Numbers of studies based on grades of students. *n/a = not applicable.

While most studies were conducted in the setting of higher education, the learning contexts were different (Fig. 6). The systematic review [23] suggested that formal learning is a type of learning arranged by institutes and guided by a curriculum that is organized and structured, in contrast with informal learning that is spontaneous, experiential, not arranged by institutes, and not guided by a curriculum. Non-formal learning means organized learning but granted no credits and not evaluated. Based on the data, over half of the studies were conducted in either non-formal (5) or informal (14) contexts. Only six studies were conducted in a formal learning context, in which the use of mobile applications was well organized and structured, also arranged by institutes, and guided by a curriculum.

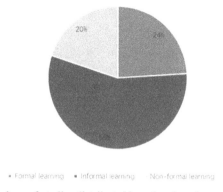

Fig. 6. Numbers of studies distributed based on learning contexts [23].

In terms of research methods, the studies reviewed adopted both qualitative and quantitative research methods. The quantitative research methods mainly included quasi-experimental designs with assessment and questionnaire surveys. Interview, observation, reflection, and transcript analysis were encompassed in qualitative research methods. As shown in the bar chart below (Fig. 7), most studies adopted quantitative research methods (11). On the contrary, qualitative research methods (1) was hardly used. Also, a growing number of researchers relied on mixed-methods (6), compared with single adoption of qualitative studies or quantitative studies. Since this study began in early 2020, the total number of studies in 2020 is relatively small. However, the general trends in the chart indicate that the number of studies on mobile apps assisted vocabulary learning is increasing especially in 2021, and the employments of all research methods are on the rise.

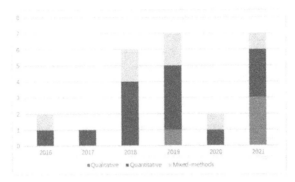

Fig. 7. Numbers of studies distributed based on research methods.

4.2 RQ2: What Are the Impacts (If Any) of using Mobile Apps in EFL Vocabulary Learning?

To figure out what the impacts of using mobile apps in EFL vocabulary learning are, Table 3 and Table 4 are provided below. Table 3 is about the application systems and related learning strategies reported in the studies reviewed. 8 categories of impacts that emerged from the study are presented in Table 4.

Table 3. Apps availability on mobile and tablet platform and related learning strategies [34].

Apps	iOS	Android	Surface App	Web based	Collaboration	Phonological Analysis	Morphological Analysis	Contextual Analysis	Game	Quiz/ Assessment
Busuu App	✓	✓	✓	✓		✓	✓	✓		✓
Baicizhan	✓	✓		✓		✓	✓	✓	✓	✓
Vocabulary Flashcards 2016	✓	✓						✓		✓
WhatsApp	✓	✓	✓	✓	✓		✓	✓		
GREvocabulary application	✓	✓	✓	✓	✓					✓
PHONE Words		✓			✓				✓	✓
Vocabulary Notebook	✓	✓	✓	✓		✓		✓		✓
Socrative			✓	✓	✓					✓
VocabGame	✓		✓			✓	✓	✓	✓	✓
Excel@EnglishPolyU				✓				✓	✓	✓
English Tody	✓	✓		✓		✓		✓		✓
Quizlet				✓		✓	✓	✓		✓
HiroTan App	✓	✓	✓	✓		✓	✓	✓		✓
My-Pet-Shop				✓				✓	✓	
Mobil2Eng	✓	✓	✓	✓	✓	✓	✓	✓		
VocUp		✓				✓	✓	✓		✓
3rd World Farmer				✓				✓	✓	

Vocabulary Acquisition and Retention. Vocabulary acquisition and retention refer to learners' ability to memorize and acquire vocabulary assisted by using the mobile application. For instance, according to researchers [20], Students' test scores improved while the mobile tool was being frequently used and failed to improve when usage subsided. It

Table 4. Impacts of using mobile applications in EFL vocabulary learning [34].

Impacts	Contents	Freq.	Studies
Vocabulary acquisition and retention	Students' ability to acquire and remember vocabulary	20	(Ma & Yodkamlue, 2019) (Andarab, 2019) (Franciosi, Yagi, Tomoshige, & Ye, 2016)
Pronunciation feature	Students use application's pronunciation model to train their English skills	11	(Demmans Epp & Phirangee, 2019) (Makoe & Shandu, 2018)
Usage Frequency	Students show different frequency in mobile applications usage	9	(Ebadi & Bashiri, 2018) (Zhang & Pérez-Paredes, 2019)
Learners' perceptions and attitudes	Learners' beliefs, perceptions and attitudes about their vocabulary competence and the use of apps	14	(Demmans Epp & Phirangee, 2019) (Makoe & Shandu, 2018) (Chen & Lee, 2018)
Motivation & Interest	Promoting learning motivation, learners' interests, engagement and confidence.	18	(Ma & Yodkamlue, 2019) (Govender & Arnedo-moreno, 2021)
Learning environment	Learning conditions that affect the behavior and development of students' learning	9	(Chen & Lee, 2018) (Ma & Yodkamlue, 2019)
Evaluation & Feedback	Quick delivery of and access to evaluation and feedback through quiz, assessment and game	10	(Makoe & Shandu, 2018) (Yarahmadzehi & Goodarzi, 2020)
Administration for learning	Pushing notification such as sending events reminders	2	(Makoe & Shandu, 2018) (Klimova & Polakova, 2020)

was indicated that students' acquisition and retention were likely to improve after using mobile applications in a high frequency. Likewise, study [47] results indicated that students with the mobile apps showed a statistically significant acquisition of words. Researchers [14] found that game-related functions of mobile applications were conducive to vocabulary acquisition. It was shown that there existed reasonable and strong correlations between learning outcomes with the usage time of gamified functions, thus improving vocabulary acquisition performance. The results of study [14] demonstrated that the participants with the mobile app were able to retain more words in their long-term memory because of spaced review and the convenience of using the mobile app to review everywhere. Similarly, compared with MEVLA-NGF (mobile English vocabulary learning apps without game-related functions), Study [44] confirmed that MEVLA-GF (mobile English vocabulary learning apps with game-related functions) achieved its educational goal and effectively assisted learners in improving their vocabulary size. Analytical results show that MEVLA-GF positively influenced learners' vocabulary acquisition and was helpful in augmenting the learners' ability to dispel the graduated interval recall hypothesis [50], thus effectively assisting learners in retaining vocabulary. However, study [16] indicated that both students in the experimental group and control group obtained a significant improvement in the performance test and revealed no significant differences between the two groups. Studies also indicated that the apps were not very supportive of their vocabulary acquisition and retention [36].

Administration for Learning. Administration for learning refers to the function of mobile applications to push notifications such as sending events reminders. Studies showed that app's notifications were helpful because it served as a constant reminder to engage in learning for most distant students who have other work to do besides studying [48]. On the contrary, study [36] indicated that students did not reach a consensus on the notification. Half of the students appreciated the notifications, which helped them study regularly, and the other half did not, which was also reflected in replies of most students that the app had a neutral effect on their study behavior. Again, this might have been caused by receiving the notifications at a not suitable time of the day.

Pronunciation Feature. One important feature of mobile applications that also seems to be connected with vocabulary is the pronunciation feature. Students used the pronunciation model of the app to train their English skills. For example, based on the questionnaire, the users found the app beneficial – they especially liked features such as listening to word pronunciation [27]. The other study showed that participants requested to include a word pronunciation feature for VocUp, a mobile application for vocabulary learning [48].

Motivation and Interest. Students' motivation, interest, engagement, and confidence improved when involving mobile applications in learning. According to researchers [19], students understand words easily with the help of a mobile application that provides a multimedia learning environment for learners to learn the target words. The mobile application also stimulates students' motivation. Results of studies indicated that assisting the lexical items with a theme or visual aids could help to motivate vocabulary learning [3]. Also, studies suggested that mobile game applications mostly had been used to improve motivation and interest to learn vocabulary [26]. Moreover, pedagogical approaches such as the communicative method (which focuses on interaction) have been shown to help with language learning but greater learner motivation and engagement have been found through gameful approach incorporation [33].

Usage Frequency. Students use mobile applications at different frequencies. For instance, one study identifies users into 4 groups: excited users, just-in-time users, responsive users, and frequent users [20]. In this study, students tended to use the application to support learning activities in more protracted sessions. These sessions were spaced over time but tended to last upwards of 40 to 50 min (students' entire spare period). This amount of focused time is well outside the range expected for microlearning activities [6, 25], especially those conducted through mobile devices [29]. Studies also suggested that the students with the new, cross-platform application exhibited a relatively significant tendency of frequent, steady, and periodical logins than those with the old, PC-only one. The analyses suggested that the cross-platform, mobile-optimized web application elicited the students' ability to regulate their everyday self-accessed online learning [27].

Learners' Perceptions and Attitudes. Learners' attitudes and perceptions towards their vocabulary competence and the usage of the mobile application is of great importance to vocabulary learning. The studies suggested that the users held positive attitudes towards the application because it influenced their learning positively and provided them with both form and meaning-focused instruction, even though they were dissatisfied with the levels and authenticity of the contents presented by the app [24]. Likewise, findings indicated that students held different attitudes towards different functions and perceived the mobile apps as facilitative for some learning actions. In addition, students would choose the implementation of the mobile app in other courses taught by teachers. Therefore, the teachers should always think about the purpose of the use of the mobile app in encouraging students' learning for generating higher learning outcomes [36].

Learning Environments. Mobile applications provide learners with learning environments different from the traditional learning context, which affect the behavior and development of students' learning. Studies found that mobile application provided a multimedia learning environment for learners to learn target words [47]. According to authors [16], incorporating digital games to support language learning provided students with an interactive environment to enrich students' learning experience. Also, ubiquitous, inexpensive, powerful, and creative learning environments can provide new and fantastic interaction opportunities and multi-synchronous modes of learning environments by employing portable social network applications such as WhatsApp and Viber [49]. In terms of mobile applications, this interaction can take place on-screen (such as when digital objects are superimposed over camera images of the real-time physical environment), or in the physical environment (for example through holographic projections of digital objects into physical space [56].There are suggestions that AR (Augmented Reality) which broadly refers to any experience in which virtual, context-sensitive information is dynamically overlaid with real-world contexts [46] can be motivating for children, due to its novelty in combining the physical and digital [10].

Feedback and Evaluation. Evaluation and feedback can be provided immediately by mobile applications. For instance, study [48] found that learners showed their satisfaction towards the fact that the mobile app was interactive in that the exercises helped them to get prompt feedback on assessing their understanding of the content. The app provided device-human interaction that facilitated feedback in the absence of human-human interaction. Students considered the feedback very strict when they made just a small mistake, such as the lack of a full stop, but they enjoyed the correction feedback of their performance. Nevertheless, the strictness of feedback is on purpose to make students realize the importance of accuracy [36]. Also, apps which create personalized path in learning and provide personalized feedback based on the tests would indicate potential weaknesses and strengths in learners' professional vocabulary. Thus, learners are helped to acquire vocabulary effectively [58].

5 Discussions

5.1 Using Context of Mobile Applications in Vocabulary Learning

Results of this study indicated that research methods of studies in mobile apps used in EFL vocabulary learning were mostly quantitative research. However, qualitative research methods can be introduced to the studies of mobile apps used in vocabulary learning. Moreover, the advantages of qualitative methods and quantitative methods can be combined to improve the validity of the study [12]. The qualitative study contributes to understanding the human condition in different contexts and a perceived situation, which can also help to study the influences of using mobile apps in different learning contexts [11].

Also, mostly the mobile apps were used in informal and non-formal learning contexts. The research found students had a waning interest in using the mobile application over the term. However, the impetus for some students to use the support tool was maintained [20]. This result implies that teachers could adapt mobile applications in a formal learning context by which can help students increase their vocabulary knowledge and give test which is closely related to what they learn from mobile vocabulary learning. As shown in the literature, test scores could be improved when mobile tools were frequently used and failed to improve when usage subsided [20].

As regards students' grades, most of the app users are college students, which is comprehensible for most university students equipped with touchscreen smartphones [60]. The second large group is primary school students. It is suggested that developing a good educational game is an optional method to arouse young children's interests in English vocabulary learning and assist them to achieve their immediate and long-term vocabulary goals. As researchers [15] mentioned, the mobile application used in vocabulary learning especially educational games, enhanced motivation in terms of the goal, feedback, autonomy, and immersion aspects. Meanwhile, the finding leads us to think about how to use mobile applications to learn vocabulary in senior high school. Most importantly, students should be given brief orientation and lectures by instructors [20]. Research implied that students were more likely to use mobile applications to assist their vocabulary learning when they spent most of their time on individual work [20]. This result suggests that providing brief instructions and facilitating students to work individually will contribute to the increasing use of mobile applications for vocabulary learning in secondary school.

5.2 Impacts of using Mobile Applications on EFL Vocabulary Learning

The studies reviewed indicate that the reason to design and implement mobile applications is mainly to improve the learners' vocabulary acquisition and retention and enhance English vocabulary teaching and learning [48]. Different studies result in different conclusions. Most studies found mobile applications are effective in promoting students' vocabulary learning [47, 49]. On the contrary, some studies imply that mobile applications are not supportive of vocabulary learning [16, 36]. These two different results demonstrate that the impacts of using mobile applications in EFL vocabulary learning still remain controversial, and more empirical studies are in need to further explore this topic. It should be cautious about adopting mobile applications in vocabulary learning. The results of this study demonstrate special features of mobile applications which can be applied by learners, teachers, software designer, and government to enhance learning outcomes.

Mobile applications provide learners with learning environments different from traditional learning contexts, which give learners much more freedom and break the boundaries of time and space. Recently, ubiquitous computing has been improved with the use of time series of contexts to organize and analyze data. A ubiquitous learning environment is any scenario in which the user can become immersed in the learning process [19]. Learners should learn to use mobile applications effectively at any time and any place with their own learning paces. However, the existence of too much freedom also challenges language learners to overcome numerous distractions [49]. Therefore, parental and teachers' supervision is of great importance to mobile application use. Students should develop the ability to be self-disciplined. In addition, different mobile applications have different features, such that learners can choose the suitable one according to their own learning style, cognitive competence, vocabulary knowledge, and the one which can best improve their learning motivation and interests.

Moreover, the results of this study indicate that the notification feature of mobile apps is supportive because it served as a constant reminder to engage learners in learning [48]. Thus, with this feature, teachers can give timely notifications to administrate distant vocabulary learning. On the other hand, students did not reach the consensus on notification mostly because receiving notifications at a not suitable timing [36], which implies that teachers should give notifications in seasonable timing and in an appropriate frequency. Otherwise, students may feel annoyed with this notification feature.

Considering the feedback and evaluation of mobile applications, teachers cannot depend much on them. While the exercises on mobile applications help learners to obtain prompt feedback on assessing their understanding of the content [46], the strictness of feedback and the absence of human-human interaction makes students feel uncomfortable [33]. Accordingly, teachers can adopt the effectiveness and accuracy of feedback and evaluation from the mobile application with the provision of formative evaluation and student-oriented feedback in the teaching and learning process. Also, the findings suggest that mobile game applications have been mostly used to improve motivation and interest to learn vocabulary [26], which can be adopted to teach lower grade students. Likewise, students' acquisition and retention are likely to improve after using mobile applications in a high frequency [20]. Thus, teachers should integrate the mobile applications in traditional and formal learning contexts, which promise students to use mobile applications frequently.

Pronunciation features, administration for learning, evaluation, and feedback, and gamification are proved to be inducive to EFL vocabulary learning. This implies that software developers ought to add these features to mobile applications. The problem of downloading mobile applications should be taken into account as some users are wary of the applications costly, others are concerned about the security of applications. Therefore, failure to study the protection and security of mobile applications can obstruct their adoption and use [48]. Although mobile applications facilitate student-content and student-device interaction where the learners appreciate the privacy of working alone, other learners may feel that they needed student-student and student-teacher interaction [48]. Hence, the developer should include interaction and collaboration components in mobile applications such as chatting and ranking.

Based on the results, the challenges of using mobile applications include phone problems, network, and connectivity, as well as a lack of familiarity [48]. One possible reason may be that not all learners possess a smartphone or the required application [41]. Hence, the government should make an effort to support the use of mobile applications officially and financially and provide an established framework. A common understanding should be reached in schools and universities, which creates a sense of trust to relieve teachers, parents, and learners' concerns. Government and schools should also work together to establish programs to train skillful teachers. Furthermore, particularly online sources, managerial cooperation, and administrative structure should be provided to control online distractions [49].

6 Conclusion and Future Directions

This study investigated Students' perceptions on how EFL vocabulary learning impacted by mobile apps by conducting qualitative and quantitative research. The results of this study indicated that mobile applications assist vocabulary learning with different features, such as feedback and evaluation. The study also reveals that students' vocabulary learning achievement and learning motivation could be improve by using the vocabulary learning apps especially the one employs game-based learning approach. However, the problems of using mobile applications still exist.

Apart from some meaningful findings, a number of factors limited the results of this study. The first limitation is concerning data sources. The reviewed studies were searched from three selected databases, and the only peer-reviewed journal articles were included. Therefore, studies from other resources were excluded, such as other databases, book chapters, dissertations, and government reports. Second, the key words for searching were "application vocabulary learning," which might exclude some studies that involved applications but defined in other ways. Third, generalizability of the findings that were developed based on only one APP *Bubei* might be limited. To avoid this, further attempts on more game- based APPs thus should be done. Last, the participants employed in the quasi-experiment are all senior two school students with limited population size. It remains largely unclear whether the findings can be generalized to larger or other grades students. Further attempts can be adopted by using the vocabulary learning app among larger or other populations to further ensure its effectiveness.

No literature searched and reviewed in this study was found exhaustive. Hence, further research should use more data sources to obtain a more holistic picture of the relation between mobile applications and vocabulary learning. According to the results of the study, more qualitative research should be conducted. Particularly, empirical research could be conducted to investigate mobile applications and vocabulary learning. First, how different elements and functions of mobile applications interact with each other to impact learning outcome and learning competence. For instance, researchers could explore how to use mobile applications as a resource to design vocabulary learning activities for students. Then observe how students develop their vocabulary acquisition and learning competence through these activities. Second, how the mobile applications can be integrated into teaching could be investigated as a context-based understanding of the educational potential of different technologies is partly determined by teachers' perceptions [8]. Third, where mobile applications might be adopted in different educational settings, and cultural contexts could be investigated. As found in this study, the use of the mobile application is limited to certain regions and learning contexts.

A Appendix: Summary of 25 Studies Reviewed

#Studies	Research question	Main findings
1(Demmans Epp & Phirangee, 2019)	How application use related to changes in student vocabulary knowledge ?	Learning is likely tied to the task design (i.e., whether it encourages deep processing) and repeated effort rather than the mobile tool's support for noticing or fast and extended mapping.
2 (Ma & Yodkamlue, 2019)	The effects of a self-developed mobile app on Chinese university EFL learners' vocabulary learning and retention.	The mobile app was feasible and effective in helping EFL learners learn more words and retain them in their long-term memory.
3 (Chen & Lee, 2018)	How application-driven model influence aspects of learning performance?	A quiz game with the support of application-driven model contributed to enhance flow experience and better learning self-regulation.
4 (Enokida, Sakaue, Morita, Kida, & Ohnishi, 2017)	How the HiroTan app assists Japanese students with effective vocabulary learning?	Users found the app beneficial and they especially like several features.
5 (Makoe & Shandu, 2018)	How to design and implement a mobile-based application aimed at enhancing English vocabulary teaching and learning?	Technological, as well pedagogical, aspects of mobile-app interventions are essential for vocabulary teaching and learning.
6 (Chih-Ming Chen, Huimei Liu, & Hong-Bin Huang, 2019)	The effects of PHONE Words, a novel mobile English vocabulary learning app (application) designed with game-related functions (MEVLA-GF) and without game-related functions (MEVLA-NGF), on learners' perceptions and learning performance.	Mobile English vocabulary learning application with game-related functions is more effective and satisfying for English vocabulary learning than without game-related functions.
7(Yarahmadzehi & Goodarzi, 2020)	Whether utilize mobile phones in EFL classroom can influence the process of vocabulary formative assessment and consequently improve vocabulary learning of Iranian pre-intermediate EFL learners or not?	Applying technology to facilitate study improves vocabulary leaning of participants better than those who are assessed formatively based on traditional way.
8 (Ebadi & Bashiri, 2018)	EFL learners' perspectives about their vocabulary learning experiences via a smartphone application.	The users held positive attitudes towards the application because it influenced their learning positively and provided them with both form and meaning-focused instruction, but they were dissatisfied with the app's levels and authenticity.
9 (Klimova & Polakova, 2020)	students' perception of the use of a mobile application aimed at learning new English vocabulary and phrases and describe its strengths and weaknesses as perceived by the students.	Students perceived the mobile app as facilitative for some learning actions but was not supportive regarding communication performance.
10 (Rosell-Aguilar, 2018)	How did users perceive the mobile app busuu?	A large proportion of users consider apps a reliable tool for language learning with vocabulary as the main area of improvement.

11 (Enokida et al., 2018)	The new, cross-platform application or the older, PC-based Web-Based Training (WBT) system, which is mor effective for vocabulary learning?	The total learning duration, the outcome, and learning efficiency are almost equivalent between the experimental and control groups. However, the cross-platform, mobile-optimized web application elicited the students' ability to regulate their everyday self-accessed online learning.
12 (Andarab, 2019)	Has humor been also extensively indicated to carry a significant role in vocabulary learning? The effect of humor-integrated pictures on vocabulary acquisition of 45 intermediate English as foreign language (EFL) learners on Quizlet.	The significant effectiveness of technology in vocabulary learning can be boosted with the help of humorous context.
13 (Mellati, Khademi, & Abolhassani, 2018)	The impact of creative interaction in social networks on learners' vocabulary knowledge in Online Mobile Language Learning (OMLL) course.	New technologies established authentic and effective interaction between human and computer in learning contexts as well as challenges that developing countries are faced with in conducting OMLL courses.
14 (Bazo, Rodríguez, & Fumero, 2016)	How Vocabulary Notebook assists vocabulary learning?	By using the application Vocabulary Notebook, the students were able to tackle the problem of incorporating specialized vocabulary derived from the use of Content and Language Integrated Learning (CLIL) in their classes.
15 (Franciosi, Yagi, Tomoshige, & Ye, 2016)	Could less complex simulation games also support the acquisition of a foreign language?	Gameplay with a simple simulation does enhance long-term vocabulary retention which may be beneficially applied in acquisition of foreign language vocabulary.
16 (Kohnke, Zhang, & Zou, 2019)	The effects of the app to enhance undergraduate students' knowledge retention of business vocabulary of different difficulty levels through extended ubiquitous learning opportunities.	Mobile gamified educational programs are a fruitful avenue for students to expand their business vocabulary knowledge and retention.
17 (Elaish, Ghani, Shuib, & Al-Haiqi, 2019)	Whether the developed VocabGame can motivate native Arab students learning the English language to achieve better performance?	VocabGame app should be part of the daily English curriculum for learning the English language. Following the feedback and statistical analysis, the features of the app can be improved in terms of designing better graphics to motivate students in their learning process.
18 (Zhang & Pérez-Paredes, 2019)	What are the uses and the motivation behind language learners' selection of mobile assisted language learning (MALL) resources?	Vocabulary development remains Chinese postgraduate EFL learners' biggest concern. Vocabulary resources, including vocabulary learning and mobile dictionary applications, are rated as Chinese postgraduate EFL learners' most favorite resources. They also prefer to take recommendations from social media and experienced experts.

19(Govender & Arnedo-moreno, 2021)	To gain a clearer picture of the developments and gaps in the digital game-based learning research in language learning.	Game element analysis reveals that the most frequently occurring elements in digital game-based language learning (DGBLL) are feedback, theme, points, narrative, and levels;More research is needed on less common design elements that have shown promise in encouraging language acquisition.
20 (Stefanovic & Klochkova, 2021)	How to develop mobile and smart platforms and systems in teaching and learning the English language for engineering professionals in different engineering study programs.	This manuscript presents software application development and its implementation in teaching English as a foreign language for engineering and technical study programs on the bachelor level. Initial results in implementation and satisfaction of end users point to the justification of implementing such solutions.
21(Li, 2021)	How the implementation of game-based vocabulary learning influences students' vocabulary learning achievement, motivation, and self-confidence?	Results demonstrated that the game-based vocabulary learning APP benefited students in vocabulary achievement, motivation, and self-confidence. Furthermore, learning self-confidence and motivation did not predict learning achievement. Implications of the study were also given.
22(Ahmed, Hassan, & Alharbi, 2021)	This study was conceived to suggest means of improving critical knowledge application.	Results showed that, the experimental group's use of mobile devices for collaboration helped them for better retention of vocabulary, postintervention, and group performance was improved drastically with more learners scoring closer to the mean value, while the control group showed no remarkable difference in performance.
23(Ishaq et al., 2021)	How different applications have been evaluated and tested at different educational levels using different experimental settings while incorporating a variety of evaluation measures?	A taxonomy has been proposed for the research work in mobile-assisted language learning, which is followed by promising future research challenges in this domain.
24(Booton et al., 2021)	The impact of features of mobile applications on children's language learning.	Real-time conversation prompts improved the quality and quantity of adult-child talk, and AR supported language learning ostensibly via increased motivation.
25(Garcia et al., 2021)	To evaluate ULearnEnglish, an open-source system to allow ubiquitous English learning focused on incidental vocabulary acquisition.	Results indicate a favorable response to the application of incidental learning techniques in com-bination with the learner context. ULearnEnglish achieved an acceptance rate of 78.66% for the perception of utility, 96% for the perception of ease of use, 86.5% for user context assessment, and 88% for ubiquity. Among its main contributions, this study demonstrates a possible tool for ubiquitous use in the future in language learning; additionally, further studies can use the available resources to develop the system.

References

1. Ahmed, A., Hassan, M., Alharbi, M.A.: MALL in collaborative learning as a vocabulary-enhancing tool for EFL learners: a study across two Universities in Saudi Arabia. SAGE Open 1–9 (2021). https://doi.org/10.1177/2158244021999062

2. Alzu'bi, M.A.M., Sabha, M.R.N.: Using mobile-based email for English foreign language learners. Turk. Online J. Educ. Technol. **12**(1), 178–186 (2013)
3. Andarab, M.S.: The effect of humor-integrated pictures using quizlet on vocabulary learning of EFL learners. J. Curric. Teach. **8**(2), 24 (2019). https://doi.org/10.5430/jct.v8n2p24
4. Baumann, J.F., Kameenui, E.J.: Vocabulary Instructin: Research to Practice. The Guildfor Press, New York (2004)
5. Bazo, P., Rodríguez, R., Fumero, D.: Vocabulary notebook: a digital solution to general and specific vocabulary learning problems in a CLIL context. In: New Perspectives on Teaching and Working with Languages in the Digital Era, pp. 269–279 (2016). https://doi.org/10.14705/rpnet.2016.tislid2014.440
6. Beaudin, J.S., Intille, S.S., Tapia, E.M., Rockinson, R., Morris, M.E.: Contextsensitive microlearning of foreign language vocabulary on a mobile device. In: Proceedings of the 2007 European Conference on Ambient Intelligence, pp. 55–72 (2007). http://dl.acm.org/citation.cfm?id=1775401.1775407
7. Bergman, E.M.L.: Finding citations to social work literature: the relative benefits of using "Web of Science", "Scopus", or "Google Scholar." J. Acad. Librariansh. **38**(6), 370–379 (2012)
8. Brown, M., Castellano, J., Hughes, E., Worth, A.: Integration of iPads into a Japanese university English language curriculum. JALT CALL J. **8**(3), 193–205 (2012)
9. Cavanaugh, C.S.: The effectiveness of interactive distance education technologies in K-12 learning: a meta-analysis. Int. J. Educ. Telecommun. **7**(1), 73–88 (2001)
10. Cerezo, R., Calderon, V., Romero, C.: A holographic mobile-based application for practicing pronunciation of basic English vocabulary for Spanish speaking children. Int. J. Hum. Comput. Stud. **124**(1), 13–25 (2019). https://doi.org/10.1016/j.ijhcs.2018.11.009
11. Creswell, J.W.: Qualitative Inquiry and Research Design: Choosing Among Five Approaches, 3rd edn. SAGE Publications, Los Angeles (2013)
12. Creswell, J.W.: Research design: qualitative, quantitative, and mixed methods approaches, 4th edn. SAGE Publications, Thousand Oaks (2014)
13. Chaffey, D.: Statistics on consumer mobile usage and adoption to inform your mobile marketing strategy mobile site design and app development, 26 October 2016. http://www.smartinsights.com/mobile-marketing/mobile-marketing-analytics/mobile-marketing-statistics/
14. Chen, C.-M., Liu, H., Huang, H.-B.: Effects of a mobile game-based English vocabulary learning app on learners' perceptions and learning performance: a case study of Taiwanese EFL learners. ReCALL **31**(2), 170–188 (2019)
15. Chen, X.-B.: Tablets for informal language learning: student usage and attitudes. Lang. Learn. Technol. **17**(1), 20–36 (2013)
16. Chen, Z.H., Lee, S.Y.: Application-driven educational game to assist young children in learning English vocabulary. Educ. Technol. Soc. **21**(1), 70–81 (2018)
17. Cho, J.Y., Lee, E.H.: Reducing confusion about grounded theory and qualitative content analysis: similarities and differences. Qual. Rep. **19**(32), 1–20 (2014)
18. Conole, G., Paredes, P.P.: An analysis of adult language learning in informal settings and the role of mobile learning. In: Yu, S., Ally, M., Tsinakos, A. (eds.) Mobile and Ubiquitous Learning, pp. 45–58. Springer, Singapore (2018). https://doi.org/10.1007/978-981-10-6144-8_3
19. da Silva, L.G., Neto, E.G.A., Francisco, R., Barbosa, J.L.V., Silva, L.A., Leithardt, V.R.Q.: ULearnEnglish: an open ubiquitous system for assisting in learning English vocabulary. Electronics **10**(14), 1692 (2021). https://doi.org/10.3390/electronics10141692Academic
20. Demmans Epp, C., Phirangee, K.: Exploring mobile tool integration: design activities carefully or students may not learn. Contemp. Educ. Psychol. **59**(July), 101791 (2019). https://doi.org/10.1016/j.cedpsych.2019.101791

21. Dörnyei, Z.: Questionnaires in Second Language Research: Construction, Administration, and Processing. Lawrence Erlbaum Associates, Mahwah (2003)
22. Dörnyei, Z.: Research Methods in Applied Linguistics: Quantitative, Qualitative, and Mixed Methodologies. Oxford University Press, Oxford (2007)
23. Eaton, S.E.: Formal, non-formal and informal learning: the case of literacy, essential skills and language learning in Canada (2010). https://eric.ed.gov/?id=ED508254
24. Ebadi, S., Bashiri, S.: Investigating EFL learners' perspectives on vocabulary learning experiences through smartphone applications. Teach. English Technol. 18(3), 126–151 (2018)
25. Edge, D., Searle, E., Chiu, K., Zhao, J., Landay, J.A.: MicroMandarin: mobile language learning in context. In: Conference on Human Factors in Computing Systems (CHI), pp. 3169–3178 (2011). https://doi.org/10.1145/1978942.1979413
26. Elaish, M.M., Ghani, N.A., Shuib, L., Al-Haiqi, A.: Development of a mobile game application to boost students' motivation in learning English vocabulary. IEEE Access 7, 13326–13337 (2019). https://doi.org/10.1109/ACCESS.2019.2891504
27. Enokida, K., Sakaue, T., Morita, M., Kida, S., Ohnishi, A.: Developing a cross-platform web application for online EFL vocabulary learning courses. In: CALL in a Climate of Change: Adapting to Turbulent Global Conditions – Short Papers from EUROCALL 2017, pp. 99–104 (2017). https://doi.org/10.14705/rpnet.2017.eurocall2017.696
28. Enokida, K., Kusanagi, K., Kida, S., Morita, M., Sakaue, T.: Tracking Online Learning Behaviour in a Cross-Platform Web Application for Vocabulary Learning Courses. Research-Publishing.Net, December 2018. http://search.ebscohost.com/login.aspx?direct=true&db=eric&AN=ED590629&lang=zh-cn&site=eds-live
29. Ferreira, D., Goncalves, J., Kostakos, V., Barkhuus, L., Dey, A.K.: Contextual experience sampling of mobile application micro-usage. In: Proceedings of the 16th International Conference on Human-Computer Interaction with Mobile Devices and Services, pp. 91–100 (2014). https://doi.org/10.1145/2628363.2628367
30. Frand, J.: The information-age mindset: changes in students and implications for higher education. EDUCAUSE Rev. 35(5), 15–24 (2000)
31. Gassler, G., Hug, T., Glahn, C.: Integrated micro learning–an outline of the basic method and first result (2004). Franciosi, S.J., Yagi, J., Tomoshige, Y., Ye, S.: The effect of a simple simulation game on long-term vocabulary retention. CALICO J. 33(3), 355–379 (2016). https://doi.org/10.1558/cj.v33i2.26063
32. Glaser, B.G., Strauss, A.L.: The Discovery of Grounded Theory: Strategies for Qualitative Research. Aldine De Gruyter, New York (1967)
33. Govender, T., Arnedo-Moreno, J.: An analysis of game design elements used in digital game-based language learning. Sustainability 13, 6679 (2021). https://doi.org/10.3390/su13126679
34. Hongjin, Z.: A Review of Empirical Studies of Effectiveness of Mobile Apps on EFL Vocabulary Learning, pp. 557–570 (2021). https://doi.org/10.5220/0010485205570570
35. Ishaq, K., Azan, N., Zin, M., Rosdi, F., Ishaq, S., Abid, A.: Mobile-based and gamification-based language learning : a systematic literature review. PeerJ Comput. Sci. 7, e496 (2021). https://doi.org/10.7717/peerj-cs.496
36. Klimova, B., Polakova, P.: Students' perceptions of an EFL vocabulary learning mobile application. Educ. Sci. 10(2) (2020). https://doi.org/10.3390/educsci10020037
37. Kohnke, L., Zhang, R., Zou, D.: Using mobile vocabulary learning apps as aids to knowledge retention: business vocabulary acquisition. J. Asia TEFL 16(2), 683–690 (2019). https://doi.org/10.18823/asiatefl.2019.16.2.16.683
38. Kukulska-Hulme, A.: Mobile usability and user experience. In: Kukulska-Hulme, A., Traxler, J. (eds.) Mobile Learning: A handbook for Educators and Trainers, pp. 45–56. Routledge, London (2005)

39. Kukulska-Hulme, A., Shield, L.: An overview of mobile assisted language learning: from content delivery to supported collaboration and interaction. ReCALL **20**(3), 271–289 (2008). https://doi.org/10.1017/S0958344008000335

40. Kukulska-Hulme, A.: Will mobile learning change language learning? ReCALL **21**(2), 157–165 (2009). https://doi.org/10.1017/S0958344009000202

41. Kukulska-Hulme, A., Lee, H., Norris, L.: Mobile learning revolution: implications for language pedagogy. In: Chapelle, C.A., Sauro, S. (eds.) The Handbook of Technology and Second Language Teaching and Learning, pp. 217–233. Wiley, Oxford (2017)

42. Lai, C., Hu, X., Lyu, B.: Understanding the nature of learners' out-of-class language learning experience with technology. Comput. Assist. Lang. Learn. **31**(1/2), 114–143 (2018). https://doi.org/10.1080/09588221.2017.1391293

43. Lander, B.: Lesson study at the foreign language university level in Japan: blended learning, raising awareness of technology in the classroom. Int. J. Lesson Learn. Stud. **4**(4), 362–382 (2015). https://doi.org/10.1108/IJLLS-02-2015-0007

44. Li, R.: Does game-based vocabulary learning APP influence Chinese EFL learners' vocabulary achievement, motivation, and self-confidence? SAGE Open 1–12 (2021). https://doi.org/10.1177/21582440211003092

45. Loewen, S., et al.: Mobile-assisted language learning: a Duolingo case study. ReCALL **31**, 1–19 (2019). https://doi.org/10.1017/S0958344019000065

46. Maas, M.J., Hughes, J.M.: Virtual, augmented and mixed reality in K–12 education: a review of the literature. Technol. Pedagog. Educ. **29**(2), 231–249 (2020). https://doi.org/10.1080/1475939X.2020.1737210

47. Ma, X., Yodkamlue, B.: The effects of using a self-developed mobile app on vocabulary learning and retention among EFL learners. PASAA **58**(December), 166–205 (2019)

48. Makoe, M., Shandu, T.: Developing a mobile app for learning english vocabulary in an open distance learning context. Int. Rev. Res. Open Distance Learn. **19**(4), 208–221 (2018). https://doi.org/10.19173/irrodl.v19i4.3746

49. Mellati, M., Khademi, M., Abolhassani, M.: Creative interaction in social networks: multi-synchronous language learning environments. Educ. Inf. Technol. **23**(5), 2053–2071 (2018). https://doi.org/10.1007/s10639-018-9703-9

50. O'Malley, C., Vavoula, G., Glew, J., Taylor, J., Sharples, M., Lefrere, P.: (2003). http://www.mobilearn.org/download/results/guidelines.pdf

51. Pachler, N., Bachmair, B., Cook, J., Kress, G.: Mobile Learning. Springer, New York (2010)

52. Pimsleur, P.: A memory schedule. Mod. Lang. J. **51**(2), 73–75 (1967). https://doi.org/10.1111/j.1540-4781.1967.tb06700.x

53. Prensky, M.: Digital natives, digital immigrants. Horizon **9**(5), 1–6 (2001). https://doi.org/10.1108/10748120110424816

54. Rosell-Aguilar, F.: Autonomous language learning through a mobile application: a user evaluation of the busuu app. Comput. Assist. Lang. Learn. **31**(8), 854–881 (2018). https://doi.org/10.1080/09588221.2018.1456465

55. Saran, M., Seferoglu, G.: Supporting foreign language vocabulary learning through multimedia messages via mobile phones. Hacettepe Univ. J. Educ. **38**, 252–266 (2010)

56. Booton, S.A., Hodgkiss, A., Murphy, V.A.: The impact of mobile application features on children's language and literacy learning: a systematic review. Comput. Assist. Lang. Learn. (2021). https://doi.org/10.1080/09588221.2021.1930057

57. Stahl, S.A., Nagy, W.E.: Teaching word meanings. Literacy Teaching Series, Lawrence Erlbaum Associates Inc., New Jersey (2006)

58. Stefanovic, S., Klochkova, E.: Digitalisation of Teaching and learning as a tool for increasing students' satisfaction and educational efficiency: using smart platforms in EFL. Sustainability **13**(9), 4892 (2021). https://doi.org/10.3390/su13094892Academic

59. Stockwell, G.: Investigating learner preparedness for and usage patterns of mobile learning. ReCALL **20**(3), 253–270 (2008). https://doi.org/10.1017/S0958344008000232

60. Tapscott, D.: Growing Up Digital: The Rise of the Net Generation. McGraw-Hill, New York (1998)

61. Wu, H.K., Lee, S.W.Y., Chang, H.Y., Liang, J.C.: Current status, opportunities and challenges of augmented reality in education. Comput. Educ. **62**, 41–49 (2013). https://doi.org/10.1016/j.compedu.2012.10.024

62. Yu, Z., Zhu, Y., Yang, Z., Chen, W.: Student satisfaction, learning outcomes, and cognitive loads with a mobile learning platform. Comput. Assist. Lang. Learn. **32**(4), 323–341 (2018)

63. Yarahmadzehi, N., Goodarzi, M.: Investigating the role of formative mobile based assessment in vocabulary learning of pre-intermediate EFL Learners in comparison with paper based assessment. Turk. Online J. Distance Educ. **21**(1), 181–196 (2020)

64. Zhang, D., Pérez-Paredes, P.: Chinese postgraduate EFL learners' self-directed use of mobile English learning resources. Comput. Assist. Lang. Learn. (2019). https://doi.org/10.1080/09588221.2019.1

Current Topics

Enhancing the Design of a Supply Chain Network Framework for Open Education

Barbara Class[1]([✉]) [iD], Sandrine Favre[2] [iD], Felicia Soulikhan[3] [iD],
and Naoufel Cheikhrouhou[3] [iD]

[1] TECFA, Faculty of Psychology and Educational Sciences, University of Geneva, Pont d'Arve 40, 1211 Geneva 4, Switzerland
Barbara.Class@unige.ch
[2] PHBern, Fabrikstrasse 8, 3012 Bern, Switzerland
sandrine.favre@phbern.ch
[3] Geneva School of Business Administration, University of Applied Sciences Western Switzerland (HES-SO), 1227 Geneva, Switzerland
{felicia.soulikhan,naoufel.cheikhrouhou}@hesge.ch

Abstract. This article addresses the issue of education in the knowledge society. More precisely, it suggests conceptualizing Open Education as a supply chain in the form of a network of responsible citizens who switch roles and participate meaningfully in all education endeavours. This results in co-designing learning paths and creating common goods in the form of knowledge commons. These insights are gathered through a reflection conducted using a method of Scholarship of Teaching and Learning, a theoretical framework based on value creation [1] and epistemologies of absences and emergences [2], and a case study [3].

Keywords: Open Education · Knowledge society · Open Science · Supply Chain

1 Introduction

The transition to the knowledge society and knowledge economy is now underway; this is an established fact. However, what are the implications of this for science and education [4, 5]? Is it clear in stakeholders' and citizens' minds that such a society and economy are goals to achieve and not accomplished states?

Knowledge is central to both Open Education (OE) and Open Science (OS), which are dedicated to co-creating and sharing public and common goods[1]. Higher education institutions are major suppliers of science and education, and they currently address the challenge of the knowledge society through internationalization [6, 7].

Within this dynamic, a growing awareness of the importance of rethinking science and education has given rise to new voices, be it from organisations like UNESCO or the League of European Research Universities that produce recommendations for policy [8–10], international researchers who share their reflections in terms of epistemologies (e.g.

---

[1] To learn about the differences between public, common and global common goods, please visit: https://www.iesalc.unesco.org/en/2022/04/10/public-goods-common-goods-and-global-common-goods-a-brief-explanation/

© Springer Nature Switzerland AG 2022
B. Csapó and J. Uhomoibhi (Eds.): CSEDU 2021, CCIS 1624, pp. 499–516, 2022.
https://doi.org/10.1007/978-3-031-14756-2_24

[2, 11, 12]), or new practices in terms of research funding (e.g. crowdfunding, citizen science). This movement highlights scientific practices that existed before copyright law [13] and warns of the ecological impacts of a digital society and economy [14].

From the perspective of supply chain management, education has typically been described as a linear endeavour from preschool on through to lifelong learning that involves the interaction of different types of resources - intellectual, human, natural, financial, physical, etc. [15]. To move away from linear processes that fail to adequately capture educational processes in a knowledge society, we have developed the concept of the Open Education Supply Chain (OESC) [16]. This concept is based on the three basic phases of supply chain management: the design phase, which consists of developing 'roads' and 'nodes' through which physical, information and financial flows are managed; the planning phase, which uses advanced systems to plan out the flows; and the control phase, in which the flows are monitored and controlled at the operational level. In the OESC, roads refer to the different type of competences and knowledge developed in institutional and certified settings as well as in non-institutional settings, whether certified or not (e.g., self-learning). Nodes refer to the different educational stakeholders providing any type of training (undergraduate, postgraduate, continuing education with or without accredited certification, etc.) in any mode (face-to-face, online or blended). The diversity of potential roads and nodes are conducive to the building of highly individual learning paths.

The purpose of this reflection, conducted in a Scholarship of Teaching and Learning approach [17] is to further develop the OESC concept with the support of a case study taking place at the lifelong learning centre of the University of Geneva [3]. We first discuss the method and the theoretical framework used for the paper. We then review the concept of OE from several perspectives and present the case study. Finally, we present our current understanding of OE conceptualized as a supply chain as the main findings of this article.

2   Method and Theoretical Framework

The methodology developed within this article is based on a Scholarship of Teaching and Learning (SoTL) approach [17–20]. It describes researchers' progress and reflection on OE conceptualized as a supply chain. Using categories from Hubball and Clarke [19, p. 4], Table 1 outlines how the outcomes shared in the present paper have been produced.

The theoretical framework guiding this reflection is composed of the value creation framework [1] (Fig. 1) on one hand and the epistemology of absences and emergences [2] on the other. As stated in Class et al. [16, p. 619], "Value is defined in terms of agency and meaningfulness of participation. More precisely, participating is perceived as conducting to a difference that matters."

The epistemology of absences and emergences is a call to consider all the knowledge that science has deliberately set aside because it has been designated as 'non-scientific knowledge'. It is a call to let knowledge express itself without filtering it through "Western-centred" glasses of what scientific knowledge is. For instance, when certain stakeholders from countries in Latin America were asked to express key concepts in their native languages, the concepts they listed were related to the elements (e.g.

Table 1. SoTL methodology using the categories put forth by Hubball and Clarke, 2010 [19].

SoTL research context	Central SoTL research question	Methodological approach	Data collection methods	General outcomes
Contribute to the design and understanding of Open Education at the levels of epistemology and praxis	How can Open Education be designed as a supply chain network?	Action and reflection are guided by progress in the understanding of the breadth and depth of both Open Education and supply chain networks	Data relating to the case study were collected through the master's thesis of Favre [3]	Enhanced understanding and visual representations of Open Education conceptualized as a supply chain network

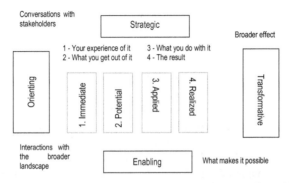

Fig. 1. Value creation according to Wenger and Wenger [1, p. 75].

"water", "fire") or to "Mother-Earth". Finally, this epistemology is a call to stop the mindset of always moving into further development (e.g. planned obsolescence; artificial scarcity) without taking into account the resources that the planet is able to produce or absorb.

Considering the current ecological crisis, and in light of the above epistemology and other voices concerned about modern scientific approach and its inevitable crash [21], it seems timely and wise to make space for ignored knowledge to emerge. This is in line with Open Science as understood by UNESCO [9, p. 15] – i.e. openness towards the "diversity of knowledge" and "the process of scientific knowledge creation and circulation".

3   Open Education

3.1   Knowledge Society

The knowledge society, as its name indicates, is based on knowledge. What is knowledge? How does it differ from information? Knowledge is defined as a cognitive capability that empowers its owners with intellectual and physical actions, whereas information is formatted and structured data that exists in the world and is activated only when actors who have the needed knowledge to process it do so [4]. Knowledge-based communities, like those of open-source software programmers, create and reproduce extensive knowledge. They develop advanced strategies for sharing and disseminating the knowledge they produce with the support of digital technologies. In contrast with private companies, which regard new knowledge as an "exclusive property" to be monetized, "sharing knowledge is their raison d'être" [4, p. 30].

3.2 Defining Openness in Open Education

Openness is defined in a special issue of the Journal of Information Technology on openness and IT as being characterized by access, participation, transparency and democracy [22]. An analysis of the relationships between openness and education reveals that depending on the perspective adopted, a myriad of interpretations of both concepts is possible. What matters are the following five essential issues regarding values, theorizing sharing, standards, deep philosophical questioning and meta-critical thinking: (1) although openness and education are positively connoted, they are not values "per se" [23, p. 5]; (2) since sharing is an essential concept for OE, extensive work must be carried out to operationalize and theorize what sharing means; (3) policies like UNESCO (2019)'s state recommendations for Open Educational Resources (OER) do exist, but the issue of standards must be addressed seriously in its full breadth and depth if guiding documents are to be adopted by practitioners. (4) Since OE draws on technologies, the "post or trans-humanist" (p. 5) complexity of IT and AI in education are philosophical questions to be debated actively. (5) Finally, in academia, it is important to foster meta-critical thinking that goes beyond current contradictions (e.g. "involution of democratic achievements in the name of democracy" (p. 6)) to lay the ground for education as a common good and let knowledge commons emerge fully [23].

Indeed, education understood from the perspective of von Humboldt is "a means of realizing individual possibility rather than a way of drilling traditional ideas into youth to suit them for an already established occupation or social role" [25]. Initiatives for open schools as well as universities were conducted in the 1960s and 1970s based on von Humboldt's ideas. Although the movement did not break through in schools, it did in universities, paving the way for OER and MOOCs. Indeed, because of these efforts toward Openness in the 1960s, OER were strongly associated with licensing and copyleft issues when they began to be designed in the 2000s. Interestingly, and as an example of the contradictions mentioned by Hug [23], MOOCs "deliberately altered the criteria for openness insofar as it was now only open (i.e., cost-free) access instead of open licenses" [26, p.5].

3.3 Open Education Invariants

It is obvious that Open Education, similar to Open Science, is in the process of being defined and can serve as an umbrella term for alternative ways of approaching education [27, 28]. However, authors agree that underlying values of OE include: (1) being geared towards humans and commoning (as opposed to profit); (2) trustworthiness; and (3) ecological systems. For Kahle [29], these values are operationalized through a design that prioritizes access (i.e. diversity of knowledge, universal design), agency (i.e. degree of user action and control on the developed artefact), ownership (i.e. making meaningful through ownership), participation (i.e. taking part in the life cycle of the artefact) and experience (human-centred design). Much research has been carried out to theorize and map OE and provide experience-based insights to advance our understanding of this endeavour (e.g. [30–42]).

3.4 Assessment in Open Education

According to the literature, assessing and certifying competences are important issues in OE which are approached from various perspectives, including open admission and open credentials (Fig. 2). Open admission is understood as changing academic policy to open up admissions to everyone without requiring any prior certifications [43]. Open competencies are related to open assessment; they take the form of a contextual catalog of competencies (i.e. in French, the so-called référentiel de compétences) that lists knowledge and skills against which open assessment is defined [42, 44]. Open assessment, in turn, is understood as assessment in which students can showcase knowledge and skills developed using Open Education Practice and Open Educational Resources [32]. Finally, Open credentials are understood as certifications issued by an accountable and authorized entity (e.g. an institution or community) within a technological infrastructure over which learners have full control [42].

In this paradigm, learners should be able to redistribute their credentials without involving third-party bodies and remix and regroup them as they wish. Essentially, learners should own and have full control over their credentials. To ensure validity, credentials must be tamper-proof and their origin must be trustworthy [42]. The aim of open credentials is to gain trust by requiring transparency [45]. To enhance transparency and thereby trust, the certifying entity must take measures to increase the visibility of its practices, for example by sharing detailed information on the competences developed, the design process, the syllabi, the assessment procedures, etc. [46].

The need is growing for alternative credentialing to document continuing education, whether completed online, face-to-face or in a blended modality and including semi-formal or informal means of learning [47]. Alternative credentialing also allows learners to receive credit for 'transversal' valued skills and knowledge that are acknowledged as part of the 21st century skillset [48, 49] but not credited in 'formal' systems [50]. One way to offer open credentials is through the use of badges. Badges are promising because they are portable and easy to share on social media (e.g. LinkedIn) [51], although currently they are only used for micro-knowledge and skills [52].

Combining Open badges and Open competencies represents an opportunity to develop and receive certification for micro-knowledge and skills based on learners' decisions. It is important that learners take the lead in their education journey [44].

With regard to the integration of Open credentials, the criteria of inviolability, controllability, verifiability, independency and transparency have been identified as a bottom line to be followed. These criteria are challenging, especially in view of achieving a fully automated solution spread at large scale [3].

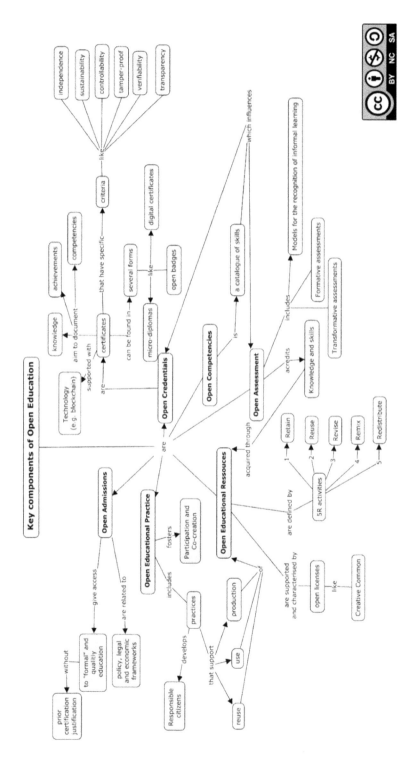

Fig. 2. Key components of assessment in Open Education - inspired from Favre [3].

Combining Open badges with blockchain technology (Fig. 3) or a similar ecological technological process [3] would provide a certification of competences and knowledge whose validity can easily be proven. Indeed, numerous entities provide open badges for a wide variety of micro-level certifications. It is thus a good solution to secure them in a backpack (Fig. 4).

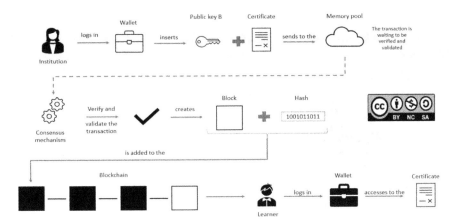

Fig. 3. Combining the Open badge with blockchain technology, inspired from Favre [3].

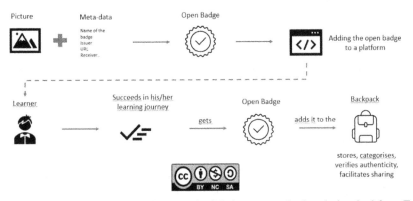

Fig. 4. Collecting Open badges or Open credentials in a secure backpack, inspired from Favre [3].

3.5 Open Ecosystem

Certification is important for several reasons, i.e. because OE is connected to and interacts with the rest of the social, economic and political world. Stacey [53] has mapped this ecosystem in the form of a tree (Fig. 5). Elements shown below ground are still being developed and are not yet at the stage of germination. Elements shown as leaves are already well developed, and elements in between are those that are germinating. OE is both an element and a central component at the heart of this ecosystem. There are at least three reasons for this: (1) OE creates new forms of each of these Open components, whether through practice, theoretical contributions, reflection or all three – i.e. Freire's [54] concept of *praxis*; (2) OE uses these forms, tries them out, questions them in light of practice and policies, etc. and refines them over time; (3) OE allows for continual monitoring of knowledge and skills through its close connections with the remaining elements of the ecosystem.

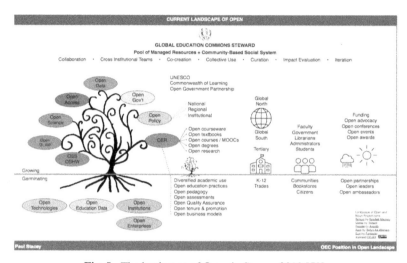

Fig. 5. The landscape of Open in Stacey, 2018 [53].

In a nutshell, the Open Ecosystem, as its name suggests, refers to the interactions and interconnections of the different components of the diverse Open landscape [53, 55, 56].

4   Actors in the Open Education Landscape

Whereas in our position paper [16], we were strongly influenced by Stacey and Hinchliff Pearson [36]'s tripartite perspective of the world – state, commons, market – we now think that it would be an error to consider the GAFAM as simply actors in the market. Google, Apple, Facebook, Amazon et Microsoft (GAFAM) are more powerful than states [57] and take decisions in all areas, be it through the direct processing of personal data from the internet or through the funding of organizations like the World Health Organization [58–60].

In addition, in light of the epistemology of absences and emergences [2], it is important to take into account ignored actors and stakeholders and bring them into the equation. Ignored actors represent the maximum of the unknowns in the equation.

We have thus revised the map of actors and stakeholders from the current situation (Fig. 6) to reflect the new situation (Fig. 7), in a dynamic projection to help us imagine what the future could look like.

Fig. 6. Commons, state, market and ignored actors wiped out by GAFAM

Fig. 7. The return of Commons, ignored actors and the state with GAFAM scaled back within the Market's prerogatives.

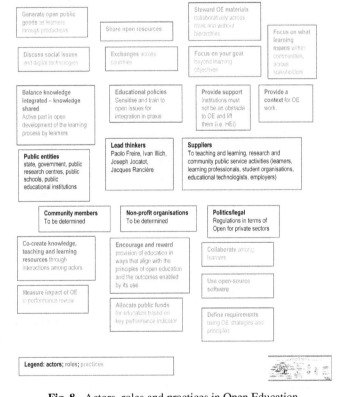

Fig. 8. Actors, roles and practices in Open Education.

At a finer granularity, the first round of a Delphi survey conducted within a current project on OE[2] has identified six actors in the Open Education landscape: public entities, lead thinkers, suppliers, community members, non-profit organizations and political/legal representatives. Some of their roles and practices are depicted below; these range from generating common goods to providing support through funding (Fig. 8).

5   Case Study

Some 15 years ago, the Swiss education system uniformized its continuing education sector [61] by creating three main certifying diplomas – Certificate, Diploma and Master of advanced studies (CAS, DAS, MAS), representing at least 10, 30 and 60 ECTS respectively. These trainings are similar to regular Bachelor and Master programmes but are oriented towards professionals who are seeking to advance their career or change their career path by adding new competencies to their CV. These training programmes are thus more practice-oriented than regular Bachelor and Master curricula. They are designed to help participants progress toward their professional goal as they move from module to module within a given programme. Although they are designed in a participatory manner with stakeholders from the private sector and from the academy, these programmes remain very closed and linear in the sense that participants must move along a given path within one preconceived programme. They cannot, for instance, mix module 1 of the CAS in digital learning with module 2 of the CAS in blockchain technology, etc. to come up with their own tailor-made CAS. Continuing education is designed by programme, where each programme is like a silo designed based on the field and type of diploma.

In 2017, a new diploma appeared in this Swiss continuing education landscape at EPFL, which was then introduced at the University of Geneva in 2020: the Certificate of Open Studies (COS) [62, 63]. This first attempt at offering OE meets the criteria for open admission [64]. However, the COS does not yet incorporate the remaining key features of OE – free access, OER, agency, empowerment, etc.

MOOCs are also readily associated with OE, but they too only meet the criteria for open admission. Certification is not free and above all, most MOOCs do not qualify as OER, as explained above, since they do not follow the principle of open licensing..

In terms of Open credentials, the conclusion of Favre's [3] study is full of insights. The proof of concept explored at the University of Geneva aims at securely distributing diplomas using a blockchain technology. However, its objective is not aligned with the recommendations for Open credentials discussed above [42] as there is no provision for allowing learners to redistribute and remix their credentials without the involvement of a third party.

6   Findings: Supply Chain Applied to Open Education

6.1   Basics of the Supply Chain

Learners taking roads through nodes (cf. introduction) generate flows, acting within a broader network and web of activities. Supply chain networks can be featured in terms

---

[2] In August 2022, at the time of proofreading this article, the project is completed and its output is available from: https://edutechwiki.unige.ch/en/Open_Education_Roadmap

of flow management, bottleneck management and queuing networks management [65]. Flow management combines innovation and value-added operations and requires digital products and services to offer new value creation through dynamic flows within the network structure [66]. Bottleneck management refers to any process activity or constraining organizational performance where the system advances more quickly than its slowest bottleneck component [67]. This consists of eliminating or acknowledging bottlenecks [68] by locating and defining their origins and causes [69]. Finally, queuing network analysis refers to identifying and modelling the performance of stochastic systems [70].

6.2   Supply Chain Concepts Applied to Open Education

Flow management in OE involves students requesting to participate in specific learning sessions to gain knowledge and skills. As intelligent agents, they choose and proactively adjust their own path based on their interactions with other intelligent agents. Dynamic and continuous flow management is thus required to deal with potential bottlenecks. Bottlenecks in the OESC may happen when the number of open positions is limited with respect to the number of learners requesting the use of a specific node. This situation requires new forms of allocating resources to accommodate the requested learning opportunities. In OESC, stochastic systems refer to competences and knowledge sought for by learners. For the same input, different outputs can be offered, e.g. different learning sources providing targeted, sought-after competences and knowledge. This is where dynamic queuing network management can help redirect learners to the most appropriate and available learning sources. Furthermore, digital technologies enhance added value for learners and other stakeholders in terms of services, decision making, visibility and prediction [71].

6.3   Four Dimensions of the Open Education Supply Chain

We have conceptualized OE as a supply chain inspired by Garay-Rondero et al.'s [66] four dimensions. The first dimension, D1, refers to components of the OESC and processes to facilitate its management (Fig. 9). These components and processes analyze data, understand learners' requests and transform this information into knowledge. For example, when a learner makes a request, processes are activated to suggest a choice between several paths and show in real time the differences between them, e.g. language, field, level, design, underpinning values, overall objective in terms of quality (understood as educating citizens for the knowledge society), etc.

The second dimension, D2, refers to OE stakeholders and needed infrastructure. On one hand, this includes core components of learning (e.g. pedagogy, resources, knowledge and skills development), learners (i.e. responsible citizens) and learning providers (e.g. institutions, communities, individuals, businesses). On the other, this includes core components of learning infrastructure (e.g. policies, legal frameworks, technological infrastructure). This is where Open Education practice comes into play, from admission and certification to open educational resources or open source software (e.g. [27, 32, 40, 42, 43]). This dimension is highly interactive and agile. Stakeholders who deliver learning, evaluate competences and knowledge, or provide certification, as well as all other stakeholders in this vast and complex network, must act in line with open values, remain accountable, and be acknowledged as competent bodies across landscapes – including market, commons, state and any emergent actors.

The third dimension, D3, refers to the Open Ecosystem. This refers to the full range of Open elements with which education interacts, including Open Science (the closest to education); open galleries, libraries, archives and museums (GLAM); open government; open institutions; and open enterprises [53].

The fourth dimension, D4, refers to digital and physical flows. It encompasses the myriad of individual learning paths supported and empowered by the three underlying dimensions described above. This flow leverages citizens in the knowledge society to contribute to the building of a collective human intelligence.

Fig. 9. A visual representation of the Open Education Supply Chain, inspired from [66].

6.4 Zooming in to Highlight the Paradigm Shift

The above has given a picture of the overall structure of the OESC. If we zoom in on its different components, from admission to certification, we can see that in the current paradigm, it is the public or private institutions accredited by the state, the market or GAFAM who hold decision-making power.

In the open paradigm, communities and ignored actors are also given power to decide, which has major implications for the entire system. Instead of having an administrative office at an institution checking whether learners have the required diplomas to start a training course, imagine that these same learners can rely on diverse communities to vouch, in the form of open credentials, that they possess specific competences and knowledge. Because they are able to transform this information into knowledge, they are responsible for enrolling in a training course if they estimate that they have the necessary prerequisites. Should that turn out not be the case once the learning journey has started, it is their responsibility to take a decision – e.g. either to find support because the learning gap is within their zone of proximal development or to change their learning route.

This is the current admission situation in MOOCs. No pass, in the form of a previous diploma, is required to attend training. It is the responsibility of learners to evaluate whether a given training course is suitable for them, decide what they want to get out of it (e.g. certification, network of interested persons, resources, etc.) and choose how they wish to proceed.

The paradigm shift occurs at precisely the level of responsibility and decision taking. It is no longer the institution that tells a learner what to learn and whether he or she is admissible. In a landscape where no pre-designed curricula exist, it is the learner's responsibility to take decisions and make choices. A second shift occurs after enrolling in a training, when learners must set clear objectives for themselves and then co-design the actual learning adventure in a participatory manner with the teaching agents.

It is important to remember that open values are about, inter alia, participation, experience, agency and empowerment. Thus, teaching agents' values in the education setting should include outstanding mastery of their respective fields, which will allow for flexibility and co-design at each step of the learning journey. The time for predefined, ready-made, single source, "educational products" is over: the maker movement is a good example of this [72]. Learners want to take control of their path to develop their knowledge and skills with full creativity and responsibility.

6.5   Visualizing the OESC at a Micro Level

To help visualize the trajectories of individuals and communities in the OESC, we have provided an initial representation of how the process works at an individual level (Fig. 10). The learner takes on several roles simultaneously in different spaces – here, as a learner of programming, a teacher of maths and a community member in a Fablab. This person is in interaction with other citizens in all these activities. Each of these individuals

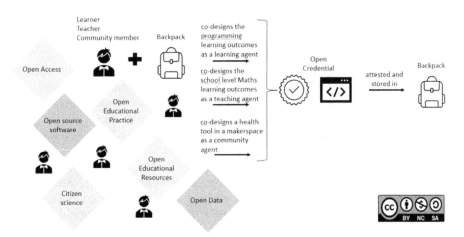

Fig. 10. Focusing on an individual trajectory in an Open Education supply chain.

evolves in an open ecosystem, using, adapting, creating and making available OER, recruiting others and being recruited for citizen science projects, and using open-source software.

Combining this individual layer with the OESC visual representation and adding in external social, economic, political and other dynamic forces produces a deeply complex network (Fig. 11) whose full dynamics cannot be captured visually at this point. Further work should be conducted in the future with relevant case studies to gain further insights.

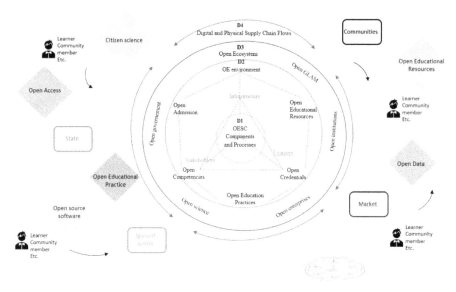

Fig. 11. A glimpse into the dynamics of the OESC network.

7   Discussion and Conclusion

A knowledge society is a society that must be co-created [12] and whose raison d'être depends on value creation in terms of knowledge. Published policies [9] have acknowledged this and can serve as guidelines for citizens. The Open movement offers a sustainable framework (e.g. Creative Commons) for citizens to take responsibility in this regard and creatively turn these policies into reality in everyday life.

In this article we have tried to contribute to the modelling of the Open Education supply chain as a network consisting of citizens and communities who co-produce, co-design and participate actively and meaningfully in education (i.e. they are not passive consumers of pre-packaged, monetized products). In doing so, they help create common goods that can take multiple forms. Ultimately, this helps build a collective human intelligence that functions as an active node in the network, demonstrating multiple skills and knowledge and flexibly switching roles in an open ecosystem.

References

1. Wenger-Trayner, E., Wenger-Trayner, B.: Learning to Make a Difference: Value Creation in Social Learning Spaces. Cambridge University Press, Cambridge (2020).https://doi.org/10.1017/9781108677431
2. Santos, B.D.S.: Epistémologies du Sud : mouvements citoyens et polémique sur la science. Desclée de Brouwer (2016)
3. Favre, S.: Une blockchain pour les compétences et connaissances certifiées ouvertes et libres ? University of Geneva, Geneva (2021). https://tecfa.unige.ch/tecfa/maltt/memoire/Favre2021.pdf
4. David, P.A., Foray, D.: Economic fundamentals of the knowledge society. Policy Futures Educ. **1**(1), 20–49 (2003). https://doi.org/10.2304/pfie.2003.1.1.7
5. Foray, D.: Editorial. Int. Soc. Sci. J. **54**(171), 1–3 (2002). https://doi.org/10.1111/1468-2451.00354
6. de Wit, H., Altbach, P.G.: Internationalization in higher education: global trends and recommendations for its future. Policy Rev. High. Educ. **5**(1), 28–46 (2021). https://doi.org/10.1080/23322969.2020.1820898
7. Jones, E., de Wit, H.: A global view of internationalisation: what Next? In: van't Land, H., Corcoran, A., Iancu, D.-C. (eds.) The Promise of Higher Education, pp. 83–88. Springer, Cham (2021). https://doi.org/10.1007/978-3-030-67245-4_13
8. Ayris, P., López de San Román, A., Maes, K., Labastida, I.: Open Science and its role in universities: a roadmap for cultural change (2018). https://www.leru.org/publications/open-science-and-its-role-in-universities-a-roadmap-for-cultural-change
9. UNESCO. Recommendation on Open Science (2021). https://unesdoc.unesco.org/ark:/48223/pf0000379949.locale=en
10. UNESCO: Recommandation sur les Ressources éducatives libres (REL) (2020b). https://unesdoc.unesco.org/ark:/48223/pf0000373755/PDF/373755eng.pdf.multi.page=11
11. Brière, L., Lieutenant-Gosselin, M., Piron, F.: Et si la recherche scientifique ne pouvait pas être neutre?. Éditions science et bien commun (2019). https://scienceetbiencommun.pressbooks.pub/neutralite/
12. Innerarity, D.: Chapitre 3. La société de la connaissance et l'ignorance. In Démocratie et société de la connaissance, pp. 47–65. Presses universitaires de Grenoble (2015). https://www.cairn.info/democratie-et-societe-de-la-connaissance--9782706122729-page-47.htm
13. Langlais, P.-C.: Quand les articles scientifiques ont-ils cessé d'être des communs? Sciences Communes (2015). https://scoms.hypotheses.org/409
14. Nardi, B., et al.: Computing within limits. Commun. ACM **61**(10), 86–93 (2018). https://doi.org/10.1145/3183582
15. Li, L.: Education supply chain in the era of industry 4.0. Syst. Res. Behav. Sci. **37**(4), 579–592 (2020). https://doi.org/10.1002/sres.2702
16. Class, B., Soulikhan, F., Favre, S., Cheikhrouhou, N.: A framework for an open education supply chain network. In: Conference on Computer Supported Education, Prague (2021)
17. Boyer, E.: Scholarship Reconsidered: Priorities of the Professoriate. The Carnegie Foundation for the Advancement of Teaching (1990)
18. Haigh, N., Withell, A.: The place of research paradigms in SoTL practice: an inquiry. Teach. Learn. Inq. **8**(2), 17–31 (2020). https://doi.org/10.20343/teachlearninqu.8.2.3
19. Hubball, H., Clarke, A.: Diverse methodological approaches and considerations for SoTL in higher education. Can. J. Scholarsh. Teach. Learn. **1**(1) (2010). https://doi.org/10.5206/cjsotl-rcacea.2010.1.2
20. Miller-Young, J., Yeo, M.: Conceptualizing and communicating SoTL: a framework for the field. Teach. Learn. Inq. ISSOTL J. **3**(2), 37–53 (2015). https://doi.org/10.2979/teachlearninqu.3.2.37

21. Latour, B.: Nous n'avons jamais été modernes: Essai d'anthropologie symétrique. La Découverte (2006)
22. Schlagwein, D., Conboy, K., Feller, J., Leimeister, J.M., Morgan, L.: "Openness" with and without information technology: a framework and a brief history. J. Inf. Technol. **32**(4), 297–305 (2017). https://doi.org/10.1057/s41265-017-0049-3
23. Hug, T.: Defining openness in education. In: Peters, M. (ed.) Encyclopedia of Educational Philosophy and Theory, pp. 1–6. Springer, Singapore (2016). https://doi.org/10.1007/978-981-287-532-7_214-1
24. UNESCO, C.: Guidelines on the development of open educational resources policies (2019). https://unesdoc.unesco.org/ark:/48223/pf0000371129
25. Wikipedia. Wilhelm von Humboldt (2021b). https://en.wikipedia.org/w/index.php?title=Wilhelm_von_Humboldt&oldid=1041591164
26. Deimann, M.: Open Education, An Overview of. In: Peters, M.A. (ed.) Encyclopedia of Educational Philosophy and Theory, pp. 1–5. Springer, Singapore (2016). https://doi.org/10.1007/978-981-287-532-7_213-1
27. Burgos, D. (ed.) Radical Solutions and Open Science : An Open Approach to Boost Higher Education. Springer, Singapore (2020).https://doi.org/10.1007/978-981-15-4276-3
28. Fecher, B., Friesike, S.: Open science: one term, five schools of thought. In: Bartling, S., Friesike, S. (eds.) Opening Science, pp. 17–47. Springer, Cham (2014). https://doi.org/10.1007/978-3-319-00026-8_2
29. Kahle: Designing open educational technology. In: Iiyoshi, T., Kumar, S. (eds.) Opening up Education, pp. 27–46. MIT Press (2008)
30. Blessinger, P., Bliss, T.J. (eds.) Open education. International Perspectives in Higher Education. Open Book Publishers (2016)
31. Conrad, D., Prinsloo, P. (eds.) Open (ing) education: theory and practice. Brill (2020). https://doi.org/10.1163/9789004422988
32. García-Holgado, A., et al.: Handbook of successful open teaching practices. OpenGame Consortium (2020). https://repositorio.grial.eu/handle/grial/2102
33. Iiyoshi, T., Kumar, S.: Opening up Education. MIT Press (2008)
34. Orr, D., Weller, M., Farrow, R.: Models for online, open, flexible and technology enhanced higher education across the globe – a comparative analysis. International Council for Distance Education (2018)
35. Pitt, R., Jordan, K., de los Arcos, B., Farrow, R., Weller, M. Supporting open educational practices through open textbooks. Distance Educ. 41(2), 303-318(2020). https://doi.org/10.1080/01587919.2020.1757411
36. Stacey, P., Hinchliff Pearson, S.: Made With Creative Commons (2017)
37. Stracke, C.: Quality frameworks and learning design for open education. Int. Rev. Res. Open Distrib. Learn. **20**(2) (2019). https://doi.org/10.19173/irrodl.v20i2.4213
38. Teixeira, A.: Widening participation in a changing landscape: how open and distance education research and policies are transforming Portuguese higher education practices. In: Huet, I., Pessoa, T., Murta, F.S. (eds.) Excellence in Teaching and Learning in Higher Education: Institutional policies, research and practices in Europe, pp. 57–74. Coimbra University Press (2021). https://doi.org/10.14195/978-989-26-2134-0
39. Weller, M.: The Battle for Open: How Openness Won and Why it Doesn't Feel Like Victory. Ubiquity Press (2014).https://doi.org/10.5334/bam
40. Weller, M.: Open and free access to education for all. In: Burgos, D. (ed.) Radical Solutions and Open Science. LNET, pp. 1–15. Springer, Singapore (2020). https://doi.org/10.1007/978-981-15-4276-3_1
41. Weller, M., Jordan, K., DeVries, I., Rolfe, V.: Mapping the open education landscape: citation network analysis of historical open and distance education research. Open Prax. **10**(2), 109–126 (2018)

42. Wiley, D.: Iterating toward openness: lessons learned on a personal journey. In: Jhangiani, R.S., Biswas-Diener, R. (eds.) Open: The Philosophy and Practices that are Revolutionizing Education and Science, pp. 195–207. Ubiquity Press (2017). https://doi.org/10.5334/bbc.o

43. Cronin, C.: Open education: walking a critical path. In: Conrad, D., Prinsloo, P. (eds.) Open (ing) Education: Theory and Practice. Brill (2020). https://doi.org/10.1163/9789004422988

44. Gama, J., Caro, R., Hernan, C., Gomez, C., Gomez, G., Mena, A.: Using input output Leontief model in higher education based on knowledge engineering and fuzzy logic. In: IEEE Global Engineering Education Conference (2016)

45. Ehrenreich, J., Mazar, I., Rampelt, F., Schunemann, I., Sood, I.: Recognition and Verification of Credentials in Open Education – OEPASS (2020). https://oepass.eu/outputs/io3/

46. dos Santos, A.I., Punie, Y., Muñoz, J.C.: Opening up education. a support framework for higher education institutions (2016). https://ec.europa.eu/jrc/en/publication/eur-scientific-and-technical-research-reports/opening-education-support-framework-higher-education-ins titutions

47. West, R.E., Newby, T., Cheng, Z., Erickson, A., Clements, K.: Acknowledging all learning: alternative, micro, and open credentials. In: Bishop, M.J., Boling, E., Elen, J., Svihla, V. (eds.) Handbook of Research in Educational Communications and Technology, pp. 593–613. Springer, Cham (2020). https://doi.org/10.1007/978-3-030-36119-8_27

48. Rios, J.A., Ling, G., Pugh, R., Becker, D., Bacall, A.: Identifying critical 21st-century skills for workplace success: a content analysis of job advertisements. Educ. Res. **49**(2), 80–89 (2020). https://doi.org/10.3102/0013189x19890600

49. WorldEconomicForum. What are the 21st-century skills every student needs? (2016). https://www.weforum.org/agenda/2016/03/21st-century-skills-future-jobs-students/

50. Finkelstein, Knight, and Manning, 2013 cited by Mathur et al. [51] (2018)

51. Mathur, A., Wood, M.E., Cano, A.: Mastery of transferrable skills by doctoral scholars: visu-alization using digital micro-credentialing. Chang. Mag. High. Learn. **50**(5), 38–45 (2018). https://doi.org/10.1080/00091383.2018.1510261

52. Halavais, A.: Microcredentials on the Open Web. AoIR Selected Papers of Internet Research 3 (2013). https://journals.uic.edu/ojs/index.php/spir/article/view/8732

53. Stacey, P.: Starting Anew in the Landscape of Open (2018). https://edtechfrontier.com/2018/02/08/starting-anew-in-the-landscape-of-open/

54. Freire, P.: Pedagogy of the oppressed. Continuum (1994)

55. FOSTER: The Open Science Training Handbook (2018). https://book.fosteropenscience.eu/en/

56. Santos-Hermosa, G.: Open Education in Europe: Overview, integration with Open Science and the Library role (2019). http://eprints.rclis.org/34423/

57. Wikipedia: GAFAM (2021a). https://fr.wikipedia.org/w/index.php?title=GAFAM&oldid=186814905

58. Crawford, J.: Does Bill Gates have too much influence in the WHO? (2021). https://www.swissinfo.ch/eng/does-bill-gates-have-too-much-influence-in-the-who-/46570526

59. McGoey, L.: No Such Thing as a Free Gift: The Gates Foundation and the Price of Philanthropy. Verso Books (2015)

60. Rogers, R.: Linsey McGoey, no such thing as a free gift: the gates foundation and the price of philanthropy. Society **53**(3), 329–330 (2016). https://doi.org/10.1007/s12115-016-0022-8

61. Swissuniversities. Further Education at Tertiary Level. https://www.swissuniversities.ch/en/topics/studying/qualifications-framework-nqfch-hs/further-education

62. EPFL-UNIL. Formation continue UNIL EPFL. https://www.formation-continue-unil-epfl.ch/en/

63. Universite-de-Genève. Statut de l'université (2020). https://www.unige.ch/files/3415/8271/1574/Statut-20fevrier2020.pdf

64. EPFL. (2018). L'EPFL remet les premiers Certificates of Open Studies de Suisse. https://actu. epfl.ch/news/l-epfl-remet-les-premiers-certificates-of-open-stu/

65. Bhaskar, V., Lallement, P.: Modeling a supply chain using a network of queues. Appl. Math. Model. **34**(8), 2074–2088 (2010). https://doi.org/10.1016/j.apm.2009.10.019

66. Garay-Rondero, C.L., Martinez-Flores, J.L., Smith, N.R., Morales, S.O.C., Aldrette-Malacara, A.: Digital supply chain model in industry 4.0. J. Manuf. Technol. Manage. (2019). https://doi.org/10.1108/Jmtm-08-2018-0280

67. Slack, N., Lewis, M.: Towards a definitional model of business process technology. Int. J. Process Manag. Benchmarking **1**(3) (2005). https://doi.org/10.1504/IJPMB.2005.006109

68. Johnston, R., Shulver, M., Slack, N., Clark, G.: Service Operations Management: Improving Service Delivery (5 ed.) Pearson (2020)

69. de Bruin, T., Rosemann, M., Freeze, R., Kulkarni, U.: Understanding the main phases of developing a maturity assessment model. In: 16th Australasian Conference on Information Systems, Sydney (2005)

70. Shortle, J., Thompson, J., Gross, D., Harris, C.: Fundamentals of Queueing Theory, 5th ed. Wiley (2017). https://doi.org/10.1002/9781119453765

71. Dinter, B.: Success factors for information logistics strategy—an empirical investigation. Decis. Support Syst. **54**(3), 1207–1218 (2013). https://doi.org/10.1016/j.dss.2012.09.001

72. Mersand, S.: The state of makerspace research: a review of the literature. TechTrends **65**(2), 174–186 (2020). https://doi.org/10.1007/s11528-020-00566-5

How Can We Recognize Formative Assessment in Virtual Environments?

Agnese Del Zozzo[1]([⊠]) [iD], Marzia Garzetti[2] [iD], Giorgio Bolondi[2] [iD], and Federica Ferretti[3] [iD]

[1] Università degli studi di Trento, Trento, Italy
agnese.delzozzo@unitn.it
[2] Free University of Bozen, Bozen, Italy
Marzia.Garzetti@education.unibz.it, Giorgio.Bolondi@unibz.it
[3] Università degli studi di Ferrara, Ferrara, Italy
federica.ferretti@unife.it

Abstract. Pandemic induced Long-Distance Learning during the first Italian lockdown has highlighted the role that digital technology-based environment can have on class interaction. This work present how formative assessment can be defined and detected in the context of virtual environment. Classroom observation has allowed us to adapt a research tool coming from a European project on formative assessment in mathematics to the virtual classroom. We developed a grid for analysis and auto-analysis of teacher practices related to this construct working on already existent tool, designed for video-analysis of physical classrooms situations. The process of adaptation to the virtual environment that led to the definition of our grid is here explained in order to function as an example for future works in the same direction, not only in relation to formative assessment.

Keywords: Long-distance learning · Formative assessment · Mathematics education

1 Introduction

The present work takes shape within a project on the practices of teaching and learning mathematics during Long-Distance Learning (LDL). The main objective is to enhance the experience gained by teachers and students during the first lockdown caused by the Covid-19 pandemic, which took place in Italy during the spring of 2020. Building on previous works [1, 2], we want to create a ground for communication, comparison and professional development that combines the experience of teachers during LDL with the results present in the literature in mathematics education related to formative assessment [3].

Through observation of the practices of some teachers who, during the LDL, have used the G-Workspace for Education[1], profound changes in the management of feedback among students and teachers have been detected. Google Classroom[2] allows for the

---

[1] https://edu.google.com/intl/it_it/products/workspace-for-education/education-fundamentals.
[2] https://support.google.com/edu/classroom/?hl=it#topic=10298088.

B. Csapó and J. Uhomoibhi (Eds.): CSEDU 2021, CCIS 1624, pp. 517–535, 2022.
https://doi.org/10.1007/978-3-031-14756-2_25

storage and duplication of student assignments and materials exchanged during teaching interactions, it also allows for communication in different environments in public and private form. This, on the one hand, has increased the workload for teachers and students; on the other hand, it has led to teachers being able to have a greater focus on individual student productions, allowing them to track individual paths within the classroom. This type of teacher attention has led to the emergence of some spontaneous classroom practices that, in [1], we have associated with the tools and techniques of formative assessment [3, 4]. Starting with some findings from the literature on formative assessment, we developed a grid to analyse the teachers' assessment practices. In our previous work the objective was to show the presence of formative assessment in correspondence with the use of certain digital technologies. For that aim, the modification of the original indicators would have been misleading. Nevertheless, the work of highlighting formative assessment has allowed us to intuit possible adaptations of the grid for use in LDL contexts and virtual classrooms.

Here we present an adaptation of the grid we have used for such analysis in [1], where relation to the LDL environment is made explicit. This adapted grid is conceived as a self-analysis tool for teachers that consider the new form of interaction and the dematerialization of didactical materials, including students' work. In this paper we want to show how an explicit adaptation of the grid to LDL context can be a tool for analysis and self-analysis, provided that we explicitly build the relationship between indicators in the physical classroom and in the virtual classroom. We show here how this adaptation was constructed: first, in Sect. 2 and 3, we frame the work done in [1], the adopted perspective and the data used, then, in Sect. 4.1 we show some significant examples of the relationship between phenomena in the physical classroom and phenomena typical of LDL context. In Sect. 4.2 we define the relationship between indicators in the two contexts using some examples taken from classroom observation. During this work we will highlight the limits of the original grid, showing the need for its adaptation to hybrid or digital contexts, increasingly common in contemporary classrooms, also because of the pandemic. Result of the work, in Sect. 5, will be the adapted grid together with the process of its definition. In Sect. 6 we set the explicit goal of making the adapted grid a tool for researchers and teachers and to foster a reflection on the educational possibilities offered by virtual environments.

2   Theoretical Framework

We can consider formative assessment as a didactical methodology, aimed at improving mathematics teaching and learning [3]. The functions, scopes and practices of assessment are closely linked to the historical context in which they occur and to the educational system of which assessment is an integral part. We embrace the shared definition of formative assessment as it is presented in the LLP-Comenius project FAMT&L - Formative Assessment for Mathematics Teaching and Learning [, p. 33]:5

The formative assessment is connected with a concept of learning according to which all students are able to acquire, at an adequate level, the basic skills of a discipline. The learning passes through the use of teaching methodologies which can respond effectively to different learning time for each student, their different learning styles, their zones of proximal development.

The formative assessment is an assessment FOR teaching and learning.

Formative assessment should be an essential part in all phases of the teaching-learning process. From this point of view, its main function is regulatory as it is aimed at modulating and continuously adapting the teaching process [6]. Specifically, one of the characteristics of formative assessment is collecting detailed information both in the initial phase and on the difficulties gradually encountered by the students during the learning in order to design targeted educational interventions and to have a constant reciprocal feedback between teacher and students [7]. In this way, on the one hand, the teacher will be able to make the appropriate changes in terms of teaching practices and, on the other hand, the student will have the opportunity to adapt their study methods and the quality of the strategies implemented in their own performance [8]. In [9] formative assessment's criteria and strategies are also highlighted:

- S1. Clarifying and sharing learning intentions and criteria for success.
- S2. Engineering effective classroom discussions and other learning tasks that elicit evidence of student understanding.
- S3. Providing feedback that moves learners forward.
- S4. Activating students as instructional resources for one another.
- S5. Activating students as the owners of their own learning.

For the sake of brevity, we will refer to these five strategies as S1, S2, S3, S4, S5.

One of the main objectives of the FAMT&L project was to improve the skills of mathematics teachers in the use of formative assessment as a tool for methodological quality in mathematics teaching [10, 11]. The main objective of the project was, ultimately, to design, implement and monitor a specific methodology for teacher professional development programs based on video analysis tools and techniques useful for promoting knowledge, attitudes and practices of formative assessment. In particular, a specific video analysis tool was designed, built and used, a structured grid – the FAMT&L Grid [12]. The FAMT&L Grid was developed starting from the international debate on formative assessment and, particularly, in line with the criteria identified by Black and Wiliam [3], and it is composed of indicators. The different indicators on assessment practices of Mathematics teachers in the grid are grouped in five macro-categories: Mathematics' contents, time of assessment, setting of assessment, kind of tools for data gathering of students' skills, phases of formative assessment. The last category gathers several kinds of behaviours and actions which are considered as indicators to be observed in the different phases of the assessment procedure.

In Bolondi et al. [1], to detect formative assessment within the LDL context, we decided to consider 72 indicators categorised according to S1, S2, S3, S4, S5. The resulting grid is available online (at https://docs.google.com/document/d/1gjlm-hSpV7t 10iNRX1FsyApzZaB6376qSy_ywt3jktU/edit?usp=sharing) and has been used to link our classroom observations to the strategies of formative assessment made explicit in

FAMT&L. The analysis conducted with the grid showed that the observed teacher was making large use of strategies that could be identified as formative assessment. Moreover, we observed how nine indicators are structural and specific to the software used for LDL, which in our case were, as stated, Google Classroom and Google Meet[3] combined. We list these nine structural indicators below (the numbers are assigned according to the order they had in the original project FAMT&L, and they can be found in [12]. In the following, T stays for Teacher:

- *1.T fixes with the students the date for the assessment.*
- *13.T distributes the text of the test/task.*
- *26. T asks questions to the whole class.*
- *27. T asks questions to a single student.*
- *36. T uses a structured tool of observation.*
- *37. T takes some record of the behaviour of one/all student/s.*
- *38. T takes some record about how much the students have achieved to handle the content of the test/task.*
- *39. T takes records from her/his desk.*
- *40. T takes records passing among the students.*

We can see how the nine indicators made possible to make student understanding visible (S2), and to clarify learning intentions and criteria for success (S1).

In Bolondi et al. [1], we conceived such indicators as being structural to some tools used for LDL, highlighting that they are sufficient to create a fruitful ground for formative assessment practices. An effective use of formative assessment then obviously depends also on other teacher's practices.

In this contribution we specifically show the use of the grid and the adaptations made to the LDL environment, but to do so we need to recall the observed episodes and our context of observation.

3 Data

During the last month of the school year (from May 10th to the first decade of June 2020) pandemic induced LDL create in Italy the opportunity to conduct an ethnographic observation in several classes, of different school grades.

We focus our analysis on one of the observed classes: namely a grade 12 class of a technical school. This class used G-Workspace for Education. It must be noted that the didactical practices that we report and analyse here were not planned or controlled interventions organized by the researchers. In particular, the teachers were not aware of the tools and techniques of formative assessment. An important point that must be remarked is that every virtual classroom in the context of LDL is dichotomous, in the sense that it has two distinct components: a synchronous one and an asynchronous one. This is also true, of course, for ordinary face-to-face classrooms, but in the context of LDL, the coordination of these two components is made easier by the fact that they develop in the same environment. On the other hand, at the time of our observation, the

---

[3] https://support.google.com/meet/?hl=en#topic=7306097.

teachers had to face a completely unexpected situation and hence also from this point of view (the coordination of the two components) the emergency can be called a "stress test" which revealed some of the beliefs behind their new and old practices.

We were provided with accesses to the synchronous component of the classroom - implemented with Google Meet - and to the asynchronous component - implemented with Google Classroom.

Two of the researchers-observers were added to the class. Regarding Google Classroom, two of us were added to the class with the role of additional teachers. In such a way, the research team was allowed to access the entire archive of materials. In particular, we were allowed to access to all the messages exchanged: both public, with the whole class, and private between the teacher and each student. The creation of this class in Google Classroom took place on February 27th and we could access all materials since then.

Therefore, our observation regarding the synchronous component of the classroom dynamics was indirect for what happened between February 27th and May 9th, and direct from May 11th until June 5th (the end of the school year). Our observation of the asynchronous component of the classroom dynamics has been complete: we could analyse the teaching flow for the whole period February 27th- June 5th.

For the sake of completeness and clarity, we briefly outline the features of the digital environment Google Classroom which define the different areas of interaction and work. These areas have different functionalities and permissions for action and access.

- Stream: it is like the homepage/message board of a specific class. Both the teacher(s) and the students can post announcements containing links, text, and attachments like images or videos. Every participant to the classroom can read the posts, everyone (both students and teachers) can delete their own posts. The teachers have the additional faculties to delete students' posts and to re-arrange the posts order.
- Classwork: whoever has a teacher role can create in this page materials and classwork, such as questions, assignments, test, for the students These materials can be organized in topics (like decimal numbers, linear systems etc.). The tasks are notified to the students which can hence enter the task activity.
- Assignments: the teacher can assign to a single student, a group of selected students,the whole class, or a set of classes an assignment. The teacher can provide instruction and additional materials needed to complete the task. A personal environment is accessible to each "student" account receiving the assignment: it contains a shared chat among all other students receiving that assignment, a private chat between the single student and the teacher, and a space to attach materials. For "teacher" type accounts, the Assignment environment consists of two areas:
- Instruction page: in this page, the teachers can see and modify the assignment and access the shared chat.
- Student work: in this page, the teachers can access the work of the students. The teacher can directly comment on the files uploaded by the student, or interact with him via the textual private chat, check the status of each student's work, grade it. Each student can access only his/her own work

We use the word "episode" to denote what happens in the virtual classroom (in both its synchronous and asynchronous components) between the assignment of a task (e.g., an Assignment on Google Classroom), and the assignment of a new task on a different set of exercises. We selected three episodes that we consider as paradigmatic of the teaching and learning practices we observed: episode 1develops between 17 and 20 March 2020, episode 2 develops between 24 and 30 March 2020, and episode 3 develops on May 18th.

3.1 Episode 1: First Assignments on Google Classroom

Episode 1 is an example of data collected before our direct observation, which by the way shows the potential (for the researchers) of data collection via a virtual environment. We consider here asynchronous exchanges on the Google Classroom at the very beginning of the LDL experience. On 17th March, a task is assigned to be completed by March 20. In this time frame various interactions between the teacher and the students occurred in the various spaces of the platform. The unfolding of episode 1 shows how the different spaces interact with each other during the different activities, framing the march of the class towards a new formative assessment practice. We distinguish among what happens on Google Classroom and what happens on Google Meet.

17th of March 2020. In Google Meet: 2-h synchronous lesson. In Google Classroom: in the instruction page a homework assignment is published with deadline 20th of March (exercises from the textbook); in the Stream, the notification of the assignment with the link to access it is published and notified.

18th and 19th of March 2020. In Google Classroom: one student communicates in her workspace that she has not the textbook. In the instruction page the teachers modifies the instructions. The scanning of the pages of the textbook is uploaded. Also, public exchanges between students and the teacher appear on this space. At the same time, in other students' workspaces we see private exchanges between students and T about the assignment. 24 students out of 29 upload their homework by sending photos of their notebooks.

20th of March 2020. In the Stream of Google Classroom, the teacher posts a reminder of the meeting on Google Meet, announcing the topic: correction of the assignment. Then, in Google Meet: 1-h synchronous lesson is conducted in which the exercises from the textbook are corrected by the teacher. After this lesson, in Google Classroom, in students' workspaces there are requests for feedback on the correction carried out synchronously during the meeting. In particular, one student uses her own workspace for reporting about her correctly completed exercises and errors. Moreover, she uploaded photos of her corrected work and outlined with red marks the corrections. From episode 2, this practice become a routine for the whole class. 2.

3.2 Episode 2: Institutionalisation of Self-correction as a Routine Practice

Like Episode 1, Episode 2 opens with an assignment on Google Classroom and develops between 24th and 30th of March. The difference with episode 1 is the institutionalisation of a process of self-correction of the assignment. This practice will be maintained until mid-May. First, the teacher asks for the task assigned to be handed in on 27th March. Then on that date, she publishes materials (in fact, videos) with the correction of the exercises. Finally, she assigns as a new task on Google Classroom the correction of the exercises of the initial assignment. Students are asked to do it on the text of the exercises already done using a different colour. Throughout the process there have been synchronous lessons and many asynchronous exchanges between the teacher and individual students, concerning both the assignment, and the correction of it.

3.3   Episode 3: Peer Assessment on Google Meet

Episode 3 occurred on 18th May in Google Meet. The teacher shares her screen for the entire duration of the episode. It is a planned assessment session in synchronous. Four students are directly involved. The session consists of three phases. In the first phase, the teacher presents as task a function study. She asks them to carry out this study in a document shared with the class. The other students, therefore, can edit the document. Each of the 4 students carries out a part of the exercise and corrects the work of the previous one. In the second phase, the teacher assigns to the whole class as a task the drawing of the graph of a function with given characteristics. The answers of the four involved students are shared and analysed. In the third and last phase, one of the exercises assigned to the whole class and conceived as preparation for the assessment period is analysed and corrected. The correction discussion involved directly the four students.

4   Codification Methodology

4.1   Redefinition of Some Classroom Practices

To start, we need to redefine some of the basic classroom practices happening in the school building in an LDL environment. Indeed, in the LDL context, many of the usual school practices have been adapted to the new context. Some are consciously adapted by the teacher, while others are the result of a process of adaptation to the software used and the new needs, limitations and opportunities. This adaptation process was done for the specific case we observed, and in this context, we try to make it explicit and generalizable. In the following, we try to give examples of some of the basic adaptations made by teachers and students during LDL, starting from the very fundamental aspects that define a classroom.

As a first step of the adaptation process, we decided to focus on the aim of basic actions in the classroom, that are: positioning in the classroom, communicative inter-action, observing student's work, collection and sending of materials. Having the aim explicit, it is possible to identify the action in the LDL classroom that has the same aim. This instantiation usually happens through a different action in the LDL classroom, but the relations among the two actions, in physical and virtual context, remains defined.

We give the emblematic examples of this process in the following. It is nevertheless important to underline that we will confront the two environments as separated: often, however, the two modes of communication and the two types of classrooms coexist, and teachers and students are accustomed to their hybrid use. In this context, the separation was given by LDL, and it is maintained to highlight the respective characteristics of the two environments.

Positioning in the Classroom. How do we identify ourselves and our classmates, or teachers in the classroom in the school building and in the virtual classroom? And how do we position in the different kinds of "classrooms" defined?

The question is not trivial, as pointed out by [13]. It must in fact be considered that individuals in the physical environment are physical people, while individuals in the virtual environment are "accounts". Physical people can implement an aware and strategic use of accounts in order to carry out some operations which have educational relevance. To clarify this statement, we can think about the class in the school building: the student's presence is detected by the fact that they are physically in the classroom, occupying a space, probably a desk, using personal or common objects in the classroom. Most often those objects (pencil, paper, computers, blackboard) are or can be used by one person at a time.

In the virtual classroom things change: the student and the teacher can have multiple accounts, they are not sharing the same physical space (at least during LDL), and their presence is detected through different means: for instance they handle homework on Google Classroom, maybe answering to teachers comments, in this case only their name and their work are visible; they participate in video call, putting their camera on or off, speaking or not, they can write on collaborative documents all together or one at a time, and so on. What is visible, and how many of these accounts are, at the same time, present in the classroom, can change a lot, and depends on the intentions of students and teachers and by the features of the software used (see Fig. 1).

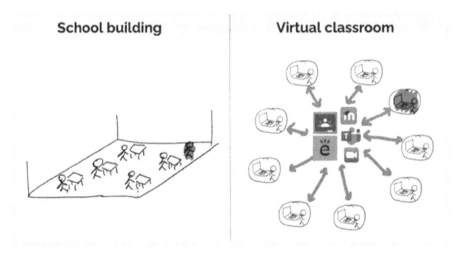

Fig. 1. Positioning in the classroom in the two environments. In green we highlight the teacher, On the right, we exemplify some software – Google Classroom, Moodle[4], Microsoft Teams[5], Zoom[6], Edmodo[7] - by their icons.

---

[4] https://moodle.org/?lang=it.

[5] https://www.microsoft.com/en-gb/microsoft-teams/log-in.

[6] https://zoom.us/.

[7] https://new.edmodo.com/.

Communicative Interaction. Students and teachers' positioning in the classroom influence the way in which interaction can be acted. Without being comprehensive, we can affirm that in the School Building participants can use written and oral communication, deictic language, gestures, etc. and that oral interactions are mostly public. The exchanges can be aimed at explaining, evaluating, assessing, affective-relational focussed, etc. but they happen mostly in a shared space. In virtual classrooms written and oral communication can have different shapes: can be done via mail, using private or public messages on different software, videoconferencing and the space of interaction can be shared or not with the whole class. Thanks to the possibility of archiving and of duplication proper to virtual environments, it becomes more immediate the exchange of documents and content, on which several students can work simultaneously or in parallel, in class subgroups, in pairs or individually. In the case of the asynchronous component in Google Classroom, any activity assignment that the teacher makes to the whole class (or to a group of students) also implies that this activity is sent to each student in the class (or belonging to the group). It appears the need, not always conscious, to mediate communicative intentions and deictic language by using "additional" devices (e.g., indicating becomes moving the mouse on the screen, or highlighting part of a document to make it visible to the other), etc.

Observing Student's Work. How can teachers observe students' work in the two environments?

In the School Building he or she can rotate between desks; call to the teacher's desk, ask students to handle or to present their written productions etc. In the virtual classroom the way teachers can observe students 'work is flexible and depend on the software used: if Google Meet is used and the students have their camera on the grid view it is possible to see students' faces, but it is also possible to watch students' work, if explicitly asked. If Google Classroom is used, the teacher can scroll through the delivered tasks, and return to them at any time, also to share them with the class or other students if needed (see Fig. 2). The use of collaborative documents and/or specific applications allow to see the work being done, but also asking for screen sharing can be an option.

Collection and Sending of Materials. Following the above paragraphs, when the aim is the collection and sending of material, we can differentiate between the school building, where students and teacher can directly exchange paper documents, or writing on the blackboard or on teacher's PC screen projection, or using textbooks and other resources, etc. and the virtual classroom, where direct paper exchange is not possible. More precisely, in the virtual classroom students and teachers can send photos of paper materials, use shared collaborative documents, share web resources via links (i.e., text string sharing, indeed, each web resource associated with a link), etc.

One important aspect of exchange in virtual classrooms is that the exchanged resource remains at disposal of both the sender and the receiver.

We have found these four aspects to be the most relevant to be discussed in the comparison among LDL and learning in the school building: many others could be listed and these same four could be further decomposed. For this work however, they seemed sufficient to justify the choice and the use of the indicators of FAMT&L project in the new environment. In the following section we specify the methodology used for the analysis, giving specific examples for each of the episodes described.

School building   **Virtual classroom**

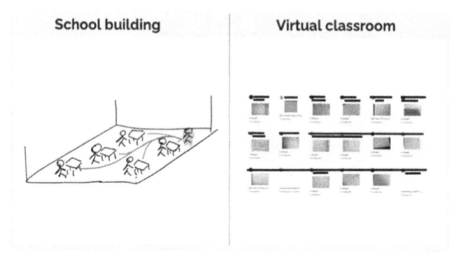

Fig. 2. Observing student's work in the two environments. On the left, we indicate the teacher in green. On the right, a screenshot of a completed assignment in Google Classroom is presented, with all the students listed (Color figure online).

4.2 Examples of Coding from Episodes 1, 2 and 3

As a second step, we analyse the three episodes described in Sect. 3 using the chosen FAMT&L indicators categorised according to the five formative assessment strategies, as described in the theoretical framework (Sect. 2). As stated, the FAMT&L table was created for the analysis of video recordings of lessons in the physical classroom. Nevertheless, we assume that given the adaptation process previously described, we can use the indicators also in LDL contexts. During the analysis we highlight the limits of some of the indicators with respect to the new context, in order to show the necessity of the final adaptation of this tool.

In Bolondi et al. [1] we observed that out of 72 considered indicators, 38 can be found in episode 1, 49 are detected in episode 2, and 53 in episode 3. The episodes were analysed by two researchers and then the resulting choices were confronted. For each one of the episodes the concordance between researchers was verified through Cohen's K [14], and good correspondence was found for the three episodes (respectively 0,69, 0,63 and 0,69 for episode 1,2 and 3). The resulting indicators are the union of the ones selected by each researcher. The indicators of each episode are well distributed along the five formative assessment strategies of Leahy et al., letting us suppose that formative assessment can be used to look at some of the emerging phenomena in LDL.

In this section we are not going to discuss the result of this work specifically, which can be found in Bolondi et. al [1], but we aim at explaining and giving examples of the specific use of the grid we adopted for the analysis, in order to propose it as a tool for future studies. To do so, we justify the choice of some of the indicators used for every described episode. To see the complete analysis you can access to https://docs.google.com/document/d/1l2Yu5Lth6QAmBMHdJxxmU1XyOfGJ1BeX lQp0nSmU6MA/edit?usp=sharing.

Episode 1. Figure 3 presents a diagram to recall the dynamics of episode 1, which is described in Sect. 3. Episode 1 started on the 17th of march with an assignment in Google Classroom of some exercises. During the two days after, students worked on their resolution to be uploaded on Google Classroom. On 20th march, the teacher conducted a synchronous lesson on Google Meet in which she corrected the assigned exercises. The same day, she asked some students via private messages if they have self-corrected their work. One of them, spontaneously, answered by uploading in Google Classroom her self-corrected notes.

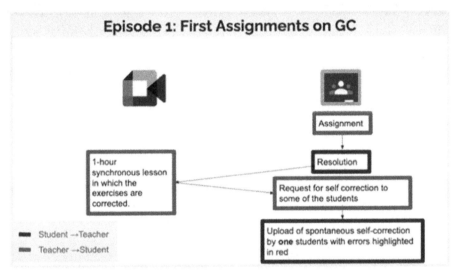

Fig. 3. Outline of episode 1. The most meaningful steps on Google Classroom and Google Meet are organised on the right and left of the image respectively.

We take advantage of this episode to exemplify the nine structural indicators, which we have already presented on Sect. 3 but that we recall here for the reader convenience:

- *1.T fixes with the students the date for the assessment*
- *13.T distributes the text of the test/task.*
- *26. T asks questions to the whole class.*
- *27. T asks questions to a single student.*
- *36. T uses a structured tool of observation*
- *37. T takes some record of the behaviour of one/all student/s*
- *38. T takes some record about how much the students have achieved to handle the content of the test/task*
- *39. T takes records from her/his desk*
- *40. T takes records passing among the students*

Let us briefly comment on them, with respect to the described episode. Firstly, every time a teacher creates an assignment in Google Classroom, it may be associated with a deadline. Thus, if a teacher establishes a deadline for the assignment in Google Classroom, a date will automatically be fixed, and all students involved in the Google Classroom virtual classroom will be aware of the deadline (n° 1).

Secondly, when in Google Classroom an assignment is sent to the whole class (n°26/13), each student receives it individually (n°27) and has at her disposal a private space of interaction with the teacher.

Finally, if we take the teacher's perspective in relation to an assignment in Google Classroom (see Fig. 3), we can say that:

- She has a synoptic overview of the whole class, depicted as the set of each student's work (n° 36).
- Such overview includes: a numerical report of the number of delivered and undelivered tasks; for each student, the number of the attachments uploaded with (if the number is not zero) a visual overview; a recap of the homework status, already handed in or not, both for each student and a global overview; furthermore, by clicking on one of the students, she has access to all details on the work of that student (n° 37 and n° 38).
- Moreover, the interface that the teacher has in front of her can be thought both as her personal desk, from which she observes who handled the homework and who did not (n°39) and as a point of access similar to the classroom "passing among students" (n°40) (Fig. 4).

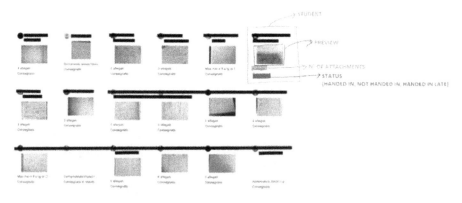

Fig. 4. A screenshot of the synoptic overview in the teacher's view of a Google Classroom assignment.

Since some of the features of Google Classroom that we have listed to justify some of the indicators are independent of the teacher's willingness to archive student materials, or to record student behaviour with respect to the assignment (handed in, not handed in, handed in late), we can say that presence of the 9 indicators previously listed is structural for a certain use of Google Classroom, and independent by this single episode. This observation was at the base of the previous work on formative assessment during LDL [1], in the present work we try to show the limitations and possibilities of this grid in the new environment.

To conclude the comment of episode 1, we provide here an example of the assignation of indicator 67:

- 67. The teacher asks every student to correct his/her own test/task

Our observed teacher, after the synchronous lesson of 20th march, took advantage of the private chat in the assignment to ask almost each student privately the correction of her homework (see Fig. 5).

Fig. 5. The exchange between one student and the teacher in relation to episode 1.

Episode 2. Figure 6 presents a diagram to recall the dynamics of episode 2, which is described in Sect. 3.

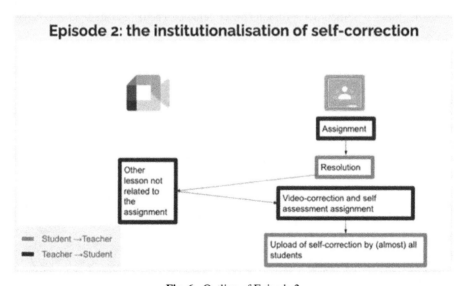

Fig. 6. Outline of Episode 2.

The episode started with an assignment published on Google Classroom the 24th of march with a deadline three days after on 27th. The same day, a synchronous lesson on Google Meet took place, which was not related to the assignment. At the end of such a synchronous lesson the teacher published on Google Classroom the video correction of the assignment, asking as homework to the students to hand in their self-correction, written in a different colour, for the 30th of march. During this process, between the first (24th) and the last day of the episode (30th), there were many private and public interactions on the platform related to the assignment.

We avoid giving new examples in relation to the 9 structural indicators commented before, and we select some other significant indicators that appeared with respect to this episode. We start with indicator 9:

- 9.T recalls the criteria to correct the test/task

This time, the teacher shares on Google Classroom the correction to the exercises as a video. More precisely, in this episode the correction is used as a new assignment designed to request the students to share their self-correction with the teacher. This request is directly connected with the indicator number 67 that we also detect in the episode 1. Nevertheless, in this case, there is a sort of institutionalisation of self-correction. Here, the quantity (and the quality) of the exchanges increases because in Episode 1 the teacher had feedback just from one student, while here she collects the self-correction from almost all of them.

In Table 1 below it is possible to see two photos of two students' notebooks with their self-correction marked with red ink. After the photos you can see the message exchange between each of them and the teacher in the private chat of Google Classroom.

The teacher's explicit request to write in red the correction makes it easier for her to detect the students' difficulties. From the observation of teacher and students exchanges we can affirm that:

Table 1. Students work and related exchanges on Google Classroom.

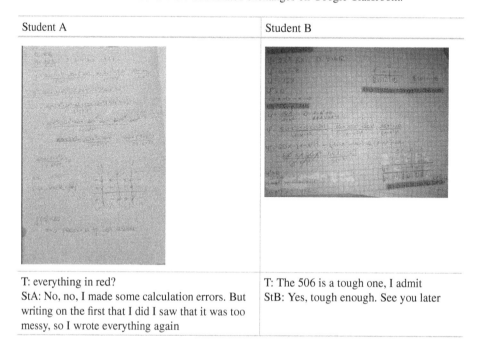

Student A	Student B
T: everything in red? StA: No, no, I made some calculation errors. But writing on the first that I did I saw that it was too messy, so I wrote everything again	T: The 506 is a tough one, I admit StB: Yes, tough enough. See you later

- 55. T delivers the results passing among the pupils' desks
- 56. T calls each student to deliver them the results
- 57. T calls each student and spares a few minutes to comment privately his/her results
- 58. T does not make negative comments on the students who failed the test/task

In particular, with respect to 55, 56, 57 the virtual environment used allows the teacher to really observe student work as if she was passing through student's desks, commenting each homework individually. Clearly, however, written chat interaction must be considered, and, as we pointed out in Sect. 4.1, the passage among desks must be redefined consistently with the characteristics of the environment.

Episode 3. The episode is described in Sect. 3 and consists of a planned moment of summative assessment involving four students. The environment of this episode is different than that of the previous two: in this case students are doing a synchronous lesson on Google Meet and at the same time using a collaborative document where the teacher has prepared some tasks. The four assessed students write on the shared document.

From this episode we recall two indicators coming from the last two columns related to strategies S4 and S5 of [9]:

- 66. T asks the student to do the corrections among themselves (cross-correction)
- 72. A student is stimulated by peer-assessment to self-assessment

Even though the structure of the described assessment already shows the presence of these indicators, we describe a particular moment of the episode that substantiates them.

At a certain moment the first student identifies $x = -1$ and $y = 1$ with the points of intersection with the axis of a given function. Prompted by the teacher, another student intervenes on her answer, commenting on the mistake done (using the notation $x = -1$ and $y = 1$ to indicate two points on the plane) and explaining that one needs two coordinates for each point (without giving the two coordinates). After the classmate's intervention, the first student corrects herself adding the appropriate notation of the two points $A(-1;0)$ and $B(0;1)$.

To conclude, we care to mention one last practice that was implemented by our observed teacher during episode three, which can be considered a practice of formative assessment but that could not be tracked with the FAMT&L indicators we were considering. In two different moments the teacher, who has at her disposal the solution of the same task as made by the students, decides to communicate with one student upon her answer using the answers of other students producing a sort of dialogue between answers. This dialogue involves the teacher, the students, and their written productions: it is a practice of formative assessment because it elicits evidence of student understanding (S2) and it activates students as instructional resources for one another (S4). However, it cannot be easily tracked by FAMT&L indicators because, as far as we noticed, it is strictly linked to the structure of a digital technology-based environment and especially storage and ease of access.

5   Results: The New Grid

In this section we propose an updated grid, emerged from the analysis done on the three episodes. LDL environment, and in general the use of virtual classrooms, make it necessary to adjust some of the indicators in order to consider the peculiar features of these environments.

The new grid is shown in Fig. 7, modifications are underlined: we discuss them here briefly, pointing as well at the redefinitions of classroom practices made in Sect. 4.1 that influence the way the grid should be used to code the observation.

As it has been highlighted in the analysis of the first episode, a structural characteristic of Google Classroom, common as well to other virtual environment, is that in some cases the difference between whole classroom and each student collapse: for instance, when the teacher sends an email to all students, or fixes the date of the assessment on the software, each student is notified on her personal device. For this reason, indicators 1 and 2 have been adjusted with respect to this duality. Indicators 26 and 27 as well reflect this duality, but they have been maintained because there are in cases in which a question can be posed to the whole classroom but not to each student: it is the case for questions written on the stream of Google Classroom or verbally during a synchronous lesson on Google Meet. There are also cases in which the question is posed to the classroom and arrives to each student, who can then answer privately: we saw an example of this in episode 2, in the context of homework assignment on Google Classroom. Then, in other cases, the teacher can decide to ask privately a question to each student in the classroom using private chat, email or other means.

Indicators 42, 43 and 44 have been modified only adding the possibility of directly sharing materials with students on the different platforms used. We can think of the publication of the correction made by the teacher in episode 2 and assigned as an homework.

Indicators 35 to 40 have not been modified, but, as shown in episode 1, reflect a structural characteristic of Google Classroom: the automatic storage and organisation of materials allowed by this software (but also from many others used in virtual environments) enable the teacher to record behaviour and achievements of the students with respect to the assignment. Moreover, the redefinition of the teacher desk in the virtual environment made in Sect. 4.1 highlights how indicators 39 and 40 can be used in the new context.

In indicators 13, 14, 55, 56, 57 has been made explicit the possibility to communicate via email or chat, also using emoticons or emoji. The text of indicator 11 has been modified for clarity.

Summarising, we rephrased some indicators where the duality of the whole *classroom/each student* collapsed; we redefined teacher's desk considering a broader definition and we have introduced the possibility to refer to asynchronous interaction, typical of some virtual environments.

a. Clarifying and sharing learning intentions and criteria for success. (S1) **Sharing learning intention**	b. Engineering effective classroom discussions and other learning tasks that elicit evidence of student understanding. (S2) **Evidence of student understanding**	c. Providing feedback that moves learners forward. (S3) **Feedback**	d. Activating students as instructional resources for one another. (S4) **Student as resources for one another**	e. Activating students as the owners of their own learning". (S5) **Owner of their learning**
1. the date for the assessment is notified to each students following teacher's communication or action on the platform 2. The teacher reminds the class and each student that today, is the day of the assessment 3. The teacher shows to the students the aims of the assessment 5. The teacher shows to the students the subject of the assessment 6. The teacher shows to the students the evaluation criteria to correct the test task 7. The teacher explains the test-task instructions 9. The teacher recalls the criteria to correct the test-task 12. The teacher makes clear the ranking for each question. 13. The teacher distributes or send the text of the test task 42. The teacher illustrates shares the results of the test task to the whole class. 43. The teacher illustrates/shares the results of the test task to the whole class. 44. The teacher illustrates /shares the results of the test task to each student. 45. The teacher gives back the results in a short time 46. The teacher describes the mistakes as an occasion to learn 50. The teacher stresses the fact that the most difficult contents will be treated again 51. The teacher avoids using marks or other kind of judgments 54. The teacher takes care not to stress the difference between high and low marks 74. The teacher explains how to perform the peer-assessment 75. The teacher explains how to perform the self-assessment 76. The teacher stresses the content focus of the peer-assessment 77. The teacher stresses the content focus of the self-assessment 78. The teacher interacts with the process of peer-assessment (by stressing the focus) 79. The teacher interacts with the process of self-assessment (by stressing the focus)	4. The teacher asks some questions to the students to verify if the students understood the aims of the assessment 8. The teacher asks some questions to the students to verify that the students understood the test-task instructions 10. The teacher discusses with the students about the above criteria 22. The teacher gives enough time so that every student can work through the test-task (without anxiety) 26. The teacher asks questions to the whole class 27. The teacher asks questions to a single student 28. The teacher does not ask "rhetorical questions" to students 29. The teacher asks a new question based on the correct answer to the previous one 30. The teacher asks a new question based on a wrong answer to the previous one 31. The teacher asks a new question based on the previous one 35. The teacher uses a narrative tool of observation 36. The teacher uses a structured tool of observation 37. The teacher takes some record of the behaviour of all student's 38. The teacher takes some record about how much the students have achieved to handle the content of the test-task 39. The teacher takes records from her/his desk 40. The teacher takes records passing among the students 62. The teacher asks the student who passed the test task to explain the correct way to do it (at the blackboard or from their seat) 63. The teacher asks the student who failed the test task to explain the correct way to do it (at the blackboard or from their seat) 68. The teacher analyses the data she ask she be collected in the classroom 69. The teacher writes a profile of every student's results with respect to knowledge 70. The teacher writes the profiles with respect to skills 71. The teacher writes the profiles with respect to skills	14. The teacher passes among the students smiling at them or sending emoticon,emoji 16. The teacher gives additional activities to students who completed the test task before time 20. The teacher provides advices or suggestions during the test-task 47. The teacher talks with calm and patience about the mistakes done 52. The teacher comments privately each student's result by voice, chat or email 58. The teacher does not make negative comments on the students who failed the test-task 59. The teacher uses the summative results to create an occasion of formative assessment 60. The teacher corrects the test-task in the classroom. 61. The teacher corrects the test-task analytically; showing the right way to do it and explaining the possible mistakes 64. The teacher takes care to write a detailed comment on the work of any student 65. The teacher uses access to value the test-task (not giving marks)	19. The teachers allows the students to collaborate among them during the test-task 21. The teacher allows the students to talk to each other during the test-task 32. The teacher "moves" the question from one student to another 32b. The teacher creates a dialogue between answers from different students. 33. One or more students take part in the answer given by another student 34. One or more students ask to intervene about the answer given by another student 48. The teacher discusses the mistakes stimulating the whole class to take part in the debate 49. The teacher generates collaboration among the students 52. The teacher stimulates the students with best results to help the ones that have had problems in the test task 66. The teacher asks the student to do the corrections among themselves (cross-correction) 73. A student is stimulated by self-assessment to peer-assessment	11. The teacher keeps care of student's observation about assessment criteria 41. The teacher urges for care and attention in the work for the test task 55. The teacher delivers the results passing among the pupils' desks, or sending them 56. The teacher delivers each student the results calling or writing her 67. The teacher asks every student to correct his/her own test task 72. A student is stimulated by peer-assessment to self-assessment *Note: the numbering is not ordered with respect to the table presented, but with respect to the FAMTAL grid*

Fig. 7. The updated grid for formative assessment practices.

6 Conclusions

The present work focus on how formative assessment can be recognized and enhanced in virtual environment. To do so we have decided to refer to well-known literature on formative assessment in our field [3–5] and to specify how virtual classrooms and digital technologies can modify the way teacher and students exchange feedbacks, store, and use materials and resources. In a previous work [1] we show that digital-technology based environment fostered the spontaneous emergence of teaching and learning practices that can be associated with formative assessment. Nevertheless, the analysis of classroom practices during LDL has made clear that some indicators needed to be reformulated to include actions specific to the new context. We ended up with the proposal of a new grid, which we describe in this contribution, that allows us to highlight some of the structural features arising from the use of certain digital technologies (e.g., Google Classroom), and aspects related to their management. The indicators of such new grid decline formative assessment practices into recognizable actions that can be observed in all kinds of classroom practices, and the structure of the grid then allows to associate these actions with specific formative assessment strategies, according to the classic framework of [3].

The teacher can use such a grid as a tool for metacognitive analysis of their own instructional actions in LDL and she is able to track them, and to consider possible, and implementable, practices of formative or summative assessment coming from the literature [15, 16]. Finally, we think that making explicit the process of adaptation to virtual environment of a validated research tool for formative assessment, and the related methodology of analysis, can be helpful for future works not only on formative assessment in LDL, but also on the adaptation of other construct of Mathematics Education in such contexts.

References

1. Bolondi, G., Del Zozzo, A., Ferretti, F., Garzetti, M, Santi, G.: Can formative assessment practices appear spontaneously during long distance learning? In Proceedings of the 13th International Conference on Computer Supported Education - Volume 1: CSEDU, pp. 625-632 (2021)
2. Ferretti, F., Santi, G.R.P., Del Zozzo, A., Garzetti, M., Bolondi, G.: Assessment practices and beliefs: teachers' perspectives on assessment during long distance learning. Educ. Sci. **11**(6), 264 (2021)
3. Black, P., Wiliam, D.: Developing the theory of formative assessment. Educ. Assess. Eval. Account. **21**(1), 5–31 (2009)
4. Burkhardt, H., Schoenfeld, A.: Formative assessment in mathematics. In: Handbook of Formative Assessment in the Disciplines, pp. 35–67. Routledge, London (2019)
5. Bolondi, G., Ferretti, F., Gimigliano, A., Lovece, S., Vannini, I.: The use of videos in the training of math teachers: formative assessment in math. In: Integrating Video into Pre-Service and In-Service Teacher Training, p. 128 (2016)
6. Tornar, C.: Il processo didattico tra organizzazione e controllo. Monolite Editrice, Roma (2001)
7. Wiliam, D.: Formative assessment in mathematics Part 2: feedback. Equals: Math. Spec. Educ. Needs **5**(3), 8–11 (1999)

8. Wiliam, D.: Formative assessment in mathematics part 3: the learner's role. Equals: Math. Spec. Educ. Needs **6**(1), 19–22 (2000)
9. Leahy, S., Lyon, C., Thompson, M., Wiliam, D.: Classroom assessment: minute-by minute and day-by-day. Educ. Leadersh. **63**(3), 18–24 (2005)
10. Ferretti, F., Michael-Chrysanthou, P., Vannini, I.: Formative Assessment for Mathematics Teaching and Learning: Teacher Professional Development Research by Videoanalysis Methodologies. FrancoAngeli, Roma (2018)
11. Gagatsis, A., et al.: Formative assessment in the teaching and learning of mathematics: teachers' and students 'beliefs about mathematical error. Sci. Paedagog. Exp. **56**(2), 145–180 (2019)
12. Franchini, E., Salvisberg, M., Sbaragli, S.: Riflessioni sulla valutazione formativa tramite l'uso di video. Linee guida per formatori. SUPSI - Dipartimento formazione e apprendimento, Locarno (2016). https://rsddm.dm.unibo.it/wp-content/uploads/2019/01/famtl-linee-guida-web.pdf
13. Del Zozzo, A., Santi, G.: Theoretical perspectives for the study of contamination between physical and virtual teaching/learning environments. Didattica Della Matematica. Dalla Ricerca Alle Pratiche d'aula (7), 9–35 (2020). https://www.journals-dfa.supsi.ch/index.php/rivistaddm/article/view/91/133
14. Landis, J.R., Koch, G.G.: The measurement of observer agreement for categorical data. Biometrics **33**, 159–174 (1977)
15. Cusi, A., Morselli, F., Sabena, C.: Promuovere strategie di valutazione formativa in Matematica con le nuove tecnologie: l'esperienza del progetto FaSMEd. Annali online della didattica e della formazione docente **9**(14), 91–107 (2017)
16. Fandiño Pinilla, M.I.: Diversi aspetti che definiscono l'apprendimento e la valutazione in matematica. Pitagora, Bologna (2020)

Lessons Learned: Design of Online Video Co-creation Workshop for ESD

Shun Arima[✉], Fathima Assilmia, Marcos Sadao Maekawa, and Keiko Okawa

Keio University, 4-1-1 Hiyoshi, Yokohama, Japan
{arima-shun,assilmia,marcos,keiko}@kmd.keio.ac.jp

Abstract. This paper discusses the design of online video co-creation workshop for ESD (education for sustainable development). Our research team designed and practiced a workshop program for 112 Japanese high school students from July to November 2020. This workshop program aims for high school students to acquire abilities, attitudes, and key competencies emphasized in ESD through co-creating short videos promoting SDGs activity. Focusing on the four teams that were particularly effective from the quantitative data of the workshop, we will analyze the work process of those teams and summarize the design recommendations for the video co-creation workshop. This research explores the possibility of online video co-creation as an ESD method and contributes to sustainable ESD research even in the era of COVID-19.

Keywords: Education for sustainable development · Competencies · Workshop design · Student-generated video · Online education

1 Introduction

Education for Sustainable Development (ESD) is defined by UNSECO as follows: "Education for Sustainable Development empowers learners to make informed decisions and responsible actions for environmental integrity, economic viability and a just society for present and future generations, while respecting cultural diversity" [39].

ESD received high praise in The United Nations Decade of Education for Sustainable Development (DESD), which was adopted by the United Nations General Assembly in 2002, and UNESCO was designated as the lead agency for promotion. The final meeting of DESD, held worldwide from 2004 to 2014, was held in Nagoya, Japan, in 2014 [41]. After The Global Action Program (GAP) on ESD from 2015, "ESD for 2030" was adopted by UNESCO in 2019 as a successor to DESD, and it has become a new framework for these activities [43]. Initially, ESD was viewed primarily as environmental education. ESD has since expanded to all forms of education, including environmental, social, ethical, and cultural aspects. Currently, the importance of ESD to involve a worldwide range of stakeholders and change behavior for a sustainable future is explicitly recognized [40].

Supported by FUJIMIGAOKA HIGH SCHOOL for GIRLS.

B. Csapó and J. Uhomoibhi (Eds.): CSEDU 2021, CCIS 1624, pp. 536–559, 2022.
https://doi.org/10.1007/978-3-031-14756-2_26

In the field of ESD, an approach called the competence approach, which focuses on learning and acquiring students' "competency," is being studied. The German concept of "Gestaltungskompetenz" ("shaping competence") [13] is one example. The concept of Gestaltungskompetenz is "key competencies [. . .] required for forward-looking and autonomous participation in shaping sustainable development" [5] and competency approaches are a major factor in Japanese ESD research by Japanese researchers influenced from this German research [25].

Japan's ESD research and concrete efforts were summarized in 2012 in the final report by the Ministry of Education, Culture, Sports, Science and Technology (MEXT) [24]. This final report designed a framework of seven abilities and attitudes to emphasize in ESD and three related key competencies (Table 1). These three key competencies cite the key competencies defined in the OECD's The Definition and Selection of Key Competencies (DeSeCo) project [29, 37].

Table 1. Seven abilities and attitudes to be emphasized in ESD and three related key competencies [24] (Some English translations are by the author).

No	Abilities and attitudes	Specific action example	Key competency
1	Ability to think critically	Carefully consider, understand and incorporate the opinions and information of others	Use tools interactively
2	Ability to plan with anticipation of a future scenario	Make a plan with a perspective and a sense of purpose	
3	Multidimensional and integrative thinking	Think of various things in relation to each other	
4	Communication skills	Incorporate the opinions of others into one's thoughts	Interact in heterogeneous groups
5	Ability to cooperate with others	Work as a team while encouraging peers	
6	Respectful of relations and connections	Realizing that we exist thanks to various things	Act autonomously
7	Proactive participation	Willing to act for others	

According to UNESCO, these types of competencies are essential to achieving sustainable development [42]. In addition, by developing the abilities necessary for the sustainable future of the knowledge society called 21st-century skills within the framework of ESD, students will be able to make informed choices and act responsibly [46, 47].

In their review paper, Gonzalez et al.(2020) conclude that the development of 21st-century skills can be achieved by creating opportunities for practice, feedback, and formative assessment in a comprehensive curriculum design based on social constructivism, including active methodologies [15]. They also state that the most effective ways to teach 21st-century skills are to participate, reflect and generate new knowledge: co-design or participatory design.

Against this background, student video co-creation is a collaborative, participatory design that fosters ICT literacy [17,49] and can be expected to be effective as one of the competency approaches in ESD. Many studies have shown the educational effect of students actually creating videos[17,32].

These key competencies should not simply create a "checklist" of competencies but should specify the desired competencies based on the context and design and practice of the educational program [25]. The educational environment has changed significantly due to the influence of COVID-19 [33]. It is necessary to design and practice an ESD education program that students can acquire these key competencies even in the COVID-19 world context. However, as far as we know, few such studies have been found.

Our research team designed an online video co-creation workshop for ESD for 112 Japanese high school students from July to December 2020. In this online video co-creation workshop, high school students were divided into groups of 5–6 people and were given the task of co-creating videos to promote the activities of the SDGs. As mentioned in our last paper, through the co-creation activities of this video, we aim the participating students to acquire the seven abilities and attitudes and three key competencies set by MEXT.

The original purpose of this online video co-creation workshop is to acquire the seven abilities and attitudes and three key competencies, but we would like to explore the wide range of educational possibilities of students-generated videos and maker-centered learning. Therefore, in this paper, we will not only decide the survey items in advance and collect data, but also discuss the recommendations for designing a more effective online video co-creation workshop by collecting a wide range of data.

In our previous paper, a preliminary study examined the effectiveness of the video co-creation workshop based on the seven abilities and attitudes [2]. The description of the paper was limited to the overview of the workshop design, and the specific description of what kind of work was made and the activities of each team of students was limited.

In this paper, from the quantitative data for each team regarding the three key competencies, we will focus on the four teams that gave positive results in acquiring key competencies through video co-creation. We will analyze in more detail what kind of online video co-creation these four teams did from the content actually created by the students, the qualitative data of students survey, and the observation data of the actual workshop, and explore the possibilities of online video co-creation for more effective ESD.

This research explores the specific possibilities of video co-creation workshop design as a new online ESD method that is sustainable even in a pandemic era such as COVID-19, and contribute to research ESD and student-generated videos.

2   Literature Review

2.1   ESD

A review of the history of ESD research reveals a difference between researchers whose primary interest is in learning and educational processes and those whose primary interest is in the role that teaching and learning ultimately play in advancing sustainable development. The former often focus on the quality of learning and on building individual capacity (e.g., the ability to think critically, systematically, and reflectively) rather than encouraging specific social or environmental outcomes. The latter tends to be the opposite [38]. The competency approach is a study in the former vein.

The conventional Competence-based approaches tend to underestimate the importance of organizational and institutional change to achieve ESD due to inadequate consideration of human independence and personal and cultural differences in the identification of competencies. Mochizuki et al. (2010) points out the need to overcome such challenges [25].

Nagata (2017) argues that there are both "shallow ESD" and "deep ESD," the former stemming primarily from a widely shared interpretation of ESD that emphasizes overlap and connections with existing school subjects and types of education. The latter is necessary to avoid the loss of this dynamism and to maximize the potential of ESD. In order to achieve this "deep ESD," a comprehensive educational approach and system-level changes are essential to replace traditional approaches [27]. This Nagata's claim is in common with Mochizuki's point, suggesting the need for a more comprehensive educational approach.

The development of major competencies is based on both cognitive and non-cognitive dispositions and requires multiple contexts. Barth et al.(2007) argues that combining formal and informal learning environments in higher education as part of a new learning culture can provide diverse contexts and facilitate competency development [5].

Lozano et al. (2013) claim universities must effectively educate students of all ages with a better understanding of the needs of current and future generations as a leader in sustainable development [22]. In addition, Shulla et al. (2020) suggest that they may facilitate the efforts of the the Sustainable Development Goals(SDGs) by providing opportunities to incorporate the concept of ESD into the mainstream of society at various levels, including multi-stakeholder collaborative learning and informal learning [36].

Our workshop design in this paper also contributes to these areas as an example of how a university collaborates with multiple stakeholders to contribute to the ESD of teenage students (high school students, not university students). In addition, ESD through video creation can be expected to extend not only to formal learning but also to informal learning, and we think it is one of the approaches with great potential. The possibility of video co-creation in learning will be further reviewed in the next section.

2.2 Student-generated Videos in Education

Student-generated content is an educational approach by making various types of content, rather than simply passively consuming teaching materials. The content made by student-generated content is diverse. For example, PeerWise, an online tool that allows students to create, answer, and explain multiple-choice questions [6], a wiki-type open architecture software [45], and written explanatory videos and literature reviews [30]. In addition, Doyle et al. (2021) Show that co-creation of Student-generated content has a statistically significant impact on academic performance [14]. These studies have shown that student-generated content can support learning by supporting collaborative learning and increasing motivation to learn.

Sener (2007) states that for student-generated content to be more widely used in future online education, the fact that student-generated content improves the quality of learning is not enough [35]. He also argues that student-generated content needs to be connected to a larger purpose that emphasizes meeting emerging educational and social needs.

Therefore, exploring the possibility of student-generated videos for ESD in this paper may contribute to the expansion of the specifications of student-generated content in the future.

This growing attention to student-generated content is similar to the growing awareness of the educational potential of maker-centered learning approaches in recent years [11,44]. The focus on video creation as an educational method has increased significantly in the last decade [17]. One of the factors is a paradigm shift in video usage (nowadays, students of almost all ages use advanced digital video tools to create their digital videos and publish them to others) [7]. YouTube and social networking services (SNS) have evolved video beyond TV broadcasting and movies. Video has become one of the crucial methods of self-expression and communication in modern youth culture [10,23].

A 2012 report by Cisco Systems noted that student-generated videos are "a powerful tool for students" and that their use will increase in the 21st century [16]. Cayari et al. (2014) found that creating music videos provides students with a new way of self-expression, enables them to become better performers and consumers of the media, and encourages students to create music videos. By doing so, students will be able to acquire lifelong music production skills [9]. Huang et al.(2020) used student-generated videos to teach math. It has been shown that this activity helps to learn math better and improves communication skills, teamwork skills, and film making skills [18].

Hawley and Allen (2018) state that making videos has several benefits for students to support the development of digital and communication skills and enhance learning effectiveness. He also argues that incorporating this technology into new educational programs that have not yet been used is attractive to students and may help improve student satisfaction [17]. There is also a learning effect of collaboration and teamwork by co-creating videos as a group [1,34].

2.3 Online Collaboration in Education

The global pandemic of COVID-19 had a profound impact on students, instructors, and educational institutions around the world [4,50]. This pandemic has shifted the learning opportunities for many students from the campus to the online environment [8,12,20]. In this context, there is a need to gain more in-depth knowledge about re-designing online lessons that are adapted to the online environment, rather than simply copying classroom lessons [3,28]. Therefore, much research has been done on online collaboration under COVID-19 in the last year or two.

Xue et al. (2021) surveyed students in a large design fundamentals class about their online collaboration experience under the COVID-19 pandemic [48]. The survey found that there were more benefits (cognitive, communication, and emotional) than online collaboration challenges. On the other hand, the most frequently mentioned online collaboration challenge for students was communication. They also emphasize the need to consider how these benefits of online collaboration can be realized in face-to-face classrooms and instruction. On the other hand, under COVID-19, teachers found challenges in online collaboration in an interview with a science teacher who remotely teaches young teens in junior high school. The authors conclude that establishing a cognitive, social, and educational presence with young teens is a major challenge in distance learning [31].

Murai et al. (2021) designed and practiced a workshop where participants could work on maker projects remotely and give critical insights into the ethical implications of biowearable devices. Their research pointed a problem in the online environment as a lack of open-minded curiosity, with no visible progress in peers' work. On the other hand, there are merits the possibility of personalized facilitation through multiple communication channels, promotion of ideas and reflections in the student environment, and joint facilitation of reflection through back-channel communication [26].

Lee et al. (2021) designed and conducted an intergenerational participatory design session using an online synchronization session using video chat. They propose a framework of elements that must be considered in online sessions and advocate the need for improvisation to anticipate and respond to confusion [21]. Jayathirtha et al.(2020) have been developing physical deliverables, schematics, coding, etc. offline, but changed them to online courses during the pandemic. The study reports that asynchronous interactions online made it difficult to share physical productions, resulting in an online course focused on schematics and coding. The lack of physical interaction also affected the depth of inquiry that teachers make learners [19].

Like these studies, ours can contribute to these research areas with insights into online collaboration, with a particular focus on video co-creation.

3   Workshop Design

This section explains the specific contents of the video co-creation workshop designed by our research team. Some of the content has been described in our previous paper, so please refer to that as well [2].

3.1   Context

The participants of this workshop were 112 first-year students from a private high school in Tokyo, Japan (Japanese high schools correspond to grades 10–12 of K-12). This private high school and our research team have been working together for several years. Our research team designs and practices educational programs (120 min per session, eight sessions per year) on SDGs, Global citizenship for the students. We conducted this video co-creation workshop using four sessions of this educational program. Our research team of designing and facilitating workshops consists of 6–10 university faculty and graduate students, some of whom have previously had video creation experience. The purpose of this educational program is to help students acquire the literacy and competencies they need for sustainable development and global citizenship through active learning about these.

This video co-creation workshop was held once a month with six sessions (One extra session. The first and sixth sessions used only part of the session time) from July to December 2020 (excluding August during the summer vacation, twice in November). Each session consisted of a 120 min class, with some teams not completing their work within the class time doing the rest of the work as out-of-class homework.

We conducted most of all workshop activities in an online environment due to the influence of the COVID-19 pandemic. However, because students were allowed to go to school while limiting the number of students, some teams' video production work, such as shooting and editing, was done offline (in high school). Participating students participated in the workshop from home or in the classroom at school (while ensuring social distance). All facilitation team members participated in the workshop from their own homes.

All participants' students own a Microsoft Surface, and most of the workshop activities were conducted using the Microsoft Surface. We also used Zoom, an online video conferencing system, for all workshop communications. Zoom has a breakout room feature that divides participants into multiple smaller groups, which we took advantage of when working in groups. Students are restricted from using smartphones in regular high school classes, but they were allowed to use them when shooting and editing videos at this workshop.

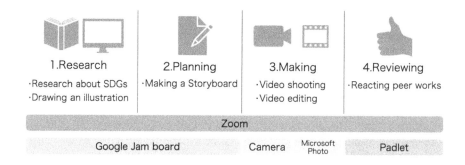

Fig. 1. Workshop flow in four phases and the tools used for each phase [2].

3.2 Goal and Flow

The goal of this workshop was to have students make short videos to promote the activities of SDGs. SDGs and ESD are closely related, and in particular, the SDGs#4 set out to provide everyone with quality education as a means of achieving the remaining SDGs. Among them, ESD is an indispensable part of Target 4.7. The multidimensional aspects of ESD related to the SDGs are strongly related to today's complex problems [36], and learning about the SDGs can be expected to become an ESD approach.

There are several other benefits to setting SDGs as a theme. One is that the SDGs have a lot of data and materials on the Internet and are a common topic worldwide.

Therefore, participants can make videos with a wide range of perspectives and abundant materials. In addition, since the SDGs have various goals, we can expect that various videos will be produced for each team.

The flow of this workshop consisted of four phases: 1. Research 2. Planning 3. Making 4. Reviewing (Fig. 1). In each phase, students need to use the tools, communicate with other team members, and complete the video within a set time limit throughout the workshop. Therefore, participating students can be expected to acquire three major abilities. The original purpose of this video co-creation workshop is to acquire three key competencies, but as mentioned in Chapter 2, we will also explore the wide range of educational possibilities of students-generated videos and maker-centered learning.

3.3 Research

In 1. Research phase, students investigate SDGs and understand the issues set by SDGs and their background. After that, as the first step of co-creation work in the group, students express concrete actions that they can realize under the title of "Summer Vacation Promise" based on the results of the investigation on SDGs by drawing an illustration in Google Jamboard (Fig. 2). This task is also an exercise in creating a storyboard for the next part. In this part, the first goal is to familiarize students with collaborative work in an online environment. students set up tasks that can be completed in a short amount of time, rather than suddenly creating a video, so that students can practice communication and collaboration.

Fig. 2. "Summer Vacation Promise" drawn by the students.

3.4 Planning

In the second phase, 2. Planning, students create a storyboard in groups as a plan to create a short video promoting the SDGs based on the information obtained from 1. Research phase. In this task, students use Google Jamboard to think about the specific content, structure, and expression of the video. Students can insert images on the Internet that can be searched by Google Image Search into Google Jamboard, so even if students are not good at drawing, they can make a storyboard.

The facilitation team showed some reference video works to help the students actually come up with video ideas. These reference works also included videos produced in a remote environment (such as Commercial Message videos produced under COVID-19).

Our facilitation team has prepared a sample storyboard on Google Jamboard to make it easier for students to work with (Fig. 3). This storyboard has a space to write the main image, the description of the scene, narration and dialogue. The sample work was made based on the famous Japanese folk tale "Momotaro".

Fig. 3. Sample storyboard on Google Jamboard.

3.5 Making

In the third phase, 3. Making, students shoot and edit videos based on the storyboard they created during 2. Planning phase. The shooting was done using a Microsoft Surface camera or a student's personal smartphone camera, and editing was done using Microsoft's "Photo" pre-installed on Microsoft Surface. Regarding how to use Photo, we used Zoom's screen sharing function and gave a lecture to students on how to use it while showing the operation screen. However, because we do not need restrictions on the use of tools, some teams used the editing software they were accustomed to (such as Inshot[1] and CapCut[2]).

As mentioned in Chap. 2 and our previous paper [2], the facilitation team decided that there was no need to explain how to use the camera, as the students have routinely experienced video shooting in their culture. Instead, we asked the students to take pictures of paper cranes (traditional Japanese paper crafts) so that they could enjoy shooting various compositions. In this work, students devised the composition of the photograph and aimed to make the folded paper cranes look smaller or larger (Fig. 4).

In the online environment using Zoom, it is difficult for facilitators to check the progress of each team after each team moves to the breakout room. Therefore, during this phase, the facilitation teams shared progress reports to support students' work. In

---

[1] https://inshot.com/.

[2] https://www.capcut.net/.

this progress report, each facilitator patrolled the breakout room of the team in charge and expressed the progress of each team in colors (5 colors according to the progress) and short sentences (Fig. 5). Based on the report, facilitators with abundant video creation experience made additional follow-ups.

As a result of this progress report, some teams were expected to not finish their work with the planned number of lesson sessions, so we provided one additional session for students.

Fig. 4. Pictures of paper cranes by students [2].

Room	Facilitator	Jamboard URL	Progress
Group 1			Editing almost finished. Already decided on the final message and continue working.
Group 2			Storyboard is finished. Plan to shoot together at school.
Group 3			Still working on the storyboard
Group 4			Storyboard is finished and already started delegating parts. Plan to shoot on Monday.
Group 5			Editing the video, while finishing the storyboard
Group 6			Will start editing soon; Previously they said they didn't have any video, so they would use voice only (?)

Fig. 5. Progress report by facilitators.

3.6 Reviewing

In the final phase, 4. Reviewing, students watch and react (posting comments or pushing the like button) to peers' videos (Fig. 6). In this phase, we asked students to post team's videos on Padlet[3] and have each student review it in an asynchronous environment. This workshop did not adopt the format of presenting the students' finished

---

[3] https://padlet.com/.

works during the session in real time. The excitement and emotion of presenting a work to friends in real time is one of the attractions of video production learning. However, in an online environment, if we play a video while sharing it in real time, it will not play smoothly. Therefore, this time we adopted the submission of works in an asynchronous environment.

Padlet allowed students to upload content, post comments, and push Like buttons without the need to register for an account. Furthermore, it is possible to make a limited share only for those who know the URL. For these reasons we used this tool in this workshop.

Fig. 6. Students' video work uploaded to Padlet [2].

4   Results: Quantitative Data

4.1   Methods

In this workshop, we surveyed quantitative data on the three key competencies of the participating students. This survey was held twice, the 1st and final session of this workshop. We got responses from 69 out of 112 participating students. The survey asked whether they had three key competencies: Use tools interactively, Interact in heterogeneous groups, and Act autonomously. Students can answer with seven frequencies, 1 is a strong disagreement, 7 is a strong agreement, and 4 is neutral (Fig. 7 and Fig. 8).

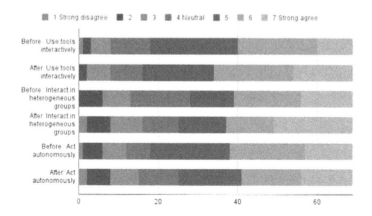

Fig. 7. Data on whether students have three key competencies before and after the workshop.

	1	2	3	4	5	6	7
Before : Use tools interactively	1	2	5	10	22	20	9
After: Use tools interactively	0	2	6	8	18	20	15
Before : Interact in heterogeneous groups	0	6	7	15	11	17	13
After: Interact in heterogeneous groups	2	6	8	9	12	12	20
Before : Act autonomously	1	5	6	6	20	19	12
After: Act autonomously	2	6	7	10	16	15	13

Fig. 8. Specific numbers of people for each answer in the graph in Fig. 7.

4.2   About Data

Figure 7 shows the responses for each competency of all students who responded to the survey. Regarding Use tools interactively, positive ratings (6–7) increased after the workshop. On the other hand, regarding Interact in heterogeneous groups, both positive and negative evaluations (1–3) increased slightly, and polarization increased. As for Act autonomously, negative evaluations increased slightly, positive evaluations decreased somewhat, and neutral responses increased somewhat. None of all changes were significant enough to make a statistically significant difference.

Next, we focused on the amount of change in each key competency before and after the workshop and summarized these data for the 20 teams that co-created the video (Fig. 9). The data showed extreme results for each team, with four teams (Team1, Team8, Team12, Team13) showing positive changes in all items and five teams (Team2, Team3, Team7, Team15, Team18) showing negative changes in all items. As shown in the Fig. 7, the overall trend was that Use tools interactively increased slightly, and Interact in heterogeneous groups and Act autonomously decreased somewhat.

From these quantitative data, we pay particular attention to the four teams (Team1, Team8, Team12, Team13) that gave particularly positive results. In the next section, we will show the work of these teams, the contents of the completed video content, and the survey by the team members as qualitative data.

5   Results: Qualitative Data

5.1   Methods

In this workshop, we surveyed qualitative data from students in open-ended questions five times after each session (excluding one additional session). The students also answered their team numbers, which allowed them to connect which team's data each survey data was. In addition, since all the storyboard data and video data described later are uploaded on the cloud, we refer to those data. This section first gives an overview of the video work of all teams and then describes the creation of each of the four teams

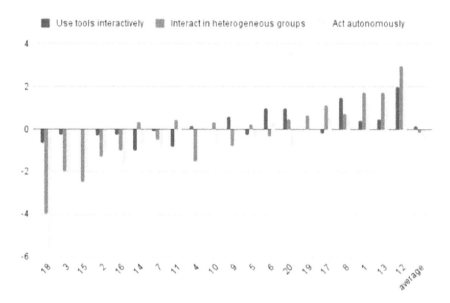

Fig. 9. Changes in each key competency before and after the workshop for each team.

mentioned above and the content of the video content in more detail. We will also explore the state of video co-creation in each team from the survey of team members.

5.2   Over View of Students Video Works

Participating students completed a total of 20 videos at this workshop. Students created each video with a different theme and production style for each team (Table 2).

Each team could select the theme of the video from the goals of SDGs. In particular, as you can see from the table, many teams chose the theme of marine pollution and plastic waste related SDGs#14. The reasons for this are that students had an opportunity to investigate and learn about marine pollution in their other classes before this workshop. Furthermore, it seems that it is related to the fact that plastic shopping bags were charged from July 2020 due to the revision of Japanese law and became a hot topic for many Japanese people. Three other teams chose LGBT. LGBT is not directly set as a goal of the SDGs, but we interpreted it as one of reducing inequality within and among countries of SDGs#10.

There were several video production styles. Online is the case where all video production work is completed online. Onsite is the case where all video production work is completed onsite. Hybrid is a combination of online and onsite work. As mentioned in Chap. 3, students were allowed to go to high school with a limited number of people, so some teams shot and edited on-site at high school.

Table 2. List of videos by students.

Team No	Video theme	Related SDGs No	Making style
1	Marine pollution	14	Online
2	Plastic waste reduction	14	Hybrid
3	Water problem	14	Hybrid
4	Marine pollution Plastic waste reduction	14	Hybrid
5	Hunger problem	2	Hybrid
6	Littering problem	12	Online
7	Water problem and Plastic waste reduction	14	Online
8	Fundraising	1	Onsite
9	Power saving	7	Onsite
10	Marine pollution	14	Online
11	Marine pollution	14	Hybrid
12	Food loss	2	Online
13	Racial equality	10	Hybrid
14	LGBT	10	Online
15	LGBT	10	Online
16	LGBT	10	Onsite
17	Plastic waste reduction	14	Online
18	Plastic waste reduction	14	Onsite
19	Fundraising	1	Onsite
20	Littering problem	12	Onsite

5.3 Team1

Fig. 10. Team1's video.

Team 1 made a video about marine pollution and plastic wasteo (Fig. 10). In this video, the video of the sea is shown first. The text would indicate the number of marine resources there, followed by the ocean where the plastic garbage floats and the text would indicate that there is more plastic garbage than the marine resources. At the middle of the video, An impressive photo of a whale that has swallowed trash and died is shown, and at the end, you see the message "Don't suffer anymore" with an original logo. In this video, the sound of quiet waves continues to flow as a sound effect. This video is a "quiet and powerful" video that simply conveys the current problem and finally conveys a message.

At the stage of making the storyboard, their team had already used the impressive photo of a whale, which is also used in the final video (Fig. 11). A survey of Team 1 members shows that there were some problems with teamwork early on. "team members turned off the camera and microphone, and when I talk something, there was no response, and I didn't know what to do..." (Team 1 student A)

Fig. 11. Team1's Storyboard.

However, when the video was finally completed, all team members recognized good teamwork like that "Our group's teamwork was good" (Team 1 student B) and "I was able to work together in a group" (Team 1 student C).

Student A, who was anxious about teamwork at the beginning, answered in a survey, "I enjoyed making group marks with everyone in the group." (Team 1 student A). This mark is an impressive mark with a message that appears at the end of the video. In Team1, the co-creation of this mark is seemed to have helped improve teamwork.

On the other hand, regarding video production, there were some comments, "I thought it was difficult to edit" (Team 1 student D) and "It was difficult to share the sound sources and images found by the members" (Team 1 student A). it seems that the members of Team1 felt difficulty overall.

5.4 Team8

Team 8 created a video on the theme of fundraising (Fig. 12). Their team did not make any significant changes to the plan from the storyboard stage, and Team8 completed it. They did all shoot at school, and in the editing, they stitched the multiple shots taken together to add music and sound effects.

Fig. 12. Team8's video.

They connected cuts taken in different places on campus to represent fundraisers in other locations by different people. The feature of this video is that the faces of the characters are not shown. This video style has a meaning as a video expression, but it also effectively solves the privacy problem described later. Team 8 was a production style categorized as Onsite, but other teams that were classified as Onsite some works that combined multiple live-action cuts like Team 8 and some video works that had a story like a drama.

The feature of this video is that the beginning part and end part of the video taken in the first-person view. In fact, in Team 8 storyboards, They decided in advance to shoot from the first-person view (Fig. 13). This first-person shoot seems to be influenced by the reference video work[4] presented by the facilitation team. In the student survey, there was a comment that "The facilitation team showed us various videos, which was very helpful." (Team8 Student A).

---

[4] https://youtu.be/qHFr1_md3Ok.

Fig. 13. Team8's Storyboard.

Also, in the Team 8 video, a donation box for tangible props is made. Not only shooting and editing but also making props, etc., may allow students who are not good at digital work to share the team's task well. Actually, as a survey of a member of Team 8, "I'm glad that all the team members were able to work in their own fields such as video editing and ideation." (Team 8 Student B)

5.5   Team12

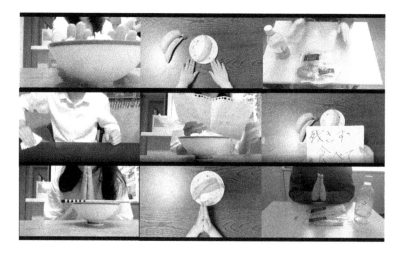

Fig. 14. Team12's video.

Team 12 filmed a meal scene at the home of each member of the team, expressing a reduction in food loss (Fig. 14). Most of the online-style teams were collages of existing materials(such as photos, videos), but Team 12 was a video production that maximized the appeal of collaboration.

This video has the following structure: each meal cuts are taken at each member's home and at each meal. At first, the members leave meals. Then one student throws a letter to the other students elsewhere. The letter contains a message (in Japanese) to eat everything to reduce food loss, and each students who receive the letter eat everything. At the end, the cuts that everyone thumbs up are projected in succession, giving a positive impression.

Team 12 also completed the video without significant changes from the storyboard (Fig. 15). The feature of Team 12's video, which is different from other teams, is that all members are shooting. For example, the Team 8, mentioned above, one person in charge of shooting can make a video. Team 1 mentioned above made a video using only existing materials without shooting a video. Team 12's all members were able to commit to shooting because all members were shooting in their own homes and school classrooms. There was a comment from a survey of a member of Team 12 that "it was interesting because it was a video taken by ourselves" (Team 12 Student A). There is a possibility that we can provide more fun to the students by all members shooting.

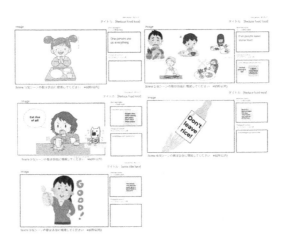

Fig. 15. Team12's Storyboard.

Fig. 16. Team13's video.

5.6   Team13

Team 13 made a video by combining the shooting in the school and the material on the net. Team 13 metaphors the difference in skin color with reversi games and piano keyboards under the theme of racial equality. First, the video showed the reversi and piano in only white and showed that we couldn't play games or play the piano in one color. After that, the video showed normal reversi and piano scenes expressing we can play reversi games and piano because we have various colors.

It then shows several images of different colors and people of different races communicating with each other and concludes with the catchphrase, "Every color is beautiful." (Fig. 16)

Team 13, like the other three teams we're focusing on this time, hasn't changed much from the storyboards (Fig. 17). In particular, They decided on specific ideas and text expressions at the storyboard stage, and it can be seen that They made the video with sufficient planning. In addition, their storyboards also included ideas (chess ideas) that were not used in the final video and prepared a wealth of ideas.

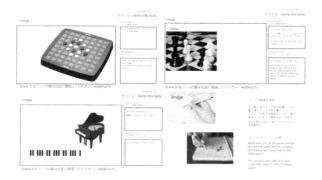

Fig. 17. Team13's Storyboard.

The video of Team 13 was very good at using different sounds. At the beginning of the video, they used the lonely sound of the piano, and in the latter half, which leads to "Every color is beautiful," I chose exhilarating music that gives a positive impression. In fact, the team members commented on the sound. "By adding sound to what we just shot, I enjoyed watching it as if I was watching a movie." (Team13 Student A)

In addition, Team 13 was the only of the four teams we focused on this time, and the work was not completed within the deadline and participated in an additional session. In fact, the members also picked up "Collecting and editing videos of all the group in time" (Team13 Student A) as a difficult thing in the workshop's task.

6   Discussions

This section discusses recommendations for designing an online video co-creation workshop for more effective ESD based on the data from the four teams mentioned above.

The four teams we focused on did not make significant changes to the storyboard until the video was completed. As mentioned in the previous paper [2], some teams started shooting and editing without completing the storyboard. Still, from this analysis, it is important to support the students to make the storyboard properly. In this workshop, the facilitation team prepared sample storyboards to help the students, but we also needed support to make the storyboards. For example, like Team13, writing a wealth of ideas at the storyboard stage, regardless of whether or not you shoot, can be expected to enhance the subsequent shooting and editing work. As you can see from the Team 8 example, ideas can come from reference works. It is conceivable to show many references works to create abundunt ideas at the stage of a storyboard.

In the context of COVID-19, technology, knowledge, and information competencies for the use of remote communication tools and co-creation in remote environments will be required. Reviewing the works, we have achieved sufficient video co-creation even in a completely online remote environment (Team1 Team12 works, etc.). Video co-creation has shown the potential to contribute to the competency of interactively using tools in the COVID-19 context.

At this workshop, all the students had a unified environment (Microsoft Surface), so there was no significant confusion. However, please note that the setting of each participant may be different when doing it in other contexts. Also, although in the same environment, some groups used the editing software they were accustomed to. The flexibility of these tools (giving them the freedom to choose which tools to use) is vital to drive student creativity. In other words, facilitators do not need to ultimately unify the tools but need to guarantee the minimum feasible environment.

As mentioned in our previous paper [2], in this workshop, we randomly created 20 groups of 5–6 people to form a group of students who met for the first time as much as possible and asked them to co-create the video. As a result, all groups have completed the video, which is expected to provide the opportunity for minimal cooperation and conflict resolution competencies. However, on the other hand, there were problems such as the work being biased toward some students (See previous paper for details).

The style in which all members shoot like Team 12's work is effective in preventing such freeride. By setting a rule that uses the material all members shot, you can expect the task to be distributed to all members. It is also possible that the facilitator will set up various types of work (not just digital work), such as making props, as in the case of Team 8. However, such tangible work is difficult to set up in a fully online environment and can only be achieved in a hybrid or onsite environment.

One of the advantages of video production is that once we create content, we can expand that content in various ways. However, with our practices this time, it was challenging to realize expanded outside this workshop.

The first reason is copyright. As you can see from the content of the students' videos, students used a great variety of media materials for their videos. Under Japanese copyright law, the copyright holder's permission is not required when using a copyrighted work for educational purposes in the classroom. However, copyright restrictions are more stringent when published outside of school. There is a dilemma that production cannot be done smoothly if copyright is a concern. This point is based on the premise that it will be released outside the school from the beginning, and it will be decided whether to use materials that do not have copyright problems or to use materials regardless of copyright restrictions instead of keeping them open to the public inside the school is required.

The second reason is privacy issues. In this video, some of the students' faces were shown, and some of their homes were shown in the video. Sharing videos within the school is fine, but if these videos were released, there would be a risk of identifying the addresses and personal names of minor students. This is also a problem caused by making it in an online environment. For example, if all the shoots are done in school like Team8's video, these risks are mitigated.

For either reason, it is suggested that publishing the work in an environment where everyone can watch it online indiscriminately is not suitable. For example, even if the works are expanded outside the school, we think it is suitable to perform it in an environment where it is possible to have a screening and interactive communication where detailed supplementary explanations are possible.

7 Conclusion

Our research team designed an online video co-creation workshop for ESD for 112 Japanese high school students from July to December 2020. This workshop program aims for high school students to acquire abilities, attitudes, and key competencies emphasized in ESD through co-creating short videos promoting SDGs activity. In this paper, from the quantitative data for each team regarding the three key competencies, we focused on the four teams that gave positive results in acquiring key competencies through video co-creation and analyzed in more detail their actual online video co-creation.

This research contributes to research ESD and student-generated videos through exploring the specific possibilities of video co-creation workshop design as a new online ESD even in a pandemic era such as COVID-19.

References

1. Alpay, E., Gulati, S.: Student-led podcasting for engineering education. Eur. J. Eng. Educ. **35**(4), 415–427 (2010)
2. Arima, S., Assilmia, F., Maekawa, M.S., Okawa, K.: Design and practice of ESD in high school in Japan through online video co-creation workshop. In: CSEDU (1), pp. 640–647 (2021)
3. Arima, S., Yasui, M., Okawa, K.: Re-design classroom into MOOC-like content with remote face-to-face sessions during the COVID-19 pandemic: a case study in graduate school. In: Proceedings of the Eighth ACM Conference on Learning @ Scale, pp. 299–302. L@S 2021, Association for Computing Machinery, New York, NY, USA (2021). https://doi.org/10.1145/3430895.3460163
4. Aristovnik, A., Keržič, D., Ravšelj, D., Tomaževič, N., Umek, L.: Impacts of the COVID-19 pandemic on life of higher education students: a global perspective. Sustainability **12**(20), 8438 (2020)
5. Barth, M., Godemann, J., Rieckmann, M., Stoltenberg, U.: Developing key competencies for sustainable development in higher education. Int. J. Sustain. High. Educ. (2007)

6. Bates, S.P., Galloway, R.K., McBride, K.L.: Student-generated content: using PeerWise to enhance engagement and outcomes in introductory physics courses. In: AIP Conference Proceedings, vol. 1413, pp. 123–126. American Institute of Physics (2012)
7. Burgess, J., Green, J.: YouTube: Online Video and Participatory Culture. Wiley, Hoboken (2018)
8. Bylieva, D., Bekirogullari, Z., Lobatyuk, V., Nam, T.: Analysis of the consequences of the transition to online learning on the example of MOOC philosophy during the Covid-19 pandemic. Humanit. Soc. Sci. Rev. **8**(4), 1083–1093 (2020)
9. Cayari, C.: Using informal education through music video creation. Gen. Music Today **27**(3), 17–22 (2014)
10. Chau, C.: Youtube as a participatory culture. New Dir. Youth Dev. **2010**(128), 65–74 (2010)
11. Clapp, E.P., Ross, J., Ryan, J.O., Tishman, S.: Maker-Centered Learning: Empowering Young People to Shape their Worlds. Wiley, Hoboken (2016)
12. Crawford, J., et al.: COVID-19: 20 countries' higher education intra-period digital pedagogy responses. J. Appl. Learn. Teach. **3**(1), 1–20 (2020)
13. De Haan, G.: The BLK '21'programme in Germany: a 'Gestaltungskompetenz'-based model for education for sustainable development. Environ. Educ. Res. **12**(1), 19–32 (2006)
14. Doyle, E., Buckley, P., McCarthy, B.: The impact of content co-creation on academic achievement. Assess. Eval. High. Educ. **46**(3), 494–507 (2021)
15. González-Salamanca, J.C., Agudelo, O.L., Salinas, J.: Key competences, education for sustainable development and strategies for the development of 21st century skills. a systematic literature review. Sustainability **12**(24), 10366 (2020)
16. Greenberg, A.D., Zanetis, J.: The impact of broadcast and streaming video in education. Cisco Wainhouse Res. **75**(194), 21 (2012)
17. Hawley, R., Allen, C.: Student-generated video creation for assessment: can it transform assessment within higher education? (2018)
18. Huang, M.C.-L., et al.: Interest-driven video creation for learning mathematics. J. Comput. Educ. **7**(3), 395–433 (2020). https://doi.org/10.1007/s40692-020-00161-w
19. Jayathirtha, G., Fields, D., Kafai, Y.B., Chipps, J.: Supporting making online: the role of artifact, teacher and peer interactions in crafting electronic textiles. Inf. Learn. Sci. (2020)
20. Jena, P.K.: Impact of COVID-19 on higher education in India. Int. J. Adv. Educ. Res. (IJAER) **5** (2020)
21. Lee, K.J., et al.: The show must go on: a conceptual model of conducting synchronous participatory design with children online. In: Proceedings of the 2021 CHI Conference on Human Factors in Computing Systems, pp. 1–16 (2021)
22. Lozano, R., Lukman, R., Lozano, F.J., Huisingh, D., Lambrechts, W.: Declarations for sustainability in higher education: becoming better leaders, through addressing the university system. J. Clean. Prod. **48**, 10–19 (2013)
23. Madden, A., Ruthven, I., McMenemy, D.: A classification scheme for content analyses of YouTube video comments. J. Documentation **69**(5), 693–714 (2013)
24. MEXT: A guide to promoting ESD (education for sustainable development) (first edition). https://www.mext.go.jp/component/english/__icsFiles/afieldfile/2016/11/21/1379653_01_1.pdf (2016). Accessed 26 Oct 2021
25. Mochizuki, Y., Fadeeva, Z.: Competences for sustainable development and sustainability: significance and challenges for ESD. Int. J. Sustain. High. Educ. (2010)
26. Murai, Y., et al.: Facilitating online distributed critical making: lessons learned. Association for Computing Machinery, New York, NY, USA (2021). https://doi.org/10.1145/3466725.3466759
27. Nagata, Y.: A critical review of education for sustainable development (ESD) in Japan: beyond the practice of pouring new wine into old bottles. Educ. Stud. Jpn. **11**, 29–41 (2017)

28. Nambiar, D.: The impact of online learning during COVID-19: students' and teachers' perspective. Int. J. Indian Psychol. **8**(2), 783–793 (2020)
29. OECD: The definition and selection of key competencies: executive summary. https://www.oecd.org/pisa/35070367.pdf (2005). Accessed 26 Oct 2021
30. Pirhonen, J., Rasi, P.: Student-generated instructional videos facilitate learning through positive emotions. J. Biol. Educ. **51**(3), 215–227 (2017)
31. Rannastu-Avalos, M., Siiman, L.A.: Challenges for distance learning and online collaboration in the time of COVID-19: interviews with science teachers. In: Nolte, A., Alvarez, C., Hishiyama, R., Chounta, I.-A., Rodríguez-Triana, M.J., Inoue, T. (eds.) CollabTech 2020. LNCS, vol. 12324, pp. 128–142. Springer, Cham (2020). https://doi.org/10.1007/978-3-030-58157-2_9
32. Reeves, T., Caglayan, E., Torr, R.: Don't shoot! understanding students' experiences of video-based learning and assessment in the arts. Video J. Educ. Pedagogy **2**(1), 1–13 (2017)
33. Reimers, F.M., Schleicher, A.: A framework to guide an education response to the COVID-19 pandemic of 2020. OECD Retrieved April, **14**(2020), 2020–2024 (2020)
34. Ryan, B.: A walk down the red carpet: students as producers of digital video-based knowledge. Int. J. Technol. Enhanced Learn. **5**(1), 24–41 (2013)
35. Sener, J.: In search of student-generated content in online education (2007)
36. Shulla, K., Filho, W.L., Lardjane, S., Sommer, J.H., Borgemeister, C.: Sustainable development education in the context of the 2030 agenda for sustainable development. Int. J. Sustain. Dev. World Ecol. **27**(5), 458–468 (2020)
37. Sleurs, W.: Competencies for ESD (Education for Sustainable Development) teachers: a framework to integrate ESD in the curriculum of teacher training institutes. CSCT-project (2008)
38. Sterling, S., Warwick, P., Wyness, L.: Understanding approaches to ESD research on teaching and learning in higher education. In: Routledge Handbook Of Higher Education for Sustainable Development, pp. 113–123. Routledge (2015)
39. UNESCO: Learning for a sustainable world: review of contexts and structures for education for sustainable development (2009)
40. UNESCO: Shaping the future we want: un decade of education for sustainable development (2005–2014). Final Report (2014)
41. UNESCO: Shaping the future we want: un decade of education for sustainable development; final report. https://sustainabledevelopment.un.org/content/documents/1682Shapingthefuturewewant.pdf (2014). Accessed 25 Jan 2021
42. UNESCO: Rethinking education. towards a global common good? (2015). http://unesdoc.unesco.org/images/0023/002325/232555e.pdf. Accessed 26 Oct 2021
43. UNESCO: Education for sustainable development: a roadmap (2020). https://unesdoc.unesco.org/ark:/48223/pf0000374802. Accessed 25 Jan 2021
44. Valente, J.A., Blikstein, P.: Maker education: where is the knowledge construction? Constructivist Found. **14**(3), 252–262 (2019)
45. Wheeler, S., Yeomans, P., Wheeler, D.: The good, the bad and the wiki: evaluating student-generated content for collaborative learning. Br. J. Edu. Technol. **39**(6), 987–995 (2008)
46. Wiek, A., Withycombe, L., Redman, C.L.: Key competencies in sustainability: a reference framework for academic program development. Sustain. Sci. **6**(2), 203–218 (2011)
47. Willard, M., Wiedmeyer, C., Warren Flint, R., Weedon, J.S., Woodward, R., Feldman, I., Edwards, M.: The sustainability professional: 2010 competency survey report. Environ. Qual. Manage. **20**(1), 49–83 (2010)
48. Xue, F., Merrill, M., Housefield, J., McNeil, T.: Beyond isolation: Benefits and challenges as perceived by students throughout online collaboration during the COVID-19 pandemic. In: Companion Publication of the 2021 Conference on Computer Supported Cooperative Work and Social Computing, pp. 195–198 (2021)

49. Zahn, C., et al.: Video clips for YouTube: collaborative video creation as an educational concept for knowledge acquisition and attitude change related to obesity stigmatization. Educ. Inf. Technol. **19**(3), 603–621 (2014)
50. Zhu, X., Liu, J.: Education in and after COVID-19: immediate responses and long-term visions. Postdigital Sci. Educ. **2**(3), 695–699 (2020). https://doi.org/10.1007/s42438-020-00126-3

Formative Assessment in LDL Workshop Activities: Engaging Teachers in a Training Program

Camilla Spagnolo[1] (✉) ⓘ, Rita Giglio[1] ⓘ, Sabrina Tiralongo[2] ⓘ, and Giorgio Bolondi[1] ⓘ

[1] Free University of Bozen, Bozen, Italy
{camilla.spagnolo,rita.giglio,giorgio.bolondi}@unibz.it
[2] MIUR, Rome, Italy
sabrina.tiralongo@libero.it

Abstract. The pandemic crisis that affected us in December 2019 is still reflected in teacher training. In this article, we describe a teacher-training experiment focused on the use of large-scale assessment in a formative perspective and in a laboratorial distance teaching setting. Between March and May 2020, during the first period of long-distance learning that affected Italy, we implemented a long-distance teacher professional development designed for teachers of all school levels. This training was structured into 16 webinars involving 2539 Italian teachers coming from different geographical areas (north, center and south Italy). At the beginning of the 2020/2021 school year, a follow-up questionnaire was developed and implemented. One of the purposes of this questionnaire was to clarify how this experience impacted teachers' beliefs and practices. As a result, we find that our long-distance teacher training helped resilient teachers understand the importance of technology and reflect on the importance of formative, not just summative, assessment.

Keywords: Mathematics education · Long-distance teaching and learning · Formative assessment

1 Introduction and Rationale of the Paper

This paper is a follow-up of a paper presented at the CSEDU 2021 conference [1] where a teacher training experimental program developed during the first months of the Covid 19 pandemic emergency was described. In this paper we reconsider that experience in a wider framework, presenting a more detailed description of the program. Moreover, we present more data, focusing on some results that can be interpreted in terms of teacher's identity. In fact. in the meanwhile, what happened all over the world has had an impact on research in Mathematics Education and a large amount of research has been performed in a completely new and unexpected setting. In particular, Mathematics teachers' identity has been re-considered as a construct that may help in describing the long-term impact of this epochal event.

© Springer Nature Switzerland AG 2022
B. Csapó and J. Uhomoibhi (Eds.): CSEDU 2021, CCIS 1624, pp. 560–576, 2022.
https://doi.org/10.1007/978-3-031-14756-2_27

Hence, this paper is positioned at the crossroad of three timely issues of Education, and Mathematics Education in particular- all of them have been heavily hit by the emergency. The first issue is institutional-systemic: how to foster the integration of the frameworks, the materials and the results of Large-Scale Assessments (LSA) into the classroom experience of the teachers [2]. This issue is also at the heart of hard debates in different countries, in particular in Italy, the context of this study [3–5]. The second issue is a research perspective. It is important for a researcher to have access to teachers' beliefs, taking into account and going beyond what is declared. *Assessment* is a key component of teachers' identity and researchers can use it as an helpful access door to teachers' choices and practices; moreover, it is perceived as such by the students. The third issue is a topical one. The pandemic fostered a forced and quick digitization of teaching and learning all over the world. Teaching and learning practices have been upset, and many theoretical constructs used to analyse, interpret and plan educational actions have been reconsidered in a completely new scale- and formative assessment (FA) in particular.

Our paper describes an in-service teaching professional development program where a "traditional" face-to-face program, based on materials from LSAs, has been switched to a LD program, and all the practical issues related to the implementation in the classrooms of the contents of the program have been reconsidered, by the participants, in the frame of the new emergency context. Our research hypothesis is that "assessment" is a crucial topic which makes "transparent" teachers' behaviours and allows teachers' attitudes and beliefs to emerge.

2 Elements for a Theoretical Background

2.1 Forced Digitization

The context of our study is Italy, a country where there is traditionally a resilience in the use of technologies in the classroom. The pandemic forced an acceleration of their use since most teachers were forced to use Long-distance learning (LDL) settings.

Long before the pandemic many teachers and researchers were aware of the fact that digital learning was reshaping education [6], but the consequences of the sudden switch to LDL on a dramatical larger scale all over the world are yet to be measured and understood [7, 8]. A huge amount of research is currently devoted to collecting experiences, descriptions, interpretation. Many journals are publishing special issues addressing specific topics related to mathematics education during Covid time ([9], and many others in press). In particular, this digitization impacted on the relationships between teachers and students, on the interaction between teachers as a social community, and in general on their identities.

2.2 Teachers' Identity

Teachers' identity is considered in research as a system of beliefs. Teachers beliefs are important since it is commonly accepted [10–15] that beliefs affect practices and behaviors. The reference is usually to beliefs about education, about mathematics, about cognitive and metacognitive aspects involved in the teaching-learning processes. More recently, Skott [16] highlighted the importance of context as crucial to the study of the relation between practices and beliefs. From this point of view, the disruption created in learning context by the pandemic must be considered a key point to consider [17].

2.3 Formative Assessment

The concept of formative assessment (term originally coined by M. Scriven in 1967) has developed over the years a long-lasting debate with some consensus point. In Italy, it was introduced by Domenici [18] in a frame where *valutazione* (the Italian term both for assessment and for evaluation) is defined as an essential part of the teaching/learning processes and the act of conferring a *value* on something or someone. Within this perspective, assessment processes must acquire more and more a formative function. Formative assessment is not only part of teaching/learning processes, but often regulates their functioning as well. The importance of assessment with formative functionality is the identification, in an analytical manner, of the strengths and weaknesses of student learning, grounding on multiple sources of references. This allows teachers to modify and evaluate teaching practices and may help in establishing dialogues between students and teachers.

In our context, several studies highlighted that despite this long tradition of research on the topic, formative assessment is neither entered in a systemic and stable way in teachers' professional development activities, nor an aware classroom practice [19].

2.4 Teacher Training Focused on the Use of Standardized Assessments

In the Italian school system assessment is entrusted to teachers, individual educational institutions and central state institutions. NVALSI (www.invalsi.it) is the research organization that, according to the legislation in force, carries out, among other tasks, periodic and systematic tests on students' knowledge and skills and on the overall quality of the educational offer, also in the context of lifelong learning. In particular, INVALSI administers every year national tests in Mathematics, in different school grades and a key point is that, following what is highlighted in the literature [2, 20] these tests try to be strictly connected to the Italian National Guidelines for the curriculum. This should allow to make these tests tools in the hands of teachers. For instance, they may help in quantifying and clarifying common difficulties or obstacles [19, 21–23]. This integration of LSA tools (frameworks, results, released items, studies…) into classroom practice cannot be achieved without a specific attention to the teacher's professional development.

2.5 Standardized Assessments from a Formative Perspective

Hence, a specific attention should be given to the information that systemic results (coming from LSA) can be add to the interpretation of local facts (in the classroom), in order to enhance teacher's actions. In particular, this can be supported by technology [24, 25].

First of all, standardized assessments can provide teachers with tools and benchmarks for their diagnostic assessment [22]. Diagnostic assessment requires a basis that is as objective and shared as possible, and standardized assessment results can provide a grounded information that can help to understand what is really being assessed and what information will actually be returned.

As a part of his/her identity each teacher has his/her own implicit epistemology and background philosophy [26] regarding the teaching/learning process and this applies to assessment as well. This personal implicit framework has a deep impact on both the definition of the actual implemented curriculum and the choice of teaching tools and practices. Standardized assessments ground and are implemented on explicit frameworks: a better knowledge of these explicit frameworks and their actual implementation through tests, items, and the related results may help in gaining awareness about these implicit factors.

Last but not least, standardized assessments can help making concrete the demands of the National Education Documents: they provide examples of standard achievements. Besides the above-mentioned effort to create a close link between the goals and objectives of the National Guidelines and the planning and organization of teaching, it also requires work on teaching materials (e.g., textbooks), which are still, in Italy, the main source from which teachers take inspiration for classroom and homework activities. We have therefore tried to build learning situations for our program from standardized assessment tests.

2.6 The "Laboratory of Mathematics"

A key element of our program was the idea of "Laboratory of Mathematics". The active involvement of the learner is taken for granted today as an essential component of the teaching-learning process, whatever the body of knowledge is involved. In mathematics this has also an epistemological grounding, that is to say it is closely related to the nature of the discipline itself (for a classical statement of this idea, one can refer to [27]. This idea is deeply-rooted in Italian school tradition, since the pioneering work of Emma Castelnuovo (see f.i. [28]). Following Decroly's approach, she formulated the iconic idea that it is the whole classroom that must be a laboratory of mathematics.

In Castelnuovo's approach, the laboratory of mathematics does not need to have its own dedicated physical space, but it must have its own well-defined time. Its goal is to get the students involved, and a necessary condition for this that the teacher must be the first to get involved. Moreover, the laboratory is a natural place where assessment is necessarily formative.

In the period of LDL many teachers abandoned laboratory teaching because they could not design it in the new teaching setting. Our course was designed to provide tools to do laboratory teaching and maintain formative assessment activities even at a distance.

It is worth fixing some characteristics of the mathematics laboratory that we used for structuring our proposal [29].

L1) *Personal involvement:* Students enter the laboratory because they *want* to understand or discover something. In a laboratory there are facts and questions to observe, study, reproduce, arrange, understand: data, facts, situations....;

L2) *Epistemological significance*: In a laboratory the starting point is a problem, not its solution as in "standard" mathematical exposition. This is a particularly crucial point for us (as mathematicians). The final point of many mathematical researches is the construction of a formal theory, with a logical-deductive organization. Concrete situations are particular cases. So, the logical organization of a mathematical theory goes usually backwards with respect to one personal's path of understanding and organizing knowledge;

L3) *Inquiry approach:* It is not possible to know a priori what will be needed in order to understand the proposed situation; this should be discovered through inquiry:

L4) *Social involvement:* In a laboratory, work is never individual and collaboration between different people and people with different roles develops by working on concrete problems, which involve the students and the teacher as real challenges;

L5) *Theory and practice* are interlaced: observations on-the-field become the starting point for a theoretical construction.

L6) *Mistakes* contribute to make sense of the problems and to build the meaning of the body of knowledge the class is working on;

L7) *Reasoning and argumentation* are embedded in processes which are also heuristic:, intuition is combined with rigor, imagination with method, inventiveness with craftsmanship. Mathematical reasoning is so formative, so important, so "beautiful" because it cannot be reduced to abstract logic.

3   The Training Program

In this section, we describe the layout of the training course, we present an example of a laboratory activity which is representative for the training course, and specify the features of the questionnaire structure administered to teachers as a follow-up.

3.1   Layout of the Long-Distance Teachers Professional Development Program

The design of the training course started from the needs illustrated above: the necessity to share the results of large-scale evaluations with Italian teachers, the necessity to understand what is and how useful is the use of formative assessment for learning, and how to do it at a LDL perspective using technologies.

The trainers involved in the LDL training had previous experience (since 2010) with in-person trainings on the same background and purposes. The course design has been improved since then through repeated implementations and validation of the original design. The challenge was to transform these experiences into a LDL course designed to support long-distance learning activities with classrooms.

The training course was designed for teachers of all school levels and focused on the functional use of standardized assessments in mathematics laboratory practices of teachers in LDL context. Different types of activities were also presented during the course giving examples for each school level in order to directly engage all teachers enrolled in the course.

This course, delivered via webinar, was structured into meetings that would highlight the following:

- theoretical references to formative assessment, large-scale assessment and their relationships and connections;

- analysis of mathematics items from INVALSI large-scale assessments and critical analysis of their relationship with the Italian national curriculum;
- analysis of assessment situations and design/implementation of mathematics laboratory teaching activities to be carried out during the lockdown period through the use of technologies (videos, platforms, padlets,…);
- Emphasis on the fact that the results of standardized assessments, which highlight macro-phenomena often already studied in the literature and referred to in the tools available to teachers, allow teachers to intervene punctually, during the teaching action, on the critical aspects of the learning process in a formative way.

The implemented distance-learning course was structured into 14 webinars involving 2539 Italian teachers and 13 trainers. The webinars were all live, with the possibility of interactions with trainers and teachers-in-training via chat. Teachers-in-training who cannot attend few meetings have received a recording (following which they could ask questions to the trainer).

For each webinar we provided different tools: from one to three videos that the teacher could propose to his/her students in synchronous or asynchronous mode; teaching sheets for the students to carry out some activities; guidelines for the teacher.

The different training sessions are summarized in Table 1 and Table 2.

Table 1. List of 14 training sessions.

Date	Title of the training session	Number of live participants
17/03/2020	*Stories, fables, tools, drawings… to make math history in primary. School*	451
30/03/2020	*Decomposing a square: towards the Pythagorean theorem*	625
30/03/2020	*What is a hexagon?*	625
08/04/2020	*Vertices, edges and faces: polyhedra and toothpicks*	634
08/04/2020	*Folds and powers*	634
17/04/2020	*Enig…matica: when puzzles meet mathematical language*	769
17/04/2020	*More and less machines*	769
23/04/2020	*What do numbers tell? Reading infographics for primary and secondary school*	788
23/04/2020	*Towards infinity and… beyond: a story of toothpicks*	788

The training experience we conducted focused on workshop activities designed from items constructed for a standardized assessment, the results of which could help teachers with formative assessment. The focus was not on standardized tests, but on materials

Table 2. List of 14 training sessions (cont.).

Date	Title of the training session	Number of live participants
30/04/2020	*Webinar for second cycle teachers*	487
08/05/2020	*Webinar for first cycle teachers. Distance learning: reflections and activities*	699
14/05/2020	*Webinar on Assessment and DAD: a change in perspective*	712
21/05/2020	*What mathematics is NOT: communication in the service of teaching*	525
29/05/2020	*AMARCORD in Covid time. Rethinking mathematics teaching in Italy in the last hundred years*	811

and teaching sequences for workshop activities designed from standardized tests. The activities were proposed, for example, through videos to be shared with students.

These activities were designed "vertically," that is, as much as possible so that teachers at all school levels could use them.

3.2 The Description of a Laboratory Activity Based on the Standardized Tasks

The standardized questions used in the webinars were a total of ten, selected from the INVALSI Gestinv test database (www.gestinv.it; [30, 31]). In the following we will describe a teaching sequence, as an example, based on one of these questions, proposed in the webinar, entitled *What is a hexagon?*.

In that webinar, held in the first Italian lockdown period, i.e. in March 2020 and mainly dedicated to teachers of the last classes of primary and secondary schools, a geometry activity was proposed having as focus that of giving names to geometric objects.

During the webinar, in addition to describing an INVALSI question and the theoretical framework in which it fits, a possible activity to be assigned to students was proposed through two videos. In the first video, through operations on figures, students are led to reason about the characteristics and properties of geometric figures and their names. Then a challenge is given to them and in the second video a possible solution is proposed.

The activity of name-geometric object association begins very early, and then becomes one of the fundamental processes of the school geometry course; it is a process that requires continuous attention from the teacher. Very often students associate the name of the geometric figure with a picture, often in a predetermined position, and not with its properties. If we were to ask a student to draw, for example, a square, they would almost certainly draw a figure with two horizontal sides and two vertical sides, because that is the image of a square they have always known and experienced.

The starting point for this webinar was the INVALSI question (see Fig. 1), administered to all Italian grade 6 students in the INVALSI mathematics test in the 2011/12 school year.

Questa è la carta politica degli Stati Uniti d'America.

Quale, tra i seguenti stati dell'Ovest, ha la forma di un esagono?

A. ☐ Colorado

B. ☐ Utah

C. ☐ Nevada

D. ☐ New Mexico

Fig. 1. Task D10, 2012 INVALSI math test, grade 6.

In this question, students should recognize among the various states the one having the shape of a hexagon, relative to the area of space and figures.

The national percentages show that 47.1% of the students answer correctly (and therefore choose answer B), 51.9% answer incorrectly, and 1% do not answer. Of the students who get it wrong, 5.5% choose option A, while over 30% choose option C Nevada, and 15.7% choose option D.

The activity proposed in the video starts precisely from this stimulus and is aimed at reasoning about how we name geometric figures and to make students reflect on the relationship between figural and conceptual aspects.

Students are invited to get a square, a ruler, a sheet of squared paper, while the presenter of the video uses the dynamic geometry software Geogebra. Obviously if students have it they can use Geogebra.

In the first phase of the activity they have to draw a rectangle. Research has shown that when faced with such a request, the majority of students draw it with horizontal and vertical sides (see Fig. 2) while it will be very unlikely that any of them will draw it with oblique sides (see Fig. 3).

The video continues by looking at the properties of the rectangles as they are drawn, while the students are invited to experiment with drawing.

The same sequence is repeated with the square, and in particular the square is reasoned about when rotated in the prototypical position of the rhombus.

Fig. 2. Rectangle in prototypical position.

Fig. 3. Rectangle turned relative to the prototypical position.

Next, students have to draw a polygon with six sides, i.e. a hexagon. Again most likely students will draw it as in Fig. 4.

Fig. 4. Regular hexagon.

Starting from this figure, we point out that the drawn hexagon is a particular hexagon since it has all sides and all angles equal.

In the second step, we get students to think about how we name figures. You draw a figure with 4 sides, call it a quadrilateral and change it until it has the properties that identify a rectangle, or a square.

Then, we move on to draw different types of hexagons: the regular hexagon rotated in all ways, a non-regular convex hexagon (Fig. 5) and finally a non-regular, non-convex hexagon (Fig. 6).

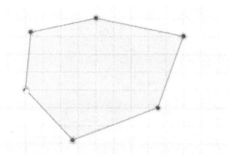

Fig. 5. Non-regular hexagon.

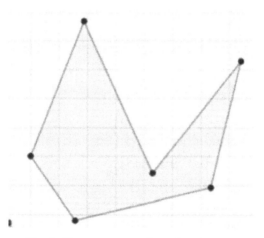

Fig. 6. Hexagon not-regular and not-convex.

In the third and last part of the video, starting from the analysis of the property of the regular hexagon to have its sides two by two parallel (therefore to have three pairs of sides parallel to each other), a challenge arises for the students: to draw a hexagon having its sides three by three parallel (therefore two triads of sides parallel to each other). The purpose of the activity and the challenge is to try to break the stereotypical image that students have for all geometric figures.

In the webinar, a second video was proposed in which possible solutions to the challenge are described and, after explaining the figure, other challenges are proposed and solved together.

To the first challenge, the possible answer proposed is the letter L: it is a polygon with six sides, therefore a hexagon, with the sides three to three to parallel (Fig. 7).

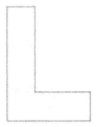

Fig. 7. Hexagon with three sides three by three parallel.

At this point you draw the letter E and ask them to describe it "using geometry language", i.e. identify it as a dodecagon with six by six parallel sides (Fig. 8).

Fig. 8. Dodecagon with six to six parallel sides.

As a final challenge we leave that of analyzing and defining the letter F as a geometric figure.

The activity proposed through the videos and explained during the webinar, leads to reflect, through play, on how we name geometric figures. Operation that in mathematics we always do, but that sometimes leads to stereotypes in the minds of students that create difficulties (e.g. the state of Utah in the INVALSI question, Fig. 1) when we have to use these shapes in non-standard situations or extend these notions to broaden the field of set of mathematical objects with which we work.

3.3 The Questionnaire

In the in-person modality, the training was conducted with small groups of teachers (20–30), who had the opportunity to take part in the activities and discuss them. Digitization made it possible to reach larger targets of teachers and have them participate at the same time, but the possibility of interaction during the training sessions was limited. At the beginning of the emergency, moreover, many teachers had a very limited knowledge of the technological tools used. We promoted discussion in the chats during the webinars and we implemented a final questionnaire that was designed to give teachers an opportunity to express their reflections, simulating what was happening in-person and for an overall validation of the course.

As described in [1], within the questionnaire there were several sections, each with a specific purpose:

Section 1: designed to investigate whether teachers were familiar with the questions from the standardized assessments used by trainers during the webinars;
Section 2: designed to investigate whether teachers knew the results of the questions from the standardized assessments used by the trainers during the webinars;
Section 3: designed to investigate whether teachers recognized the evidence of macro-phenomena highlighted by the results of the standardized assessments;
Section 4: aimed at investigating whether teachers had used some of the suggested activities in the classroom to explicite the presence of some of the misconceptions highlighted by the results of the standardized tests;
Section 5: aimed at investigating whether the reflections that emerged during the webinars had changed some of the teachers' practices or some of the ideas related to assessment.

The questionnaire consists of 19 closed-ended questions.

The questionnaire was sent to all the teachers who took part in the webinars and we received responses from 509 teachers. In the chat recorded some hundreds of comments and remarks were collected.

4 Elements that Emerged During the Program

M of the teachers involved in this experiment in teacher education had never been interested in distance learning courses before this event, hence data collected necessarily reflect this situation.

In the following, we comment on two types of results collected: the items that emerged in the chat (during the webinars) and the results of the questionnaire (post webinar). We recall here some data from [1] and we add other data that improve that presentation.

4.1 Chat: Elements that Emerged During the Webinars

This experience challenged some beliefs that teachers had about distance learning that also emerged from some of the comments written in chat by teachers during the webinars.

One teacher writes: *"I was sure I couldn't do remote lab activities and now I've changed my mind!"*. Many teachers, during the period of distance learning, began to work in the new situation simply reproducing in front of a camera lectures because they were convinced that they could not do laboratory activities. The course gave them the opportunity to see that they could do laboratory activities and with what tools.

The chat comments reveal another of the teachers' concerns during this period: assessment. A common initial belief among teachers was the idea that assessment should ultimately be nothing more than a score. So "formative assessment" was just a formal expression in official documents, with little relation to "real" assessment (which is, in their idea, summative assessment). This is reported in several comments, including the following: *"How can we assess our students without limiting ourselves to formative assessment? I currently put positive or negative notes, but never assessments because I think it is impossible to objectively assess at a distance"*. Let us recall that summative assessment, in Italian schools, is carried out mainly through oral questions and individual written open-ended tasks. Teachers' concerns on the issue of assessment have been many, since these modalities are indeed difficult to implement in a distance teaching situation.

4.2   Questionnaire: Elements Emerged Post Webinar

A total of 351 elementary school teachers (from grade 1 to 5), 106 secondary school teachers (from grade 6 to 8) and 52 secondary school teachers (from grade 9 to 13) answered the questionnaire.

Within the questionnaire, one of the sections was devoted entirely to reflections related to the INVALSI items proposed during the webinars.

We initially asked whether participants were familiar (prior to attending the webinars) with the items presented. Specifically, we distinguished knowledge of the questions, from knowledge of the results of the questions and thus of whether important macro-phenomena were present.

When asked *"Did you know the INVALSI items that were discussed?"* the response rates were as follows (Fig. 9):

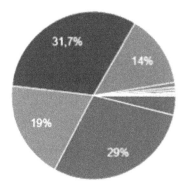

Fig. 9. Results of the question "Did you know the INVALSI items that were discussed?".

The 2.1% of teachers answered "Yes, all of them (including those in other school levels)", 29% answered "Yes, but only all of those in my school level", 19% answered "Yes, many", 31.7% answered "Yes, only some", 14% answered "No", while 4.2% answered "Other".

These results are to be related to the answers to the question *"Did you know the results of the INVALSI items discussed?"*, whose response percentages are shown in the following graph (Fig. 10):

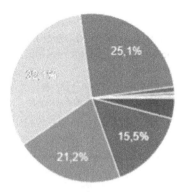

Fig. 10. Results of the question " Did you know the results of the INVALSI items discussed?".

Only 1.2% of teachers answered "Yes, the results of all questions," 15, 5% answered "Yes, all the results of the questions in my grade level," 21.2% answered "Yes, many," 32.1% answered "Yes, only some," 25.1% answered "No," and 4.9% answered "Other." It is apparent that the majority did not know all of the questions submitted, let alone the results. In addition, the majority of teachers who not only knew the questions at their own school level, but also knew the results were from Primary School (almost 75%).

Despite teachers' initial difficulties, the course also had a strong impact on the implementation of distance teacher education. This is evident through teachers' responses to some of the questions in the questionnaire.

One question asked, *"Prior to participating in our webinars, had you ever used video to implement classroom activities?"*, since the training program stressed the importance of such a tool, especially in a LDL setting.

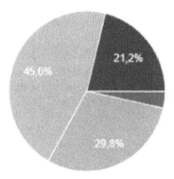

Fig. 11. Results of the question "Prior to participating in our webinars, had you ever used video to implement classroom activities?". Source [1].

The results, shown in Fig. 11, show that 3.4% of teachers responded "Systematically," 21.2% responded "Often," 45.6% responded "Rarely," and 29.8% responded "Never."

All those who responded "Systematically" were found to be secondary school teachers (grade 6 through grade 13). In fact, all of the primary teachers and most of the secondary teachers enrolled in the webinars had not previously used videos or other similar materials prior to the training.

Another important question was, *"When you return fully in attendance do you plan to continue using the materials presented during the webinars or elements of this experience?"* The results are shown in Fig. 12.

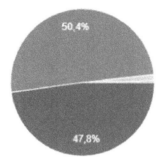

Fig. 12. Results of the question "When you return fully in attendance do you plan to continue using the materials presented during the webinars or elements of this experience?". Source [1].

A total of 47.8% of the teachers responded "Definitely yes", 50.4% responded "Probably yes", 1.4% responded "Probably not", and no one responded "Definitely not".

One of the conclusions is that these teachers did not use such materials before the course, while after the course they say that even when they return to presence they will probably or definitely use and continue to use these materials. So they think that the use of these materials can be valid also in presence and not replace the teaching in presence, which gives other things.

One of the aspects highlighted during the webinars was the development of synchronous and a-synchronous activities. In presence a teacher can only do synchronous activities, while (with these materials developed in a-synchronous) the activity can continue outside the classroom. This aspect is one of those highlighted as a positive of distance learning.

5   Conclusions

We described and analyzed an experience that highlighted how to include items and results from large-scale assessments in teacher training to create laboratory activities that can be used in LDL. The questionnaire answers we analyzed allow us to claim that the activities presented during the webinars provide opportunities for LDL activities with students in formative assessment.

The course, initially designed and tested in presence, has highlighted how it is possible to provide tools to do laboratory activities and support formative assessment even at a distance.

The project was developed in an emergency situation and in a context in which the use of technology for distance learning was very limited in Italy and the assessment substantially summative in practice. Working together with teachers to provide tools and resources with a theoretical background of research on mathematics education in order to realize LDL workshops (always in emergency) with particular attention to dynamics of formative assessment that could be applied in this situation has allowed to highlight connections with concrete teaching situations.

References

1. Spagnolo, C., Giglio, R., Tiralongo, S., Bolondi, G.: Formative assessment in LDL: a teacher-training experiment. In: Proceedings of the 13th International Conference on Computer Supported Education (CSEDU 2021), vol. 1, pp. 657–664 (2021). https://doi.org/10.5220/0010496006570664

2. Looney, J.W.: Integrating Formative and Summative Assessment: Progress Toward a Seamless System? OECD Education Working Papers, No. 58. OECD Publishing, Paris (2011)

3. Calvani, A.: Criticità e potenzialità nel cammino della valutazione. Form@re -Open J. per la formazione in rete **14**(4), 20–33 (2014)

4. Trinchero, R.: Il Servizio Nazionale di Valutazione e le prove Invalsi. Stato dell'arte e proposte per una valutazione come agente di cambiamento. Form@re -Open J. per la formazione in rete **14**(4), 34–49 (2014)

5. Pastori, G., Pagani, V.: What do you think about INVALSI tests? School directors, teachers and students from Lombardy describe their experience. J. Educ. Cult. Psychol. Stud. **13**, 97–117 (2016)

6. Borba, M.C., Askar, P., Engelbrecht, J., Gadanidis, G., Llinares, S., Aguilar, M.S.: Blended learning, e-learning and mobile learning in mathematics education. ZDM Math. Educ. **48**(5), 589–610 (2016). https://doi.org/10.1007/s11858-016-0798-4

7. Mulenga, E.M., Marbán, J.M.: Is COVID-19 the gateway for digital learning in mathematics education. Contemp. Educ. Techno. **12**(2), ep269 (2020)

8. Mulenga, E.M., Marbán, J.M.: Prospective teachers' online learning mathematics activities in the age of COVID-19: a cluster analysis approach. EURASIA J. Math. Sci. Technol. Educ. **16**(9), em1872 (2020)

9. Chan, M.C.E., Sabena, C., Wagner, D.: Mathematics education in a time of crisis—a viral pandemic. Educ. Stud. Math. **108**(1–2), 1–13 (2021). https://doi.org/10.1007/s10649-021-10113-5

10. Green, T.F.: The Activities of Teaching. McGraw-Hill, New York (1971)

11. Golding, G.A.: Affect, meta-affect, and mathematical belief structures. In: Leder, G.C., Pehko-nen, E., Törner, G. (eds.) Beliefs: A Hidden Variable in Mathematics Education?, pp. 59–72. Kluwer Academic Publishers, Dordrecht (2002)

12. Swan, M.: The impact of task-based professional development on teachers' practices and beliefs: a design research study. J. Math. Teach. Educ. **10**, 217–237 (2007). https://doi.org/10.1007/s10857-007-9038-8

13. Liljedahl, P.: Noticing rapid and profound mathematics teacher change. J. Math. Teach. Educ. **13**(5), 411–423 (2010). https://doi.org/10.1007/s10857-010-9151-y

14. Beswick, K.: Teachers' beliefs about school mathematics and mathematicians' mathematics and their relationship to practice. Educ. Stud. Math. **79**(1), 127–147 (2012). https://doi.org/10.1007/s10649-011-9333-2

15. Wong, N.Y., Ding, R., Zhang, Q.P.: From classroom environment to conception of mathematics. In: King, R.B., Bernardo, A.B.I. (eds.) The Psychology of Asian Learners, pp. 541–557. Springer, Singapore (2016). https://doi.org/10.1007/978-981-287-576-1_33
16. Skott, J.: Contextualising the notion of 'belief enactment.' J. Math. Teach. Educ. **12**, 27–46 (2006). https://doi.org/10.1007/s10857-008-9093-9
17. Ferretti, F., Santi, G.R.P., Del Zozzo, A., Garzetti, M., Bolondi, G.: Assessment practices and beliefs: teachers' perspectives on assessment during long distance learning. Educ. Sci. **11**, 264 (2021). https://doi.org/10.3390/educsci11060264
18. Domenici, G.: Manuale della valutazione scolastica. Editori Laterza, Bari (2003)
19. Bolondi, G., Ferretti, F., Giberti, C.: Didactic contract as a key to interpreting gender differences in maths. J. Educ. Cult. Psychol. Stud. **18**, 415–435 (2018)
20. Meckes, L.: Evaluación y estándares: logoros y desafíos para incrementar el impacto en calidad Educativa. Rev. Pensamiento Educativo **40**(1), 351–371 (2007)
21. Bolondi, G.: What can we learn from large-scale surveys about our students learning of maths? AAPP Atti della Accademia Peloritana dei Pericolanti, Classe di Scienze Fisiche, Matematiche e Naturali, vol. 992021, Article number 99S1A4 (2021)
22. Bolondi, G., Ferretti, F.: Quantifying solid findings in mathematics education: loss of meaning for algebraic symbols. Int. J. Innov. Sci. Math. Educ. **29**(1), 1–15 (2021)
23. Ferretti, F., Bolondi, G.: This cannot be the result! The didactic phenomenon 'the age of the earth.' Int. J. Math. Educ. **52**(2), 1–14 (2019)
24. Barana, A., Marchisio, M., Sacchet, M.: Interactive feedback for learning mathematics in a digital learning environment. Educ. Sci. **11**, 279 (2021). https://doi.org/10.3390/educsci11060279
25. Barana, A., Marchisio, M., Miori, R.: MATE-BOOSTER: design of tasks for automatic formative assessment to boost mathematical competence. In: Lane, H.C., Zvacek, S., Uhomoibhi, J. (eds.) CSEDU 2019. CCIS, vol. 1220, pp. 418–441. Springer, Cham (2020). https://doi.org/10.1007/978-3-030-58459-7_20 ISBN 978-3-030-58458-0
26. Speranza, F.: Scritti di Epistemologia della Matematica. Pitagora, Bologna (1997)
27. Courant, R., Robbins, H.: What is Mathematics? An Elementary Approach to Ideas and Methods. Oxford University Press, London (1941)
28. Castelnuovo, E.: Didattica della Matematica. La Nuova Italia, Firenze (1963)
29. Bolondi, G.: Metodologia e didattica: il laboratorio. Rassegna **29**, 59–63 (2006)
30. Ferretti, F., Gambini, A., Santi, G.: The gestinv database: a tool for enhancing teachers professional development within a community of inquiry. In: Borko, H., Potari, D. (eds.) Pre-Proceedings of the Twenty-fifth ICMI Study School Teachers of Mathematics Working and Learning in Collaborative Groups, pp. 621–628. University of Lisbon, Portugal (2020)
31. Bolondi, G., Ferretti, F., Gambini, A.: Il database GESTINV delle prove standardizzate INVALSI: uno strumento per la ricerca. In: Falzetti, P. (ed.) I dati INVALSI: uno strumento per la ricerca, pp. 33–42. FrancoAngeli, Milano (2017)

Author Index

~formation can be obtained
~testing.com
~ USA
~280822
~04B/80